Cooperative Control of Distributed Multi-Agent Systems

Cooperative Control of Distributed Multi-Agent Systems

Edited by

Jeff S. Shamma
Georgia Institute of Technology, USA

BICENTENNIAL 1807 WILEY 2007 BICENTENNIAL

John Wiley & Sons, Ltd

Copyright © 2007 John Wiley & Sons Ltd, The Atrium, Southern Gate, Chichester, West Sussex PO19 8SQ, England

Telephone (+44) 1243 779777

Email (for orders and customer service enquiries): cs-books@wiley.co.uk
Visit our Home Page on www.wileyeurope.com or www.wiley.com

Other Wiley Editorial Offices

John Wiley & Sons Inc., 111 River Street, Hoboken, NJ 07030, USA

Jossey-Bass, 989 Market Street, San Francisco, CA 94103-1741, USA

Wiley-VCH Verlag GmbH, Boschstr. 12, D-69469 Weinheim, Germany

John Wiley & Sons Australia Ltd, 42 McDougall Street, Milton, Queensland 4064, Australia

John Wiley & Sons (Asia) Pte Ltd, 2 Clementi Loop #02-01, Jin Xing Distripark, Singapore 129809

John Wiley & Sons Canada Ltd, 6045 Freemont Blvd, Mississauga, Ontario, L5R 4J3, Canada

Wiley also publishes its books in a variety of electronic formats. Some content that appears in print may not be available in electronic books.

Anniversary Logo Design: Richard J. Pacifico

British Library Cataloguing in Publication Data

A catalogue record for this book is available from the British Library

ISBN 978-0-470-06031-5

Typeset in 10/12 Times by Laserwords Private Limited, Chennai, India
Printed and bound in Great Britain by Antony Rowe Ltd, Chippenham, Wiltshire
This book is printed on acid-free paper responsibly manufactured from sustainable forestry in which at least two trees are planted for each one used for paper production.

Contents

List of Contributors

Alessandro Arsie Laboratory for Information and Desision Systems, Massachusetts Institute of Technology, Cambridge, MA

Gurdal Arslan Department of Electrical Engineering, University of Hawaii

Gary J. Balas Department of Aerospace Engineering and Mechanics, University of Minnesota, Minneapolis, MN

Francesco Borrelli Dipartimento di Ingegneria, Università degli Studi dell Sannio, Italy

Georgios C. Chasparis Department of Mechanical and Aerospace Engineering, University of California Los Angeles, CA

Daniel Choukroun Department of Mechanical Engineering, Ben-Gurion University of the Negev, Israel

T.C. Collier University of California Los Angeles, CA

Munther A. Dahleh Laboratory for Information and Decision Systems, Massachusetts Institute of Technology, Cambridge, MA

Raffaello D'Andrea Mechanical and Aerospace Engineering Department, Cornell University, New York

Domitilla Del Vecchio Department of Electrical Engineering and Computer Science, University of Michigan

Jan DeMot School of Aerospace, Georgia Institute of Technology, AL

William B. Dunbar Department of Computer Engineering, University of California Santa Cruz, CA

Matthew G. Earl BAE Systems, Advanced Information Technologies, Burlington, MA

John J. Enright Department of Mechanical and Aerospace Engineering, University of California Los Angeles, CA

Eric Feron School of Aerospace, Georgia Institute of Technology, Atlanta, GA

Emilio Frazzoli Laboratory of Information and Decision Systems, Massachusetts Institute of Technology, Cambridge, MA

Sonja Glavaški Honeywell International Inc., Minneapolis, MN

S. Goldfarb Cornell University, New York

Tamás Keviczky Delft Center for Systems and Control, Delft University of Technology, The Netherlands

Y. Lee University of California Los Angeles, CA

Nuno C. Martins Department of Electrical Engineering, University of Maryland

Brandon J. Moore Department of Electrical and Computer Engineering, The Ohio State University, OH

Richard M. Murray Control and Dynamical Systems, California Institute of Technology, Pasadena, CA

Kevin M. Passino Department of Electrical and Computer Engineering, The Ohio State University, OH

V.G. Rao Cornell University, New York

Reza Olfati-Saber Dartmouth College, Thayer School of Engineering

Tariq Samad Honeywell International Inc., Minneapolis, MN

Vincent Pei-wen Seah Department of Mechanical and Aerospace Engineering, University of California Los Angeles, CA

Jeff S. Shamma School of Electrical and Computer Engineering, Georgia Institute of Technology, GA

Jason L. Speyer Department of Mechanical and Aerospace Engineering, University of California Los Angeles, CA

E.P. Stabler University of California Los Angeles, CA

Ashitosh Swarup Department of Mechanical and Aerospace Engineering, University of California Los Angeles, CA

C.E. Taylor University of California Los Angeles, CA

Anca Williams FLIR Systems, Beaverton, OR

Preface

Cooperative control involves a collection of decision-making components with limited processing capabilities, locally sensed information, and limited inter-component communications, all seeking to achieve a collective objective. Examples include autonomous vehicle teams, mobile sensor networks, data network congestion control and routing, transportation systems, multi-processor computing, and power systems. The distributed nature of information processing, sensing, and actuation makes these applications a significant departure from the traditional centralized control system paradigm.

There has been substantial and increasing interest in recent years in cooperative control systems. Indications of the level of interest include several multi-year/multi-university research projects, calls for proposals, journal special issues, and specialty conferences.

This volume represents an effort to recognize certain themes that have emerged from recent cooperative control research. The themes, or 'dimensions', we will use are: (1) Distributed control and computation; (2) Adversarial interactions; (3) Uncertain evolution; and (4) Complexity management. Of course, these themes do not constitute a 'partition' of cooperative control research, and alternative headings could have been used. Furthermore, research results typically fall under more than one dimension. Nonetheless, it is instructive to impose some structure on the broad scope of research that has emerged in cooperative control.

Many of the contributions in this volume were the outcome of a multi-university collaboration between UCLA, Caltech, Cornell, and MIT sponsored by the Air Force Office of Scientific Research, whose support is gratefully acknowledged.

Jeff S. Shamma

Part I

Introduction

1

Dimensions of cooperative control

Jeff S. Shamma and Gurdal Arslan

1.1 WHY COOPERATIVE CONTROL?

1.1.1 Motivation

Cooperative control is concerned with engineered systems that can be characterized as a collection of decision-making components with limited processing capabilities, locally sensed information, and limited inter-component communications, all seeking to achieve a collective objective. Examples include mobile sensor networks, network congestion control and routing, transportation systems, autonomous vehicle systems, distributed computation, and power systems. Areas of research that are related to cooperative control include 'multi-agent control', 'distributed systems', 'networked control', as well as 'team theory' or 'swarming'. Regardless of the nomenclature, the central challenge remains the same. That is, to derive desirable collective behaviors through the design of individual agent control algorithms.

The primary distinguishing feature of a cooperative control system is distribution of information. As opposed to 'centralized' solutions, no one decision-maker has access to the information gathered by all agents. Furthermore, there is typically a communication cost in distributing locally gathered information.

A secondary distinguishing feature is complexity. Even if information were centrally available, the inherent complexity of the decision problem makes a centralized solution computationally infeasible. Therefore, one looks to 'decomposition' approaches as in cooperative control.

Both these features result in the distributed decision architecture of cooperative control. The potential benefits of such architectures include the opportunity for real-time adaptation (or self-organization) and robustness to dynamic uncertainties such as individual component failures, non-stationary environments, and adversarial elements. These benefits come with significant challenges, such as the complexity associated with a potentially large number of interacting agents and the analytical difficulties of dealing with overlapping and partial information.

Cooperative Control of Distributed Multi-Agent Systems edited by Jeff S. Shamma

1.1.2 Illustrative example: command and control of networked vehicles

We will use the vehicle-target assignment problem, described below, as an illustrative scenario for the discussion of many of the concepts in this chapter.

Recent years have seen a boom of research motivated by the application of mobile unmanned systems that operate with either complete autonomy or semi-autonomy, i.e., given high level commands from a remote human operator. In fact, the broader topic of 'network-centric operations' was the subject of a very recent National Research Council committee's study on Network Science for Future Army Applications.[1]

A representative scenario is the vehicle-target assignment problem (Ahuja *et al.* 2003; Beard *et al.* 2002; Murphey 1999), illustrated in Figure 1.1. A collection of vehicles seeks to visit a collection of targets. Each vehicle has a limited range of reachable targets. There can be an advantage in having multiple vehicles visit a single target in that vehicle capabilities are complementary. However, there is also a diminishing return having redundancy in vehicle-target assignments. An additional complication is to allow vehicles to visit multiple targets in succession.

Vehicle-target assignment also encompasses the problem of multiple vehicle motion planning. In an obstacle-filled congested environment, vehicles need to get to their destination while avoiding obstacles and avoiding each other, yet having only limited communications with neighboring vehicles.

Again, both distinguishing features of cooperative control problems are apparent. Because of the spatial distribution of vehicles, any information gathered by a single vehicle does not reach other vehicles without explicit communication. This adds a dimension of what to communicate and to whom. Furthermore, the vehicle-target assignment problem is known to be NP-complete (Murphey 1999), and so even a centralized solution is computationally prohibitive for large numbers of vehicles.

An additional possible complication is the presence of an intelligent adversary. Most work on vehicle-target assignment assumes an asymmetry between vehicle and target capability – i.e., targets have simple scripted behavior. The problem becomes much more

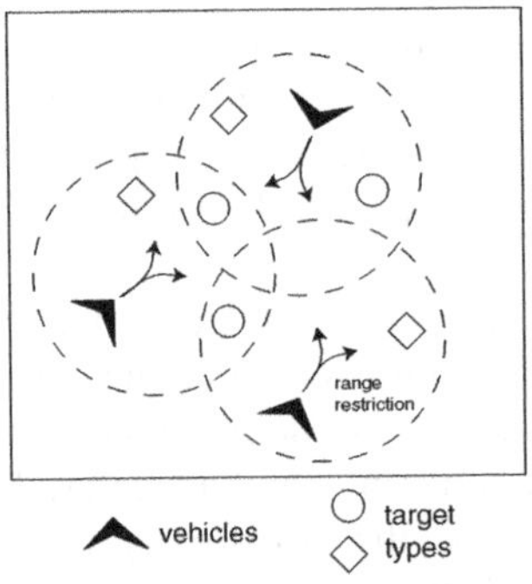

Figure 1.1 Vehicle target assignment.

[1] The NRC report, *Network Science*, is available online at http://darwin.nap.edu/books/0309100267/html/.

interesting if targets also have strategic capability, e.g., mobile targets with stealth modes of operation. In this case, the problem resembles a sort of networked team-vs-team encounter.

1.2 DIMENSIONS OF COOPERATIVE CONTROL

It is helpful to impose some sort of structure on the broad scope of issues that can emerge in cooperative control. With this motivation, we will discuss cooperative control problems in terms of the following 'dimensions': (1) distributed control and computation; (2) adversarial interactions; (3) uncertain evolution; and (4) complexity management.

In the following subsections, we will describe each of these dimensions in general and in terms of the above illustrative example. The discussion will largely be in terms of chapters in this volume and selected immediately relevant results. Other reviews of cooperative control and related research may be found in Murray (2006) and Panai and Luke (2005).

1.2.1 Distributed control and computation

This is the primary defining characteristic of cooperative control. The distribution of information necessitates the distributed control and computation among interacting components. Not surprisingly, this dimension represents the highest number of contributions to the volume.

1.2.1.1 Multivehicle motion planning

Multivehicle motion planning arises in vehicle-target assignment in congested environments. The objective is to generate real-time trajectories to guide a collection of vehicles to their destination while avoiding obstacles and each other. Destinations may be specified in terms of absolute locations, relative locations, coverage patterns, or even general direction of flow. Example scenarios include rendezvous or formation keeping.

The distributed nature of multivehicle motion planning stems from each vehicle determining its own trajectory while having only partial measurements of only neighboring vehicles.

Chapter 2[2] focuses on a 'swarm' of vehicles, i.e., a large collection of autonomous vehicles that need not maintain a regular formation. An important underlying technique in this problem is that of 'consensus algorithm', which was introduced in Tsitsiklis *et al.* (1986) and has received considerable renewed interest in recent years (Blondel *et al.* 2005; Jadbabaie *et al.* 2003; Kashyap *et al.* 2006; Marden *et al.* 2007; Moreau 2004; Olfati-Saber and Murray 2003; Olfati-Saber *et al.* 2007; Xiao and Boyd 2004, 2005). The general form of the algorithm can be described as follows. Let $i \in \{1, 2, \ldots, \mathcal{I}\}$

[2] R. Olfati-Saber, 'Design of behavior of swarms: from flocking to data fusion using microfilter networks'.

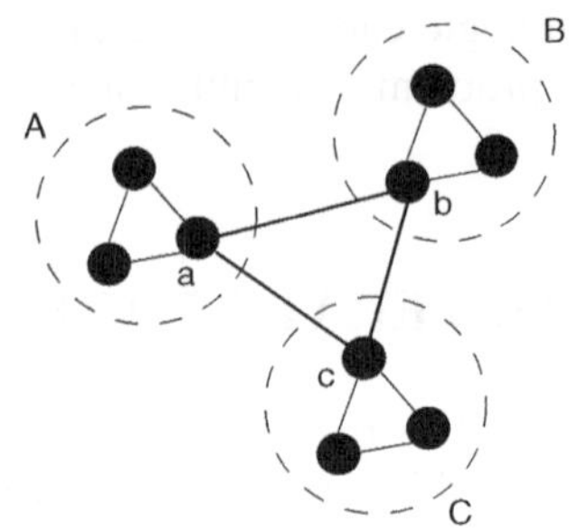

Figure 1.2 Hierarchical formation.

denote the index of the ith agent in a collection $\{1, 2, \ldots, \mathcal{I}\}$. Let $\theta_i(t)$, $t \in \{0, 1, 2, \ldots\}$, denote a time-varying scalar quantity associated with the ith agent. Now define

$$\theta_i(t+1) = \sum_{j \in N_i(t)} a_{ij}(t)\theta_j(t), \tag{1.1}$$

where $N_i(t) \subset \{1, 2, \ldots, \mathcal{I}\}$ denotes the set of 'neighbors' of the ith agent at time t and $a_{ij}(t)$ is the relative weight that agent i places on agent j. Under suitable assumptions, one can show that all agents asymptotically synchronize, i.e., for all $i \neq j$,

$$\lim_{t \to \infty} \theta_i(t) = \lim_{t \to \infty} \theta_j(t).$$

For motion planning problems, the consensus algorithm has been used as an approach to synchronize heading angles, velocities, or positions, using only 'local' information characterized by $N_i(t)$ and in the face of time-varying network topologies.

Chapter 3[3] also addresses multivehicle motion planning, but the emphasis is shifted away from swarms in favor of formations. That is, each vehicle seeks to hold a specified distance from a subset of other vehicles. In particular, this chapter considers 'hierarchical formations'. A very simple example (Figure 1.2) consists of three groups of vehicles, say, group A, B, and C. Group A seeks to maintain a relative formation, as do groups B and C. Furthermore, there is a single vehicle in each group, say, vehicles a, b, and c, that seek to maintain a relative formation with each other. In this way, the overall formation is induced by the intergroup formation and intragroup formation.

1.2.1.2 Distributed predictive control

Receding horizon control or model predictive control (e.g., Bemporad and Morari 1999) is a well-known approach to nonlinear control design through successive online optimizations. Distributed control and computation present an obstacle to implementing predictive control in a multiagent system.

[3] S. Glavaški, A. Williams, and T. Samad, 'Connectivity and convergence of formations'.

The 'single agent' version can be described briefly as follows. Assume state dynamics of the form

$$\dot{x} = f(x, u).$$

Let $u^*(\cdot)$ be the optimal control trajectory for the optimization problem

$$\min_{u(\cdot)} \int_t^{t+T} g(x(\tau), u(\tau), \tau)d\tau + G(x(t+T))$$

subject to the above dynamics with initial condition $x(t)$. In model predictive control, the initial portion of the optimal control is implemented, i.e.,

$$u(\tau) = u^*(\tau), \quad \tau \in [t, t+\delta)$$

for some $\delta > 0$. The optimization is then resolved at time $t + \delta$ using the state $x(t + \delta)$ as the initial condition, and the process is repeated.

Two complications arise in a multiagent framework. For simplicity, consider the case with two agents. One complication arises when the systems are dynamically coupled, as in

$$\dot{x}_1 = f_1(x_1, x_2, u_1)$$
$$\dot{x}_2 = f_2(x_2, x_2, u_2).$$

Another complication arises when the systems are not dynamically coupled, as in

$$\dot{x}_1 = f_1(x_1, u_1) \tag{1.2a}$$
$$\dot{x}_2 = f_2(x_2, u_2), \tag{1.2b}$$

but their objective functions are coupled, as in

$$\min_{u_1(\cdot)} \int_t^{t+T} g_1(x_1(\tau), x_2(\tau), u_1(\tau), \tau)d\tau + G_1(x_1(t+T)) \tag{1.3a}$$
$$\min_{u_2(\cdot)} \int_t^{t+T} g_2(x_1(\tau), x_2(\tau), u_2(\tau), \tau)d\tau + G_2(x_2(t+T)). \tag{1.3b}$$

In either case, the requirement of distributed control and computation prohibits an immediate application of receding horizon control, the main reason being that neither agent is in control of the other. Accordingly, neither agent is able to control or predict the trajectory of the other agent.

Chapters 4[4] and 5[5] present complementary approaches to multiagent versions of predictive control. These chapters address how to introduce additional constraints on the control trajectories, such as $u_i(\cdot) \in \mathcal{U}_i$, to guarantee stability and performance. As one

[4] W.B. Dunbar, 'Distributed receding horizon control: stability via move suppression'.
[5] T. Keviczky, F. Borrelli, and G.J. Balas, 'Distributed predictive control: synthesis, stability and feasibility'.

would expect, an important underlying concern of any single agent is the deviation of the trajectories of other agents from their predicted values. While the problem set-up is motivated by multivehicle motion planning, the methods and concepts carry over to more general settings.

1.2.1.3 Task assignment

The multivehicle motion planning component of vehicle-target assignment addresses issue of 'how to get there?'. This was the main subject of the Chapters 2–5.

A higher level strategic decision is 'where to go'? This is the problem of task assignment. Let $\mathcal{V} = \{1, 2, \ldots, n_V\}$ denote a set of vehicles and $\mathcal{T} = \{1, 2, \ldots, n_T\}$ denote a set of targets. Let a_i, $i \in \mathcal{V}$ denote the assignment of the ith vehicle. For example, $a_1 = 2$ means that vehicle #1 is assigned to target #2. The collection of assignment profiles is denoted by the vector a. In general, these assignments are constrained (e.g., by vehicle range limitations), and so each vehicle's assignment must belong to a specified set, denoted by $\mathcal{A}_i$ for vehicle #i.

In static assignment problems, there is a scoring function that indicates the utility of an assignment profile, e.g.,

$$U_g(a) = U_g(a_1, a_2, \ldots, a_{n_V}).$$

The task assignment problem is then the combinatorial optimization

$$\max_{a_i \in \mathcal{A}_i, i=1,2,\ldots,n_V} U_g(a_1, a_2, \ldots, a_{n_V}).$$

The situation is more complicated in dynamic assignment problems. In these problems, vehicle locations, target locations, and target constraint sets are all time-varying, which induces time-variations in the overall utility function. The problem is then to design dynamic vehicle-target assignment policies whose performance is measured by time-aggregated behavior such as the rate of successful target assignments.

This problem becomes one of distributed computation and control because individual vehicles need not be aware of the entirety of target locations. Furthermore, there may be an incentive for vehicles to be assigned to the same target, and so vehicles must coordinate to complement the actions of other vehicles. Typically, one assumes some sort of communication graph over which vehicles message each other in an effort to compute an assignment profile. This brings in the additional dimension of what to communicate as part of the decision problem. Another reason for a distributed approach is complexity. Even if a single decision-maker had access to all target locations to determine an assignment profile, the resulting optimization is known to be NP-complete (cf., Murphey 1999). Imposing a distributed architecture does not alleviate the inherent complexity, per se, but represents an attempt to compute effective solutions with low computational cost.

Chapters 6–7 address different aspects of the task assignment problem. Chapter 6[6] investigates the one-to-one assignment problem with static targets. Vehicles can communicate through a specified communication topology, but these communications introduce delays in determining an assignment profile. Furthermore, the assignment utility function changes as vehicles move in the environment en route to their interim assignments.

[6] B.J. Moore and K.M. Passino, 'Task assignment for mobile agents'.

Vehicles use a modified distributed auction algorithm to determine the action profile, and the performance is measured through Monte Carlo simulation.

Chapter 7[7] also considers mobile vehicles and static targets, but targets are regenerated by a stochastic process. The performance measure is to minimize the average time to visit a target once it has appeared. This performance objective is an example of the aforementioned time-aggregated behavior (as opposed to optimal assignment for a specific fixed target scenario.) The emphasis in Chapter 7 is to investigate the role of communication. Namely, to what degree do explicit vehicle communications affect performance? Surprisingly, the result in this formulation is that asymptotic performance is not affected. The intuition is that vehicles implicitly communicate when one vehicle reaches a target before another vehicle.

Another approach to task assignment is reported by Arslan *et al.* (2006). This paper takes a learning in games (Fudenberg and Levine 1998; Young 2006) approach to the many-to-one static assignment problem. Since many vehicles can visit the same target, there is an assumed communication topology where two vehicles can communicate if and only if they share a common target, i.e., $\mathcal{A}_i \cap \mathcal{A}_j \neq \emptyset$.

The method proposed by Arslan *et al.* (2006) is twofold. The first element is design of local utility functions. Individual vehicles are endowed with a 'local' utility function that reflects a vehicle's local influence. A particularly appealing local utility construction is the marginal contribution utility (or wonderful life utility (Wolpert and Tumer 1999)) defined as follows. We will write an assignment profile as $a = (a_i, a_{-i})$ where a_i is the assignment of the ith vehicle, and a_{-i} is the assignment of all *other* vehicles. Let $T_\emptyset$ denote the 'null' target, i.e., no assignment, and assume that $T_\emptyset \in \mathcal{A}_i$ for all agents. Now define the local utility function

$$U_i(a_i, a_{-i}) = U_g(a_i, a_{-i}) - U_g(T_\emptyset, a_{-i}).$$

In words, $U_i(\cdot)$ reflects the change in utility that the ith vehicle can affect. Any sort of geographical distribution of targets and vehicles reflected in the global utility is automatically reflected in this definition.

The second element is the design of negotiation protocols. Vehicles that share common targets communicate repeatedly and propose self-assignments to each other. The negotiation protocol determines how the new proposals are generated as a (stochastic) function of the history of prior proposals. In particular, Arslan *et al.* (2006) present different protocols and show that these protocols are guaranteed to converge to a Nash equilibrium of the resulting multi-player game. The performance of these protocols is compared through Monte Carlo simulations.

1.2.1.4 Inducing coordination

The problem of inducing coordination is to design individual agent control laws that use very limited descriptions of the behaviors of other agents. In terms of the discussion on predictive control, assume that two agents are not dynamically coupled but are coupled

[7] A. Arsie, J.J. Enright, and E. Frazzoli, 'On the value of information in dynamic multiple-vehicle routing problems'.

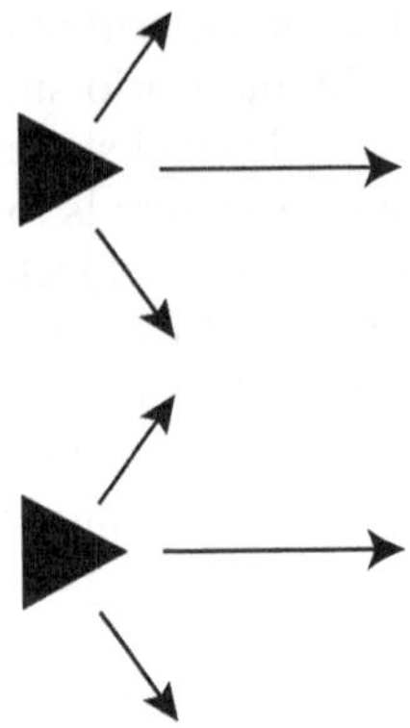

Figure 1.3 Multivehicle search.

through objective functions, as in Equations (1.2)–(1.3). A centralized control policy would have each agent's control law as a function of the state of both agents, e.g.,

$$u_1 = g_1(x_1, x_2),$$
$$u_2 = g_2(x_1, x_2)$$

The objective in inducing coordination is to derive control laws that only use a suitably defined aggregate measure, denoted by $m_i(\cdot)$, of the behavior of the other agent, as in

$$u_1 = g_1(x_1, m_1(x_2)),$$
$$u_2 = g_2(x_2, m_2(x_1)).$$

Chapter 8[8] considers the following scenario, illustrated in Figure 1.3. Two agents are traversing an unknown (gridded) map that models terrain with obstructed views. Each agent can measure the cost of travel for many steps ahead in a forward direction but has very limited view of the cost of travel in diagonal direction. The objective is for both vehicles to cross the map with minimum cost. This brings up the well-known common tension between exploration and exploitation. One the one hand, vehicles should explore the terrain in search of low cost paths. On the other hand, if vehicles are too far apart, then neither vehicle cannot take advantage of a low cost path found by the other. Chapter 8 computes the optimal policy in this scenario and shows that the optimal policy is characterized by an ideal separation. In terms of the above discussion, the separation factor serves as the 'aggregate measure' of the other vehicle's performance. If a vehicle exceeds this separation, then it must be that it has found a path whose cost is below a threshold value.

Chapter 9[9] investigates a linear version of the vehicle-target assignment problem illustrated in Figure 1.4 motivated by the 'RoboFlag drill' of Earl and D'Andrea (2002). As in Chapter 7, targets are regenerated stochastically. The approach here is that each agent makes a simplified model of the other agent in the form of a probability of being on

[8] E. Feron and J. DeMot, 'Optimal agent cooperation with local information'.
[9] V.P.-W. Seah and J.S. Shamma, 'Multiagent cooperation through egocentric modeling'.

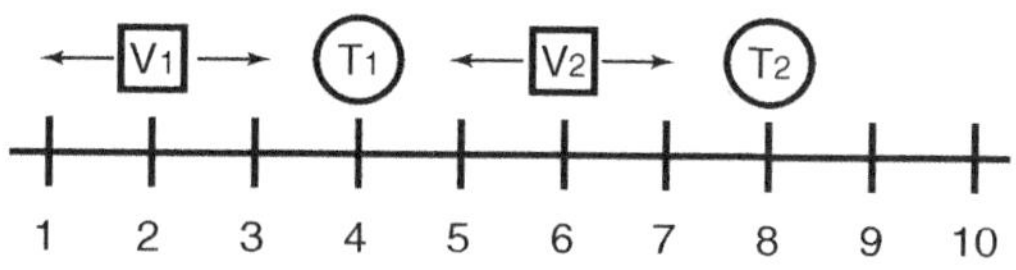

Figure 1.4 Linear vehicle-target assignment.

the left versus being on the right. This 'probability' becomes the aggregate measure of the behavior of the other agent. A semi-decentralized policy is computed through the following iterative process. First, an agent forms an empirical model of the behavior of the other agent. Second, an agent designs an optimal policy given this model. Once agents deploy the new policy, they remodel each other's behavior and redesign their individual policies. The implication of convergence is a consistency condition. Namely, each agent's behavior is consistent with how the agent is modeled by others. Furthermore, each agent's local strategy is optimal with respect to how it models other agents. Chapter 9 derives conditions for convergence for the scenario in Figure 1.4, presents illustrative simulations, and establishes a connection to the learning in games approach.

1.2.2 Adversarial interactions

An important application for cooperative control is autonomous vehicle systems operating in a possibly hostile environment. Adversarial interactions highlight that these systems must address the possibility of other hostile vehicles that are also capable of strategic planning. For example in the vehicle-target assignment problems previously discussed, targets exhibited only scripted as opposed to strategic behavior. Targets were either stationary or they regenerated stochastically. In adversarial situation, targets may make strategic decisions related to mobility or visibility.

1.2.2.1 Trajectory planning with adversaries

Both Chapters 10[10] and 11[11] look at a particular version of vehicle target assignment called 'Roboflag drill' (see Earl and D'Andrea 2002) illustrated in Figure 1.5. Mobile attackers seek to penetrate a defense zone. However, an attacker is disabled if it is 'tagged' by a defender. Suppose that attacker paths were known in advance. Then it is possible to set up an optimization problem to compute defender trajectories to intercept attacker positions. The general form of this optimization is a mixed integer linear program (MILP). MILP optimization is similar to linear programming, except that some (or all) variables are constrained to be integers. An illustrative form is

$$\max c^{\mathrm{T}} x$$

$$\text{subject to } Ax \le b,\ x_i \in \{0, 1\}\,.$$

[10] M.G. Earl and R. D'Andrea, 'Multi-vehicle cooperative control using mixed integer linear programming'.
[11] G.C. Chasparis and J.S. Shamma, 'LP-based multi-vehicle path planning with adversaries'.

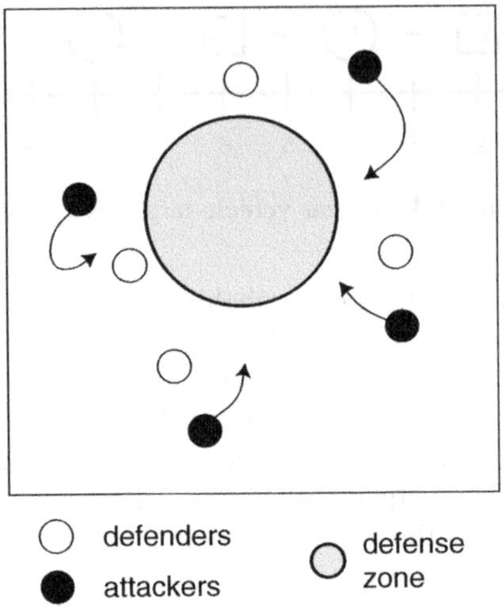

Figure 1.5 RoboFlag drill.

This problem would be a linear program if the constraint set on x were changed to the interval condition $x_i \in [0, 1]$. MILP optimization can be significantly more computationally demanding than LP optimization. However, there are effective commercial solvers. The integer constraints in Roboflag drill stem from certain binary states in the problem set-up, such as whether or not an attacker has been intercepted. Chapter 10 goes beyond the assumption of a known attacker trajectory. Rather, the assumption is a known *strategy*, which can be viewed as a guidance law, of the attackers.

Chapter 11 addresses a very similar framework. However, the approach taken is to avoid MILP formulations in favor of LP formulations. This is achieved by a predictive control implementation of the defender trajectory planning. The procedure is as follows. First, a simplified model for trajectory planning is constructed. In this simplification, certain discrete elements of the model are ignored, thereby allowing an LP computation of a defender trajectories. An initial phase of the computed trajectory is implemented, at which point the process is repeated. As in Chapter 11, there is an assumed strategy of the attackers. However, this strategy is based on myopic predictions of the attacker trajectories and updated as part of the predictive control iteration. Since the method is based on computationally simpler LP, it allows one to endow both attackers and defenders with equal strategic capability. That is, both attackers and defenders optimize online based on revised strategy predictions of their opponent.

1.2.2.2 Adversary forecasting

An important consideration in both of the above approaches was the lack of knowledge, and hence assumptions on, the strategy, or 'intent', of the opponent. In order to compute an optimal response, one must make some sort of model of opponent strategies. Recent work (Mannor *et al.* 2006) addresses this issue in the framework of learning in games. The set-up is as follows. Two players $\mathcal{P}_1$ and $\mathcal{P}_2$ repeatedly play a matrix game over times $t = \{0, 1, 2, \ldots\}$. Player $\mathcal{P}_i$ selects its action $a_i(t)$ from its action set $\mathcal{A}_i$. Each player receives a utility (reward) $U_i(a_i(t), a_{-i}(t))$. (Recall that $-i$ denotes the 'other' player.) An important assumption is that players do *not* know each other's utility function. In particular, a zero-sum structure as in

$$U_1(a_1, a_2) = -U_2(a_2, a_1)$$

is not assumed. This restriction can be interpreted as a lack of knowledge of the intent of the other player. To deal with this lack of knowledge, players make forecasts of each other's actions. More precisely, given the action history $(a_2(0), a_2(1), \ldots, a_2(t-1))$ of player $\mathcal{P}_2$, player $\mathcal{P}_1$ constructs a forecast $f_2(t)$ which can be interpreted as a probability distribution on the anticipated $a_2(t)$. Similarly, player $\mathcal{P}_2$ creates a forecast, $f_1(t)$, of $a_1(t)$. The action of each player is then to maximize its expected utility based on this forecast, i.e.,

$$a_1(t) = \max_{a_1 \in \mathcal{A}_1} \mathbf{E}_{a_2(t) \sim f_2(t)}[U_1(a_1, a_2(t))]$$

$$a_2(t) = \max_{a_2 \in \mathcal{A}_2} \mathbf{E}_{a_1(t) \sim f_1(t)}[U_2(a_2, a_1(t))].$$

We use the notation $\sim$ to denote the probability distribution over which the expectation operation is performed. Of course, this procedure relies heavily on how players forecast each other's actions. The forecasting approach in (Mannor *et al.* 2006) is one of 'calibrated forecasting'. In brief, calibrated forecasts are a 'universal' approach to construct statistically consistent (i.e., calibrated) forecasts of a discrete sequence in the absence of a model of the generator of the sequence (Foster and Vohra 1997). Unfortunately, these forecasting schemes, while universal, are not computationally feasible. In (Mannor *et al.* 2006), universality is sacrificed for computational efficiency. An alternative forecasting scheme, termed 'tracking forecast', is constructed that results in calibrated forecasts for special classes of sequence generators. This special class include many forms of learning algorithms. The final result is the ability in repeated play to optimally respond to an opponent based on observations of past play but without knowledge of the opponent's utility function, and hence, intent.

1.2.2.3 Effects of information patterns

Chapter 12[12] considers a very different problem in adversarial interactions that is at the heart of the distributed information patterns prevalent in cooperative control problems. The problem set-up one of linear systems with a combination of stochastic and deterministic inputs. The system dynamics are of the form

$$\dot{x} = Ax + B_0 d + B_1 w + B_2 u$$

$$y = Cx + n$$

The inputs d and n represent stochastic process noise and measurement noise, respectively. The inputs w and u are opposing adversarial inputs, where w seeks to maximize

[12] A. Swarup and J.L. Speyer, 'Characterization of LQG differential games with different information patterns'.

a specified cost function that u seeks to minimize. The control u has access to the output measurement y. A standard assumption in worst case paradigms such as robust control is that the adversary, w, being 'nature', is all knowing. That is, it has full access to the state and to the measurement of the controller. Chapter 12 departs from this set-up by assuming that the maximizer, w, can measure the state x but *not* the noisy output y. It turns out that such seemingly innocuous differences can significantly alter the resulting constructions of optimal controllers. (A classic illustration is the Witsenhausen example (Ho 1980; Witsenhausen 1968).) For example, under standard assumptions, optimal controllers are of the same dimension as the state dynamics. Under non-standard assumptions, such as the above setting, it is not known whether optimal controllers are even finite dimensional, let alone the same order as the state dynamics. Chapter 12 shows that under mild technical assumptions, controllers remain finite dimensional even in this non-standard information scenario.

1.2.3 Uncertain evolution

An operating environment, even if not hostile, introduces significant uncertainty in the operation of autonomous vehicle systems. Uncertain evolution means that cooperative control solutions may encompass both estimation and adaptation methods to overcome environmental uncertainty.

1.2.3.1 Hybrid mode estimation

Let us take the position of an evading mobile target. The construction of an evasion trajectory would benefit from knowledge of the target assignments of vehicles (cf., the previous discussion on unknown opponent strategies and intent). Since there is no explicit communication between vehicles and targets, a target must estimate a vehicle's assignment based on some sort of assumed model. In this setting, and for other cooperative control problems as well, such an underlying model often combines both continuous dynamics (e.g., vehicle motions) and discrete dynamics (e.g., target assignments), resulting in the now standard 'hybrid' system form.

The general assumed model is

$$x(t+1) = A(q(t))x(t) + Bw(t)$$

$$y(t) = Cx(t) + n(t)$$

where w and n are stochastic process and measurement noise, respectively. The parameter $q(t)$ takes its values from a finite discrete set $\mathcal{Q}$. The problem of mode estimation is to infer $q(t)$ based on observations $y(t)$.

Chapters 13 and 14 provide complementary approaches to address this problem. Chapter 13[13] takes an information theoretic approach. It is assumed that the mode switches, $q(t)$, are described by a Markov process with associated transition probabilities. The connection to information theory is that the mode history can be viewed as a 'message',

[13] N.C. Martins and M.A. Dahleh, 'Modal estimation of jump linear systems: an information theoretic viewpoint'.

and the system dynamics and observations can be viewed as a 'channel'. The information theoretic connection allows one to provide bounds on mode estimation errors. Chapter 14[14] takes a more explicit approach. The evolution dynamics of the mode is augmented onto the state dynamics, thereby producing a higher order nonlinear system. Approximate nonlinear filtering methods can then be applied to the augmented system to derive an online mode estimate.

1.2.3.2 Autonomous evolution of languages

The a long-term goal of cooperative control is to enable agents to explore unknown environments, communicate their findings to one another, and ultimately report back to human observers. A particularly ambitious goal is that this exploration will occur in highly unknown 'first encounter' environments. Such a scenario can be contrasted with a more standard search and rescue operation, where the uncertainty is much more structured. There is an unknown location to be specified, but the general environment of the search is well understood. By contrast, in first encounter environments, many basic elements of the environment are not known, and so agents must learn to communicate about findings that were not 'pre-programmed' into their communication protocol.

Chapter 15[15] makes an interesting connection between this scenario and the evolution of languages. Different agents will encounter different discoveries, thereby altering their internal representations of the environment. The problem of language evolution involves determining whether initially homogenous languages will evolve into mutually unintelligible dialects or stay mutually understandable. The underlying graphical structure of communication among agents plays an important role in determining the ultimate outcome of this process. Interestingly, there are strong connections to the aforementioned consensus problem as well, cf. Equation (1.1).

1.2.4 Complexity management

Cooperative control problem formulations, if taken at face value, often result in computationally prohibitive solutions. Complexity management seeks to alleviate the inherent computational complexity through exploitation of special structures or the introduction of effective approximations.

Many of the problems discussed have some element of complexity management by virtue of an imposed distributed structure. For example, much of the task assignment work involved deriving effective solutions while bypassing an explicit optimization of the assignment problem. Chapters 16–17 take a more explicit view at complexity management.

Chapter 16[16] revisits the problem of hybrid mode estimation, but for multiple agents. A motivating problem is estimating a *collective* assignment profile through observations of a group of vehicles. A complementary approach to stochastic methods in estimation

[14] D. Choukroun and J.L. Speyer, 'Conditionally-linear filtering for mode estimation in jump-linear systems'.
[15] Y. Lee, T.C. Collier, C.E. Taylor, and E.P. Stabler, 'Cohesion of languages in grammar networks'.
[16] D. Del Vecchio and R.M. Murray, 'Complexity management in the state estimation of multi-agent systems'.

is set-valued estimation (e.g., Milanese and Vicino 1991; Schweppe 1973; Shamma and Tu 1997, 1999). The main idea is to compute the entire *set* of values (as opposed to a single estimate) of an unknown quantity that are consistent with available observations to date. For example, for linear systems with bounded process disturbances and measurement noises, this amounts to constructing the set of all possible states based on output measurements. Not surprisingly, the exact characterization of such a set in the absence of any additional structure can be computationally prohibitive. Chapter 16 addresses this problem by superimposing a lattice structure on the state dynamics. The benefit of this lattice structure is that set-valued estimates can be expressed as generalized intervals, as in $\mathrm{lb(t)} \leq \mathrm{x(t)} \leq \mathrm{ub(t)}$, where $\mathrm{lb}(t)$ and $\mathrm{ub}(t)$ denote lower bounds and upper bounds from the lattice structure, respectively. As new measurements are taken, these lower and upper bounds are recomputed. The approach is demonstrated on the problem of estimating an assignment profile based on vehicle motions. This also falls under the heading of hybrid mode estimation, but without a stochastic structure imposed on mode dynamics.

Chapter 17[17] discusses complexity management from the perspective of scenario representation. The objective is to derive representations of a multivehicle scenario for a human operator that are both descriptive and not overly detailed. The challenge is to find an appropriate level of granularity and to represent this level of detail in a graphically appealing manner. The specific framework taken here is abstraction-based patch models. Rather than specify locations and trajectories of vehicles, patch models specify sets of possibilities. The tradeoff is that the omission of details results in the admission of auxiliary scenarios. By having a variable level of resolution, auxiliary scenarios can be pruned. In many ways, the framework resembles the set-valued estimation of Chapter 16. The main difference is that the set of possibilities is deliberately constructed as a representation device as opposed to the outcome of an estimation process.

1.3 FUTURE DIRECTIONS

We conclude this opening chapter with some comments on future directions in cooperative control. It should be clear from the present discussion that the scope of cooperative control admits multiple problem formulations for addressing multiple aspects of the underlying issues. And so, rather than suggest specific open questions, we will offer some general directions that have received relatively limited attention.

- *Strategic decision-making*: Much of the effort in cooperative control has been devoted to multivehicle trajectory planning and motion execution, such as formation keeping, rendezvous, or area coverage. Less attention has been given to *strategic* decision-making. Strategic decision-making entails determining, coordinating and executing a plan of attack in response to real-time data as events unfold. In terms of a feedback hierarchy, strategic decision-making involves higher level planning, while multivehicle motion planning is at a lower level (with conventional vehicular control at the lowest level). Some of the work in vehicle-task assignment falls under strategic decision-making, but in more general settings, one may not have an underlying geometric framework to exploit.

[17] V. G. Rao, S. Goldfarb and R. D'Andrea, 'Abstraction-based command and control with patch models'.

- *Increasing role of game theory*: The use of game theory in cooperative control has largely been limited to the concept of Nash equilibrium in zero sum or general sum games. Game theory, by definition, is concerned with the interactions between multiple decision-makers, and hence it is clearly relevant to cooperative control. There are a variety of important concepts, such as mechanism design, bargaining, coalition theory, and correlated equilibrium that are not yet widely known. A challenge in exploiting these concepts is that they were developed with the intention of being models of social phenomena, whereas cooperative control applications involve engineered systems and require prescriptive synthesis tools.
- *Benchmark scenarios and problems*: A collection of benchmark scenarios would help to compare different approaches in the broad scope of cooperative control research. The terminology benchmark 'scenario' rather than benchmark 'problem' is deliberate here. For example, we have used vehicle target assignment as a benchmark scenario for much of the discussion in this chapter. Various research results were discussed in terms of the relevant component of this benchmark scenario. Benchmark problems are much more structured, where competing methodologies can be compared quantitatively. Benchmark problems, stemming from benchmark scenarios would also be very valuable. Here, one faces the usual challenge of defining specific problems that are neither trivially concocted nor intractable.

ACKNOWLEDGEMENTS

Research supported by AFOSR/MURI grant #F49620-01-1-0361, NSF grant #ECS-0501394, and ARO grant #W911NF–04–1–0316.

REFERENCES

Ahuja RK, Kumar A, Jha K and Orlin JB 2003 Exact and heuristic methods for the weapon-target assignment problem. Technical Report #4464-03, MIT, Sloan School of Management Working Papers.

Arslan G, Marden JR and Shamma JS 2007 Autonomous vehicle-target assignment : A game theoretical formulation. *ASME Journal of Dynamic Systems, Measurement and Control* **129**(5) 584–596.

Beard RW, McLain TW, Goodrich MA and Anderson EP 2002 Coordinated target assignment and intercept for unmanned air vehicles. *IEEE Transactions on Robotics and Automation* **18**(6), 911–922.

Bemporad A and Morari M 1999 Robust model predictive control: A survey. In *Robustness in Identification and Control* (ed. Garully A, Tesi A and Vicino A), vol. **245**. Springer-Verlag, Berlin, Lecture Notes in Control and Information Sciences.

Blondel VD, Hendrickx JM, Olshevsky A and Tsitsiklis J 2005 Convergence in multiagent coordination, consensus, and flocking. In *44th IEEE Conf. on Decision and Control*, pp. 2996–3000.

Earl M and D'Andrea R 2002 A study in cooperative control: The Roboflag drill. In *Proceedings of the American Control Conference*, Anchorage, Alaska.

Foster D and Vohra R 1997 Calibrated learning and correlated equilibrium. *Games and Economic Behavior* **21**, 40–55.

Fudenberg D and Levine D 1998 *The Theory of Learning in Games*. MIT Press, Cambridge, MA.

Ho YC 1980 Team decision theory and information structures. *Proceedings of the IEEE* **68**(6), 644–654.

Jadbabaie A, Lin J and Morse AS 2003 Coordination of groups of mobile autonomous agents using nearest neighbor rules. *IEEE Trans. on Automatic Control* **48**(6), 988–1001.

Kashyap A, Basar T and Srikant R 2006 Consensus with quantized information updates. In *45th IEEE Conference on Decision and Control*, pp. 2728–2733.

Mannor S, Shamma JS and Arslan G 2007 Online calibrated forecasts: Memory efficiency versus universality for learning in games. *Machine Learning*. Special issue on Learning and Computational Game Theory **129**(5) 585–596.

Marden JR, Arslan G and Shamma JS 2007 Connections between cooperative control and potential games illustrated on the consensus problem. In *Proceedings of the European Control Conference*, 4604–4611.

Milanese M and Vicino V 1991 Optimal estimation theory for dynamic systems with set membership uncertainty: An overview. *Automatica* **27**, 997–1009.

Moreau L 2004 Stability of continuous-time distributed consensus algorithms. In *43rd IEEE Conference on Decision and Control*, pp. 3998–4003.

Murphey R 1999 Target-based weapon target assignment problems. In *Nonlinear Assignment Problems: Algorithms and Applications* (ed. Pardalos P and Pitsoulis L). Kluwer Academic Publishers, Dord recht, pp. 39–53.

Murray R 2006 Recent research in cooperative control of multi-vehicle systems. Submitted to *ASME Journal of Dynamic Systems, Measurement, and Control.*

Olfati-Saber R and Murray RM 2004 Consensus problems in networks of agents with switching topology and time-delays. *IEEE Trans. on Automatic Control* **AC–49**(9), 1520–1533.

Olfati-Saber R, Fax JA and Murray RM 2007 Consensus and cooperation in networked multi-agent systems. In *Proceedings of the IEEE* **95**(1), 215–233.

Panai L and Luke S 2005 Cooperative multi-agent learning: The state of the art. *Autonomous Agents and Multi-Agent Systems* **11**, 387–434.

Schweppe F 1973 *Uncertain Dynamic Systems*. Prentice-Hall, Englewood Cliffs, NJ.

Shamma J and Tu KY 1997 Approximate set-valued observers for nonlinear systems. *IEEE Transactions on Automatic Control* **AC–42**(5), 648–658.

Shamma J and Tu KY 1999 Set-valued observers and optimal disturbance rejection. IEEE Transactions on Automatic Control **AC–44**(2), 253–264.

Tsitsiklis JN, Bertsekas DP and Athans M 1986 Distributed asynchronous deterministic and stochastic gradient optimization algorithms. *IEEE Transactions on Automatic Control* **35**(9), 803–812.

Witsenhausen H 1968 A counterexample in stochastic optimum control. *SIAM Journal of Control* **6**(1), 131–147.

Wolpert D and Tumer K 1999 An overview of collective intelligence In *Handbook of Agent Technology* (ed. Bradshaw JM) AAAI Press/MIT Press, Cambridge, MA.

Xiao L and Boyd S 2004 Fast linear iterations for distributed averaging. *Systems and Control Letters* **53**(1): 65–78.

Xiao L and Boyd S 2005 A scheme for robust distributed sensor fusion based on average consensus. In *Information Processing in Sensor Networks*, pp. 63–70.

Young H 2006 *Strategic Learning and its Limits*. Oxford University Press, Oxford.

Part II

Distributed Control and Computation

2

Design of behavior of swarms: From flocking to data fusion using microfilter networks

Reza Olfati-Saber

2.1 INTRODUCTION

Complex networks of interacting *agents*, dynamic systems, and humans appear in broad applications in engineering, biological, and social systems. We refer to these multi-agent complex networks as *swarms*. Swarms can form by interconnections of homogeneous or heterogeneous agents. In many engineering applications, the information flow among the agents is specified by *complex networks*.[1] In science, the researchers try to understand why certain types of collective behavior of swarms emerge. Our objective is to design the local interaction rules for swarms so that the engineered swarm exhibits a desired collective behavior.

In this chapter, we focus on design and analysis of two categories of collective behavior of swarms: (1) flocking for swarms of mobile agents (e.g. robots and unmanned vehicles): and (2) distributed data fusion in sensor networks using swarms of microfilters.

At first, sensor networks and flocks do not seem to be fundamentally related as motion coordination for flocks is a robotics problem, whereas data fusion in sensor networks is a form of distributed computing or signal processing problem. We demonstrate that there is a common thread that connects alignment and spatial self-organization in flocks to information fusion in sensor networks. This thread turns out to be consensus problems for networked dynamic systems that have recently emerged as the key in design of distributed algorithms for both motion coordination tasks such as rendezvous by Cortés *et al.* (2006), flocking by Olfati-Saber (2006), and formation control by Fax (2001); Fax and Murray (2004); Olfati-Saber and Murray (2002) as well as information processing tasks in sensor networks by Olfati-Saber (2005a); Olfati-Saber and Shamma (2005); Spanos *et al.* (2005a, b).

[1] See Newman (2003) for a survey.

Cooperative Control of Distributed Multi-Agent Systems edited by Jeff S. Shamma

The outline of this chapter is as follows. Some background on consensus problems for networked dynamic systems is provided in Section 2.2. A detailed discussion on flocking and its role in distributed area coverage with mobile agents is presented in Section 2.3. Swarms of cooperative microfilters for data fusion in sensor networks are discussed in Section 2.4. Several examples and simulation results will be presented throughout the chapter.

2.2 CONSENSUS PROBLEMS

Consider a group of agents with opinion dynamics

$$\dot{x}_i = u_i$$

where every node only interacts with its neighbors $N_i = \{j : (i, j) \in E\}$ on a graph $G = (V, E)$ as shown in Figure 2.1. In general, G is a directed graph with the set of nodes $V = \{1, , \ldots, n\}$, the set of edges $E \subset V \times V$, and the adjacency matrix $A = [a_{ij}]$ with non-negative elements. In reaching a consensus, the objective of the agents is to converge to the following space

$$x_1 = x_2 = \cdots = x_n$$

where the opinion of all agents is the same. This can be achieved by applying a linear consensus protocol

$$\dot{x}_i = \sum_{j \in N_i} a_{ij}(x_j - x_i). \tag{2.1}$$

For undirected graphs, it is easy to see that the linear system in Equation (2.1) is identical to a gradient system

$$\dot{x} = -\nabla \Phi_G(x)$$

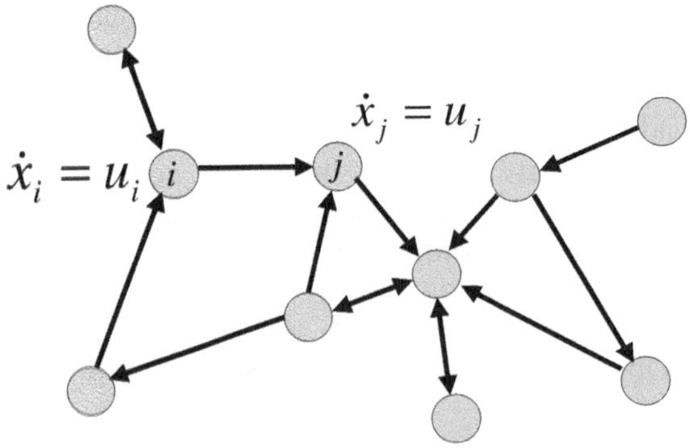

Figure 2.1 A network of dynamic systems $\dot{x}_i = u_i$. Reprinted from R. Olfati-Saber, J. A. Fax, and R. M. Murray, "Consensus and cooperation in networked multi-agent systems," Proceedings of the IEEE, vol. 95, no. 1, pp. 215–233, Jan. 2007. ©2007 IEEE.

with a potential function

$$\Phi_G(x) = \frac{1}{2}\sum_{i<j} a_{ij}(x_j - x_i)^2 \tag{2.2}$$

called the disagreement function. The disagreement function quantifies the total disagreement for a group of agents with opinion vector x and information flow G. The consensus dynamics in Equation (2.1) can be rewritten as

$$\dot{x} = -Lx$$

where $L = D - A$ is the graph Laplacian matrix, $D = \mathrm{diag}(A\mathbf{1})$ is the degree matrix with diagonal elements $d_i = \sum_j a_{ij}$, and $\mathbf{1} = (1, \ldots, 1)^T$ is the vector of ones. By definition, $L\mathbf{1} = 0$ and thus, graph Laplacian always has a zero eigenvalue $\lambda_1 = 0$. If G is a strongly connected graph,[2] or has a directed spanning tree,[3] $\mathrm{rank}(L) = n - 1$ and all other eigenvalues of L are nonzero. In other words, an equilibrium of system Equation (2.1) is a state in the form $x^* = (\alpha, \ldots, \alpha)^T = \alpha\mathbf{1}$ where all nodes agree.

Theorem 2.2.1 *(Average-Consensus Theorem by Olfati-Saber and Murray (2004)) Let G be a strongly connected digraph. Then, all agents asymptotically reach a consensus opinion α. Moreover, assuming that the digraph G is balanced (i.e. $\sum_j a_{ij} = \sum_j a_{ji}, \forall i$), a consensus to $\alpha = \overline{x}(0) = 1/n\sum_{i=1}^{n} x_i(0)$ is reached exponentially fast with a rate greater or equal to $\lambda_2((L + L^T)/2)$, i.e. the second-largest eigenvalue, or algebraic connectivity, of the symmetric part of the Laplacian of the digraph G).*

The above theorem allows one to use the consensus protocol in Equation (2.1) for two purposes: (1) creating alignment in flocks of mobile agents, or agreement in social networks: and (2) distributed computation of averages, multiplications, and aggregate function of the sensed measurements in sensor networks by Olfati-Saber and Shamma (2005); Olfati-Saber *et al.* (2006); Spanos *et al.* (2005a). We intend to use extensions of this basic consensus algorithm to achieve both objectives.

In complex networked systems, some adversarial phenomena such as packet-loss, node/link failures, and presence of obstacles that cut communication links between unmanned vehicles can be detrimental to stability and performance of the overall system. All these effects in a network of agents can be modeled as changes in the topology of the network that results in multi-agent systems with *switching networks*. A dynamic graph $G(t)$ can be represented as $G(t) : [0, \infty) \to \mathcal{G}$ with $\mathcal{G}$ being the superset of all directed graphs with edges (i, j) in $V \times V$. Let $L(t)$ denote the graph Laplacian of $G(t)$. Then, the following linear switching system

$$\dot{x} = -L(t)x$$

is capable of capturing the model of consensus in networks with packet-loss and link/node failures. For example, loss of a packet sent from node j to node i can be represented as

[2] See Olfati-Saber and Murray (2004).

[3] See Ren and Beard (2005).

the loss of the link a_{ij} in the network. The loss of a node in a network with topology G can be modeled as the loss of a subgraph G_i of G that is induced by node i and its neighbors N_i.

Consensus in networks with switching topology is considered by several researchers including Jadbabaie *et al.* (2003); Moreau (2005); Olfati-Saber and Murray (2004); Ren and Beard (2005). For a survey on this subject, the reader can refer to Olfati-Saber *et al.* (2007). The alignment rules analyzed in the work of Jadbabaie *et al.* (2003); Moreau (2005); Ren and Beard (2005) are all special cases of the protocol introduced in Fax (2001) for multi-vehicle formation stabilization. Fax's algorithm is in turn a special case of the consensus algorithm Equation (2.1) by Olfati-Saber and Murray (2004). Furthermore, contrary to popular belief and the assertions in Jadbabaie *et al.* (2003), none of the above alignment algorithms is due to Vicsek *et al.* (1995).

Remark 2.2.1 The correct form of the alignment rule by Vicsek *et al.* (1995) is a nonlinear consensus protocol given in Equation (8) of Moreau (2005). The term *Vicsek Model* that has often appeared in recent control literature[4] is a 'myth' that has to nothing with the model studied numerically by Vicsek and co-workers in Vicsek *et al.* (1995). Unfortunately, this myth has propagated through many papers that refer to Fax and Murray's alignment rule applied to networks with n self-loops as Vicsek's rule. It turns out that both of these rules are equivalent to the flock centering rule of Reynolds (1987).

Some of the most recent advances in consensus theory include consensus on random graphs by Hatano and Mesbahi (2005); ultrafast consensus in small-world networks of Watts and Strogatz (1998) and Ramanujan graphs by Olfati-Saber (2005b, 2007); χ-consensus[5] in its most general form by Cortés (2006a, b); consensus with quantized states by Kashyap *et al.* (2006); and modeling consensus problems using potential games by Marden *et al.* (2007).

A comprehensive introduction on discrete-time consensus problems is provided in the survey by Olfati-Saber *et al.* (2007). In discrete-time, the Perron matrix for a dynamic graph G_k with n self-loops is defined as

$$P_k = I_n - \varepsilon_k L(G_k) \tag{2.3}$$

with $0 < \varepsilon_k < 1/\max_i d_i(G_k)$. This matrix plays a key role in specifying the iterative form of consensus algorithm as

$$x(k+1) = P_k x(k).$$

Remark 2.2.2 In the work of Fax and Murray (2004), Jadbabaie *et al.* (2003) and Moreau (2005), and Ren and Beard (2005), the Perron matrix is used with a fixed $\varepsilon_k = 1$ and a Laplacian $L(G_k) = I_n - D(G_k)^{-1}A(G_k)$ that results in the Perron matrix $P_k = D(G_k)^{-1}$ $A(G_k)$.

[4] See Blondel *et al.* (2005); Jadbabaie *et al.* (2003).

[5] By χ-consensus, we mean that all agents reach an agreement regarding the value of a multi-variable function $\chi(x)$ that depends on the state of all agents. See Olfati-Saber and Murray (2004).

By definition, the Perron matrix P of a graph G with Laplacian L is a stochastic matrix satisfying $P\mathbf{1} = \mathbf{1}$. Thus, the convergence of the consensus algorithm

$$x_i(k+1) = x_i(k) + \varepsilon_k \sum_{j \in N_i} a_{ij}(x_j(k) - x_i(k)) \tag{2.4}$$

requires verification of the convergence of infinite products of Perron stochastic matrices. Independent of recent advances in consensus problems for networked dynamic systems, Tsitsiklis (1984) and Tsitsiklis *et al.* (1986) studied the convergence of products of stochastic matrices of this type in the context of convergence of stochastic gradient-descent algorithms based on the properties of weakly ergodic Markov chains.[6] From the original work of Tsitsiklis *et al.* (1986), it turns out that asynchronous consensus is equivalent to consensus on a switching network that satisfies an appropriate form of collective connectivity, e.g. the union of the set of edges of all graphs on an interval of length T must be strongly connected,[7] or have a directed spanning tree.[8]

2.3 FLOCKING BEHAVIOR FOR DISTRIBUTED COVERAGE

The availability of unmanned vehicles and mobile sensors allows deployment of large numbers of such mobile agents in an environment for the purpose of information gathering, surveillance, search and rescue, disaster relief, and combat. All of these tasks require distributed coverage of an environment – the main focus of this section. Networks of many mobile agents constitute swarms. Due to mobility of the agents, the resulting swarm has a dynamic topology that varies as a function of the position of the agents. We call this spatially-induced network a proximity graph[9] (see Cortés *et al.* (2006); Olfati-Saber (2006)). The proximity subgraph of agent i is depicted in Figure 2.2. The proximity graph is the union of the proximity subgraph of all agents. Formally speaking, let $q_i \in \mathbb{R}^m$ denote the position of agent i in an m-dimensional space, then the set of neighbors of i

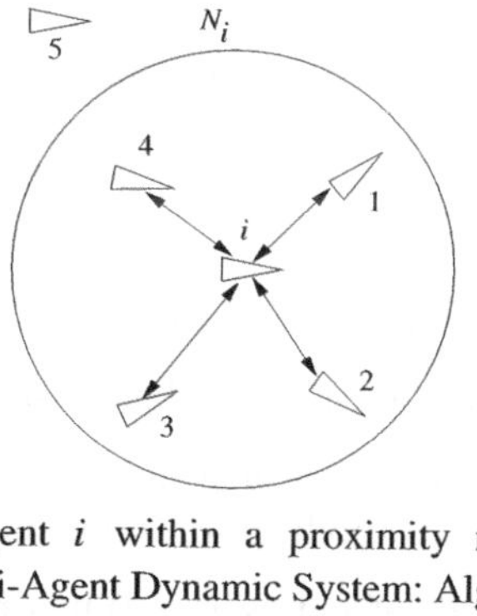

Figure 2.2 The neighbors of agent i within a proximity radius of $r > 0$. Reprinted from Olfati-Saber, R., "Flocking for Multi-Agent Dynamic System: Algorithms and Theory," IEEE Trans. on Automatic Control, vol. 51, no. 3, pp. 401–420, March 2006. ©2006 IEEE.

[6] See Tsitsiklis and Bertsekas (2006).
[7] See Tsitsiklis and Bertsekas (2006); Tsitsiklis *et al.* (1986).
[8] See Moreau (2005).
[9] These graphs are also known as 'state-dependent graphs' in the work of Mesbahi (2005).

are $N_i(q) = \{j : |q_j - q_i| < r, j \neq i\}$ and the set of vertices and edges of the proximity graph $G(q) = (V, E(q))$ are defined as

$$\begin{aligned} V &= \{1, 2, \ldots, n\} \\ E(q) &= \{(i, j) \in V \times V : |q_j - q_i| < r, j \neq i\} \end{aligned} \tag{2.5}$$

where $|\cdot|$ denotes the 2-norm in $\mathbb{R}^m$ and $q = \text{col}(q_1, \ldots, q_n)^T$.

Inspired by cohesive collective behavior of flocks of birds and schools of fish in nature, we propose a distributed algorithm for area coverage that relies on creation of flocking behavior for swarms. The fundamental challenge in this approach is to provide a mathematical model of flocks and a formal representation of key features in flocking behavior. In addition, one needs to develop a meaningful notion of 'stability of flocking motion' that can be verified for a proposed flocking algorithm – this is rather similar to defining when a nonlinear state feedback $u = k(x)$ stabilizes a nonlinear control system $\dot{x} = f(x, u)$ in the sense of Lyapunov.

In the past, three 'heuristic rules' of flocking were introduced by Reynolds (1987) called *cohesion, separation*, and *alignment* that led to creation of the first animation of flocking. Recently, a formal analysis of novel flocking algorithms for a particle-based model of swarms that encompass these rules was introduced by Olfati-Saber (2006). Reynolds rules for continuum model of flocks are formally analyzed by Raymond and Evans (2006). Some notable studies on numerical analysis of behavior of particle flocks include the work of D'Orsogna *et al.* (2006), Helbing *et al.* (2000), Levine *et al.* (2001), and Vicsek *et al.* (1995). Distributed coverage of a restricted area without flocking was also studied in the past by Cortés *et al.* (2004).

Let us consider a group of mobile agents (particles) with the following dynamics

$$\begin{cases} \dot{q}_i = p_i, \\ \dot{p}_i = u_i, \end{cases} \tag{2.6}$$

where $q_i, p_i, u_i \in \mathbb{R}^m$ are position, velocity, and control input of agent i, respectively. We would like to create flocks with local interactions that their emergent behavior satisfies the following conditions in steady-state:

(i) *Alignment*: the velocity of the agents is aligned.

(ii) *Spatial self-organization*: the agents form lattice-shaped meshes/structures in space.

(iii) *Collision and obstacle avoidance*: nearby agents have a tendency to avoid collision with each other and obstacles (both in transient-state and steady-state).

(iv) *Cohesion*: there exists a ball of radius $R > 0$ that contains all the agents.

(v) *Connectivity*: the agents self-assemble connected networks of mobile agents.

The above conditions have nothing to do with Reynolds flocking rules. These conditions are desired features that define flocking as a form of collective behavior of particle-based model of swarms.[10] For simplicity, assume that our objective is to avoid collision with spherical obstacles and straight walls (i.e. convex obstacles with smooth boundaries).

Remark 2.3.1 In the existing control literature, the term 'flocking' is often misused. Anything from formation control for multi-robot systems to alignment and synchronization of oscillator networks is incorrectly referred to as flocking. From the above conditions, it is clear that it takes more than just alignment or a simple consensus to achieve flocking behavior, i.e. flocking is a multifaceted form of collective behavior of swarms that is more complex than synchronization of oscillators and alignment.

To achieve velocity alignment, one can use consensus protocols on velocity of the agents. For collision avoidance, the use of attractive/repulsive potentials is common. However, no obvious solutions come to mind to achieve the remaining conditions of flocking – namely, conditions (ii), (iv), and (v). It turns out that the answer to the remaining questions is almost entirely hidden in the design of 'collective potential' of flocks and introduction of 'virtual agents' that are embedded in the computing infrastructure of physical agents.

To separate the physical agents from virtual agents, we refer to the members of a swarm as *α-agents*. Later, we introduce two types of virtual agents: (1) *β-agents* that move on the boundary of obstacles and are obtained from the projection of α-agents on their neighboring obstacles: and (2) *γ-agents* that are embedded in α-agents as an internal dynamics and contain the information regarding a desired trajectory for the center of the mass of the entire flock for migration/tracking. Figure 2.3 illustrates the role of each type of agent.

2.3.1 Collective potential of flocks

The following set of pairwise distance-based algebraic constraints

$$\|q_j - q_i\| = d, \quad \forall j \in N_i(q) \tag{2.7}$$

defines a class of lattice-shaped configurations q called an *α-lattice*. Examples of conformations of α-lattices are shown in Figure 2.4. A configuration q that approximately

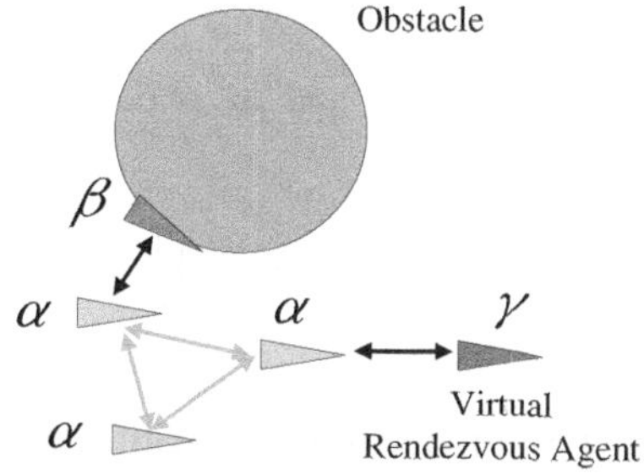

Figure 2.3 The role of α, β, and γ agents.

[10] The notion of flocking behavior is formally defined in Olfati-Saber (2006).

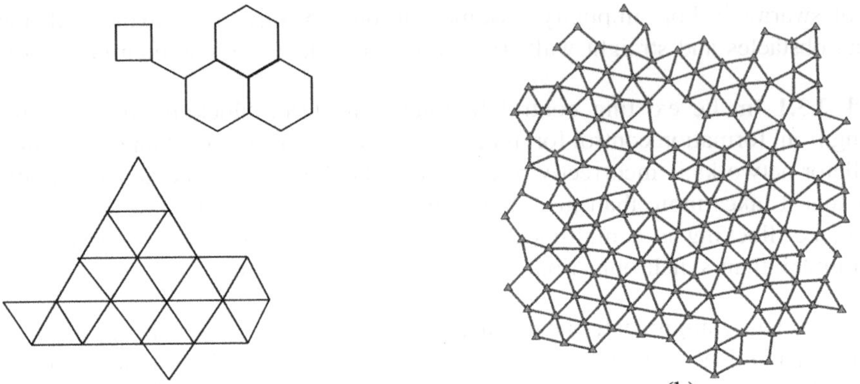

(a) (b)

Figure 2.4 Examples of α-lattices and quasi α-lattices: (a) a 2-D α-lattice with a disconnected proximity net and (b) a 2-D quasi α-lattice with $n = 150$ nodes. Reprinted from Olfati-Saber, R., "Flocking for Multi-Agent Dynamic System: Algorithms and Theory," IEEE Trans. on Automatic Control, vol. 51, no. 3, pp. 401–420, March 2006. ©2006 IEEE.

satisfies the above algebraic constraints is called a 'quasi α-lattice', i.e. $-\delta < |q_j - q_i| - d < \delta$. Parameters d and $\kappa = r/d$ are called 'scale' and 'ratio' of an α-lattice, respectively.

We find it convenient to define a map[11] $\|\cdot\||: \mathbb{R}^m \to \mathbb{R}_{\geq 0}$ called *σ-norm* for an m-vector z as

$$\|z\| = \frac{1}{\varepsilon}[\sqrt{1+\varepsilon|z|^2} - 1] \tag{2.8}$$

where $\varepsilon > 0$ is a parameter of the σ-norm and the gradient of this map is given by

$$\sigma_\varepsilon(z) = \frac{z}{\sqrt{1+\varepsilon|z|^2}} = \frac{z}{1+\varepsilon\|z\|}. \tag{2.9}$$

The benefit of defining σ-norm is that $\|z\|$ is differentiable everywhere, but the 2-norm $|z|$ is not differentiable at $z = 0$. Later, this feature becomes useful in introducing smooth gradient-based flocking algorithms. Using a smooth bump function $\rho_h : \mathbb{R} \to [0, 1]$, one can define smooth spatial adjacency (or interaction) coefficients $a_{ij}(q)$ for a swarm of particles as

$$a_{ij}(q) = \rho_h(\|q_j - q_i\|/r_\alpha) \in [0, 1], \quad j \neq i \tag{2.10}$$

where $r_\alpha = \|r\|$, $h \in (0, 1)$ is a parameter, and ρ_h is defined by

$$\rho_h(z) = \begin{cases} 1, & z \in [0, h) \\ \frac{1}{2}\left[1 + \cos\left(\pi \frac{(z-h)}{(1-h)}\right)\right], & z \in [h, 1] \\ 0 & \text{otherwise.} \end{cases} \tag{2.11}$$

One can show that $\rho_h(z)$ is a C^1-smooth function with the property that $\rho'_h(z) = 0$ over the interval $[1, \infty)$ and $|\rho'_h(z)|$ is uniformly bounded in z.

[11] This is not a norm but has similarities to a norm.

Now, the smooth collective potential function $V(q)$ of a flock can be expressed as

$$V(q) = \frac{1}{2}\sum_{i}\sum_{j\neq i}\psi_\alpha(\|q_j - q_i\|_\sigma) \tag{2.12}$$

where $\psi_\alpha(z)$ is a smooth pairwise attractive/repulsive potential with a finite cut-off at $r_\alpha = \|r\|$ and a global minimum at $z = d_\alpha = \|d\|$. This potential takes its minimum at a configuration q that is an α-lattice, i.e. the set of q satisfying

$$\|q_j - q_i\| = d_\alpha, \quad \forall j \in N_i(q). \tag{2.13}$$

To construct a smooth pairwise potential $\psi_\alpha(z)$ with a finite cut-off, we integrate an action function $\phi_\alpha(z)$ that vanishes for all $z \geq r_\alpha$. Define this action function as

$$\begin{aligned} \phi_\alpha(z) &= \rho_h(z/r_\alpha)\phi(z - d_\alpha) \\ \phi(z) &= \tfrac{1}{2}[(a+b)\sigma_1(z+c) + (a-b)] \end{aligned} \tag{2.14}$$

where $\sigma_1(z) = z/\sqrt{1+z^2}$ and $\phi(z)$ is an uneven sigmoidal function with parameters that satisfy $0 < a \leq b$, $c = |a-b|/\sqrt{4ab}$ to guarantee $\phi(0) = 0$. The pairwise attractive/repulsive potential $\psi_\alpha(z)$ in Equation (2.12) is defined as

$$\psi_\alpha(z) = \int_{d_\alpha}^{z}\phi_\alpha(s)\,ds. \tag{2.15}$$

2.3.2 Distributed flocking algorithms

We propose the following distributed algorithm for flocking

$$u_i = u_i^\alpha + u_i^\beta + u_i^\gamma \tag{2.16}$$

with interactions between α-agents and their neighboring α- and β-agents and also their internal γ-agent. The terms of this algorithm was originally introduced in Equation (59) of Olfati-Saber (2006). The first term is

$$u_i^\alpha = \underbrace{\sum_{j\in N_i}\phi_\alpha(\|q_j - q_i\|)\mathbf{n}_{ij}}_{\text{gradient-based term}} + \underbrace{\sum_{j\in N_i} a_{ij}(q)(p_j - p_i)}_{\text{consensus term}} \tag{2.17}$$

where $\mathbf{n}_{ij} = \sigma_\varepsilon(q_j - q_i) = (q_j - q_i)/\sqrt{1+\varepsilon|q_j - q_i|^2}$ is a vector along the line connecting q_i to q_j and $\varepsilon \in (0, 1)$ is a fixed parameter of the σ-norm. The gradient-based term is equal to $-\nabla_{q_i}V(q)$. The term u_i^α represents all the interactions among α-agents, u_i^β is a repulsive force that keeps an α-agent away from its projection on obstacles (or β-agents), and

$$u_i^\gamma = -c_1(q_i - q_r) - c_2(p_i - p_r); \quad c_1, c_2 > 0$$

is the tracking control that allows α-agents to track their internal γ-agents with dynamics

$$\begin{cases} \dot{q}_r = p_r, \\ \dot{p}_r = f_r(q_r, p_r), \end{cases} \tag{2.18}$$

that are initialized to $(q_r(0), p_r(0)) = (q_d, p_d)$ for all γ-agents.

If one is not interested in tracking or obstacle avoidance for flocks, one can use two simpler algorithms $u_I = u_i^\alpha$ and $u_{II} = u_i^\alpha + u_i^\gamma$. In Olfati-Saber (2006), it was demonstrated that Algorithm I encompasses all three rules of Reynolds and leads to fragmentation (and not flocking). Moreover, Algorithm II leads to flocking behavior. By 'fragmentation', we mean that small flocks of particles form that incohesively disperse in random directions. Contrary to popular belief, Reynolds rules only create flocking behavior under restricted initial conditions of the particles, whereas Algorithm II (or $u = u_i^\alpha + u_i^\gamma$) generically leads to the creation of flocking.

According to flocking Algorithm III of Olfati-Saber (2006) involving all three species of agents, flocks need no leaders for migration. This is consistent with earlier studies on schools of fish by animal behavior experts including Shaw (1975) and Partridge (1982).

2.3.3 Stability analysis for flocking motion

Let q_c and p_c denote the position and velocity of the center of mass of the swarm of particles. Define $x_i = q_i - q_c$ and $v_i = p_i - p_c$. This allows one to express the structural dynamics of a swarm applying Algorithms I and II as nonlinear systems Σ_1 and Σ_2 in the following:

$$\Sigma_1 : \begin{cases} \dot{x} &= v \\ \dot{v} &= -\nabla V(x) - \hat{L}(x)v \end{cases} \tag{2.19}$$

with a multi-dimensional Laplacian matrix[12] $\hat{L}(x)$. In comparison, the structural dynamics of a group of agents applying Algorithm II is in the form

$$\Sigma_2 : \begin{cases} \dot{x} &= v \\ \dot{v} &= -\nabla U_\lambda(x) - D(x)v \end{cases} \tag{2.20}$$

where $U_\lambda(x)$ is called the aggregate potential function and is defined by

$$U_\lambda(x) = V(x) + \lambda J(x). \tag{2.21}$$

The map $J(x) = \frac{1}{2}\sum_{i=1}^{n} \|x_i\|^2$ is the moment of inertia of all particles and $\lambda = c_1 > 0$ is a parameter of u_i^γ. The damping matrix $D(x)$ is given by $D(x) = c_2 I + \hat{L}(x)$ and is a

[12] Let $L(x)$ be the Laplacian of a spatially-induced graph $G(x)$ with adjacency matrix $A(x) = [a_{ij}(x)]$. Then, $\hat{L}(x) = L(x) \otimes I_m$ where $\otimes$ is the Kronecker product of matrices.

positive definite matrix. The structural Hamiltonians of systems Σ_1 and Σ_2 are

$$\begin{aligned} H(x, v) &= V(x) + K(v), \\ H_\lambda(x, v) &= U_\lambda(x) + K(v), \end{aligned} \tag{2.22}$$

where $K(v) = \frac{1}{2}\sum_i \|v_i\|^2$ is the velocity mismatch function of the entire group.

We say a swarm of particles is a flock if the proximity graph of the agents is connected. A swarm of particles is called cohesive over an interval $[t_0, t_f)$ if there exists a ball of radius $R > 0$ at the center of mass of the swarm that contains the agents over that interval. The main stability results regarding these two nonlinear systems are proven in Olfati-Saber (2006) and rephrased in the following:

Theorem 2.3.1 *Consider a group of α-agents applying protocol Equation* (2.17) *(Algorithm I) with structural dynamics Σ_1. Let $\Omega_c = \{(x, v) : H(x, v) \leq c\}$ be a level-set of the Hamiltonian $H(x, v)$ of Σ_1 such that for any solution starting in Ω_c, the agents form a cohesive flock $\forall t \geq 0$. Then, the following statements hold:*

(i) Almost every solution of the structural dynamics converges to an equilibrium $(x^*, 0)$ with a configuration x^* that is an α-lattice.

(ii) All agents asymptotically move with the same velocity.

(iii) Given $c < c^* = \psi_\alpha(0)$, no inter-agent collisions occur for all $t \geq 0$.

The above theorem does not guarantee that the proximity network induced by x^* is a connected graph. In fact, our observations show that generically this proximity network is disconnected.

Theorem 2.3.2 *Consider a group of α-agents applying Algorithm II with $c_1, c_2 > 0$ and structural dynamics Σ_2. Assume that the initial velocity mismatch $K(v(0))$ and inertia $J(x(0))$ are finite. Then, the following statements hold:*

(i) The group of agents remain cohesive for all $t \geq 0$.

(ii) Almost every solution of Σ_2 asymptotically converges to an equilibrium point $(x^*_\lambda, 0)$ where x^*_λ is a local minima of $U_\lambda(x)$.

(iii) All agents asymptotically move with the same velocity.

(iv) Assume the initial structural energy of the particle system is less than $(k + 1)c^*$ with $c^* = \psi_\alpha(0)$ and $k \in \mathbb{Z}_+$. Then, at most k distinct pairs of α-agents could possibly collide ($k = 0$ guarantees a collision-free motion).

Under some additional assumptions, this theorem can be improved so that Algorithm II can be used for self-assembly of connected networks of mobile agents.

Theorem 2.3.3 *Consider a group of α-agents applying Algorithm II with $c_1, c_2 > 0$ and structural dynamics Σ_2. Assume the initial structural energy is finite and the interaction range satisfies $r/d = 1 + \varepsilon$ ($\varepsilon \ll 1$). If conjectures 1 and 2 by Olfati-Saber (2006) hold, then almost every solution of Σ_2 asymptotically converges to an equilibrium point $(x^*_\lambda, 0)$ where x^*_λ is a quasi α-lattice and a flock is asymptotically self-assembled.*

2.3.4 Simulations of flocking

Figure 2.5 shows the consecutive snapshots of flocking behavior for 150 mobile agents that are moving in an environment with six obstacles. In many surveillance and search and rescue operations, unmanned vehicles need to fly in lower altitudes that increases the possibility of collision with buildings and natural obstacles like trees and mountains. A flocking-based distributed coverage has the benefit of spanning a large region over a relatively small period, particularly if multiple mobile sensing agents are necessary to carry out a distributed information gathering task. The agents asymptotically self-assemble a connected mobile network (i.e. a flock).

Figure 2.6 illustrates the task of squeezing maneuver for $n = 150$ agents. The squeezing maneuver is the result of flocking in presence of obstacles that are relatively close to each other. This a special case of escape panic phenomenon that is studied by Helbing *et al.* (2000). The initial positions of the agents are chosen at random in a square. Again, the agents asymptotically self-assemble a flock and achieve self-alignment. Furthermore, the conformation of the agents is close to an α-lattice in both cases.

Fragmentation phenomenon is shown in Figures. 2.7 (a) through (d). It can be observed that two agents which belong to two different components of the proximity net, move further apart from each other as time goes by. Interestingly, for search and rescue operations, fragmentation phenomenon might be highly desirable as it allows distribution of the resources.

2.4 MICROFILTER NETWORKS FOR COOPERATIVE DATA FUSION

In this section, we discuss a class of swarms that are capable of performing cooperative information processing in sensor networks. In this approach, a microfilter (to be defined) is embedded in every node of a sensor network. This network of microfilters is an information processing swarm that acts as the 'distributed mind' of the sensor network for data fusion.

Similar to flocks of mobile agents, this swarm of microfilters is a networked system with the same communication topology (or overlay network) G as the sensor network. For simplicity of the presentation, we assume G is a fixed graph. In general, a wireless network could be subject to failures such as packet-loss and node/link failures that render the network topology dynamic.

A fundamental distributed estimation and data fusion problem for sensor networks is to develop a distributed algorithm for Kalman filtering.[13] Decentralized Kalman filters with all-to-all communication links were proposed by Rao *et al.* (1993) and Speyer (1979). The communication complexity of these algorithms is $O(n^2)$ which makes them non-scalable in n (the size of the network). Recently, a Distributed Kalman Filtering (DKF) algorithm was introduced by Olfati-Saber (2005a) that relied on embedded consensus filters for distributed computation of averaged measurements and inverse-covariance matrices of the

[13] See Anderson and Moore (1979).

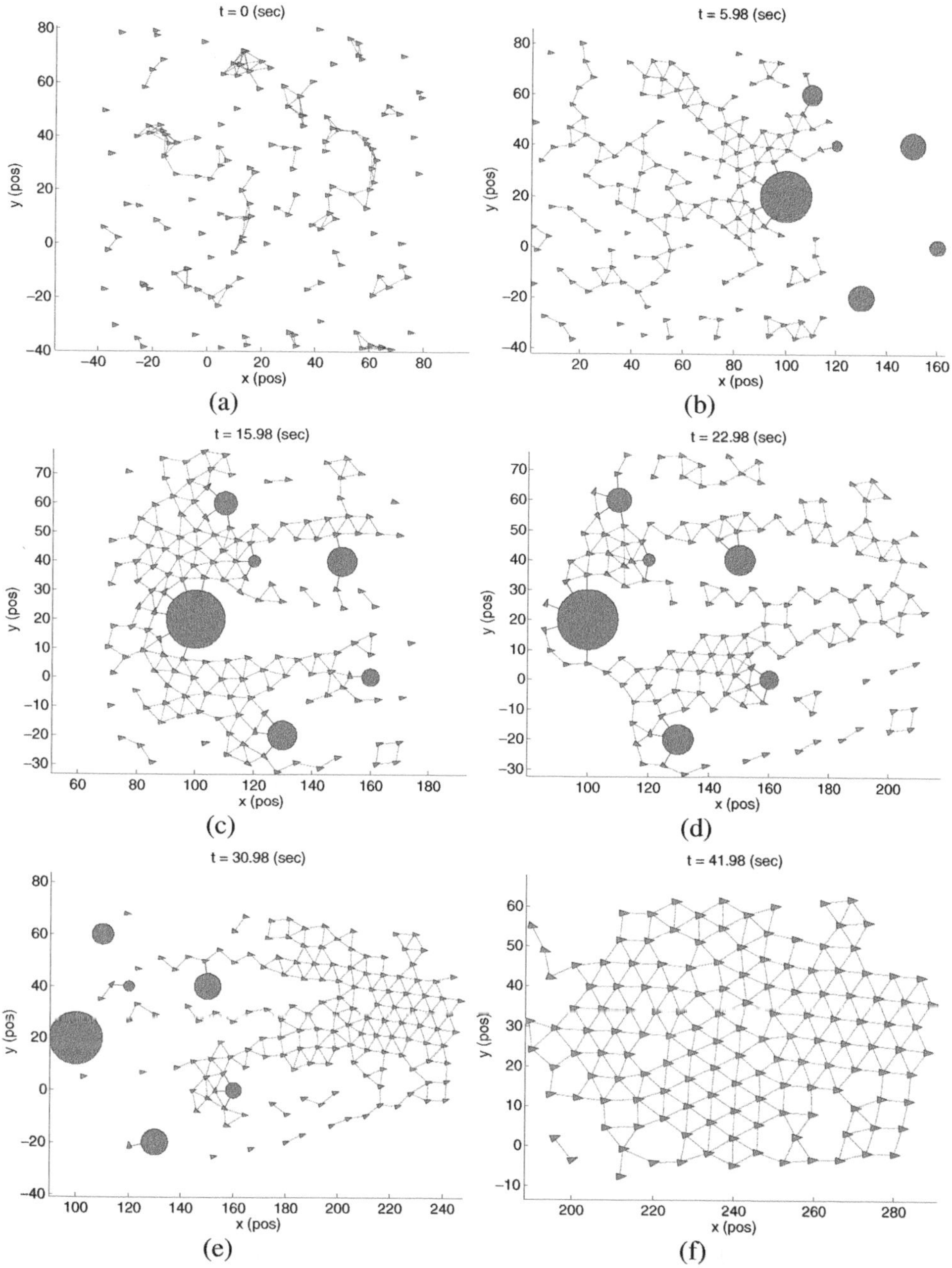

Figure 2.5 The split/rejoin maneuver for $n = 150$ agents using the flocking algorithm in Olfati-Saber (2006). Reprinted from Olfati-Saber, R., "Flocking for Multi-Agent Dynamic System: Algorithms and Theory," IEEE Trans. on Automatic Control, vol. 51, no. 3, pp. 401–420, March 2006. ©2006 IEEE.

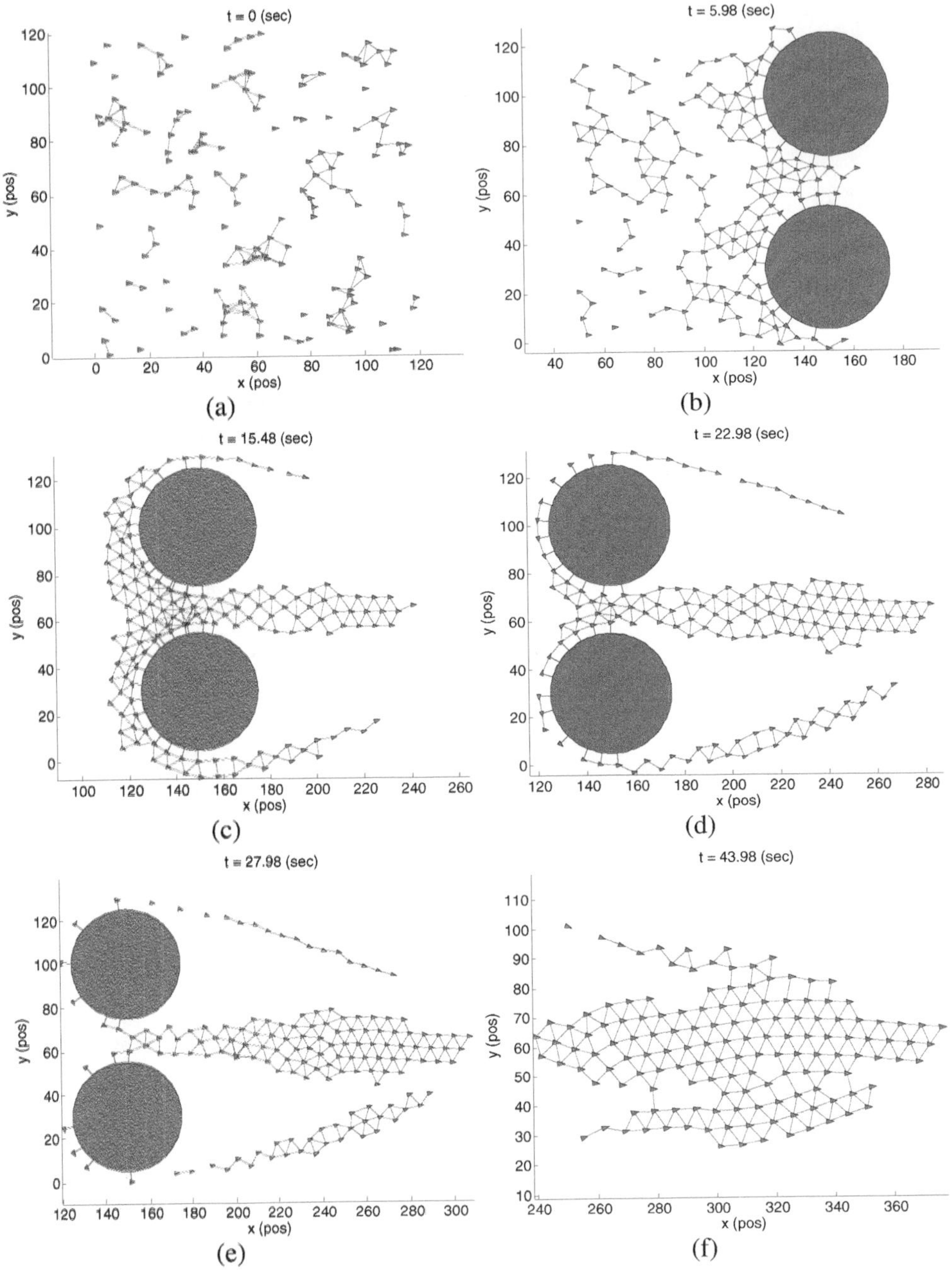

Figure 2.6 The squeezing maneuver for $n = 150$ agents using the flocking algorithm in Olfati-Saber (2006). Reprinted from Olfati-Saber, R., "Flocking for Multi-Agent Dynamic System: Algorithms and Theory," IEEE Trans. on Automatic Control, vol. 51, no. 3, pp. 401–420, March 2006. ©2006 IEEE.

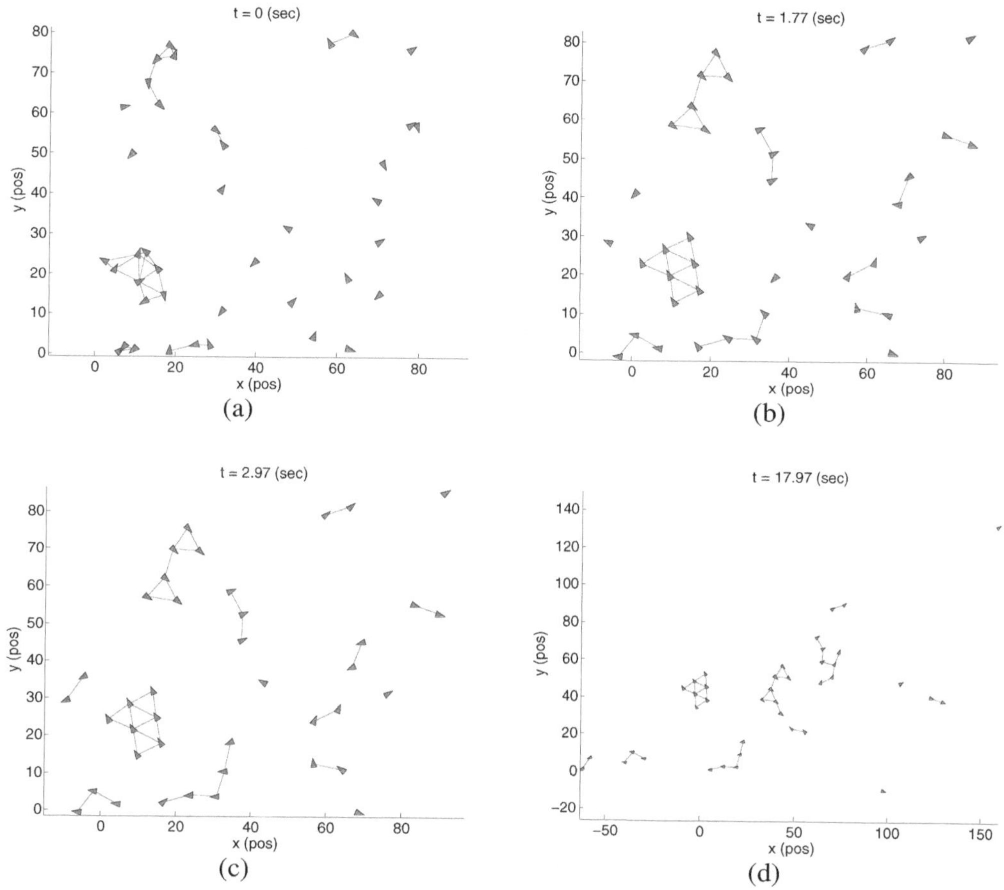

Figure 2.7 Fragmentation phenomenon for 40 agents applying Algorithm I of Olfati-Saber (2006). Algorithm I encompasses all three rules of Reynolds that are clearly insufficient for the creation of flocking behavior without fragmentation. Reprinted from Olfati-Saber, R., "Flocking for Multi-Agent Dynamic System: Algorithms and Theory," IEEE Trans. on Automatic Control, vol. 51, no. 3, pp. 401–420, March 2006. ©2006 IEEE.

nodes of a sensor network. These consensus filters are extensions of the basic consensus algorithm in Equation (2.1) and are introduced by Olfati-Saber and Shamma (2005) and Spanos *et al.* (2005a) in two forms: a low-pass filter and a high-pass filter.

The low-pass consensus filter is in the following form:

$$\dot{x}_i = \sum_{j \in N_i} a_{ij}(x_j - x_i) + \sum_{j \in J_i} a_{ij}(u_j - x_i) \tag{2.23}$$

and is a key element of the DKF algorithm. In the last equation, $J_i = \{i\} \cup N_i$ denotes the set of inclusive neighbors of node i on graph G. The convergence analysis for this consensus filter is performed for inputs $u_i = r(t) + v_i(t)$ that are noisy versions of a signal $r(t)$.

Consider a sensor network of size n observing a moving *target* with discrete-time model

$$x_{k+1} = A_k x_k + B_k w_k.$$

Assume the sensing model of the nodes are in the form

$$z_i(k) = H_i x(k) + v_i(k) \tag{2.24}$$

with the noise statistics given by

$$\begin{aligned} E(w_k w_l') &= Q_k \delta_{kl}, \\ E(v_i(k) v_i(l)') &= R_i(k) \delta_{kl}, \\ x_0 &= \mathcal{N}(\overline{x}_0, P_0). \end{aligned} \tag{2.25}$$

Let us define the following $m \times m$ average inverse-covariance matrix S as

$$S = \frac{1}{n} H_c' R_c^{-1} H_c = \frac{1}{n} \sum_{i=1}^{n} H_i' R_i^{-1} H_i \tag{2.26}$$

and weighted-average measurement y

$$y_i = H_i' R_i^{-1} z_i, \; y = \frac{1}{n} \sum_{i=1}^{n} y_i. \tag{2.27}$$

Then, according to the following result that is rephrased from Olfati-Saber (2005a), a group of microfilters (shown in Figure 2.8) with local communication are capable of computing the state estimate $\hat{x} = E(x_k | Z_k)$ with $Z_k = \{z_0, z_1, \ldots, z_k\}$ obtained by a central Kalman filter in a distributed way provided that two average-consensus problems can be solved.

Theorem 2.4.1 *(Distributed Kalman filter) Consider a sensor network with n sensors and topology G that is a connected graph observing a process of dimension m using $p \leq m$ sensor measurements. Assume the nodes of the network solve two consensus problems that allow them to calculate average inverse-covariance S and average measurements y at every iteration k. Then, every node of the network can calculate the state estimate $\hat{x}$ at iteration k using the update equations of its micro-Kalman filter (or μKF iterations)*

$$M_\mu = (P_\mu^{-1} + S)^{-1}, \tag{2.28}$$

$$\hat{x} = \overline{x} + M_\mu (y - S\overline{x}), \tag{2.29}$$

$$P_\mu^+ = A M_\mu A' + n B Q B', \tag{2.30}$$

$$\overline{x}^+ = A\hat{x}. \tag{2.31}$$

This gives an estimate identical to the one obtained via a central Kalman filter (note that the time indices are dropped for clarity and the plus superscripts mean updated values).

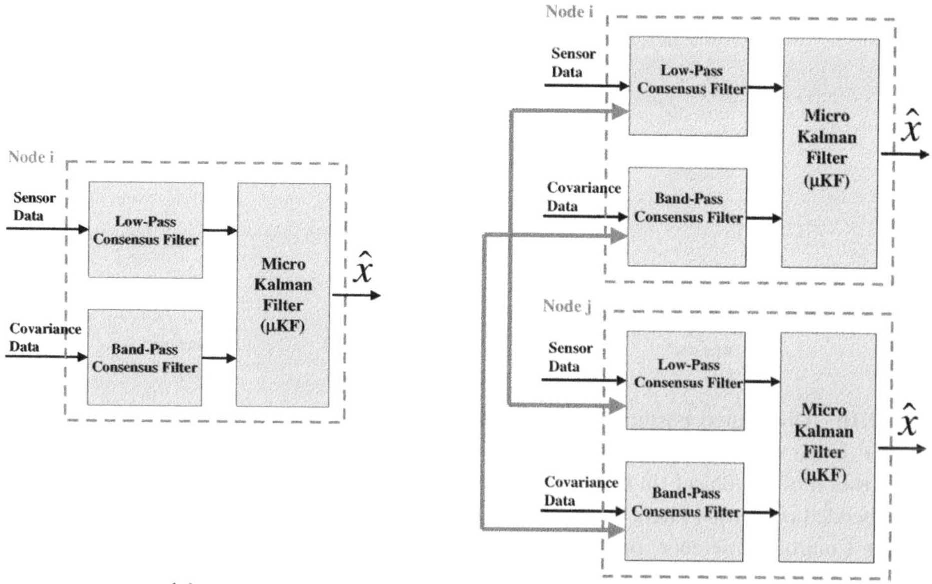

(a) (b)

Figure 2.8 Elements of the DKF algorithm: (a) a microfilter and (b) a network of microfilters with information exchange between embedded low-pass and band-pass consensus filters of neighboring microfilters. Reprinted from Olfati-Saber, R., "Distributed Kalman filter with embedded consensus filters," 44th IEEE Conference on Decision and Control, 2005 and 2005 European Control Conference, pp. 8179–8184, Dec. 2005. ©2005 IEEE.

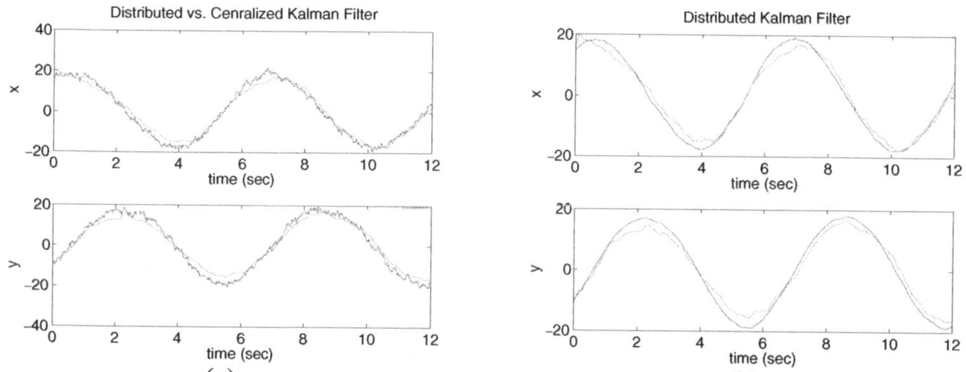

(a) (b)

Figure 2.9 Distributed position estimation for a moving object by node $i = 100$: (a) DKF vs. KF (DKF is the smooth curve in red) and (b) Distributed Kalman filter estimate (in red) vs. the actual position of the object (in blue). Reprinted from Olfati-Saber, R., "Distributed Kalman filter with embedded consensus filters," 44th IEEE Conference on Decision and Control, 2005 and 2005 European Control Conference, pp. 8179–8184, Dec. 2005. ©2005 IEEE.

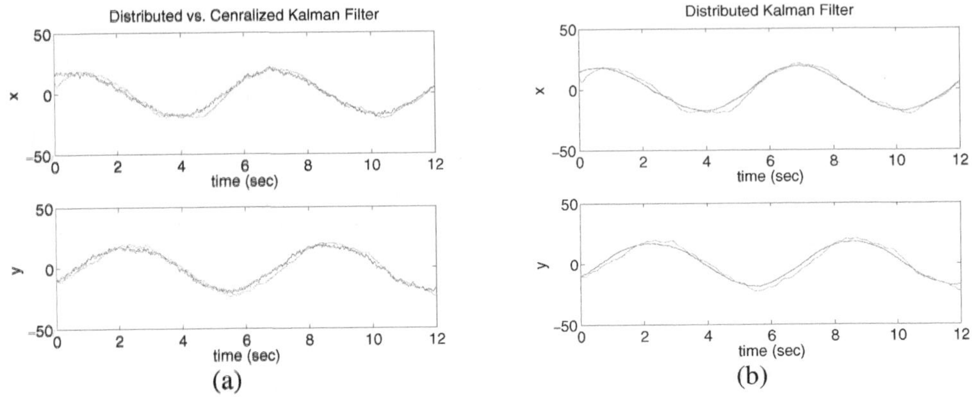

Figure 2.10 Distributed position estimation for a moving object by node $i = 25$: (a) DKF vs. KF (DKF is the smooth curve in red) and (b) Distributed Kalman filter estimate (in red) vs. the actual position of the object (in blue). Reprinted from Olfati-Saber, R., "Distributed Kalman filter with embedded consensus filters," 44th IEEE Conference on Decision and Control, 2005 and 2005 European Control Conference, pp. 8179–8184, Dec. 2005. ©2005 IEEE.

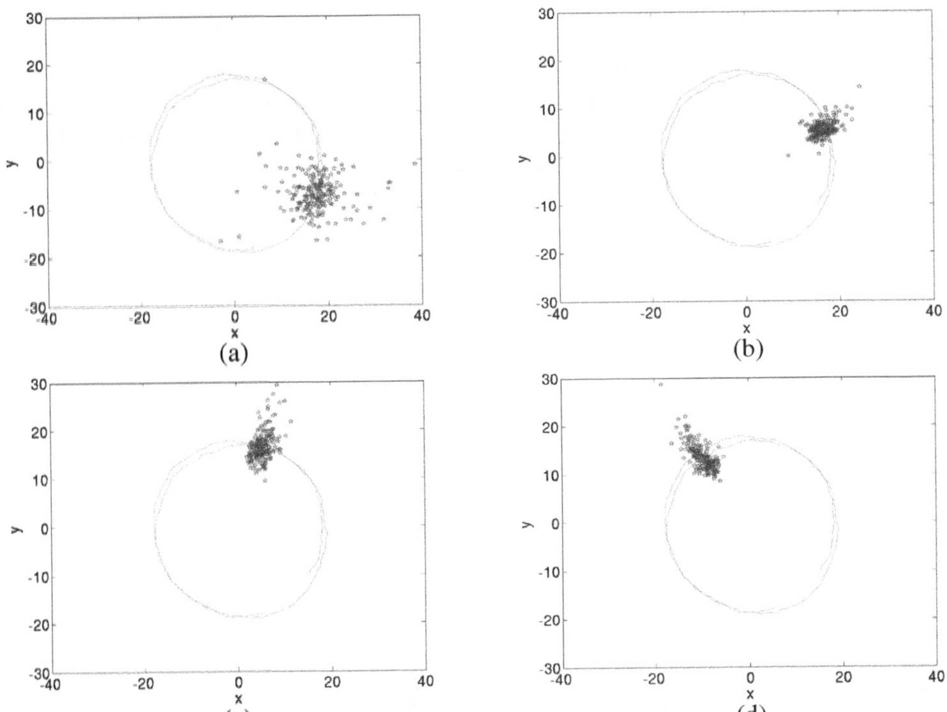

Figure 2.11 Snapshots of the estimates of all nodes regarding the position of a moving target that goes on circles. Reprinted from Olfati-Saber, R., "Distributed Kalman filter with embedded consensus filters," 44th IEEE Conference on Decision and Control, 2005 and 2005 European Control Conference, pp. 8179–8184, Dec. 2005. ©2005 IEEE.

To avoid waiting for an average-consensus algorithm to finish calculating S and y at every step between two consecutive updates of the micro-Kalman filter, we use a combination of low-pass and high-pass consensus filters as shown in Figure 2.8. This allows considerably faster distributed computation of the averaged quantities S and y.

We call a dynamic system that embodies a micro-Kalman filter and its associated consensus filters a microfilter. The cooperative network of microfilters embedded in the computing units of the sensors form an information fusion swarm.

Complex networks of microfilters embedded in sensor nodes constitute information processing swarms. The convergence analysis of this distributed Kalman filtering algorithm and its improvement is the subject of ongoing research. Simulation results for a sensor network with 200 nodes tracking a moving target are shown in Figures 2.9 through 2.11.

ACKNOWLEDGEMENTS

The author would like to thank Richard Murray and Jerrold Marsden for discussions on flocking behavior. Also, many thanks go to Jeff Shamma, Jason Speyer, Emilio Frazzoli, Demetri Spanos, and Elisa Franco for our discussions on distributed Kalman filters.

REFERENCES

Anderson BDO and Moore JB 1979 *Optimal Filtering*. Prentice-Hall, Englewood Cliffs, NJ.

Blondel V, Hendrickx JM, Olshevsky A and Tsitsiklis JN 2005 Convergence in multiagent coordination, consensus, and flocking. In *Proceedings of the 44th IEEE Conference on Decision and Control, 2005 and 2005 European Control Conference (CDC-ECC '05)*, pp. 2996–3000.

Cortés J 2006a Analysis and design of distributed algorithms for χ-consensus. In *Proceedings of the 45th IEEE Conference on Decision and Control*, pp. 3363–3368.

Cortés J 2006b Distributed algorithms for reaching consensus on arbitrary functions. *Automatica (submitted)*.

Cortés J, Martínez S and Bullo F 2006 Robust rendezvous for mobile autonomous agents via proximity graphs in arbitrary dimensions. *IEEE Trans. on Automatic Control* **51**(8), 1289–1298.

Cortés J, Martínez S, Karatas T and Bullo F 2004 Coverage control for mobile sensing networks. *IEEE Trans. on Robotics and Automation* **20**(2), 243–255.

D'Orsogna MR, Chuang YL, Bertozzi AL and Chayes L 2006 Self-propelled particles with soft-core interactions: patterns, stability, and collapse. *Physical Review Letters* **96**.

Fax JA 2001 Optimal and cooperative control of vehicle formations, PhD thesis, Control and Dynamical Systems, California Institute of Technology, Pasadena, CA.

Fax JA and Murray RM 2004 Information flow and cooperative control of vehicle formations. *IEEE Trans. on Automatic Control* **49**(9), 1465–1476.

Hatano Y and Mesbahi M 2005 Agreement over random networks. *IEEE Trans. on Automatic Control* **50**(11), 1867–1872.

Helbing D, Farkas I and Vicsek T 2000 Simulating dynamical features of escape panic. *Nature* **407**, 487–490.

Jadbabaie A, Lin J and Morse AS 2003 Coordination of groups of mobile autonomous agents using nearest neighbor rules. *IEEE Trans. on Automatic Control* **48**(6), 988–1001.

Kashyap A, Basar T and Srikant R 2006 Consensus with quantized information updates. In *Proceedings of the 45th IEEE Conference on Decision and Control*, pp. 2728–2733.

Levine H, Rappel WJ and Cohen I 2001 Self-organization in systems of self-propelled particles. *Phys, Rev. E* **63**, 017101.

Marden JR, Arslan G and Shamma JS 2007 Connections between cooperative control and potential games illustrated on the consensus problem. *The 2007 European Control Conference (submitted).*

Mesbahi M 2005 On state-dependent dynamic graphs and their controllability properties. *IEEE Trans. on Automatic Control* **50**(3), 387–392.

Moreau L 2005 Stability of multiagent systems with time-dependent communication links. *IEEE Trans. on Automatic Control* **50**(2), 169–182.

Newman MEJ 2003 The structure and function of complex networks. *SIAM Review* **45**, 167–256.

Olfati-Saber R 2005a Distributed Kalman filter with embedded consensus filters. In *Proceedings of the 44th IEEE Conference on Decision and Control, 2005 and 2005 European Control Conference (CDC-ECC '05)*, pp. 8179–8184.

Olfati-Saber R 2005b Ultrafast consensus in small-world networks. In *Proceedings of the 2005 American Control Conference*, pp. 2371–2378.

Olfati-Saber R 2006 Flocking for multi-agent dynamic systems: algorithms and theory. *IEEE Trans. on Automatic Control* **51**(3), 401–420.

Olfati-Saber R 2007 Algebraic connectivity ratio of Ramanujan graphs. In *Proceedings of the 2007 American Control Conference*.

Olfati-Saber R and Murray RM 2002 Distributed cooperative control of multiple vehicle formations using structural potential functions. *The 15th IFAC World Congress*, Barcelona, Spain.

Olfati-Saber R and Murray RM 2004 Consensus problems in networks of agents with switching topology and time-delays. *IEEE Trans. on Automatic Control* **49**(9), 1520–1533.

Olfati-Saber R and Shamma JS 2005 Consensus filters for sensor networks and distributed sensor fusion. In *Proceedings of the 44th IEEE Conference on Decision and Control, 2005 and 2005 European Control Conference (CDC-ECC '05)*, pp. 6698–6703.

Olfati-Saber R, Fax JA and Murray RM 2007 Consensus and cooperation in networked multi-agent systems. In *Proceedings of the IEEE* **95**(1), 215–233.

Olfati-Saber R, Franco E, Frazzoli E and Shamma JS 2006 Belief consensus and distributed hypothesis testing in sensor networks. In *Network Embedded Sensing and Control* (ed. Antsaklis PJ and Tabuada P), vol. LNCIS 331, pp. 169–182. Springer-Verlag, Berlin.

Partridge BL 1982 The structure and function of fish schools. *Scientific American* **246**(6), 114–123.

Rao BSY, Durrant-Whyte HF and Sheen JA 1993 A fully decentralized multi-sensor system for tracking and surveillance. *Int. Journal of Robotics Research* **12**(1), 20–44.

Raymond JR and Evans MR 2006 Flocking regimes in a simple lattice model. *Physical Review E* **73**, 036112.

Ren W and Beard RW 2005 Consensus seeking in multiagent systems under dynamically changing interaction topologies. *IEEE Trans. on Automatic Control* **50**(5), 655–661.

Reynolds CW 1987 Flocks, herds, and schools: a distributed behavioral model. *Computer Graphics (ACM SIGGRAPH '87 Conference Proceedings)* **21**(4), 25–34.

Shaw E 1975 Fish in schools. *Natural History* **84**(8), 40–45.

Spanos D, Olfati-Saber R and Murray RM 2005a Dynamic consensus on mobile networks. *The 16th IFAC World Congress*, Prague, Czech.

Spanos DP, Olfati-Saber R and Murray RM 2005b Approximate distributed Kalman filtering in sensor networks with quantifiable performance. *Fourth International Symposium on Information Processing in Sensor Networks*, pp. 133–139.

Speyer JL 1979 Computation and transmission requirements for a decentralized linear-quadratic-Gaussian control problem. *IEEE Trans. on Automatic Control* **24**(2), 266–269.

Tsitsiklis JN 1984 Problems in decentralized decision making and computation, PhD thesis, Department of Electrical Engineering and Computer Science, Laboratory for Information and Decision Systems, Massachusetts Institute of Technology, Cambridge, MA.

Tsitsiklis JN and Bertsekas DP 2007 Comment on 'Coordination of groups of mobile autonomous agents using nearest neighbor rules'. *IEEE Trans. on Automatic Control* **AC-52**(5), 968–969.

Tsitsiklis JN, Bertsekas DP and Athans M 1986 Distributed asynchronous deterministic and stochastic gradient optimization algorithms. *IEEE Trans. on Automatic Control* **31**(9), 803–812.

Vicsek T, Czirоók A, Ben-Jacob E, Cohen I and Shochet O 1995 Novel type of phase transition in a system of self-driven particles. *Phys. Rev. Letters* **75**(6), 1226–1229.

Watts DJ and Strogatz SH 1998 Collective dynamics of 'small-world' networks. *Nature* **393**, 440–442.

3

Connectivity and convergence of formations

Sonja Glavaški, Anca Williams and Tariq Samad

3.1 INTRODUCTION

Formation control for collections of vehicles is a problem that has recently generated much interest in the research community, motivated largely by applications to autonomous vehicles (terrestrial and aerial). Many of the feedback schemes investigated are inspired by the motion of aggregates of individuals in nature. For example, flocks of birds and schools of fish achieve coordinated motions without the use of a central controlling mechanism (Okubo 1986). The notion of a communication graph in formation stability is introduced in (Fax 2001), and an averaging feedback law is proposed based on the flow of information. The convergence of such a formation is further explored in (Glavaški *et al.* 2003) for the case where the transmission connections between vehicles break randomly.

Converging to the same velocity is also referred to as *formation flocking*. Jadbabaie *et al.* (2003) and Tanner *et al.* (2003) investigate the motions of vehicles modeled as double integrators. Their goal is for the vehicles to achieve a common velocity while avoiding collisions. The control laws involve graph Laplacians for an associated undirected (neighborhood) graph but also nonlinear terms resulting from artificial potential functions. Rather than reaching a predetermined *formation*, the vehicles converge to an equilibrium formation that minimizes all individual potentials. Ren and Beard (2004a) present a special case of some of the results shown here (among other results). Different motions of the actual vehicle formation are considered in (Veerman *et al.* 2005; Williams *et al.* 2005b).

A related area of research is consensus seeking by autonomous agents (de Castro and Paganini 2004; Olfati-Saber and Murray 2004; Ren and Beard 2004a). In this case, n agents achieve consensus if their associated variables converge to a common value. The vector x of variables satisfies a differential (or difference) equation: $\dot{x} = -L(x)$. Ren and Beard (2004a) show that a necessary and sufficient condition for achieving consensus is

*At the time this research was developed Anca Williams was a student intern with Honeywell Inc.

Cooperative Control of Distributed Multi-Agent Systems edited by Jeff S. Shamma

that the communication directed graph admits a directed spanning tree. Olfati-Saber and Murray (2003, 2004) only prove sufficiency and assume that the graph is either undirected or strongly connected. The same condition of the existence of a rooted directed spanning tree is proved necessary and sufficient in (Lafferriere *et al.* 2005) for the existence of certain decentralized feedback laws that control the vehicles to arbitrary formations.

In this chapter we discuss our recent research in formation control, specifically on the convergence and stability of multivehicle formations. We adopt the feedback scheme of Fax and Murray (2002a) and show that under some assumptions, communication mechanisms and control laws can be constructed that guarantee such convergence and stability. Control to formation is in a way a consensus problem, since in order to converge to formation the vehicles have to achieve, among other things, the same velocity. In addition, the vehicle dynamics in the formation problem are governed by second-order dynamics. Key features of our development include that the vehicle controls are decentralized, and each vehicle computes its own control based on information available locally to it. The communication among vehicles can be sparse and formation stability is maintained. We also define the minimum communication requirements necessary to find a stabilizing feedback. Building on our earlier work (Williams *et al.* 2004), we define formation hierarchy and show that our convergence and stability results apply in this case too. Only subformation leaders need to communicate with the higher level of the hierarchy. In all these cases, the communication graph can be independent of the desired formation geometry. This allows, for example, the same communication links to be used for different formation geometries. As a natural next step in our development, we investigate stability of formations when the spatial and communication structures are allowed to vary over time (Williams *et al.* 2005a).

Below we first present our problem formulation, defining what we mean by formation convergence and stability, presenting some basic notation, and identifying the key technical problem to be solved. Our research approach is based on concepts from algebraic graph theory; relevant results from this discipline are reviewed in Section 3.3. The main results are proved in Sections 3.4, 3.5, and 3.6. In Section 3.7 we discuss, and present some simulation results related to, the issue of ensuring connectivity in moving formations with neighborhood-based communication schemes.

3.2 PROBLEM FORMULATION

Before we proceed, let us define a moving formation as it will be used in this chapter. We assume given N homogeneous vehicles with the following discrete-time dynamics:

$$x_i(k+1) = A_{veh}x_i(k) + B_{veh}u_i(k)$$

where $i = 1, 2, \ldots, N$ and the entries of x_i represent the n configuration variables for vehicle i, referred to as position-like variables, and their derivatives, referred to as velocity-like variables, and u_i is the control input for vehicle i.

For each vehicle we use the error signal $z_i(k)$ for coordination:

$$z_i(k) = \sum_{j \in J_i(k)} (x_i(k) - x_j(k)) - (h_i - h_j)$$

where $J_i(k)$ is the set of neighbors of vehicle i at time k, and h is defined below.

Definition 3.2.1 *(Lafferriere et al. 2005) A* ***formation*** *of N vehicles is given by an offset vector* $h = h_p \otimes \begin{pmatrix} 1 \\ 0 \end{pmatrix} \in R^{2nN}$. *The N vehicles are said to be in formation at time k if there exist* R^n *valued vectors q and w such that* $(x_p)_i(k) - (h_p)_i = q$ *and* $(x_v)_i(k) = w$, *for* $i = 1, 2, \ldots, N$, *where the subscript p refers to the position components of* x_i *and the subscript v refers to the corresponding velocities. The vehicles converge to formation h if there exist real valued functions* $q(\cdot)$ *and* $w(\cdot)$ *such that* $(x_p)_i(k) - (h_p)_i - q(k) \to 0$ *and* $(x_v)_i(k) - w(k) \to 0$, *as* $k \to \infty$ *for* $i = 1, 2, \ldots, N$.

Note that our definition covers both moving ($|w| > 0$) and static, or hovering ($w = 0$) formations. Figure 3.1 illustrates the interpretation of vectors in the definition.

Our interest is mainly in the interdependence of formation stability and the underlying communication structure. At any time k, the information exchange between vehicles is captured by a communication graph, $G(k)$, where the N vertices represent the vehicles, and the edges represent the communication links. We call vehicle j a neighbor of vehicle i at time k if $G(k)$ has the edge (i, j). It is important to note that if a vehicle is a neighbor of vehicle i at time k, it is not guaranteed that the same vehicle will be a neighbor of i at time $k + 1$. In fact, depending on the formation requirements, vehicles may move in such a fashion so that they change neighbors often, while converging to formation.

For simplicity of the analysis, we consider a formation of homogeneous vehicles and we look for a stabilizing feedback matrix F_{veh} for each vehicle system. The system of N-homogeneous vehicles is given by:

$$x(k+1) = Ax(k) + Bu(k)$$

$$z(k) = L(k)(x(k) - h)$$

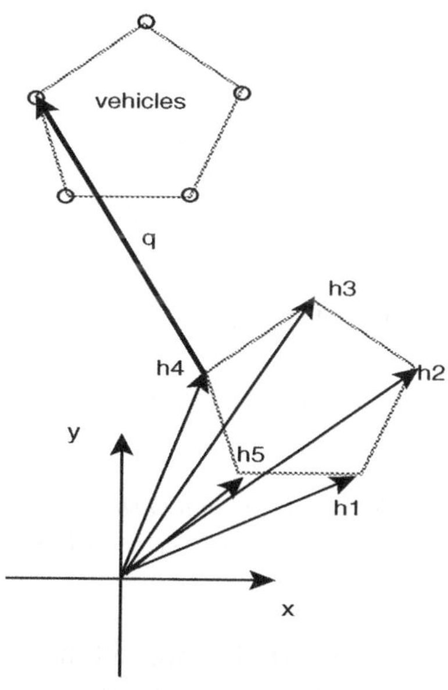

Figure 3.1 Example of a pentagon formation with the corresponding offset vectors.

where x is the augmented state vector, $A = I_N \otimes A_{veh}$, $B = I_N \otimes B_{veh}$, and $L(k) = L_G(k) \otimes I_{2n}$, where the matrix $L_G(k)$ belongs to a finite set of (constant) matrices, $\mathbb{L}$ (see below). Assuming that each vehicle has a stable control law already designed, we are interested in defining the control law at the formation level and the intervehicle communication requirements so that formation convergence and stability can be guaranteed. The coordinated control is implemented using a decentralized feedback control matrix $F = diag[F_{veh}]$ such that if:

$$u(k) = Fz(k)$$

then the vehicles converge to formation. We remark here that this feedback matrix F should be the same regardless of the communication graph between vehicles.

The closed loop system becomes:

$$x(k+1) = Ax(k) + BFL(k)(x(k) - h)$$

In the analysis that follows, we use algebraic graph theory to analyze the stability of a dynamical system that is of the form:

$$x(k+1) = (A + BFL(k))x(k)$$

Let $\sigma : \mathbb{N} \rightarrow \mathbb{L}$ be a piecewise constant function that determines the L matrix at each time k, $\sigma(k) = L_s$, where $L_s \in \mathbb{L}$. Then the system defined above can be considered as a switched dynamical system. Its stability depends in part on the stability of the individual systems:

$$x(k+1) = (A + BFL_s)x(k)$$

but it is not guaranteed by that. We will use the properties of the eigenvalues of matrices L_s to prove the stability of a related system matrix. We consider a matrix to be *stable* in the discrete Hurwitz sense: a matrix is stable if and only if all its eigenvalues are located in the interior of the circle of radius 1 in the complex plane.

3.3 ALGEBRAIC GRAPH THEORY

In this section we introduce some of the necessary algebraic graph theory concepts required for our analysis.

Definition 3.3.1 *A* ***directed graph*** *G consists of a vertex set $V(G)$ and an edge set $E(G) \subseteq V(G) \times V(G)$. For an edge $e = (u, v) \in E(G)$, u is called the head vertex of e and v is called the tail vertex of e.*

If $(u, v) \in E(G)$ for all $(v, u) \in E(G)$, then we call the graph undirected. We call the graph simple if there are no edges of the form (u, u) for $u \in V(G)$. Let G be a graph representing the communication links between a set of vehicles. The properties of such a communication graph are captured by its adjacency matrix $A_d(G)$ defined by:

Definition 3.3.2 *The* ***adjacency matrix*** *of a graph G, denoted* $A_d(G)$, *is a square matrix of size* $|V(G)| \times |V(G)|$ *defined as follows:*

$$A_d(G)_{ij} = \begin{cases} 1 & \text{if } (u_i, u_j) \in E(G) \\ 0 & \text{otherwise.} \end{cases}$$

where $u_i, u_j \in V(G)$.

Definition 3.3.3 *The* ***indegree of a vertex*** *u is the number of edges that have u as their head vertex. The indegree matrix of a graph G, denoted D(G), is a diagonal matrix of size* $|V(G)| \times |V(G)|$ *defined as follows:*

$$D(G)_{ii} = \text{indegree}(u_i)$$

where $u_i \in V(G)$.

The outdegree of a vertex u for a graph G is defined analogously, as the number of edges having u as the tail vertex.

Definition 3.3.4 *Given a graph G, the* ***Laplacian matrix*** *associated with it is given by*

$$L_G = D(G) - A_d(G)$$

where D(G) is the indegree matrix and $A_d(G)$ *is the adjacency matrix associated with the graph.*

The diagonal entry d_{ii} of the Laplacian matrix is then the indegree of vertex i and the negative entries in row i correspond to the neighbors of vertex i. For any communication graph considered, the row sums of the corresponding Laplacian matrix are always zero and hence the all ones vector is always an eigenvector corresponding to the zero eigenvalue for any Laplacian matrix considered. In the case of undirected graphs, the Laplacian matrix is symmetric and all its eigenvalues are non-negative (Chung 1994; Godsil and Royle 2001); the smallest eigenvalue of the Laplacian, λ_1, is zero; and its multiplicity equals the number of connected components of the graph. The second eigenvalue, λ_2, is directly related to the connectivity of the graph (Fiedler 1973). This is also the case for directed graphs.

Definition 3.3.5 *The edge (vertex)* ***connectivity*** *of a graph G is the smallest number of edge (vertex) deletions sufficient to disconnect G.*

There is a close relationship between the two quantities. The vertex connectivity is always no smaller than the edge connectivity, since deleting one vertex incident on each edge in a cut set succeeds in disconnecting the graph. Of course, smaller vertex subsets may be possible. The minimum vertex degree is an upper bound on both the edge and vertex connectivity, since deleting all its neighbors (or the edges to all its neighbors) disconnects the graph into one big and one single-vertex component. In this chapter we refer to the communication graph connectivity as the vertex connectivity of the defined communication graph.

The properties of the eigenvalues of the Laplacian matrices in the case of directed graphs are explored by Lafferriere *et al.* (2005).

Definition 3.3.6 *Given a directed graph G, $P = (u_1, \ldots, u_k)$ is a* **directed path** *in G if for every $1 \le i < k$ there exists edge $(u_i, u_{i+1}) \in E(G)$.*

Definition 3.3.7 *Given a directed graph G, we call T a* **rooted directed spanning tree** *of G, if T is a subgraph of G, it has no cycles and there exists a (directed) path from at least one vertex, the root, to every other vertex in T.*

All graphs in this chapter are directed graphs, unless mentioned otherwise. One property of the eigenvalues for the Laplacian matrix of such a communication graph can be derived from Gershgorin's Theorem. Since all diagonal entries of the Laplacian are the indegrees of the corresponding vertex, it follows that all its eigenvalues will be located in the disc centered at $d = max_i(\text{indegree}(i))$ of radius d, so for any eigenvalue λ of the Laplacian

$$|\lambda| \le 2d$$

Further properties of the eigenvalues of such a graph are explored by Veerman *et al.* (2005).

3.4 STABILITY OF VEHICLE FORMATIONS IN THE CASE OF TIME-INVARIANT COMMUNICATION

First we consider the case where the communication graph $G(k)$ is the same at every instance of time. Consider the system of N vehicles described in Section 3.2:

$$x(k+1) = (A + BFL_s)x(k)$$

This is the case where the matrix L_s is time invariant, i.e. σ is a constant function. The stability of this type of system has been analyzed by Lafferriere *et al.* (2005). The authors show that a necessary and sufficient condition for this system to converge to formation is that the communication graph has a rooted directed spanning tree. This ensures that the algebraic multiplicity of the zero eigenvalue of L_s is one, which in turn guarantees that the system above is formation stable.

3.4.1 Formation hierarchy

Before we proceed to analyze the stability of the time-variant communication case, we extend the concept of formations to hierarchical formations and we consider formations with a special case of hierarchical communication that is time invariant. Given several stable formations that communicate with each other, can the results from formation convergence and stability be extended to stability of formations?

Definition 3.4.1 *Given n graphs, we call $G = G_1 \times G_2 \times \cdots \times G_n$ their* **hierarchical product graph** *if the vertices of G_{i+1} are replaced by a copy of G_i such that only the vertex labeled 1 from each G_i replaces each of the vertices of G_{i+1} for all $1 \le i \le n-1$.*

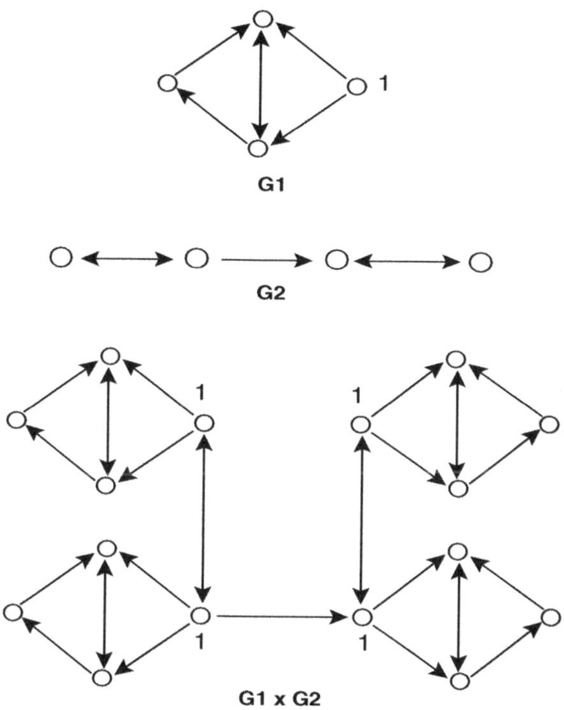

Figure 3.2 Hierarchical communication graph.

Figure 3.2 has an example of a two-layer hierarchical communication graph. The adjacency and the indegree matrices of the $G_1 \times G_2$ graph are given by:

$$A_d = I_4 \otimes A_{d_1} + A_{d_2} \otimes One$$

$$D = I_4 \otimes D_1 + D_2 \otimes One$$

where A_{d_i}, and D_i, are the adjacency, and indegree matrices respectively, of the graph G_i and One is a 4×4 matrix with only the first entry equal to 1, as defined prior. The Laplacian L is given by:

$$L = I_4 \otimes L_1 + L_2 \otimes One$$

Definition 3.4.2 *Given graphs $G_1, \ldots, G_n$ with vertex sets $V_i = V(G_i) = \{1_{G_i}, \ldots, n_{G_i}\}$ distinct for all i and edge sets $E_i = E(G_i)$, we call a graph G their* ***sum graph*** *$G_1 + \cdots + G_n$ if there exists a map $f : V(G) \to \cup_{i=1}^{n} V_i$ such that 1) $V(G) = \cup_{i=1}^{n} V_i$ and 2) v_i and v_k are adjacent in G if either $f(v_i)$ and $f(v_k)$ are adjacent in some graph G_l or $f(v_i) = 1_{G_i}$ and $f(v_k) = 1_{G_k}$.*

The graph G can be thought of as a graph with vertices v_i replaced by the vertices labeled 1_{G_i} of graphs G_i. See Figure 3.3.

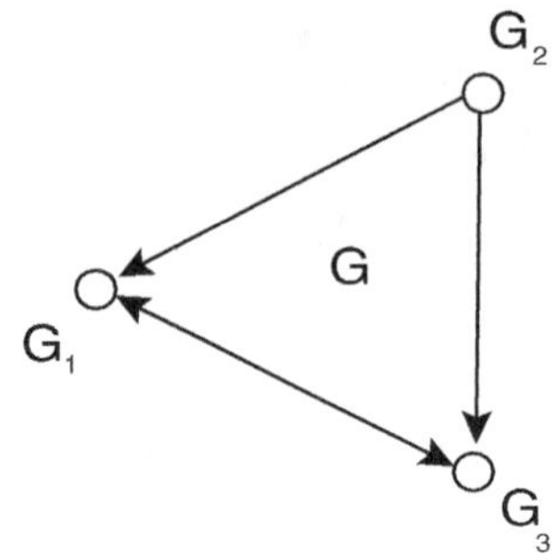

Figure 3.3 Sum of hierarchical communication graphs.

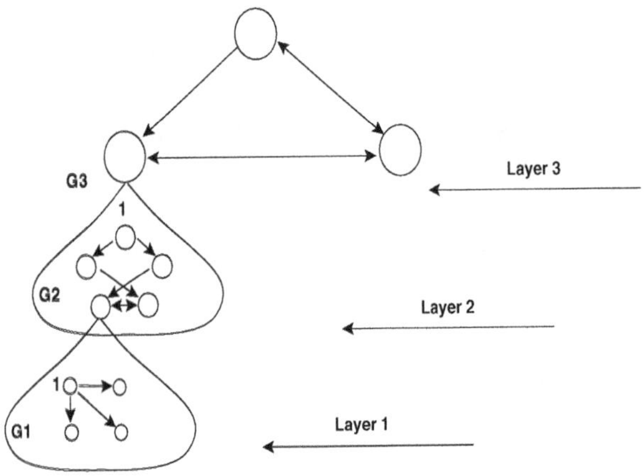

Figure 3.4 Three layer hierarchy.

Definition 3.4.3 *A group of N vehicles is said to be in a* ***hierarchical formation*** *if at all times they are the vertices of a pre-specified geometrical shape and the vehicles' communication graph is a sum of hierarchical products of graphs.*

At each layer in the hierarchy the vertices represent a subformation (see Figure 3.4.). Each subformation may be constructed of other subformations, but we also allow the subformation at layer 1 to be constructed of individual vehicles. The subformation (or vehicle) labeled 1 is assumed to be the 'leader' of its subformation and is the only one that can receive or transmit information to other subformations. For the rest of this section we assume that only the 'leader' of each subformation communicates with vehicles at the next layer and the rest of the vertices in its corresponding graph have positive indegrees. Practically, this is interpreted as follows: the non-leader subformations receive information from at least one other subformation in their group. In the subformation graph this is captured by having the first diagonal entry of the Laplacian equal to zero and all other diagonal entries positive. In this section it will become necessary to work with Laplacian matrices from the set:

$$\mathcal{S} = \{I_n \otimes L_1 + L_2 \otimes One\}$$

where $I_n \in R^{n\times n}$ is identity matrix, $L_1 \in R^{m\times m}$ is a Laplacian matrix, $L_2 \in R^{n\times n}$ is a Laplacian matrix and $One \in R^{m\times m}$ is a matrix with the first entry equal to 1 and all other entries equal to 0.

For a matrix $L \in \mathcal{S}$ its eigenvalues depend on the Laplacians L_1 and L_2. In the case of undirected graphs, some bounds for the eigenvalues of the matrix L are derived as follows. First, notice that the graph G, corresponding to the Laplacian L, has maximum vertex indegree

$$d = \max\{d_1, indegree(1) + d_2\}$$

where d_1 is the maximum indegree of the graph G_1 corresponding to Laplacian L_1, $indegree(1)$ is the indegree of the vertex labeled 1 in G_1 and d_2 is the maximum vertex indegree of the graph G_2, corresponding to the Laplacian L_2. So for all eigenvalues λ of L:

$$|\lambda| \leq 2\max\{d_1, indegree(1) + d_2\}$$

Other bounds can also be found; the reader is directed to (Anderson and Morley 1985; Fiedler 1973; Godsil and Royle 2001; Guattery and Miller 2000). In the case of directed graphs we derive a bound for the eigenvalues of L as follows.

Lemma 3.4.4 *Let $L_1 \in R^{m\times m}$ and $L_2 \in R^{n\times n}$ be Laplacian matrices of two directed graphs and suppose L_1 has its first diagonal entry equal to zero and all other diagonal entries positive. If $L = I_n \otimes L_1 + L_2 \otimes One \in \mathcal{S}$, then the eigenvalues of L are those of L_2 and the non-zero eigenvalues of L_1 repeated n times.*

Proof. We first notice that since the first entry of L_1 is zero, the indegree of vertex 1 is zero, so the first row of L_1 must have only zero entries. Also, due to the assumptions on the graph and since no other diagonal entries are zero, the multiplicity of eigenvalue zero of L_1 is 1. Then the matrix L has the following form:

$$L_1 = \begin{pmatrix} 0 & 0 & \cdots & 0 \\ \star & & & \\ \vdots & & \tilde{L_1} & \\ \star & & & \end{pmatrix}$$

The non-zero eigenvalues of L_1 are the eigenvalues of $\tilde{L_1}$. Linear transformation can be used to show that L is similar to a matrix of the form:

$$\begin{pmatrix} L_2 & 0 & \cdots & 0 \\ \star & \tilde{L_1} & 0 & 0 \\ \vdots & 0 & \ddots & 0 \\ \star & 0 & 0 & \tilde{L_1} \end{pmatrix}$$

that has $n+1$ blocks on the diagonal and whose eigenvalues are the eigenvalues of L_2 and the eigenvalues of $\tilde{L_1}$ repeated n times. We notice that if the n L_1 matrices were in fact distinct, then the eigenvalues of L would be the set of all non-zero eigenvalues of the matrices replacing the matrix L_1 and the eigenvalues of matrix L_2.

We now present a stability result for hierarchical formations: *hierarchical formation stability*.

Theorem 3.4.5 *Suppose we have a two-layer formation hierarchy, with n vehicles at the subformation level and m subformations. Let each subformation be represented by the following controllable system:*

$$x_{k+1} = A_s x_k + B_s u_k$$
$$z_k = L_s(x_k - h_s)$$
$$u_k = F_s z_k$$

where A_s and B_s are the individual subformation dynamics, L_s is the information exchange graph Laplacian, F_s is the subformation control feedback matrix and h_s is the corresponding formation offset vector. With F_s such that each subformation is stable, if G is a graph on m vertices whose Laplacian eigenvalues λ_f have the property that $A + \lambda_f BF$ is stable, then G can be used as the communication infrastructure for the second layer of the formation so that the entire system converges to a new overall stable formation.

Proof. For simplicity we assume that the subformation dynamics are identical, however, this is not necessary. Let h be the vector of offsets that specify the overall formation. We let h_s^i be the offsets for each subformation i. Consider the system of all $N = nm$ vehicles:

$$x_{k+1}^1 = A_s x_k^1 + B_s F_s L_s (x_k^1 - h_s^1)$$
$$\cdot = \cdot$$
$$\cdot = \cdot$$
$$\cdot = \cdot$$
$$x_{k+1}^m = A_s x_k^m + B_s F_s L_s (x_k^m - h_s^m)$$

Equivalently, the m systems can be captured by the following:

$$x_{k+1} = Ax_k + Bu_k \tag{3.1}$$

where $A = I_m \otimes A_s$, $B = I_m \otimes B_s$, F is a new feedback matrix and L is the Laplacian for the entire information exchange graph. As in the example, the overall Laplacian, L, takes the following form:

$$L = I_n \otimes L_s + L_f \otimes One$$

where L_f is the Laplacian for the graph G. For the purpose of analyzing stability, we let $h = 0$. Let $F = I_m \otimes F_s$. Then

$$u_k = \begin{pmatrix} \boxed{F_s} & & 0 \\ & \ddots & \\ 0 & & \boxed{F_s} \end{pmatrix} L \begin{pmatrix} x_1 \\ \vdots \\ x_n \end{pmatrix}_k$$

We will show that with this feedback control matrix F, all vehicles achieve a stable hierarchical formation, or equivalently the system

$$x_{k+1} = (A + BFL)x_k$$

is *formation stable*. To prove stability one must show that $A + \lambda BF$ is stable for all eigenvalues λ of L. From Lemma 3.4.4, the set of eigenvalues of L is equal to the set $\{\lambda_s$, non-zero eigenvalues of $L_s\} \cup \{\lambda_f$, eigenvalues of $L_f\}$. Clearly $A + \lambda_s BF$ is stable for the chosen matrix F. All that remains to do in order to guarantee stability is to choose L_f for which $A + \lambda_f BF$ is also stable.

We note that when converging to formation, the vehicles achieve consensus, as introduced by Fax and Murray (2002b), on the location of the center of the final formation. This consensus is achieved at both the second layer level, i.e. the subformation leaders, as well as at the first layer levels. This is done independently, in the sense that the relative position of each subformation vehicle only depends on the positions of the vehicles in its group. In the case when the subformations are not identical, i.e. the communication graph is a sum of hierarchical graphs, the choice of L_f should be such that each of $A + \lambda BF$ should be stable. An example of such is the case when at all levels, the information exchange graphs can be modeled using directed tree graphs.

A simulation of hierarchical formation control with the above control law is shown in Figure 3.5. In this example, the formation has two levels, with four vehicles in the first layer and five subformations in the second layer. As can be seen, the twenty vehicles achieve a stable hierarchical formation.

In essence, Theorem 3.4.5 shows that, given a local feedback controller that makes the subformations stable, there exists an entire class of communication infrastructures at the next hierarchy level for which we can guarantee stability for the entire formation.

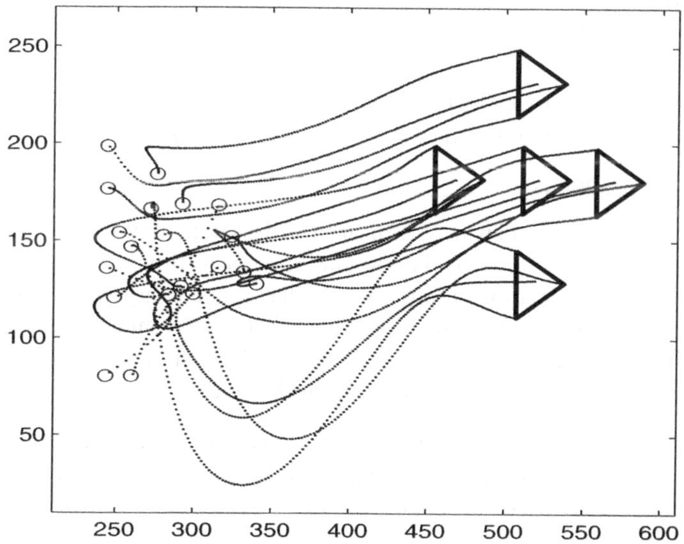

Figure 3.5 Twenty vehicle formation convergence.

Conjecture 3.4.1 *The assumptions that the subformations have a leader is sufficient, but not necessary.*

Corollary 3.4.6 *Theorem 3.4.5 can be extended to an m-layer hierarchy.*

Proof. We use strong mathematical induction as follows: Suppose layers $1, 2, \ldots, k$ have information graphs given by $G_1, G_2, \ldots, G_k$ and are stable, altogether forming a system Σ. Then at layer $k+1$ we have stable subformations given by Σ. We construct F as above and choose a communication graph that is co-spectral with one of the graphs G_i, $1 \leq i \leq k$. Then, the entire system is stable.

This allows us to construct a communication structure and indirectly a decentralized control for the next layer of the hierarchy based on the feedback controls used at the previous layer. We can thus build as large a hierarchy as required.

This section shows how we can use the connectivity of a communication graph to ensure stability of a hierarchy of formations. We have allowed the information exchange graph to be as general as possible, our only assumption being the existence of subformation leaders and that all non-leader vehicles should be able to receive information from at least one other vehicle.

3.5 STABILITY OF VEHICLE FORMATIONS IN THE CASE OF TIME-VARIANT COMMUNICATION

We now consider the case where the communication graph is time-variant. We will investigate under which conditions such a switched system of N vehicles converges to formation. Section 3.4 gave the necessary and sufficient conditions for which the individual systems are formation stable.

Proposition 3.5.1 *Let $M = \{C_1, \ldots, C_m\}$ and consider the switched linear system*

$$x(k+1) = C_{i(k)}x(k), C_{i(k)} \in M$$

If all matrices in M have a spectral radius less than 1 and the Lie algebra associated to M is solvable, then the system has a common quadratic Lyapunov function.

Proof. (Blondel *et al.* 2004)

The switching function σ can be given arbitrarily, or it may depend on the physical location of the vehicles. In some of the simulations that follow we impose an additional condition on the communication graph: we say that a necessary condition for vehicle j to be a neighbor of vehicle i is that the distance between vehicles i and j is less than or equal to ε_i, for some given set of ε_i, $i = 1, \ldots, N$. This corresponds to the case of limited sensor range. In that case, we need the following:

Proposition 3.5.2 *The vehicles are in formation if and only if there exists a communication graph G with Laplacian matrix L such that for communication graph Laplacian matrices $L_G(k)$ we have the following conditions:*

$$L_G(k) \to L$$

$$(L \otimes I_{2n})(x - h) = 0$$

Proof. If such a constant limit Laplacian L exists, then the distances between vehicles must be fixed. On the other hand, the vehicles are in a (possibly moving) formation if the distances between them are fixed and therefore the communication graph becomes independent of time. Let L be the Laplacian matrix for this graph. Also, as the vehicles are in formation, there exist $q(\cdot)$ and $w(\cdot)$ such that $(x_p)_i(k) - (h_p)_i = q(k)$ and $(x_v)_i(k) = w(k)$, $i = 1, 2, \ldots, N$. Then

$$
\begin{aligned}
&(L \otimes I_{2n})(x(k) - h) = \\
&(L \otimes I_{2n})\left(1_N \otimes \left((x_p)_i \otimes \begin{pmatrix} 1 \\ 0 \end{pmatrix} + (x_v)_i \otimes \begin{pmatrix} 0 \\ 1 \end{pmatrix}\right)\right) - \\
&(L \otimes I_{2n})\left(h_p \otimes \begin{pmatrix} 1 \\ 0 \end{pmatrix}\right) = \\
&(L \otimes I_{2n})\left(1_N \otimes \left(q(k) \otimes \begin{pmatrix} 1 \\ 0 \end{pmatrix} + w(k) \otimes \begin{pmatrix} 1 \\ 0 \end{pmatrix}\right)\right) = \\
&(L \otimes I_{2n})(1_N \otimes \alpha(k)) = 0
\end{aligned}
$$

We now provide a sufficient condition for convergence to formation in the case of time-dependent communication graphs. First we establish a lemma.

Lemma 3.5.3 *Consider a finite collection of matrices $M = \{P(k) : k = 1, \ldots, m\}$, each of the form $P(k) = \begin{pmatrix} P_1(k) & P_2(k) \\ 0 & P_3(k) \end{pmatrix}$. Let A be the Lie algebra generated by M and let B be the Lie algebra generated by the matrices $P_3(k)$, for $k = 1, \ldots, m$. Then B is isomorphic to a quotient Lie algebra of A. In particular, if A is solvable, then so is B.*

Proof. First notice that $\left[\begin{pmatrix} P_1 & P_2 \\ 0 & P_3 \end{pmatrix}, \begin{pmatrix} Q_1 & Q_2 \\ 0 & Q_3 \end{pmatrix}\right] = \begin{pmatrix} [P_1, Q_1] & * \\ 0 & [P_3, Q_3] \end{pmatrix}$. Consider the set H of matrices of the form $\begin{pmatrix} H_1 & H_2(k) \\ 0 & 0 \end{pmatrix}$ (where the dimensions correspond to those of the blocks in M). A direct calculation shows that $H \cap A$ is an ideal of the Lie algebra A. It is easy to see that $A/(H \cap A)$ is isomorphic to B.

Theorem 3.5.4 *The vehicles converge to formation if at each time k, the communication graph $G(k)$ has a rooted directed spanning tree and the Lie algebra of matrices $A + BFL(k)$ is solvable.*

Proof. We note that this is a sufficient condition, but not necessary. It was shown by Lafferriere *et al.* (2005) that a graph has a rooted directed spanning tree if and only if zero is an eigenvalue of multiplicity one of the corresponding Laplacian matrix. By analogy with (Lafferriere *et al.* 2005) we consider the extended discrete-time system:

$$
\begin{aligned}
x(k+1) &= Ax(k) + BFL(k)x(k) - BFL(k)\left(I_{nN} \otimes \begin{pmatrix} 1 \\ 0 \end{pmatrix}\right) h_p \\
h_p(k+1) &= h_p(k)
\end{aligned}
$$

where $h_p(k) = h_p, \forall k$. Let $y(k) = \begin{pmatrix} x(k) \\ h_p(k) \end{pmatrix}$. Equivalently, the system becomes:

$$
y(k+1) = M(k)y(k)
$$

where $M(k) = \begin{pmatrix} A + BFL(k) & -BFL(k)\left(I_{nN} \otimes \begin{pmatrix} 1 \\ 0 \end{pmatrix}\right) \\ 0 & I_{nN} \end{pmatrix}$. The eigenvalues of $M(k)$ consist of those of $A + BFL(k)$ and those of I_{nN}. Consider now the subspace S of R^{3nN}:

$$S = \left\{ \begin{pmatrix} x \\ h_p \end{pmatrix} : L(k)(x - h) = 0 \right\}.$$

While the matrices $L(k)$ depend on k (and, in fact on x), the space S is independent of k since, for every k, the nullspace of $L(k)$ is spanned by the all-ones vector $\mathbf{1}_N$ (here we used our connectivity assumption on the graphs $G(k)$). Indeed, a basis of S is given by:

$$B = \left\{ \begin{pmatrix} 1_N \otimes e_i \\ 0 \end{pmatrix} : e_i \in R^{2n}, i = 1, \ldots, 2n \right\}$$
$$\bigcup \left\{ \begin{pmatrix} e_j \otimes \begin{pmatrix} 1 \\ 0 \end{pmatrix} \\ e_j \end{pmatrix} : e_j \in R^{nN}, j = 1, \ldots, nN \right\}$$

Lafferriere *et al.* (2005) showed that for a fixed L, the space S is M-invariant. Here we show that S is $M(k)$-invariant, regardless of k. Let $y(k) \in S$, for some k. Then,

$$y(k) = \begin{pmatrix} 1_N \otimes \alpha \\ 0 \end{pmatrix} + \begin{pmatrix} \beta \otimes \begin{pmatrix} 1 \\ 0 \end{pmatrix} \\ \beta \end{pmatrix}$$

$$M(k)y(k) = \begin{pmatrix} 1_N \otimes A_{veh}\alpha \\ 0 \end{pmatrix} + \begin{pmatrix} \beta \otimes \begin{pmatrix} 1 \\ 0 \end{pmatrix} \\ \beta \end{pmatrix}$$

and hence, $M(k)y(k) \in S$. The last equation comes from basic rules of multiplication of Kronecker products and the specific form of A_{veh}. Therefore, for all k, S is $M(k)$-invariant. So, the linear transformation matrix induced by every $M(k)$ on S is constant of the form: $\begin{pmatrix} A & 0 \\ 0 & I_{nN} \end{pmatrix}$. Extending the above basis to a basis of R^{3nN} we can write all the matrices $M(k)$ in block upper triangular form

$$M(k) = \begin{pmatrix} \begin{pmatrix} A & 0 \\ 0 & I_{nN} \end{pmatrix} & * \\ 0 & Q(k) \end{pmatrix}.$$

The matrices $Q(k)$ become the defining matrices of the induced dynamical system on the quotient space R^{2nN}/S. The eigenvalues of $Q(k)$ are those of $A + \lambda(k)BF$, for $\lambda(k)$ any nonzero eigenvalue of $L(k)$.

Until now we assumed that a feedback matrix F that stabilizes each matrix $A + \lambda(k)BF$ is already given. The construction of such a matrix in the case when vehicles have simple dynamics is shown in the section that follows. From our assumptions and Lemma 3.5.3, the Lie algebra generated by the matrices $Q(k)$ is solvable. Therefore the quotient system above is asymptotically stable (Blondel *et al.* 2004). Moreover, stability in the quotient is equivalent to having $L(k)(x(k) - h) \to 0$, which in turn is equivalent to convergence to formation.

3.6 STABILIZING FEEDBACK FOR THE TIME-VARIANT COMMUNICATION CASE

It remains to show that under the connectivity assumption for the time-variant communication graph, a stabilizing feedback matrix F exists. In the proof of the following proposition, we will show how such a matrix can be constructed.

Proposition 3.6.1 *For the double integrator discrete-time vehicle model, a stabilizing feedback matrix F_{veh} exists.*

Proof. We are looking at stabilizing a matrix of the form

$$\begin{pmatrix} 1 + \lambda f_1 \frac{dt^2}{2} & dt + \lambda f_2 \frac{dt^2}{2} \\ \lambda f_1 dt & 1 + \lambda f_2 dt \end{pmatrix}$$

The characteristic polynomial of this matrix is of the form $p(x) = x^2 - sx + p$, where s and p can be complex numbers. We are interested in deriving conditions that will ensure that the roots of this polynomial are in the interior of a circle of radius 1 centered at the origin. By using the transformation $x = \frac{1+w}{1-w}$ and tools developed in Lafferriere *et al.* (2005), we can derive the necessary and sufficient conditions. As these are rather lengthy inequalities, we illustrate here the sufficient conditions in the case when the communication graph is undirected:

$$f_1 < 0$$

$$f_2 < 0$$

$$f_2 - f_1 < -\frac{4}{\lambda dt^2}$$

for λ a nonzero eigenvalue of the communication graph. If we allow $f_1 = f_2$, then the last condition above becomes:

$$f_2 > -\frac{2}{\lambda dt}$$

In our case, the communication graph is state-dependent, so the nonzero eigenvalues of the communication graph will vary with k. Since the number of possible graphs on N vertices is finite, and in particular the set of nonzero eigenvalues is finite, we can choose negative f_1 and f_2 that satisfy the required inequalities. The resulting feedback matrix

$$F = \left(I_{2nN} \otimes \begin{pmatrix} f_1 & f_2 \end{pmatrix}\right)$$

will stabilize all matrices of the form $A + BFL(k)$, as required in the previous section. Using similar methods, conditions can be deduced for the case of directed graphs.

3.7 GRAPH CONNECTIVITY AND STABILITY OF VEHICLE FORMATIONS

In the previous sections we showed that stability of the vehicle formation is guaranteed in the case when the communication graph contains a rooted directed spanning tree at all times k. The issue of graph connectivity has been considered in previous research, however, determining criteria for the communication graph to remain connected in the case when the control law depends on the proximity of neighbors remains an open problem.

Consider again the system defined previously:

$$y(k+1) = M(k)y(k)$$

By selecting a matrix F that stabilizes each $M(k)$ in the quotient space R^{2nN}/S, we therefore find that the induced matrix on this space has eigenvalues in the interior of the circle of radius 1, centered at the origin, and therefore each $M(k)$ must be a contraction mapping on this space. The challenge that remains in this case is to find the connection between of the relative distance between vehicles and the function defined above. In particular, one would like to say that if at time k the communication graph has a rooted directed spanning tree, then one must also exist at time $k+1$. For the case when $h = 0$, the condition should be: if two vertices are adjacent at time k, then they must be adjacent at time $k+1$. This not only depends on the vehicle dynamics given, but also on the eigenvalues of the Laplacian matrices from the set considered. We illustrate this via some examples.

We consider N vehicles with dynamics given by the discrete-time double integrator vehicle model. The initial states are fixed or randomly selected. We consider a proximity radius for vehicle i to be a value ε such that if another vehicle, j, is within that proximity radius, there is a link in the communication graph from j to i. We choose a proximity radius that would guarantee that in the final formation required every vehicle has at least one neighbor. We illustrate the connection between the proximity radius, the required formation shape, the feedback control matrix, and the initial conditions. In Figure 3.6, we have six vehicles converging to a stationary hexagonal formation, starting from given

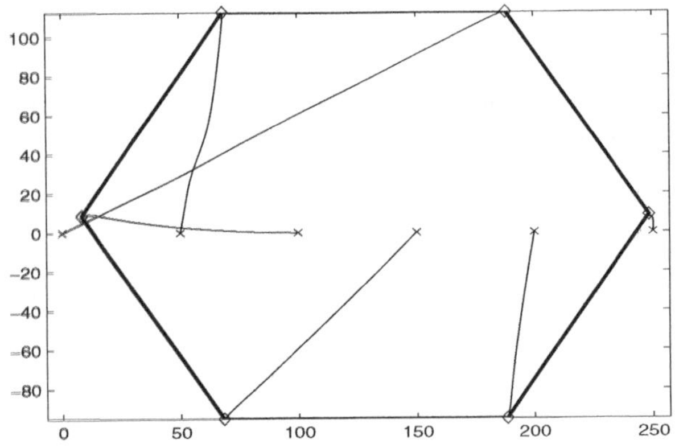

Figure 3.6 Stationary hexagonal formation.

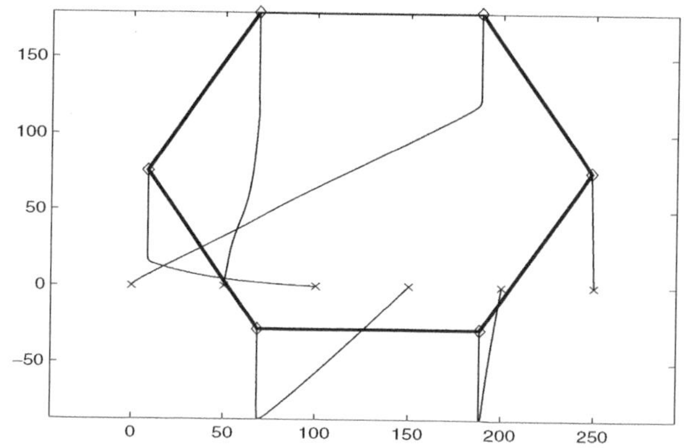

Figure 3.7 A moving hexagonal formation.

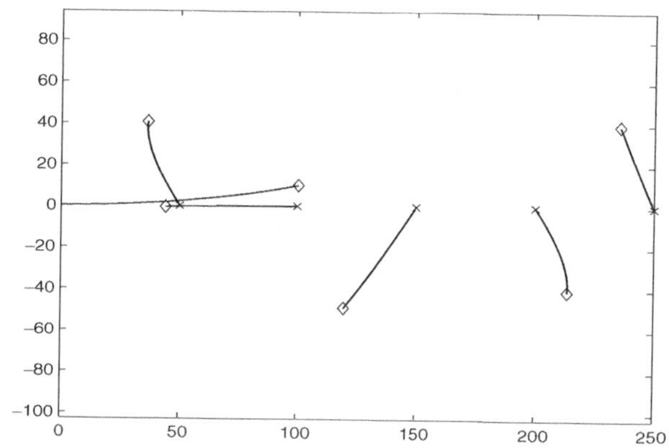

Figure 3.8 Same proximity radius and initial conditions, but different feedback.

initial conditions. The communication between vehicles is neighbor dependent, in the sense that each vehicle receives position and velocity information only from the vehicles that are in its proximity. Both in Figure 3.6 and Figure 3.7 the communication graph is undirected. The values of the gains in this case are $f_1 = -3$ and $f_2 = -3$.

In Figure 3.8, we illustrate the six vehicles, with the same initial conditions as in the vehicles in Figure 3.7, but with different feedback matrix values: $f_1 = -1$ and $f_2 = -1$.

Keeping the same gain values and detection radius, we show in Figure 3.9 that other initial conditions can still lead to a convergent formation. In this case the initial conditions were chosen randomly from a given set.

In all of the above simulations we notice that for small enough initial separation between vehicles, if the graph is connected at time k, it will remain connected at time $k + 1$.

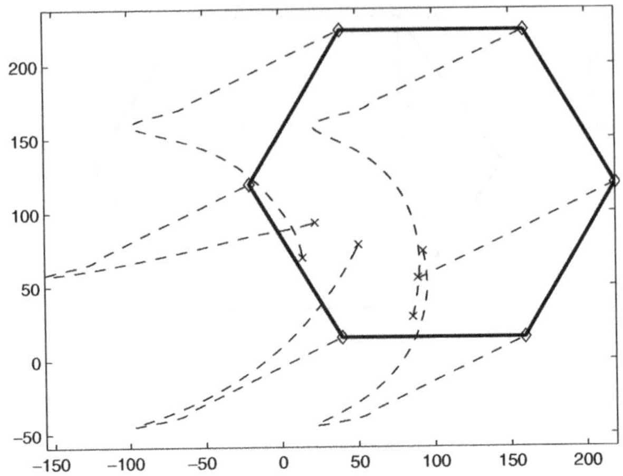

Figure 3.9 Stationary hexagonal formation with random initial conditions.

3.8 CONCLUSION

Our objective in this chapter was to illustrate some of the issues that arise in the analysis of vehicle formation stability under different communication structures. Communication schemes we have analyzed include hierarchical communication with both static and dynamic connectivity graphs. In the latter case, the communication architecture can vary based on the relative physical locations of vehicles or based on other criteria. Although we do not consider any communication delays or vehicle dynamics nonlinearities like actuator saturations, under the assumptions of homogeneous vehicles and linear discrete-time dynamics we prove that formation stability can be achieved both in the cases where the communication structure is fixed and where it is time-dependent. In addition, we investigate how the connectivity of the communication graph can be maintained using the proposed stabilizing feedback. It is clear in this case that the connectivity of the formation, and hence the stability of the formation, depends not only on the feedback gains but also on the proximity radius defined for every vehicle, the initial conditions, and ultimately on the formation structure required. It is sufficient for one of these parameters to change in order to make a stable formation unstable.

ACKNOWLEDGEMENTS

This work was supported in part by the U.S. Defense Advanced Research Porject Agency (DARPA) and the U.S. Air Force Research Laboratory under Contract No. F33615-01-C-1848.

REFERENCES

Anderson W and Morley T 1985 Eigenvalues of the Laplacian of a graph. *Lin. Multilin. Algebra* **18**, 141–145.

Blondel V, Nesterov Y and Theys J 2004 Approximations of the rate of growth of switched linear systems. *Lecture Notes in Computer Science: Hybrid Systems: Computation and Control HSCC.*

Chung F 1994 Spectral graph theory. In *AMS Regional Conference Series in Mathematics*. American Mathematical Society.

de Castro GA and Paganini F 2004 Convex synthesis of controllers for consensus. In *Proc. of American Control Conference*, pp. 4933–4938.

Fax A 2001 Optimal and cooperative control of vehicle formations, PhD thesis, California Institute of Technology, Pasadena, CA.

Fax JA and Murray RM 2002a Graph Laplacians and stabilization of vehicle formations. *15th IFAC World Congress*. Barcelona.

Fax JA and Murray RM 2002b Information flow and cooperative control of vehicle formations. **AC-49**(a), 1465–1476.

Fiedler M 1973 Algebraic connectivity of graphs. *Czechoslovak Mathematical Journal*, **23**(98), 289–305.

Glavaški S, Chavez M, Day R, Nag P, Williams A and Zhang W 2003 Vehicle networks: Achieving regular formation. In *Proceedings of American Control Conference*, pp. 4095–4100.

Godsil C and Royle G 2001 *Algebraic Graph Theory*. Springer Verlag, Berlin.

Guattery S and Miller G 2000 Graph embeddings and Laplacian eigenvalues. *SIAM J. Matrix Anal. Appl.* **21**(3), 703–723.

Jadbabaie A, Lin J and Morse A 2003 Coordination of groups of mobile autonomous agents using nearest neighbor rules. *IEEE Transactions on Automatic Control* **48**(6), 988–1001.

Lafferriere G, Williams A, Caughman J and Veerman J 2005 Decentralized control of vehicle formations. *Systems and Control Letters* **54**, 899–910.

Okubo A 1986 Dynamical aspects of animal grouping: swarms, schools, flocks and herds. *Advances in Biophysics* **22**, 1–94.

Olfati-Saber R and Murray R 2003 Agreement problems in networks with directed graphs and switching topology. In *Proc. of IEEE Conference on Decision and Control*, pp. 4126–4132.

Olfati-Saber R and Murray R 2004 Consensus problems in networks of agents with switching topology and time-delays. *IEEE Trans. Autom. Control* **AC-49**(9), 1520–1533.

Ren W and Beard R 2004a A decentralized scheme for spacecraft formation flying via the virtual structure approach. *AIAA Journal of Guidance, Control and Dynamics* **27**, 73–82.

Ren W and Beard R 2005 Consensus seeking in multiagent systems under dynamically charging interaction topologies. In *IEEE Transactions on Automatic Control* **50**(5), 655–661.

Tanner H, Jadbabaie A and Pappas G Flocking in fixed and switching networks. *IEEE Transactions on Automatic Control* **52**(5), 863–868.

Veerman P, Lafferriere G, Caughman J and Williams A 2005 Flocks and formations. *Special Issue of the Journal of Statistical Physics*.

Williams A, Glavaški S and Samad T 2004 Formations of formations: Hierarchy and stability. In *Proc. of IEEE American Control Conference*, pp. 2992–2997.

Williams A, Glavaški S and Samad T 2005a Connectivity and convergence of formations. In *Proc. of American Control Conference*, pp. 1691–1696.

Williams A, Lafferriere G and Veerman P 2005b Stable motions of vehicle formations. In *Proc. of IEEE Conference on Decision and Control*, pp. 72–77.

4

Distributed receding horizon control: stability via move suppression

William B. Dunbar

4.1 INTRODUCTION

This chapter considers the problem of controlling a set of dynamically decoupled agents that are required to perform a cooperative task. An example of such a situation is a group of autonomous vehicles cooperatively converging to a desired formation, as explored in (Dunbar and Murray 2002; Leonard and Fiorelli 2001; Olfati-Saber *et al.* 2003; Ren and Beard 2004). One control approach that accommodates a general cooperative objective is receding horizon control (RHC), also known as model predictive control. In RHC, the current control action is determined by solving a finite horizon optimal control problem at each sampling instant. In continuous time formulations, each optimization yields an open-loop control trajectory and the initial portion of the trajectory is applied to the system until the next sampling instant. A survey of RHC is given by Mayne *et al.* (2000).

For the problem of interest here, cooperation between agents can be incorporated in the optimal control problem by including coupling terms in the cost function, as done Dunbar and Murray (2006) and Olfati-Saber *et al.* (2003). Agents that are coupled in the cost function are referred to as *neighbors*. When the agents are operating in a real-time distributed environment, as is typically the case with multi-vehicle systems, a centralized RHC implementation is generally not viable, due to the computational requirements of solving the centralized problem at every update. In this chapter, a *distributed implementation* of RHC is presented in which each agent is assigned its own optimal control problem, optimizes only for its own control at each update, and exchanges information with neighboring agents. The work presented here is a continuation of a recent work (Dunbar and Murray 2006). While communication network issues (such as limited bandwidth and delay) are important for many multiagent problems, these issues are not addressed here.

Cooperative Control of Distributed Multi-Agent Systems edited by Jeff S. Shamma

Most theory for distributed RHC is for coupled linear agent dynamics (Acar 1995; Jia and Krogh 2001; Motee and Sayyar-Rodsari 2003; Venkat *et al.* 2004, 2005). In the context of multiple autonomous vehicle missions, several researchers have recently proposed hierarchical distributed RHC schemes. In (Keviczky *et al.* 2004, 2005), agents optimize for themselves and all others to whom they are coupled. A directed acyclic graph defines who computes for whom to avoid deadlock in the control update of any agent. In (Richards and How 2004a, b, 2005; Wesselowski and Fierro 2003), a leader-follower hierarchical structure on control computations is also required. In contrast to hierarchical methods, no hierarchical assignment is required and agents compute their own control in parallel for the framework presented here. As RHC updates are done in parallel, each agent is required to assume a trajectory for neighbors. Control autonomy of the agents makes the approach scalable, since the size of each optimal control problem is independent of the number of neighbors for each agent. Another recent work in which agents update in parallel is Franco *et al.* (2005), and while the work is novel in that inter-agent communication delay is admitted, a conservative small-gain condition is required. A novelty of the approach here is use of a move suppression term in each local cost function, to penalize the deviation of the computed state trajectory from the assumed state trajectory. Closed-loop stability follows if the weight on the move suppression term is larger than a parameter that bounds the amount of coupling in the cost function between neighbors. While move suppression terms are traditionally on the rate of change of the control inputs, specifically in discrete-time applications of model predictive process control, the move suppression term here involves the state trajectory.

The chapter is organized as follows. Section 4.2 begins by defining the subsystem dynamics and constraints, and the generic form of coupling cost function. In Section 4.3, distributed optimal control problems are defined for each subsystem, and the distributed RHC algorithm is stated. Feasibility and stability results are then given in Section 4.4. Finally, Section 4.5 provides conclusions.

4.2 SYSTEM DESCRIPTION AND OBJECTIVE

In this section, the system dynamics and control objective are defined. We make use of the following notation. The symbol $\mathbb{R}_+$ represents the set of non-negative reals. The symbol $\|\cdot\|$ denotes the Euclidean norm in $\mathbb{R}^n$, and dimension n follows from the context. For any vector $x \in \mathbb{R}^n$, $\|x\|_P$ denotes the P-weighted 2-norm, given by $\|x\|_P^2 = x^T P x$, and P is any positive definite real symmetric matrix. Also, $\lambda_{\max}(P)$ and $\lambda_{\min}(P)$ denote the largest and smallest eigenvalues of P, respectively. Often, the notation $\|x\|$ is understood to mean $\|x(t)\|$ at some instant of time $t \in \mathbb{R}$.

In the control problem below, subsystems will be coupled in an integrated cost function of an optimal control problem. For example, vehicles i and j might be coupled in the integrated cost by the term $\|q_i - q_j + d_{ij}\|^2$, where $q_{(\cdot)}$ is the position of the vehicle, and d_{ij} is a constant desired relative position vector that points from i to j. The purpose of the distributed RHC approach is to decompose the overall cost so that, in this example, i and j would each take a fraction of the term $\|q_i - q_j + d_{ij}\|^2$ (among other terms) in defining their local cost functions. Then, i and j update their RHC controllers in parallel, exchanging information about each other's anticipated position trajectory so that each local cost can be calculated. More generally, the coupling cost terms may not be quadratic; the assumption is that the coupled cost terms are non-separable, so that agents i and j must exchange trajectory information if they are coupled via the cost function. Examples of quadratic and non-quadratic coupling cost functions are examined in this chapter. The concept of non-separability is now formally defined.

A non-negative function $g : \mathbb{R}^n \times \mathbb{R}^n \to \mathbb{R}_+$ is called *non-separable* in $x \in \mathbb{R}^n$ and $y \in \mathbb{R}^n$ if g is *not additively separable* for all $x, y \in \mathbb{R}^n$. That is, g cannot be written as the sum of two non-negative functions $g_1 : \mathbb{R}^n \to \mathbb{R}_+$ and $g_2 : \mathbb{R}^n \to \mathbb{R}_+$ such that $g(x, y) = g_1(x) + g_2(y)$ for all $x, y \in \mathbb{R}^n$. By this definition, note that g is non-separable in vectors x and y even if only one of the components of y is coupled non-separably to any of the components of x.

The objective is to stabilize a collection of subsystems, referred to as *agents*, to an equilibrium point using RHC. In addition, each agent is required to cooperate with a set of other agents, where cooperation refers to the fact that every agent has incentive to optimize the collective cost function that couples their state to the states of other agents. For each agent $i \in \mathcal{V} \triangleq \{1, \ldots, N_a\}$, the state and control vectors are denoted $z_i(t) \in \mathbb{R}^n$ and $u_i(t) \in \mathbb{R}^m$, respectively, at any time $t \geq t_0 \in \mathbb{R}$. The decoupled, time-invariant nonlinear system dynamics are given by

$$\dot{z}_i(t) = f_i(z_i(t), u_i(t)), \quad t \geq t_0. \tag{4.1}$$

While the system dynamics can be different for each agent, the dimension of every agent's state (control) is assumed to be the same, for notational simplicity and without loss of generality. Each agent i is also subject to the decoupled state and input constraints,

$$z_i(t) \in \mathcal{Z}, \quad u_i(t) \in \mathcal{U}, \quad \forall t \geq t_0,$$

where $\mathcal{Z}$ and $\mathcal{U}$ are also assumed to be common to every agent i for notational simplicity and without loss of generality. The Cartesian product is denoted $\mathcal{Z}^{N_a} = \mathcal{Z} \times \cdots \times \mathcal{Z}$. The concatenated vectors are denoted $z = (z_1, \ldots, z_{N_a})$ and $u = (u_1, \ldots, u_{N_a})$. In concatenated vector form, the system dynamics are

$$\dot{z}(t) = f(z(t), u(t)), \quad t \geq t_0, \tag{4.2}$$

where $f(z, u) = (f_1(z_1, u_1), \ldots, f_{N_a}(z_{N_a}, u_{N_a}))$. The desired equilibrium point is the origin, and some standard assumptions regarding the system are now stated

Assumption 4.2.1 *The following holds for every $i \in \mathcal{V}$:*

(a) $f_i : \mathbb{R}^n \times \mathbb{R}^m \to \mathbb{R}^n$ is continuous, $0 = f_i(0, 0)$, and f_i is locally Lipschitz in z_i.

(b) $\mathcal{Z}$ is a closed connected subset of $\mathbb{R}^n$ containing the origin in its interior.

(c) $\mathcal{U}$ is a compact, convex subset of $\mathbb{R}^m$ containing the origin in its interior.

It is assumed that a single collective cost function $L : \mathbb{R}^{nN_a} \to \mathbb{R}_+$ is provided that couples the states of the agents, and that each agent has incentive to minimize this function with respect to their own state. For each $i \in \mathcal{V}$, let $\mathcal{N}_i \subseteq \mathcal{V} \setminus \{i\}$ be the set of other agents whose states are coupled non-separably to z_i in $L(z)$. By definition, $j \in \mathcal{N}_i$ if and only if $i \in \mathcal{N}_j$, for all distinct $i, j \in \mathcal{V}$. Denote $N_i = |\mathcal{N}_i|$ and let $z_{-i} = (z_{j_1}, \ldots, z_{j_{N_i}})$ be the collective vector of states coupled non-separably to z_i in $L(z)$.

Assumption 4.2.2 *The function $L : \mathbb{R}^{nN_a} \rightarrow \mathbb{R}_+$ is continuous, positive definite and can be decomposed as follows: for every $i \in \mathcal{V}$ there exists an integer $N_i \in \{1, \ldots, N_a - 1\}$ and a continuous and non-negative function $L_i : \mathbb{R}^n \times \mathbb{R}^{nN_i} \rightarrow \mathbb{R}_+$, not identically zero, such that*

(a) $L_i(z_i, z_{-i})$ is non-separable in $z_i \in \mathbb{R}^n$ and $z_{-i} \in \mathbb{R}^{nN_i}$.

(b) $\sum_{i \in \mathcal{V}} L_i(z_i, z_{-i}) = L(z)$.

(c) there exists a positive constant $c_i \in (0, \infty)$ such that

$$L_i(x, y) \leq L_i(x, w) + c_i \|y - w\|, \quad \forall x \in \mathbb{R}^n, \quad \forall y, w \in \mathbb{R}^{nN_i}.$$

The constant c_i is referred to as the ***strength-of-coupling parameter****, and the term $c_i \| \cdot \|$ is referred to as the* ***cost coupling bound****.*

In the example cost functions below, the cost coupling bound is shown to hold if states are assumed to remain bounded. That states remain bounded is also a prerequisite for any implemented nonlinear optimization algorithm (Polak 1997).

4.2.0.1 Example cost functions

In the following examples for the coupling cost L, the structure required by Assumption 4.2.2 is demonstrated.

Example 4.2.1 (Quadratic Multi-Vehicle Formation Cost) *The integrated cost for multi-vehicle formation stabilization considered in Dunbar and Murray (2006) is given by*

$$L(z) = \|(q_1 + q_2 + q_3)/3 - q_\Sigma\|^2 + \sum_{(i,j) \in \mathcal{E}} \|q_i - q_j + d_{ij}\|^2 + \sum_{i \in \mathcal{V}} \|\dot{q}_i\|^2,$$

for vehicles with position and velocity states $(q_i, \dot{q}_i) \in \mathbb{R}^{2n}$. For stabilization to the origin, the state for the dynamics of each i is defined as $z_i = (q_i - q_i^c, \dot{q}_i)$, where the desired equilibrium in this context is a constant position $q_i = q_i^c$ for every $i \in \mathcal{V}$, rather than having all vehicles fly to the origin. The set $\mathcal{E}$ is all pair-wise neighbors, and $\mathcal{D} = \{d_{ij} \in \mathbb{R}^n | (i, j) \in \mathcal{E}\}$ is the set of constant relative vectors between neighboring vehicles. For any two neighbors i and j, $q_i^c + d_{ij} = q_j^c$, and $i \in \mathcal{N}_j$ iff $j \in \mathcal{N}_i$. It is assumed at the outset that $\mathcal{E}$ and $\mathcal{D}$ (and therefore L) are provided by some supervisory mechanism. To make L positive definite, the first term is included, where $q_\Sigma = (q_1^c + q_2^c + q_3^c)/3$ (see Dunbar and Murray (2006) for details). The decomposition required in Assumption 4.2.2 (b) can be satisfied for each $i \in \mathcal{V}$ by taking

$$L_i(z_i, z_{-i}) = \sum_{j \in \mathcal{N}_i} \frac{1}{2} \|q_i - q_j + d_{ij}\|^2 \; + \; \|\dot{q}_i\|^2,$$

and $1/3$ of the q_Σ term in L is taken if $i \in \{1, 2, 3\}$. Now, the expression for c_i is derived to satisfy Assumption 4.2.2 (c). Since L_i is quadratic, it is straightforward to define a matrix

$Q_i = Q_i^T \geq 0$ *such that* $L_i(z_i, z_{-i}) = \|(z_i, z_{-i})\|^2_{Q_i}$. *For any* $x \in \mathbb{R}^n$ *and* $y, w \in \mathbb{R}^{nN_i}$,

$$L_i(x, y) - L_i(x, w) = \left\| \begin{matrix} x \\ y \end{matrix} \right\|^2_{Q_i} - \left\| \begin{matrix} x \\ w \end{matrix} \right\|^2_{Q_i} = (y - w)^T Q_{i,c}(y + w) + 2x^T Q_{i,b}(y - w),$$

$$\text{with } Q_i = \begin{bmatrix} Q_{i,a} & Q_{i,b} \\ Q_{i,b}^T & Q_{i,c} \end{bmatrix}.$$

If every state z_i *is assumed to remain bounded as* $\|z_i\| \leq \rho$ *for some* $\rho \in (0, \infty)$, *then*

$$\begin{aligned}(y - w)^T Q_{i,c}(y + w) + 2x^T Q_{i,b}(y - w) \ &\leq \ \|y - w\| 2\rho \\ &\left\{\lambda_{\max}(Q_{i,c})N_i + \lambda_{\max}^{1/2}(Q_{i,b}^T Q_{i,b})\right\}.\end{aligned}$$

In Dunbar (2004), Lemma A.1 proves that $\lambda_{\max}(Q_i) \geq \lambda_{\max}^{1/2}(Q_{i,b}^T Q_{i,b})$ *and* $\lambda_{\max}(Q_i) \geq \lambda_{\max}(Q_{i,c})$. *Thus,*

$$L_i(x, y) - L_i(x, w) \leq c_i \|y - w\|,$$

with $c_i = 2\rho\lambda_{\max}(Q_i)(N_i + 1)$. □

Example 4.2.2 (Full Quadratic Coupling) *Let* $L(z) = \|z\|^2_Q$, *where* $Q = Q^T > 0$ *is full rank and* $N_i = N_a - 1$ *for every* $i \in \mathcal{V}$. *In this case, one could simply choose* $L_i(z_i, z_{-i}) = \|z\|^2_Q / N_a$. *To satisfy Assumption 4.2.2 (c), for any* $i \in \mathcal{V}$, *one can shuffle submatrices and vector components so that* z_i *appears as the first subvector. That is, there exists a matrix* $Q_i = Q_i^T > 0$ *such that* $L_i(z_i, z_{-i}) = \|(z_i, z_{-i})\|^2_{Q_i} = \|z\|^2_Q / N_a$ *for each* $i \in \mathcal{V}$. *Following the logic in Example 4.2.1,* $c_i = 2\rho\lambda_{\max}(Q_i)N_a$ *ensures Assumption 4.2.2 (c) holds.* □

Example 4.2.3 (Non-Quadratic Multi-Vehicle Formation Cost) *In Olfati-Saber et al. (2003), local cost functions are defined to meet multi-vehicle formation stabilization and tracking objectives. The terms in the cost function are designed to penalize deviations of relative distances between neighboring vehicles, as opposed to penalizing relative vectors as in Examples 4.2.1 and 4.2.2. Specifically, each local cost* L_i *contains terms in the form* $[\|q_i - q_j\| - \alpha]^2/2$, *where* α *is the desired scalar distance between* i *and* j. *The cost is non-quadratic. Still, the cost functions can be shown to satisfy the conditions of Assumption 4.2.2. Part (a) is trivial, as in the previous example, since each* i *can take a fraction of each coupling term, e.g.,* $[\|q_i - q_j\| - \alpha]^2/4$. *Part (c) follows by employing the triangle inequality and by assuming that each state remains bounded. Specifically,*

$$[\|x - y\| - \alpha]^2/2 - [\|x - w\| - \alpha]^2/2 \leq (3\rho + \alpha)\|y - w\|,$$

for any $x, y, w \in \mathbb{R}^n$, *and* ρ *is the bound on any n-dimensional state.* □

The same logic in Example 4.2.1 can be used to show that generic partial quadratic coupling terms satisfy the conditions of Assumption 4.2.2. From Examples 4.2.1 and 4.2.2, c_i is proportional to $(N_i + 1)$ for quadratic coupling costs. In the limit that there is no coupling with a quadratic cost L, $c_i = 0$ trivially satisfies Assumption 4.2.2 (c). For the example cost functions above, more coupling terms (i.e., more neighbors) implies a larger c_i value.

Remark 4.2.1 Observe that, from the analysis in the examples above, the coupling cost functions *do not* satisfy $L_i(x, y) \leq L_i(x, w) + c_i\|y - w\|^2$ in general, for any positive constant c_i, while they *do* satisfy the form of cost coupling bound stated in part (c) of Assumption 4.2.2. For this reason, as demonstrated in the analysis in Section 4.4, the move suppression term in the cost function of each local optimal control problem also takes the form $b_i\|\cdot\|$ for some positive weighting constant b_i.

4.3 DISTRIBUTED RECEDING HORIZON CONTROL

In this section, N_a separate optimal control problems and the distributed RHC algorithm are defined. In every distributed optimal control problem, the same constant prediction horizon $T \in (0, \infty)$ and constant update period $\delta \in (0, T]$ are used. The receding horizon update times are denoted $t_k = t_0 + \delta k$, where $k \in \mathbb{N} = \{0, 1, 2, \ldots\}$. In the following implementation, every distributed RHC law is updated globally synchronously, i.e., at the same instant of time t_k for the kth-update.

At each update, every agent optimizes only for its own open-loop control, given its current state and that of its neighbors. Since each cost $L_i(z_i, z_{-i})$ depends upon the neighboring states z_{-i}, each agent i must presume some trajectories for z_{-i} over each prediction horizon. To that end, prior to each update, each agent i receives an assumed state trajectory from each neighbor. Likewise, agent i transmits an assumed state to all neighbors prior to each optimization. To distinguish the different trajectories, we introduce the following notation for each agent $i \in \mathcal{V}$:

$$\begin{aligned} z_i(t) &: \text{ the actual state trajectory, at any time } t \geq t_0, \\ z_i^p(\tau; t_k) &: \text{ the predicted state trajectory, } \tau \in [t_k, t_k + T],\ k \in \mathbb{N}, \\ \hat{z}_i(\tau; t_k) &: \text{ the assumed state trajectory, } \tau \in [t_k, t_k + T],\ k \in \mathbb{N}. \end{aligned}$$

For the RHC implementation here, $z_i(t) = z_i^p(t; t_k)$ for all $t \in [t_k, t_{k+1}]$ and any $i \in \mathcal{V}$. The predicted and assumed control trajectories are likewise denoted $u_i^p(\tau; t_k)$ and $\hat{u}_i(\tau; t_k)$, respectively. Let $\hat{z}_{-i}(\tau; t_k)$ be the vector of assumed state trajectories of the neighbors of i, corresponding to current time t_k. At time t_k, the cost function for the optimal control problem for each agent $i \in \mathcal{V}$ is

$$\begin{aligned} J_i(z_i(t_k), u_i^p(\cdot; t_k)) = \int_{t_k}^{t_k+T} \Big[& L_i(z_i^p(s; t_k), \hat{z}_{-i}(s; t_k)) + \|u_i^p(s; t_k))\|_{R_i}^2 \\ & + b_i\|z_i^p(s; t_k) - \hat{z}_i(s; t_k)\| \Big] ds \; + \; \|z_i^p(t_k + T; t_k)\|_{P_i}^2, \end{aligned} \tag{4.3}$$

given constant $b_i \in [0, \infty)$, and matrices $R_i = R_i^T > 0$ and $P_i = P_i^T > 0$. The cost term $b_i\|z_i^p(s; t_k) - \hat{z}_i(s; t_k)\|$ in Equation (4.3) is a state move suppression term. It is a way of penalizing the deviation of the computed predicted state trajectory from the assumed trajectory, which neighboring agents are presuming from that agent. In previous works, this term was incorporated into the distributed RHC framework as a constraint, called a consistency constraint (Dunbar and Murray 2006; Dunbar 2007). The formulation presented here is an improvement over these past formulations, since the move suppression cost formulation yields an optimization problem that is much easier to solve, and allows a greater degree of freedom to the calculated RHC control law. Note the move suppression term is in the form $b_i\|\cdot\|$, and not $b_i\|\cdot\|^2$. The reason for this is directly related to the form of the cost coupling bound made in part (c) of Assumption 4.2.2, which takes the form $c_i\|\cdot\|$ and not $c_i\|\cdot\|^2$ (see Remark 4.2.1 for justification). The connection between the move suppression term and the cost coupling bound will be clarified in the stability analysis provided in Section 4.4.

RHC stability results sometimes rely on the calculation of the optimal cost at each RHC update, e.g., (Jadbabaie *et al.* 2001; Mayne *et al.* 2000). To relax this requirement here, each cost is minimized while subject to the improvement constraint $J_i(z_i(t_k), u_i^p(\cdot; t_k)) \leq J_i(z_i(t_k), \hat{u}_i(\cdot; t_k))$. The cost $J_i(z_i(t_k), \hat{u}_i(\cdot; t_k))$ is the same as in Equation (4.3), but replacing $(z_i^p(s; t_k), u_i^p(s; t_k))$ with $(\hat{z}_i(s; t_k), \hat{u}_i(s; t_k))$. As will be shown in the coming sections, a feasible solution to this constraint is always available, and the resulting distributed RHC law is stabilizing even without computation of the optimal cost, i.e, feasibility is sufficient for stability. A fundamental reason to use an improvement constraint for stability, instead of requiring optimality, is that for nearly all nonlinear optimal control problems, optimal solutions can never be exactly calculated in finite time. The improvement constraint formulation, on the other hand, is always computationally tractable and feasible. Other (centralized) RHC methods that also rely on feasibility for stability, instead of optimality, are discussed (Chen and Allgöwer 1998; Michalska and Mayne 1993). The collection of distributed optimal control problems is now defined.

Problem 4.3.1 *For each agent $i \in \mathcal{V}$ and at any update time t_k, $k \in \mathbb{N}$:*
Given: $z_i(t_k)$, $\hat{u}_i(\tau; t_k)$, $\hat{z}_i(\tau; t_k)$ and $\hat{z}_{-i}(\tau; t_k)$ for all $\tau \in [t_k, t_k + T]$,
Find: a state and control pair $(z_i^p(\tau; t_k), u_i^p(\tau; t_k))$ that minimizes $J_i(z_i(t_k), u_i^p(\cdot; t_k))$ subject to the constraints

$$J_i(z_i(t_k), u_i^p(\cdot; t_k)) \leq J_i(z_i(t_k), \hat{u}_i(\cdot; t_k)) \tag{4.4}$$

$$\left.\begin{aligned} \dot{z}_i^p(\tau; t_k) &= f_i(z_i^p(\tau; t_k), u_i^p(\tau; t_k)) \\ u_i^p(\tau; t_k) &\in \mathcal{U} \\ z_i^p(\tau; t_k) &\in \mathcal{Z} \end{aligned}\right\} \quad \tau \in [t, t+T],$$

$$z_i^p(t_k + T; t_k) \in \Omega_i(\varepsilon_i) := \{z \in \mathbb{R}^n \mid \|z\|_{P_i}^2 \leq \varepsilon_i\}, \tag{4.5}$$

with $z_i^p(t_k; t_k) = z_i(t_k)$, and given the constant $\varepsilon_i \in (0, \infty)$. ■

As stated, the constraint (4.4) is used to guarantee stability. Minimization of the cost J_i is done solely for performance purposes. In practice, the resulting computed control must be feasible, but need not be optimal. The closed-loop system for which stability is to be guaranteed is

$$\dot{z}(t) = f(z(t), u_{\mathrm{RH}}(t)), \quad \tau \geq t_0, \tag{4.6}$$

with the applied *distributed RHC law*

$$u_{\mathrm{RH}}(t) = (u_1^p(t; t_k), \ldots, u_{N_a}^p(t; t_k)),$$

for $t \in [t_k, t_{k+1})$ and any $k \in \mathbb{N}$. As with most nominally stabilizing RHC formulations (Chen and Allgöwer 1998; Dunbar and Murray 2006; Jadbabaie *et al.* 2001; Michalska and Mayne 1993), observe that, under the closed-loop distributed RHC law, the predicted state $z_i^p(t; t_k)$ for every $i \in \mathcal{V}$ is equal to the actual state $z_i(t)$ for all $t \in [t_k, t_{k+1})$ and any $k \in \mathbb{N}$. Before stating the control algorithm formally, which in turn defines the assumed trajectories for each agent, decoupled terminal controllers associated with each terminal cost and terminal constraint (4.5) are required.

Assumption 4.3.1 *For every agent $i \in \mathcal{V}$, there exists a (possibly nonlinear) state feedback controller $\kappa_i(z_i)$ and a constant $\varepsilon_i \in (0, \infty)$ such that:*

(a) $\kappa_i(z_i) \in \mathcal{U}$ for all $z_i \in \Omega_i(\varepsilon_i)$,

(b) $\Omega_i(\varepsilon_i) \subset \mathcal{Z}$, and

(c) the function $V(z) = \sum_{i \in \mathcal{V}} \|z_i\|_{P_i}^2$ satisfies

$$\frac{d}{dt}V(z) \leq -\left[L(z) + \sum_{i \in \mathcal{V}} \|\kappa_i(z_i)\|_{R_i}^2\right], \quad \forall z \in \Omega_1(\varepsilon_1) \times \cdots \times \Omega_{N_a}(\varepsilon_{N_a}).$$

The assumption provides sufficient conditions under which the closed-loop system $\dot{z}_i(t) = f_i(z_i(t), \kappa_i(z_i(t)))$ is asymptotically stabilized to the origin, with constraint feasible state and control trajectories, and positively-invariant region of attraction $\Omega_i(\varepsilon_i)$. Variants of this assumption are common in the RHC literature (Mayne *et al.* 2000). For example, by assuming stabilizable and C^2 dynamics f_i for each agent i, a feasible local linear feedback $u_i = K_i z_i$ which stabilizes each linearized and nonlinear subsystem (4.1) in $\Omega_i(\varepsilon_i)$ can be constructed (Chen 1997; Chen and Allgöwer 1998; Michalska and Mayne 1993). Moreover, with this linear feedback control, one can show that ε_i exists for each $i \in \mathcal{V}$ when L is quadratic (Michalska and Mayne 1993), in which case the assumption can be converted into an existence lemma.

Some comments about the parameters required to solve Problem 4.3.1 are in order. For the cost function J_i in (4.3) for any $i \in \mathcal{V}$, each function L_i is assumed given, and one is free to choose the input weighting matrix R_i. The terminal cost weight P_i, and the terminal state constraint set parameter ε_i in (4.5), are determined so that Assumption 4.3.1 is satisfied, as shown for quadratic L_i with linear terminal controllers in (Chen and Allgöwer 1998; Dunbar 2007; Michalska and Mayne 1993). The parameters T and δ may be freely chosen, with the choice usually dictated by the application. Lastly, conditions on the move suppression weight parameter b_i in (4.3) will be derived in Section 4.4 to ensure that the distributed RHC algorithm is stabilizing.

As a result of Assumption 4.3.1, $\Omega_i(\varepsilon_i)$ is a positively invariant region of attraction for (4.1) with $u_i = \kappa_i(z_i)$. The decoupled feedback controllers are therefore referred to as *terminal controllers*, since they are associated with the terminal state constraint set. With the terminal controllers defined, the assumed trajectories can now also be defined, given by

$$\hat{z}_i(t; t_k) = \begin{cases} z_i^p(t; t_{k-1}), & t \in [t_k, t_{k-1} + T) \\ z_i^\kappa(t; t_{k-1} + T), & t \in [t_{k-1} + T, t_k + T] \end{cases} \tag{4.7}$$

$$\hat{u}_i(t; t_k) = \begin{cases} u_i^p(t; t_{k-1}), & t \in [t_k, t_{k-1} + T) \\ \kappa_i(z_i^\kappa(t; t_{k-1} + T)), & t \in [t_{k-1} + T, t_k + T] \end{cases} \tag{4.8}$$

where $z_i^{\kappa}(\cdot; t_{k-1} + T)$ is the solution to

$$\dot{z}_i^{\kappa}(t) = f_i(z_i^{\kappa}(t), \kappa_i(z_i^{\kappa}(t))), \tag{4.9}$$

with initial condition $z_i^{\kappa}(t_{k-1} + T; t_{k-1} + T) = z_i^p(t_{k-1} + T; t_{k-1})$. By construction, each assumed state and control trajectory is the remainder of the previously predicted trajectory, concatenated with the closed-loop response under the terminal controller. As stated in the examples in Section 4.2, an assumption on the boundedness of the computed state trajectories is required, for each optimal control problem to be computationally well posed, and for Assumption 4.2.2 to hold.

Assumption 4.3.2 *For every agent $i \in \mathcal{V}$ and for any update t_k, $k \in \mathbb{N}$, there exists a positive constant $\rho \in (0, \infty)$ such that $\|z_i^p(t; t_k)\| \leq \rho$ for all $t \in [t_k, t_k + T]$.*

Since $\Omega_i(\varepsilon_i)$ is compact, and therefore bounded, boundedness of $z_i^p(t; t_k)$ is sufficient to guarantee boundedness of $\hat{z}_i(t; t_k)$. For each $i \in \mathcal{V}$, let $Z_i \subset \mathbb{R}^n$ denote the bounded set of initial states $z_i(t)$ which can be steered to $\Omega_i(\varepsilon_i)$ by a piecewise right-continuous control $u_i^p(\cdot; t) : [t, t + T] \to \mathcal{U}$, with the resulting trajectory $z_i^p(\cdot; t) : [t, t + T] \to \mathcal{Z}$. Note that at initial time, $\hat{z}_i(t; t_0)$ cannot be calculated as a function of the previously predicted trajectory, since no such trajectory exists. Therefore, a different means of computing $\hat{z}_i(t; t_0)$ must be defined so that the distributed RHC algorithm can be initialized. The procedure for initialization is incorporated into the control algorithm.

When every agent is in its terminal state constraint set, all agents synchronously switch to their terminal controllers. The terminal controllers are then employed for all future time, resulting in asymptotic stabilization. Switching from RHC to a terminal controller is known as *dual-mode* RHC (Michalska and Mayne 1993). To determine if all agents are in their terminal sets, a simple protocol is used. If an agent has reached its terminal set at an update time, it sends a flag message to all other agents. If an agent sends a flag and receives a flag from all other agents, that agent switches to its terminal controller, since this will only happen if all agents have reached their terminal set. The control algorithm is now stated.

Algorithm 4.3.1 *[Distributed RHC Algorithm] For any agent $i \in \mathcal{V}$, the distributed RHC law is computed as follows:*

Data: $z_i(t_0) \in Z_i$, $T \in (0, \infty)$, $\delta \in (0, T]$.

Initialization: At time t_0, if $z_i(t_0) \in \Omega_i(\varepsilon_i)$, transmit a flag message. If a flag is sent and a flag is received from all other agents, employ the terminal controller κ_i for all future time $t \geq t_0$.

Otherwise, solve a modified Problem 4.3.1, setting $b_i = 0$ in (4.3), $\hat{z}_{-i}(t; t_0) = z_{-i}(t_0)$ *for all $t \in [t_0, t_0 + T]$, and removing the constraint* (4.4). *Proceed to controller step 1.*

Controller:

1. *Over any interval $[t_k, t_{k+1})$, $k \in \mathbb{N}$:*
   (a) *Apply $u_i^p(\tau; t_k)$, $\tau \in [t_k, t_{k+1})$.*
   (b) *Compute $\hat{z}_i(\cdot; t_{k+1})$ according to* (4.7) *and $\hat{u}_i(\cdot; t_{k+1})$ according to* (4.8).
   (c) *Transmit $\hat{z}_i(\cdot; t_{k+1})$ to every neighbor $j \in \mathcal{N}_i$. Receive $\hat{z}_j(\cdot; t_{k+1})$ from every neighbor $j \in \mathcal{N}_i$, and assemble $\hat{z}_{-i}(\cdot; t_{k+1})$. Proceed to step 2.*

2. *At any update time* t_{k+1}, $k \in \mathbb{N}$:

   (a) *Measure* $z_i(t_{k+1})$.

   (b) *If* $z_i(t_{k+1}) \in \Omega_i(\varepsilon_i)$, *transmit a flag to all other agents. If, in addition, a flag is received from all other agents, employ the terminal controller* κ_i *for all future time* $t \geq t_{k+1}$. *Otherwise, proceed to step (c).*

   (c) *Solve Problem 4.3.1 for* $u_i^p(\cdot; t_{k+1})$, *and return to step 1.* ■

At initialization of Algorithm 4.3.1, if all agents are not in their terminal sets, a modified Problem 4.3.1 is solved in which neighbors are assumed to remain at their initial conditions. While this choice facilitates initialization, it is known that neighbors will not remain at their initial conditions. As such, for performance reasons, it be may preferable to reduce the weight of the L_i term in (4.3), so that the cost J_i would be dominated by the terminal state cost and integrated control input cost terms. If an iterative procedure can be tolerated, the computed initial $z_i^p(\cdot; t_0)$ and $u_i^p(\cdot; t_0)$ could be defined as the assumed trajectories $\hat{z}_i(\cdot; t_0)$ and $\hat{u}_i(\cdot; t_0)$, with $\hat{z}_i(\cdot; t_0)$ transmitted to neighbors, and the modified Problem 4.3.1 resolved again. While the move suppression cost is removed at initialization, by setting $b_i = 0$, the value of b_i will be nonzero at every subsequent update t_k, $k \geq 1$, with conditions on permissible values for b_i defined in the next section. While each agent must communicate (flag messages) with all other agents at initialization and in step 2(b) of the controller algorithm, all other inter-agent communication is local, i.e., agents communicate only with their neighbors.

4.4 FEASIBILITY AND STABILITY ANALYSIS

This section demonstrates the feasibility and stability properties of the distributed RHC algorithm. For computational reasons, it is desirable for any RHC algorithm to have a feasible solution to the optimal control problem at every update, since the optimization algorithm can then be preempted at each RHC update. A feasible solution at time t_k is defined as a state and control pair $(z_i^p(\cdot; t_k), u_i^p(\cdot; t_k))$ that satisfies all constraints, and results in bounded cost, in Problem 4.3.1. According to Algorithm 4.3.1, a modified problem is solved at time t_0 (initialization). The remainder of this section assumes that a solution to this problem can be found for every $i \in \mathcal{V}$. Assuming initial feasibility is standard in the RHC literature (Chen and Allgöwer 1998; Michalska and Mayne 1993; Mayne *et al.* 2000). The first result of this section is to show that, if a feasible solution to the modified problem can be found at t_0, then there is a feasible solution to Problem 4.3.1 at every subsequent RHC update time t_k, $k \geq 1$.

Lemma 4.4.1 *Suppose Assumptions 4.2.1–4.2.2 and Assumptions 4.3.1–4.3.2 hold, and* $z_i(t_0) \in Z_i$ *for every* $i \in \mathcal{V}$. *For every agent* $i \in \mathcal{V}$, *suppose that a feasible solution to the modified Problem 4.3.1 is computed at initial time* t_0, *with the modified problem defined in the initialization step of Algorithm 4.3.1. Then, for every agent* $i \in \mathcal{V}$, *there exists a feasible solution to Problem 4.3.1 at every subsequent update time* t_k, $k \geq 1$.

Proof. The following analysis holds for any agent $i \in \mathcal{V}$. For time t_0, it is assumed that a solution $(z_i^p(\tau; t_0), u_i^p(\tau; t_0))$ to the modified Problem 4.3.1 is found. The assumed state and control pair $(\hat{z}_i(\cdot; t_1), \hat{u}_i(\cdot; t_1))$ is constructed according to Equations (4.7)–(4.8). To show this pair is a feasible solution to Problem 4.3.1, observe first that since $z_i^p(t_1; t_0) = z_i(t_1)$, $\hat{z}_i(t_1; t_1) = z_i(t_1)$. By construction, the assumed trajectories also satisfy the dynamic equation. The cost improvement is trivially satisfied since $J_i(z_i(t_k), \hat{u}_i(\cdot; t_k)) \leq J_i(z_i(t_k), \hat{u}_i(\cdot; t_k))$, and the resulting cost is bounded using Assumption 4.3.2 and the fact that $\Omega_i(\varepsilon_i)$ and $\mathcal{U}$ are compact. By the properties on the terminal controller and terminal state constraint set stated in Assumption 4.3.1, the state and control pair $(\hat{z}_i(t; t_1), \hat{u}_i(t; t_1)) \in \mathcal{Z} \times \mathcal{U}$, for all $t \in [t_1, t_1 + T]$, and $\hat{z}_i(t_1 + T; t_1) \in \Omega_i(\varepsilon_i)$. Therefore, the assumed state and control pair $(\hat{z}_i(\cdot; t_1), \hat{u}_i(\cdot; t_1))$ is a feasible solution to Problem 4.3.1 at time t_1.

Setting $(z_i^p(\cdot; t_1), u_i^p(\cdot; t_1)) = (\hat{z}_i(\cdot; t_1), \hat{u}_i(\cdot; t_1))$ as the feasible solution at update time t_1, the assumed state and control pair $(\hat{z}_i(\cdot; t_2), \hat{u}_i(\cdot; t_2))$ is constructed, again according to Equations to (4.7)–(4.8). Following the same logic above, this pair is a feasible solution for update time t_2. Moreover, a feasible solution $(z_i^p(\cdot; t_k), u_i^p(\cdot; t_k))$, for any $k \geq 0$, implies that the pair $(\hat{z}_i(\cdot; t_{k+1}), \hat{u}_i(\cdot; t_{k+1}))$ is a feasible solution at time t_{k+1}. Therefore, since an initially feasible solution $(z_i^p(\tau; t_0), u_i^p(\tau; t_0))$ is assumed to be available, a feasible solution to Problem 4.3.1 at every update time t_k, $k \geq 1$ is also available. ■

The lemma proof shows that the assumed trajectories, which neighbors rely on in computing their own local solutions to Problem 4.3.1, can be used as a warm-start for computing a solution to Problem 4.3.1 at each update. Next, the stability of the closed-loop system (4.6) is analyzed.

Theorem 4.4.2 *Suppose Assumptions 4.2.1–4.2.2 and Assumptions 4.3.1–4.3.2 hold, and $z_i(t_0) \in Z_i$ for every $i \in \mathcal{V}$. Suppose also that a solution to the modified Problem 4.3.1 is computed at initial time t_0 for every agent $i \in \mathcal{V}$. Then, by application of Algorithm 4.3.1 for all time $t \geq t_0$, the closed-loop system* (4.6) *is asymptotically stabilized to the origin, provided the move suppression weight b_i in the cost function* (4.3) *satisfies*

$$b_i \geq \sum_{j \in \mathcal{N}_i} c_j, \quad \forall i \in \mathcal{V}, \tag{4.10}$$

where c_i is the strength-of-coupling parameter, defined in Assumption 4.2.2.

Equation (4.10) says that the weight b_i on the move suppression term is bounded from below by the sum of the strength-of-coupling parameters c_j of the neighbors of i. Each strength-of-coupling parameter c_j is a (possibly conservative) measure of how much net coupling there is between j and *all* other neighbors $\mathcal{N}_j$. The larger c_j, and hence the more net coupling, the more dependent j is on the assumed trajectories of neighbors in the term L_j in the optimal control cost function.[1] So, another way of interpreting Equation (4.10) is that, the more the neighbors of i rely on assumed trajectories in their own coupled cost term L_j, the more i must adhere to its own assumed trajectory in the move suppression cost term $b_i \|\cdot\|$. In this way, Equation (4.10) is a way of ensuring consistency between what neighbors assume an agent does, and what the agent actually does.

[1] While c_j may be large, it may only be a subset of neighbors $\mathcal{N}_j$ in L_j that are dominating the cost, and j would be more dependent on those neighbors assumed trajectories. Still, we can leave this detail aside to interpret Equation (4.10).

Proof. As with most stability results in RHC theory, a non-negative value function is shown to be strictly decreasing for states outside the terminal constraint sets. Define the value function

$$J(t_k) := \sum_{i\in\mathcal{V}} J_i(z_i(t_k), u_i^p(\cdot; t_k)).$$

By application of Algorithm 4.3.1, if $z_i(t_k) \in \Omega_i(\varepsilon_i)$ for all $i \in \mathcal{V}$ at any update time t_k, the terminal controllers take over and stabilize the system to the origin. Therefore, it remains to show that, by application of Algorithm 4.3.1, the closed-loop system (4.6) is driven to the set $\Omega_1(\varepsilon_1) \times \cdots \times \Omega_{N_a}(\varepsilon_{N_a})$ in finite time.

Suppose the closed-loop system (4.6) does not enter set $\Omega_1(\varepsilon_1) \times \cdots \times \Omega_{N_a}(\varepsilon_{N_a})$ in finite time. Then, for any $k \geq 0$, $z_i(t_k) \notin \Omega_i(\varepsilon_i)$ and $z_i(t_{k+1}) \notin \Omega_i(\varepsilon_i)$ for all $i \in \mathcal{V}$. ¿From the cost improvement constraint (4.4), $J(t_{k+1}) \leq \sum_{i\in\mathcal{V}} J_i(z_i(t_{k+1}), \hat{u}_i(\cdot; t_{k+1}))$, thus, for any $k \in \mathbb{N}$,

$$\begin{aligned}
J(t_{k+1})-J(t_k) \ \leq\ & -\int_{t_k}^{t_{k+1}} \sum_{i\in\mathcal{V}} \left[L_i(z_i^p(s; t_k), \hat{z}_{-i}(s; t_k)) + \|u_i^p(s; t_k))\|_{R_i}^2\right] ds \\
& + \int_{t_{k+1}}^{t_k+T} \sum_{i\in\mathcal{V}} L_i(\hat{z}_i(s; t_{k+1}), \hat{z}_{-i}(s; t_{k+1})) - L_i(z_i^p(s; t_k), \hat{z}_{-i}(s; t_k))\, ds \\
& - \int_{t_{k+1}}^{t_k+T} \sum_{i\in\mathcal{V}} b_i\|z_i^p(s; t_k) - \hat{z}_i(s; t_k)\|\, ds + V(\hat{z}(t_{k+1}+T; t_{k+1})) \\
& + \int_{t_k+T}^{t_{k+1}+T} L(\hat{z}(s; t_{k+1})) + \sum_{i\in\mathcal{V}} \|\kappa_i(\hat{z}_i(s; t_{k+1}))\|_{R_i}^2\, ds - V(\hat{z}(t_k+T; t_{k+1})),
\end{aligned}$$

where $L(z) = \sum_{i\in\mathcal{V}} L_i(z_i, z_{-i})$ and $V(z) = \sum_{i\in\mathcal{V}} \|z_i\|_{P_i}^2$. From Assumption 4.3.1(c),

$$\begin{aligned}
&\int_{t_k+T}^{t_{k+1}+T} L(\hat{z}(s; t_{k+1})) + \sum_{i\in\mathcal{V}} \|\kappa_i(\hat{z}_i(s; t_{k+1}))\|_{R_i}^2\, ds \\
&\qquad + V(\hat{z}(t_{k+1}+T; t_{k+1})) - V(\hat{z}(t_k+T; t_{k+1})) \leq 0,
\end{aligned}$$

since $\hat{z}_i(s; t_{k+1}) \in \Omega_i(\varepsilon_i)$ for all $s \in [t_k+T, t_{k+1}+T]$ and every $i \in \mathcal{V}$. Since $\hat{z}_i(s; t_{k+1}) = z_i^p(s; t_k)$ for all $s \in [t_{k+1}, t_k+T]$, from Assumption 4.2.2,

$$\begin{aligned}
&\int_{t_{k+1}}^{t_k+T} \sum_{i\in\mathcal{V}} L_i(\hat{z}_i(s; t_{k+1}), \hat{z}_{-i}(s; t_{k+1})) - L_i(z_i^p(s; t_k), \hat{z}_{-i}(s; t_k))\, ds \\
&\qquad \leq \int_{t_{k+1}}^{t_k+T} \sum_{i\in\mathcal{V}} c_i\|\hat{z}_{-i}(s; t_{k+1}) - \hat{z}_{-i}(s; t_k)\|\, ds.
\end{aligned}$$

Therefore, the cost difference satisfies

$$J(t_{k+1}) - J(t_k) + \eta_k$$

$$\leq \int_{t_{k+1}}^{t_k+T} \sum_{i\in\mathcal{V}} c_i \|\hat{z}_{-i}(s;t_{k+1}) - \hat{z}_{-i}(s;t_k)\| - b_i \|z_i^p(s;t_k) - \hat{z}_i(s;t_k)\| \,\mathrm{d}s, \qquad (4.11)$$

with

$$\eta_k := \int_{t_k}^{t_{k+1}} \sum_{i\in\mathcal{V}} \left[L_i(z_i^p(s;t_k), \hat{z}_{-i}(s;t_k)) + \|u_i^p(s;t_k))\|_{R_i}^2\right] \mathrm{d}s.$$

Next, it is shown that the term on the right of the inequality in Equation (4.11) is non-positive, provided each b_i satisfies the inequality (4.10). From the triangle inequality,

$$\sum_{i\in\mathcal{V}} c_i \|\hat{z}_{-i}(s;t_{k+1}) - \hat{z}_{-i}(s;t_k)\| \leq \sum_{i\in\mathcal{V}} c_i \sum_{j\in\mathcal{N}_i} \|\hat{z}_j(s;t_{k+1}) - \hat{z}_j(s;t_k)\|$$
$$= \sum_{i\in\mathcal{V}} \left(\|\hat{z}_i(s;t_{k+1}) - \hat{z}_i(s;t_k)\| \sum_{j\in\mathcal{N}_i} c_j\right),$$

and the equality follows since the term $\|\hat{z}_j(s;t_{k+1}) - \hat{z}_j(s;t_k)\|$ is present in the overall summation N_j times, each with a coefficient c_l, $l \in \mathcal{N}_j$. If each b_i satisfies (4.10), the cost difference becomes

$$J(t_{k+1}) - J(t_k) + \eta_k \leq \int_{t_{k+1}}^{t_k+T} \sum_{i\in\mathcal{V}} \left[\sum_{j\in\mathcal{N}_i} c_j - b_i\right] \|z_i^p(s;t_k) - \hat{z}_i(s;t_k)\| \,\mathrm{d}s \leq 0.$$

If $z_i(t_k) \notin \Omega_i(\varepsilon_i)$ and $z_i(t_{k+1}) \notin \Omega_i(\varepsilon_i)$ for all $i \in \mathcal{V}$, for any $k \geq 0$, then $\inf_k \eta_k > 0$. Thus, from the inequality above, if $z_i(t_k) \notin \Omega_i(\varepsilon_i)$ and $z_i(t_{k+1}) \notin \Omega_i(\varepsilon_i)$ for all $i \in \mathcal{V}$, there exists a constant $\eta \in (0, \inf_k \eta_k]$ such that $J(t_{k+1}) \leq J(t_k) - \eta$ for any $k \geq 0$. From this inequality, it follows by contradiction that there exists a finite update time t_l such that $z_i(t_l) \in \Omega_i(\varepsilon_i)$ for all $i \in \mathcal{V}$. If this were not the case, the inequality implies $J(t_k) \to -\infty$ as $k \to \infty$. However, $J(t_k) \geq 0$; therefore, there exists a finite time such that the closed-loop system (4.6) is driven to the set $\Omega_1(\varepsilon_1) \times \cdots \times \Omega_{N_a}(\varepsilon_{N_a})$, concluding the proof. ■

Note that the move suppression term $b_i\|\cdot\|$ in each cost function $J_i(z_i(t_k), u_i^p(\cdot;t_k))$ is not differentiable at the origin. As this can play havoc on optimization algorithms, it is preferable in practice to replace $b_i\|\cdot\|$ with a smooth alternative that still exhibits linear growth everywhere outside of a neighborhood of the origin. Recall that linear growth is required due to the form of the cost coupling bound, $c_i\|\cdot\|$, in Assumption 4.2.2(c). A suitable smooth function that exhibits linear growth everywhere outside of a neighborhood of the origin is the C^∞, bounded and positive definite cost function $\phi : \mathbb{R}^n \to \mathbb{R}_+$ defined by $\phi(x) = \sqrt{a_1\|x\|^2 + a_2^2} - a_2$, with constants $a_1 > 1$ and $a_2 > 0$. The function ϕ has bounded gradients, another nice feature for optimization algorithms, and grows linearly, satisfying $\|x\| \leq \phi(x)$ for all $\|x\| \geq 2a_2/(a_1 - 1)$ (Olfati-Saber *et al.* 2003). In theory, to replace $b_i\|\cdot\|$ with $b_i\phi(\cdot)$ requires that one choose a_1, a_2 such that $\|z_i^p(s;t_k) - \hat{z}_i(s;t_k)\| \geq 2a_2/(a_1 - 1)$ for all $s \in [t_{k+1}, t_k + T]$, whenever the closed-loop system is outside the terminal constraint sets. Practically, since this is not an easy condition to check, one would simply choose a_1, a_2 such that $2a_2/(a_1 - 1)$ is small.

4.5 CONCLUSION

In this chapter, a distributed implementation of receding horizon control for coordinated control of multiple agents has been formulated. A generic integrated cost function that couples the states of the agents is first defined, where the coupling terms in the cost are required to satisfy a linear growth bound. Example cost functions for formation stabilization of multiple autonomous vehicles were shown to satisfy the required bound, provided all state trajectories remain bounded. One aspect of the generality of the approach is that agent's dynamics are nonlinear and heterogeneous. The coupling cost is decomposed and distributed optimal control problems are then defined. Each distributed problem is augmented with a move suppression cost term, which is a central element in the stability analysis by ensuring that actual and assumed responses of each agent are not too far from one another. Stability of the closed-loop system is proven when the move suppression weight parameter is large enough, in comparison to the growth parameters that bound the amount coupling in the cost function. The resulting distributed RHC algorithm is computationally scalable, since the size of each local optimal control problem is independent of the number of neighbors for each agent, as well as the total number of agents. The implementation also does not require agents to update sequentially, as in hierarchical methods; agents update their control in parallel. This is likely to be an advantage when the agents are operating in time-critical networked environments, in which turn-taking could be too taxing to manage.

ACKNOWLEDGEMENT

Research was supported in part by NIH grant #K25 HG004035-01.

REFERENCES

Acar L 1995 Boundaries of the receding horizon control for interconnected systems. *Journal of Optimization Theory and Applications*, **84**(2), 251–271.

Chen H 1997 Stability and robustness considerations in nonlinear MPC, PhD thesis, University of Stuttgart.

Chen H and Allgöwer F 1998 A quasi-infinite horizon nonlinear model predictive scheme with guaranteed stability. *Automatica* **14**(10), 1205–1217.

Dunbar WB 2004 Distributed receding horizon control of multiagent systems, PhD thesis, California Institute of Technology Pasadena, CA.

Dunbar WB 2007 Distributed receding horizon control of dynamically coupled nonlinear systems. *IEEE Trans. Auto. Contr.* **52**(7) 1249–1263.

Dunbar WB and Murray RM 2002 Model predictive control of coordinated multi-vehicle formations. *Proceedings of the 41st, IEEE Conference on Decision and Control*, pp. 4631–4636, Las Vegas, NV.

Dunbar WB and Murray RM 2006 Distributed receding horizon control for multi-vehicle formation stabilization. *Automatica* **42**(4), 549–558.

Franco E, Parisini T and Polycarpou MM 2005 Cooperative control of distributed agents with nonlinear dynamics and delayed information exchange: a stabilizing receding-horizon approach, In

Proc. of the 44th IEEE Conference on Decision and Control / IEE European Control Conference, Seville, Spain, pp. 2206–2211.

Jadbabaie A, Yu J and Hauser J 2001 Unconstrained receding horizon control of nonlinear systems. *IEEE Trans. Auto. Contr.* **46**(5), 776–783.

Jia D and Krogh BH 2001 Distributed model predictive control. In *Proceedings of the IEEE American Control Conference*, pp. 2767–2772.

Keviczky T, Borrelli F and Balas GJ 2004 Model predictive control for decoupled systems: A study of decentralized schemes *Submitted to the American Control Conference*, Boston, MA.

Keviczky T, Borrelli F and Balas GJ 2005 Stability analysis of decentralized RHC for decoupled systems. *Proc. of the 44th IEEE Conference on Decision and Control / IEE European Control Conference*, Seville, Spain, pp. 1689–1694.

Leonard NE and Fiorelli E 2001 Virtual leaders, artificial potentials and coordinated control of groups. In *40th Conference on Decision and Control*, Florida, pp. 2968–2973.

Mayne DQ, Rawlings JB, Rao CV and Scokaert POM 2000 Constrained model predictive control: Stability and optimality. *Automatica* **36**, 789–814.

Michalska H and Mayne DQ 1993 Robust receding horizon control of constrained nonlinear systems. *IEEE Trans. Auto. Contr.* **38**, 1623–1632.

Motee N and Sayyar-Rodsari B 2003 Optimal partitioning in distributed model predictive control. In *Proceedings of the IEEE American Control Conference*, pp. 5300–5305.

Olfati-Saber R, Dunbar WB and Murray RM 2003 Cooperative control of multi-vehicle systems using cost graphs and optimization. *Proceedings of the IEEE American Control Conference*, Denver, CO.

Polak E 1997 *Optimization: Algorithms and Consistent Approximations*. Springer, New York.

Ren W and Beard R 2004 A decentralized scheme for spacecraft formation flying via the virtual structure approach. *AIAA Journal of Guidance, Control and Dynamics* **27**(1), 73–82.

Richards A and How J 2004a A decentralized algorithm for robust constrained model predictive control. *Proc. American Control Conference*, Boston, MA, pp. 4261–4266.

Richards A and How J 2004b Decentralized model predictive control of cooperating UAVs. In *Proc. 43rd IEEE Conf. on Decision and Control*, Bahamas, pp. 4286–4291.

Richards A and How J 2005 Robust model predictive control with imperfect information. In *Proc. American Control Conference*, pp. 268–273.

Venkat AN, Rawlings JB and Wright SJ 2004 Plant-wide optimal control with decentralized MPC. In *Proc. IFAC Dynamics and Control of Process Systems Conference*, Boston, MA.

Venkat AN, Rawlings JB and Wright SJ 2005 Stability and optimality of distributed model predictive control. In *Proc. 44th IEEE Conference on Decision and Control / IEE European Control Conference*, Seville, Spain, pp. 6680–6685.

Wesselowski K and Fierro R 2003 A dual-mode model predictive controller for robot formations. In *Proc. 42nd IEEE Conf. on Decision and Control*, Maui, HI, pp. 3615–3620.

5

Distributed predictive control: synthesis, stability and feasibility

Tamás Keviczky, Francesco Borrelli and Gary J. Balas

5.1 INTRODUCTION

Research on decentralized control dates back to the pioneering work of Wang and Davison (1973) and since then, the interest has grown significantly. Decentralized and distributed control techniques today can be found in a broad spectrum of applications ranging from robotics and formation flight to civil engineering. Approaches to distributed control design differ from each other in the assumptions they make on: (1) the kind of interaction between different systems or different components of the same system (dynamics, constraints, objective); (2) the model of the system (linear, nonlinear, constrained, continuous-time, discrete-time); (3) the model of information exchange between the systems; and (4) the control design technique used.

Dynamically coupled systems have been the most studied (Camponogara *et al.* 2002; D'Andrea and Dullerud 2003; Rotkowitz and Lall 2002; Vadigepalli and Doyle 2003; Wang and Davison 1973). In particular Camponogara *et al.* (2002) and Jia and Krogh (2002) consider distributed min-max model predictive control problems for large-scale systems with coupled linear time-invariant dynamics and propose a scheme using stability-constraints and assuming a one-step communication delay. Sufficient conditions for stability with information exchange between the local controllers are given by treating neighboring subsystem states as diminishing disturbances.

In this work, we focus on *decoupled systems*. Our interest in distributed control for dynamically decoupled systems arises from the abundance of networks of independently actuated systems and the necessity of avoiding centralized design when this becomes computationally prohibitive. Networks of vehicles in formation, production units in a power plant, network of cameras at an airport, mechanical actuators for deforming surface are just a few examples.

In a descriptive way, the problem of distributed control for decoupled systems can be formulated as follows. A dynamical system is composed of (or can be decomposed into)

Cooperative Control of Distributed Multi-Agent Systems edited by Jeff S. Shamma

distinct dynamical subsystems that can be independently actuated. The subsystems are dynamically decoupled but have common objectives and constraints which make them interact between each other. Typically the interaction is local, i.e., the objective and the constraints of a subsystem are function of only a subset of other subsystems' states. The interaction will be represented by an 'interaction graph', where the nodes represent the subsystems and an edge between two nodes denotes a coupling term in the objectives and/or in the constraints associated with the nodes. Also, typically it is assumed that the exchange of information has a special structure, i.e., it is assumed that each subsystem can sense and/or exchange information with only a subset of other subsystems. Often the interaction graph and the information exchange graph coincide. A distributed control scheme consists of distinct controllers, one for each subsystem, where the inputs to each subsystem are computed only based on local information, i.e., on the states of the subsystem and its neighbors.

In this work we make use of Receding Horizon Control (RHC) schemes. The main idea of RHC is to use the model of the plant to predict the future evolution of the system, and based on current measurements repeatedly optimize over future control signals, which are implemented until the next measurement becomes available (Mayne *et al.* 2000). A receding horizon controller where the finite-time optimal control law is computed by solving an on-line optimization problem with regular measurement updates is usually referred to as *Model Predictive Controller* (MPC). In this case, at each time step t we optimize a certain performance index under operating constraints with respect to a sequence of future input moves. The first of such optimal moves is the control action applied to the plant at time t. At time $t + 1$, a new optimization is solved over a shifted prediction horizon. We will use 'MPC' to refer to model predictive control and model predictive controller interchangeably.

In this chapter we describe a rigorous mathematical framework for designing distributed model predictive controllers (DMPCs) in discrete time. In our framework a centralized MPC is broken into distinct MPCs of smaller sizes. Each MPC is associated with a different node and computes the local control inputs based only on the states of the node and of its neighbors. We take explicitly into account constraints and use the model of the neighbors to predict their behavior. We analyze the properties of the proposed scheme and introduce sufficient stability conditions. Such conditions (1) highlight the role of prediction errors between neighbors in the stability of the aggregate system; (2) are local to each node and function only of neighboring nodes that can be reached through at most two edges, thus leading to complexity reduction for interconnection graphs of large diameter; and (3) help understand the importance of information exchange between neighbors and its role in stabilizing the entire system. In Section 5.2, we focus on linear systems and show how to recast the stability conditions into a set of semi-definiteness tests. In Section 5.3 we describe a simple way of constructing stabilizing distributed controllers in continuous time by solving one local LQR problem if the subsystems are in addition identical and unconstrained. Finally, in Section 5.5, we provide an overview of various methodologies and avenues of research for addressing the feasibility issues arising in DMPC schemes. Results presented in this work are based on a collection of our recent publications (Borrelli and Keviczky 2006b; Borrelli *et al.* 2005b; Keviczky 2005; Keviczky *et al.* 2006a, 2007).

5.2 PROBLEM FORMULATION

We consider a set of N_v decoupled dynamical systems, the i-th system being described by the discrete-time time-invariant state equation:

$$x_{k+1}^i = f^i(x_k^i, u_k^i), \tag{5.1}$$

where $x_k^i \in \mathbb{R}^{n^i}$, $u_k^i \in \mathbb{R}^{m^i}$, $f^i : \mathbb{R}^{n^i} \times \mathbb{R}^{m^i} \to \mathbb{R}^{n^i}$ are state, input and state update function of the i-th system, respectively. Let $\mathcal{X}^i \subseteq \mathbb{R}^{n^i}$ and $\mathcal{U}^i \subseteq \mathbb{R}^{m^i}$ denote the set of feasible states and inputs of the i-th system, respectively:

$$x_k^i \in \mathcal{X}^i, \quad u_k^i \in \mathcal{U}^i, \quad k \geq 0, \tag{5.2}$$

where $\mathcal{X}^i$ and $\mathcal{U}^i$ are given polytopes.

We will refer to the set of N_v constrained systems as the *aggregate system*. Let $\tilde{x}_k \in \mathbb{R}^{\tilde{n}}$ with $\tilde{n} = \sum_i n^i$ and $\tilde{u}_k \in \mathbb{R}^{\tilde{m}}$ with $\tilde{m} = \sum_i m^i$ be the vectors which collect the states and inputs of the aggregate system at time k, i.e., $\tilde{x}_k = [x_k^1, \ldots, x_k^{N_v}]$, $\tilde{u}_k = [u_k^1, \ldots, u_k^{N_v}]$, with

$$\tilde{x}_{k+1} = f(\tilde{x}_k, \tilde{u}_k), \tag{5.3}$$

where $f(\tilde{x}_k, \tilde{u}_k) = [f^1(x_k^1, u_k^1), \ldots, f^{N_v}(x_k^{N_v}, u_k^{N_v})]$ and $f : \mathbb{R}^{\tilde{n}} \times \mathbb{R}^{\tilde{m}} \to \mathbb{R}^{\tilde{n}}$. We denote by (x_e^i, u_e^i) the equilibrium pair of the i-th system and $(\tilde{x}_e, \tilde{u}_e)$ the corresponding equilibrium for the aggregate system. The state update functions f^i are assumed to be continuous and stabilizable at the equilibrium for all $i = 1, \ldots, N_v$.

So far the subsystems belonging to the aggregate system are completely decoupled. We consider an optimal control problem for the aggregate system where cost function and constraints couple the dynamic behavior of individual systems. We use a graph structure to represent the coupling in the following way. We associate the i-th system to the i-th node of the graph, and if an edge (i, j) connecting the i-th and j-th node is present, then the cost and the constraints of the optimal control problem will have coupling terms, which are functions of both x^i and x^j. The graph $\mathcal{G}$ will be defined as

$$\mathcal{G} = (\mathcal{V}, \mathcal{A}) \tag{5.4}$$

where $\mathcal{V}$ is the set of nodes (or vertices) $\mathcal{V} = \{1, \ldots, N_v\}$ and $\mathcal{A} \subseteq \mathcal{V} \times \mathcal{V}$ the set of edges (i, j) with $i \in \mathcal{V}$, $j \in \mathcal{V}$. The degree d_j of a graph vertex j is the number of edges which start from j. Let $d_{max}(\mathcal{G})$ denote the maximum vertex degree of the graph $\mathcal{G}$.

We denote by $\mathbf{A}(\mathcal{G})$ the $(0, 1)$ adjacency matrix of the graph $\mathcal{G}$. Let $\mathbf{A}_{i,j} \in \mathbb{R}$ be its i, j element, then $\mathbf{A}_{i,i} = 0$, $\forall\, i = 1, \ldots, N_v$, $\mathbf{A}_{i,j} = 0$ if $(i, j) \notin \mathcal{A}$ and $\mathbf{A}_{i,j} = 1$ if $(i, j) \in \mathcal{A}$, $\forall\, i, j = 1, \ldots, N_v$, $i \neq j$.

Remark 5.2.1 Often the graph edges are chosen to be time-varying, based on a particular neighbor selection policy. For instance in the case of formation flight, the interconnection graph is full due to collision avoidance (since each vehicle has to avoid every other), but it is usually replaced with a time-varying 'closest spatial neighbor' relationship. For simplicity, a time-invariant graph structure is assumed throughout the manuscript. References to distributed MPC using time-varying graphs can be found in Borrelli *et al.* (2005b); Keviczky *et al.* (2007).

Once the graph structure has been fixed, the optimization problem is formulated as follows. Denote with $\tilde{x}^i$ the states of all neighboring systems of the i-th system, i.e., $\tilde{x}^i = \{x^j \in \mathbb{R}^{n^j} | (i, j) \in \mathcal{A}\}$, $\tilde{x}^i \in \mathbb{R}^{\tilde{n}^i}$ with $\tilde{n}^i = \sum_{j|(i,j)\in\mathcal{A}} n^j$. Analogously, $\tilde{u}^i \in \mathbb{R}^{\tilde{m}^i}$ denotes the inputs to all the neighboring systems of the i-th system, and $(\tilde{x}^i_e, \tilde{u}^i_e)$ represent their equilibria. Let

$$g^{i,j}(x^i, x^j) \leq 0 \tag{5.5}$$

define the coupling constraints between the i-th and the j-th systems, where $(i, j) \in \mathcal{A}$, with $g^{i,j} : \mathbb{R}^{n^i} \times \mathbb{R}^{n^j} \to \mathbb{R}^{nc^{i,j}}$ being a continuous function. We will often use the following shorter form of the coupling constraints defined between the i-th system and all its neighbors:

$$g^i(x^i, \tilde{x}^i) \leq 0, \tag{5.6}$$

with $g^i : \mathbb{R}^{n^i} \times \mathbb{R}^{\tilde{n}^i} \to \mathbb{R}^{nc^i}$, where $nc^i = \sum_{j|(i,j)\in\mathcal{A}} nc^{i,j}$.

Consider the following cost:

$$l(\tilde{x}, \tilde{u}) = \sum_{i=1}^{N_v} l^i(x^i, u^i, \tilde{x}^i, \tilde{u}^i), \tag{5.7}$$

where $l^i : \mathbb{R}^{n^i} \times \mathbb{R}^{m^i} \times \mathbb{R}^{\tilde{n}^i} \times \mathbb{R}^{\tilde{m}^i} \to \mathbb{R}$ is the cost associated with the i-th system and is a function only of its states and the states of its neighbor nodes.

$$l^i(x^i, u^i, \tilde{x}^i, \tilde{u}^i) = l^{i,i}(x^i, u^i) + \sum_{(i,j)\in\mathcal{A}} l^{i,j}(x^i, u^i, x^j, u^j) \tag{5.8}$$

where $l^{i,i} : \mathbb{R}^{n^i} \times \mathbb{R}^{m^i} \to \mathbb{R}$ and $l^{i,j} : \mathbb{R}^{n^i} \times \mathbb{R}^{m^i} \times \mathbb{R}^{n^j} \times \mathbb{R}^{m^j} \to \mathbb{R}$ is the cost function for two adjacent nodes. We assume throughout this work that $l^{i,i}$ and $l^{i,j}$ are positive convex continuous functions such that $l^{i,j}(x^i, u^i, \tilde{x}^i, \tilde{u}^i) \geq c\|(x^i, u^i, \tilde{x}^i, \tilde{u}^i)\|_2$, $c > 0$ and that $l^i(x^i_e, u^i_e, \tilde{x}^i_e, \tilde{u}^i_e) = 0$.

We design a model predictive controller by repeatedly solving finite time optimal control problems in a receding horizon fashion as described in Section 5.1. Assume at time t the current state $\tilde{x}_t$ to be available. Consider the following constrained finite time optimal control problem:

$$\tilde{J}^*_N(\tilde{x}_t) \triangleq \min_{\tilde{U}_t} \sum_{k=0}^{N-1} l(\tilde{x}_{k,t}, \tilde{u}_{k,t}) + l_N(\tilde{x}_{N,t}) \tag{5.9a}$$

$$\text{subj. to } x^i_{k+1,t} = f^i(x^i_{k,t}, u^i_{k,t}), \quad i = 1, \ldots, N_v, \quad k \geq 0$$

$$x^i_{k,t} \in \mathcal{X}^i, \quad u^i_{k,t} \in \mathcal{U}^i, \quad i = 1, \ldots, N_v, \quad k = 1, \ldots, N-1$$

$$g^{i,j}(x^i_{k,t}, x^j_{k,t}) \leq 0, \quad i = 1, \ldots, N_v, \quad (i, j) \in \mathcal{A}, \qquad k = 1, \ldots, N-1 \tag{5.9b}$$

$$\tilde{x}_{N,t} \in \mathcal{X}_f, \tag{5.9c}$$

$$\tilde{x}_{0,t} = \tilde{x}_t, \tag{5.9d}$$

where N is the prediction horizon, $\mathcal{X}_f \subseteq \mathbb{R}^{\tilde{n}}$ is a terminal region, l_N is the cost on the terminal state. In (5.9) $\tilde{U}_t \triangleq [\tilde{u}_{0,t}, \ldots, \tilde{u}_{N-1,t}] \in \mathbb{R}^s$, $s \triangleq \tilde{m}N$ denotes the optimization vector, $x^i_{k,t}$ denotes the state vector of the i-th node predicted at time $t+k$ obtained by starting from the state x^i_t and applying to system (5.1) the input sequence $u^i_{0,t}, \ldots, u^i_{k-1,t}$.

Let $\tilde{U}^*_t = [\tilde{u}^*_{0,t}, \ldots, \tilde{u}^*_{N-1,t}]$ be an optimal solution of (5.9) at time t. Then, the first sample of $\tilde{U}^*_t$ is applied to the aggregate system (5.3)

$$\tilde{u}_t = \tilde{u}^*_{0,t}. \tag{5.10}$$

The optimization (5.9) is repeated at time $t+1$, based on the new state $\tilde{x}_{t+1}$.

It is well known that stability is not ensured by the MPC law (5.9)–(5.10). Usually the terminal cost l_N and the terminal constraint set $\mathcal{X}_f$ are chosen to ensure closed-loop stability. A treatment of sufficient stability conditions goes beyond the scope of this work and can be found in the surveys (Chen and Allgöwer 1998; Magni *et al.* 2003; Mayne *et al.* 2000). We assume that the reader is familiar with the basic concept of MPC and its main issues, we refer to the above references for a comprehensive treatment of the topic. In general, the optimal input u^i_t to the i-th system computed by solving (5.9) at time t, will be a function of the overall state information $\tilde{x}_t$. In the next section we propose a way to distribute the MPC problem defined in (5.9)–(5.10).

5.3 DISTRIBUTED MPC SCHEME

We address the complexity associated with a centralized optimal control design for the class of large-scale decoupled systems described in the previous section by formulating N_v distributed finite time optimal control problems, each one associated with a different node. Each node has information about its current states and its neighbors' current states. Based on such information, each node computes its optimal inputs and its neighbors' optimal inputs. The input to the neighbors will only be used to predict their trajectories and then discarded, while the first component of the optimal input to the node will be implemented where it was computed.

A more formal description follows. Let the following finite time optimal control problem $\mathcal{P}_i$ with optimal value function $J^{i*}_N(x^i_t, \tilde{x}^i_t)$ be associated with the i-th system at time t:

$$\min_{\tilde{U}^i_t} \sum_{k=0}^{N-1} l^i(x^i_{k,t}, u^i_{k,t}, \tilde{x}^i_{k,t}, \tilde{u}^i_{k,t}) + l^i_N(x^i_{N,t}, \tilde{x}^i_{N,t})$$

$$\text{subj. to } x^i_{k+1,t} = f^i(x^i_{k,t}, u^i_{k,t}), \tag{5.11a}$$

$$x^i_{k,t} \in \mathcal{X}^i, \quad u^i_{k,t} \in \mathcal{U}^i, \quad k = 1, \ldots, N-1$$

$$x^j_{k+1,t} = f^j(x^j_{k,t}, u^j_{k,t}), \quad (i, j) \in \mathcal{A},, \quad k = 1, \ldots, N-1 \tag{5.11b}$$

$$x^j_{k,t} \in \mathcal{X}^j, \quad u^j_{k,t} \in \mathcal{U}^j, \quad (i, j) \in \mathcal{A}, \quad k = 1, \ldots, N-1$$

$$g^{i,j}(x^i_{k,t}, u^i_{k,t}, x^j_{k,t}, u^j_{k,t}) \leq 0, \quad (i, j) \in \mathcal{A}, \quad k = 1, \ldots, N-1 \tag{5.11c}$$

$$x^i_{N,t} \in \mathcal{X}^i_f, \qquad x^j_{N,t} \in \mathcal{X}^j_f, \quad (i, j) \in \mathcal{A} \tag{5.11d}$$

$$x^i_{0,t} = x^i_t, \quad \tilde{x}^i_{0,t} = \tilde{x}^i_t, \tag{5.11e}$$

where $\tilde{U}^i_t \triangleq [u^i_{0,t}, \tilde{u}^i_{0,t}, \ldots, u^i_{N-1,t}, \tilde{u}^i_{N-1,t}] \in \mathbb{R}^{s^i}$, $s^i \triangleq (\tilde{m}^i + m^i)N$ denotes the optimization vector, $x^i_{k,t}$ denotes the state vector of the i-th node predicted at time $t + k$ obtained by starting from the state x^i_t and applying to system (5.1) the input sequence $u^i_{0,t}, \ldots, u^i_{k-1,t}$. The tilded vectors denote the prediction vectors associated with the neighboring systems by starting from their states $\tilde{x}^i_t$ and applying to their models the input sequence $\tilde{u}^i_{0,t}, \ldots, \tilde{u}^i_{k-1,t}$. Denote by $\tilde{U}^{i*}_t = [u^{*i}_{0,t}, \tilde{u}^{*i}_{0,t}, \ldots, u^{*i}_{N-1,t}, \tilde{u}^{*i}_{N-1,t}]$ an optimizer of problem $\mathcal{P}_i$.

Note that problem $\mathcal{P}_i$ involves only the state and input variables of the i-th node and its neighbors at time t. We will define the following distributed MPC scheme. At time t

(i) Each node i solves problem $\mathcal{P}_i$ based on measurements of its state x^i_t and the states of all its neighbors $\tilde{x}^i_t$.

(ii) Each node i implements the first sample of $\tilde{U}^{i*}_t$

$$u^i_t = u^{*i}_{0,t}. \tag{5.12}$$

(iii) Each node repeats steps (i) to (iii) at time $t + 1$, based on the new state information $x^i_{t+1}, \tilde{x}^i_{t+1}$.

In order to solve problem $\mathcal{P}_i$ each node needs to know its current states, its neighbors' current states, its terminal region, its neighbors' terminal regions and models and constraints of its neighbors. Based on such information each node computes its optimal inputs and its neighbors' optimal inputs. The input to the neighbors will only be used to predict their trajectories and then discarded, while the first component of the i-th optimal input of problem $\mathcal{P}_i$ will be implemented on the i-th node. The solution of the i-th subproblem will yield a control policy for the i-th node of the form $u^i_t = c^i(x^i_t, \tilde{x}^i_t)$, where $c^i : \mathbb{R}^{n^i} \times \mathbb{R}^{\tilde{n}^i} \to \mathbb{R}^{m^i}$ is a time-invariant feedback control law implicitly defined by the optimization problem $\mathcal{P}_i$.

Remark 5.3.1 In the formulation above, a priori knowledge of the aggregate system equilibrium $(\tilde{x}_e, \tilde{u}_e)$ is assumed. The equilibrium could be defined in several other different ways. For instance, in a problem involving mobile agents we can assume that there is a leader (real or virtual) which is moving and the equilibrium is given in terms of distances of each agent from the leader. Also, it is possible to formulate the equilibrium by using relative distances between agents and signed areas. The approach of this work does not depend on the way the aggregate system equilibrium is defined, as long as this is known a priori. In some distributed control schemes, the equilibrium is not known a priori, but is the result of the evolution of distributed control laws such as in Jadbabaie *et al.* (2003). The approach of this chapter is not applicable to such schemes.

Remark 5.3.2 The problem formulation of (5.11) lends itself to generalization and is flexible enough to describe additional characteristics of a particular application. For instance, delayed information and additional communication between neighbors can be incorporated in the formulation as well. Additional terms that represent coupling between any two neighbors of a particular node can be included in the local cost function (5.8) as well. This leads to a better representation of the centralized problem and based on our numerical simulations, more accurate predictions regarding the behavior of neighbors. The derivation of results presented in this chapter can be found in Keviczky *et al.* (2006b) for such a case. One can also assume that terminal set constraints of neighbors are not known exactly. For the sake of simplicity, we will not consider and describe all these possible modifications and focus on problem (5.11).

Even if we assume N to be infinite, the distributed MPC approach described so far does not guarantee that solutions computed locally are globally feasible and stable. The reason is simple: at the i-th node the prediction of the neighboring state x^j is done independently from the prediction of problem $\mathcal{P}_j$. Therefore, the trajectory of x^j predicted by problem $\mathcal{P}_i$ and the one predicted by problem $\mathcal{P}_j$, based on the same initial conditions, are different (since in general, $\mathcal{P}_i$ and $\mathcal{P}_j$ will be different). This will imply that constraint fulfillment will be ensured by the optimizer u_t^{*i} for problem $\mathcal{P}_i$ but not for the centralized problem involving the states of all nodes.

Stability and feasibility of distributed RHC schemes are currently active research areas (Camponogara *et al.* 2002; Dunbar and Murray 2006; Keviczky *et al.* 2004b; Richards and How 2004a). In the following section the stability of the distributed MPC scheme given in (5.11) and (5.12) is analyzed in detail.

5.4 DMPC STABILITY ANALYSIS

Without loss of generality, we assume the origin to be an equilibrium for the aggregate system. In this section, we rely on the general problem formulation introduced in Section 5.2 and focus on systems *with* input and state constraints, *no* coupling constraints and terminal point constraint to the origin $\mathcal{X}_f^i = \mathbf{0}$. Thus the distributed finite time optimal control problem associated with the i-th node at time t will have the following form:

$$\min_{\tilde{U}_t^i} \sum_{k=0}^{N-1} l^i(x_{k,t}^i, u_{k,t}^i, \tilde{x}_{k,t}^i, \tilde{u}_{k,t}^i)$$

$$\text{subj. to } x_{k+1,t}^i = f^i(x_{k,t}^i, u_{k,t}^i), \tag{5.13a}$$

$$x_{k,t}^i \in \mathcal{X}^i, \quad u_{k,t}^i \in \mathcal{U}^i, \quad k = 1, \ldots, N-1, \tag{5.13b}$$

$$x^j_{k+1,t} = f^j(x^j_{k,t}, u^j_{k,t}), \quad (i, j) \in \mathcal{A}, \tag{5.13c}$$

$$x^j_{k,t} \in \mathcal{X}^j, \quad u^j_{k,t} \in \mathcal{U}^j, \quad (i, j) \in \mathcal{A}, \quad k = 1, \ldots, N-1, \tag{5.13d}$$

$$x^i_{N,t} = \mathbf{0}, \quad x^j_{N,t} = \mathbf{0}, \quad (i, j) \in \mathcal{A} \tag{5.13e}$$

$$x^i_{0,t} = x^i_t, \quad \tilde{x}^i_{0,t} = \tilde{x}^i_t. \tag{5.13f}$$

We will make the following assumption on the structure of individual cost functions:

Assumption 5.4.1 *The cost term l^i in (5.8) associated with the i-th system can be written as follows*

$$\begin{aligned} l^i(x^i, u^i, \tilde{x}^i, \tilde{u}^i) &= \|Qx^i\|_p + \|Ru^i\|_p \\ &+ \sum_{j|(i,j)\in\mathcal{A}} \|Qx^j\|_p + \sum_{j|(i,j)\in\mathcal{A}} \|Ru^j\|_p \\ &+ \sum_{j|(i,j)\in\mathcal{A}} \|Q(x^i - x^j)\|_p. \end{aligned} \tag{5.14}$$

where $\|Mx\|_p$ denotes the p-norm of the vector Mx if $p = 1, \infty$ or $x'Mx$ if $p = 2$.

Remark 5.4.1 The cost function structure in Assumption 5.4.1 can be used to describe several practical applications including formation flight, paper machine control and monitoring network of cameras (Borrelli *et al.* 2005b). The relative state term in (5.14) also assumes $n^i = \tilde{n}/N_v$ for all $i = 1, \ldots, N_v$.

In classical RHC schemes, stability and feasibility are proven by using the value function as a Lyapunov function. We will investigate two different approaches to analyzing and ensuring stability of the aggregate system:

(i) Use of individual cost functions as Lyapunov functions for each node (Sections 5.4.1 and 5.4.2).

(ii) Exchange of optimal solutions between neighbors (Section 5.4.3).

Remark 5.4.2 If we consider the sum of individual cost functions as a Lyapunov function for the entire system, one might obtain a less restrictive stability condition for the price of increased complexity when testing such conditions (Keviczky 2005; Keviczky *et al.* 2006a).

The following notation will be used to describe state and input signals. For a particular variable, the first superscript refers to the index of the corresponding system, the second superscript refers to the location where it is computed. For instance the input $u^{i,j}$ represents the input to the i-th system calculated by solving problem $\mathcal{P}_j$. Similarly, the state variable $x^{i,j}$ stands for the states of system i predicted by solving $\mathcal{P}_j$. The lower indices conform to the standard time notation of MPC schemes. For example, variable $x_{k,t}$ denotes the k-step ahead prediction of the states computed at time instant t.

5.4.1 Individual value functions as Lyapunov functions

In order to illustrate the fundamental issues regarding stability in a simple way, we first consider two systems ($N_v = 2$). The general formulation for an arbitrary number of nodes is treated later in this section. We consider two distributed MPC problems $\mathcal{P}_1$ and $\mathcal{P}_2$ according to (5.13). In order to simplify notation, we define

$$\ell^1(x_t^1, U_t^{1,1}, x_t^2, U_t^{2,1}) = \sum_{k=0}^{N-1} l^1(x_{k,t}^{1,1}, u_{k,t}^{1,1}, x_{k,t}^{2,1}, u_{k,t}^{2,1}), \tag{5.15}$$

where x_t^1 and x_t^2 are the initial states of systems 1 and 2 at time t, and $U_t^{1,1}$, $U_t^{2,1}$ are the control sequences for node 1 and 2, respectively, calculated by node 1. Let $[U_t^{1,1*}, U_t^{2,1*}]$ be an optimizer of problem $\mathcal{P}_1$ at time t:

$$U_t^{1,1*} = \left[u_{0,t}^{1,1}, \ldots, u_{N-1,t}^{1,1}\right], \; U_t^{2,1*} = \left[u_{0,t}^{2,1}, \ldots, u_{N-1,t}^{2,1}\right], \tag{5.16}$$

and

$$\mathbf{x}_t^{1,1} = \left[x_{0,t}^{1,1}, \ldots, x_{N,t}^{1,1}\right], \quad \mathbf{x}_t^{2,1} = \left[x_{0,t}^{2,1}, \ldots, x_{N,t}^{2,1}\right],$$

be the corresponding optimal state trajectories of node 1 and 2 predicted at node 1 by $\mathcal{P}_1$.

Analogously, let $[U_t^{1,2*}, U_t^{2,2*}]$ be an optimizer of problem $\mathcal{P}_2$ at time t:

$$U_t^{1,2*} = \left[u_{0,t}^{1,2}, \ldots, u_{N-1,t}^{1,2}\right], \; U_t^{2,2*} = \left[u_{0,t}^{2,2}, \ldots, u_{N-1,t}^{2,2}\right], \tag{5.17}$$

and

$$\mathbf{x}_t^{1,2} = \left[x_{0,t}^{1,2}, \ldots, x_{N,t}^{1,2}\right], \quad \mathbf{x}_t^{2,2} = \left[x_{0,t}^{2,2}, \ldots, x_{N,t}^{2,2}\right],$$

be the corresponding optimal state trajectories of node 1 and 2 predicted at node 2 by $\mathcal{P}_2$. By hypothesis, neighboring systems either measure or exchange state information, so the initial states for both problems are the same at each time step, i.e., $x_{0,t}^{1,1} = x_{0,t}^{1,2}$ and $x_{0,t}^{2,1} = x_{0,t}^{2,2}$.

Remark 5.4.3 It should be noted that although the two problems $\mathcal{P}_1$ and $\mathcal{P}_2$ involve the same subsystems, multiple optima can arise from non-strictly convex cost functions. Furthermore, in a more general setting, for larger number of nodes with an arbitrary graph interconnection, adjacent nodes have different set of neighbors and thus are solving different subproblems $\mathcal{P}_i$. Non-convex coupling constraints would be a source of multiple optimal solutions as well. These factors lead to different optimal solutions for neighboring problems and warrant distinguishing between $U^{1,1*}$, $U^{1,2*}$ and $U^{2,1*}$, $U^{2,2*}$ in (5.16) and (5.17).

We denote the set of states of node i at time k feasible for problem $\mathcal{P}_i$ by

$$\mathcal{X}_k^i = \{x^i \mid \exists u^i \in \mathcal{U}^i \text{ such that } f^i(x^i, u^i) \in \mathcal{X}_{k+1}^i\} \cap \mathcal{X}^i,$$
$$\text{with } \mathcal{X}_N^i = \mathcal{X}_f^i. \tag{5.18}$$

Since we are neglecting coupling constraints, the set of feasible states for the distributed MPC scheme described by (5.13) and (5.12) applied to the aggregate system is the cross-product of the feasible set of states associated with each node:

$$\mathcal{X}_k = \prod_{i=1}^{N_v} \mathcal{X}_k^i, \tag{5.19}$$

where the symbol $\prod$ denotes the standard Cartesian product of sets.

Denote with

$$c(\tilde{x}_k) = \left[u_{0,k}^{1,1*}(\tilde{x}_k), u_{0,k}^{2,2*}(\tilde{x}_k)\right], \tag{5.20}$$

the control law obtained by applying the distributed MPC policy in (5.13) and (5.12) with cost function (5.15), when the current state is $\tilde{x}_k = [x_k^1, x_k^2]$. Consider the aggregate system model (5.3) consisting of two nodes ($N_v = 2$), and denote with

$$\tilde{x}_{k+1} = f\left(\tilde{x}_k, c(\tilde{x}_k)\right), \tag{5.21}$$

the closed-loop dynamics of the entire system. In the following theorem we state sufficient conditions for the asymptotic stability of the closed-loop system.

Theorem 5.4.1 *Assume that*

(A0) $Q = Q' \succ 0$, $R = R' \succ 0$ *if* $p = 2$ *and* Q, R *are full column rank matrices if* $p = 1, \infty$.

(A1) The state and input constraint sets $\mathcal{X}^1, \mathcal{X}^2$ *and* $\mathcal{U}^1, \mathcal{U}^2$ *contain the origin in their interior.*

(A2) The following inequality is satisfied for all $x_t^i \in \mathcal{X}_0^i$, $x_t^j \in \mathcal{X}_0^j$ *with* $i = 1$, $j = 2$ *and* $i = 2$, $j = 1$:

$$\varepsilon \le \|Qx_t^i\|_p + \|Qx_t^j\|_p + \|Q(x_t^i - x_t^j)\|_p + \|Ru_{0,t}^{i,i}\|_p + \|Ru_{0,t}^{j,i}\|_p, \tag{5.22}$$

where

$$\varepsilon = \sum_{k=1}^{N-1}\left(2\|Q(x_{k,t}^{j,j} - x_{k,t}^{j,i})\|_p + \|R(u_{k,t}^{j,j} - u_{k,t}^{j,i})\|_p\right). \tag{5.23}$$

Then, the origin of the closed loop system (5.21) is asymptotically stable with domain of attraction $\mathcal{X}_0^1 \times \mathcal{X}_0^2$.

Proof. The proof can be found in Keviczky *et al.* (2006a). It involves showing that the value function of each individual node is a Lyapunov function, which decreases along the closed-loop trajectories at each time step. □

Theorem 5.4.1 highlights the relationship between the stability of the distributed scheme given in (5.13) and (5.12), and the allowable prediction mismatch at all points in the state space of the aggregate system. The term ε in inequality (5.22) is a function of the error between the trajectories of node 2 predicted by node 1 and the one predicted by node 2 itself. The smaller the error, the larger the set of initial states for which the value function will decrease along the aggregate system trajectories.

Similar ideas can be used if instead of a terminal point constraint, nonzero terminal cost and terminal set constraints $\mathcal{X}_f \neq \mathbf{0}$ are used. In this case, the terminal set has to be control invariant and the terminal cost is chosen as a control Lyapunov function (Mayne *et al.* 2000).

5.4.2 Generalization to arbitrary number of nodes and graph

The results discussed so far carry over to any number of nodes and general graph structure. Let us denote the distributed model predictive control law for the aggregate system with

$$c(\tilde{x}_k) = \left[u_{0,k}^{1,1*}(x_k^1, \tilde{x}_k^1), \ldots, u_{0,k}^{N_v,N_v*}(x_k^{N_v}, \tilde{x}_k^{N_v})\right], \tag{5.24}$$

obtained by applying the distributed MPC policy of each subproblem $\mathcal{P}_i$ described in (5.13) and (5.12) when the current state is $\tilde{x}_k = [x_k^1, \ldots, x_k^{N_v}]$. Note that since there are no coupling constraints, the feasible states for the aggregate system is the cross-product of the feasible states associated with each node as defined in (5.18) and (5.19). Consider the system model (5.3) and denote by

$$\tilde{x}_{k+1} = f\left(\tilde{x}_k, c(\tilde{x}_k)\right), \tag{5.25}$$

the closed-loop dynamics of the aggregate system. Sufficient conditions for asymptotic stability of the closed-loop system are given next.

Theorem 5.4.2 *Assume*

(A0) $Q = Q' \succ 0$, $R = R' \succ 0$ *if* $p = 2$ *and* Q, R *are full column rank matrices if* $p = 1, \infty$.

(A1) The state and input constraint sets $\mathcal{X}^i$ *and* $\mathcal{U}^i$ *contain the origin for each node in their interior.*

(A2) The following inequality is satisfied for each node and all $x_t^i \in \mathcal{X}_0^i$:

$$\sum_{j|(i,j)\in\mathcal{A}} \varepsilon^{i,j} \leq J_0^{i*}, \tag{5.26}$$

where

$$\varepsilon^{i,j} = \sum_{k=1}^{N-1}\left(2\|Q(x_{k,t}^{j,j} - x_{k,t}^{j,i})\|_p + \|R(u_{k,t}^{j,j} - u_{k,t}^{j,i})\|_p\right), \tag{5.27}$$

and

$$J_0^{i*} = \|Qx_t^i\|_p + \|Ru_{0,t}^{i,i}\|_p + \sum_{j|(i,j)\in\mathcal{A}}(\|Qx_t^j\|_p + \|Ru_{0,t}^{j,i}\|_p) + \sum_{j|(i,j)\in\mathcal{A}}\|Q(x_t^i - x_t^j)\|_p. \tag{5.28}$$

Then, the origin of the closed loop system (5.25) is asymptotically stable with domain of attraction $\prod_{i=1}^{N_v} \mathcal{X}_0^i$.

Proof. It can be found in Keviczky *et al.* (2006a).□

Remark 5.4.4 Assumption *(A2)* of Theorem 5.4.2 extends Assumption *(A2)* of Theorem 5.4.1 to arbitrary graphs and number of nodes. In general, nodes may have multiple neighbors, which lead to additional terms in inequality (5.26) compared to (5.22).

Theorem 5.4.2 presents a sufficient condition for testing the stability of the distributed scheme introduced in Section 5.3. It involves local conditions to be tested at each individual node and requires bounding the prediction mismatch between neighboring subsystems. These results follow in the footsteps of one of the stability analysis methods presented in the survey paper of Sandell *et al.* (1978), where stability tests for large-scale interconnected systems are formulated in terms of the individual Lyapunov functions and bounds on subsystem interconnections. In our work, the concept of these bounds has an exact relationship with the prediction mismatch between neighboring subsystems. It is clear that for large-scale systems, the stability condition (5.26) leads to complexity reduction. The formulation of these local stability tests is highlighted in Section 5.4.4 for the case of linear systems.

5.4.3 Exchange of information

Stability conditions derived in the previous sections show that it is the mismatch between the predicted and actual control solutions of neighbors that plays a central role in the stability problem. Therefore we are prompted to investigate how sufficient conditions for stability could be improved by allowing the exchange of optimal solutions between neighbors. Examining condition (5.22) from this standpoint, we can immediately make two general observations:

(i) Using bounds on the mismatch between the predicted and actual inputs and states of neighbors, the stability condition (5.22) could be made less restrictive by reducing the terms, which adversely affect the value function variation. In other words, using a coordination scheme based on information exchange, it may be possible to reduce the size of ε to decrease the left side of inequality (5.22).

(ii) One can observe that as each node is getting closer to its equilibrium (in our example the origin), the right side of inequality (5.22) starts to diminish, which leads to more stringent restrictions on the allowable prediction mismatch between neighbors, represented by the left side of the inequality.

These observations suggest that information exchange between neighboring nodes has a beneficial effect in proving stability, *if it leads to reduced prediction mismatch.* As each system converges to its equilibrium, assumptions on the behavior of neighboring systems should become more and more accurate to satisfy the stability condition (5.22).

In fact, as system (5.21) approaches its equilibrium, the right-hand side of inequality (5.22) decreases. In turn, the left-hand side of inequality (5.22) has to diminish as well. This leads to the counter-intuitive conclusion that an increasing information exchange rate between neighbors might be needed when approaching the equilibrium. These conclusions are in agreement with the stability conditions of a distributed RHC scheme proposed in Dunbar and Murray (2006), where it is shown that convergence to a smaller neighborhood of the system equilibrium requires more frequent updates. However, our simulation examples (Borrelli *et al.* 2005b; Keviczky *et al.* 2004b; Online 2006) suggest that the prediction errors between neighbors tend to disappear as each node approaches its equilibrium, and the prediction mismatch converges to zero at a faster rate than the decay in the right-hand side. A different, sequential information exchange scheme can be found in Richards and How (2004a, b), which is valid for a special graph structure based on a leader-follower architecture.

5.4.4 Stability analysis for heterogeneous unconstrained LTI subsystems

The stability condition (5.26) developed in Section 5.4.2 for a particular node i involves the states of its neighbors $x^{j,j}_{k,t}$, $(i, j) \in \mathcal{A}$ predicted at time t by the neighbor j itself. A predicted neighboring state $x^{j,j}_{k,t}$ is a function of the input sequence $u^{j,j}_{k,t}$ computed at node j, which is then a function of the *initial* states of all the neighbors of node j. This implies that the test for node i involves the states of node i, all its neighbors' states, and the states of the neighbors of neighbors. Thus the dimension of the local stability tests are limited by the maximum size of any subgraph with diameter less than or equal to four. This leads to an immediate complexity reduction in testing the stability of the large-scale control system. For heterogeneous unconstrained linear time-invariant (LTI) subsystems, with state matrices $A^i \in \mathbb{R}^{n\times n}$, $B^i \in \mathbb{R}^{n\times m}$, $i = 1, \ldots, N_v$, the stability condition (5.26) when $p = 2$ is used in Assumption 5.4.1 leads to testing the semi-definiteness of N_v matrices. In this section, we will describe these local semi-definiteness tests deriving from (5.26). An example of the case of identical subsystems can be found in Keviczky *et al.* (2006b).

Since the local MPC problems are time-invariant, without loss of generality we set the generic initial time to $t = 0$ for notational simplicity. Consider node i with initial state x^i_0. As defined in Section 5.2, $\tilde{x}^i_0$ denotes the states of its neighbors at the same time instant. We denote the states of the neighbors of neighbors to node i (which are not connected to the i-th node) by $\breve{x}^i_0 = \{x^q \in \mathbb{R}^{n^q} | \exists j\ (j, q) \in \mathcal{A}, (i, j) \in \mathcal{A}, (i, q) \notin \mathcal{A}\}$. We use $\overline{x}^i_0 = [x^i_0, \tilde{x}^i_0, \breve{x}^i_0]$ to denote the collection of self, neighboring and two-step neighboring states of node i. Using the notation $u^{i,i}_{[0,N-1],0} = [u^{i,i}_{0,0}, \ldots, u^{i,i}_{N-1,0}]$ for time sequences, the solution to problem (5.13) associated with node i can be expressed as

$$\begin{bmatrix} u^{i,i}_{[0,N-1],0} \\ \tilde{u}^{i,i}_{[0,N-1],0} \\ \breve{u}^{i,i}_{[0,N-1],0} \end{bmatrix} = \begin{bmatrix} K^i_{11} & K^i_{12} & \mathbf{0} \\ K^i_{21} & K^i_{22} & \mathbf{0} \\ \mathbf{0} & \mathbf{0} & \mathbf{0} \end{bmatrix} \begin{bmatrix} x^i_0 \\ \tilde{x}^i_0 \\ \breve{x}^i_0 \end{bmatrix} = \begin{bmatrix} K^i_1 \\ K^i_2 \\ \mathbf{0} \end{bmatrix} \overline{x}^i_0, \tag{5.29}$$

when the subsystems (5.1) are unconstrained LTI systems. Based on (5.29), we will use the following notation for k-step ahead predicted input values:

$$u_{k,0}^{i,i} = \underbrace{\left[K_{11,k}^{i} \;\; K_{12,k}^{i} \;\; \mathbf{0}\right]}_{K_{1,k}^{i}} \bar{x}_0^{i}, \quad u_{k,0}^{j,i} = \underbrace{\left[K_{21,k}^{j,i} \;\; K_{22,k}^{j,i} \;\; \mathbf{0}\right]}_{K_{2,k}^{j,i}} \bar{x}_0^{i}, \tag{5.30}$$

where $K_{1,k}^{i}$ and $K_{2,k}^{j,i}$ are submatrices of K_1^i and K_2^i, respectively.

We use the above notation to express the solution $u_{[0,N-1],0}^{j,j}$ to problem (5.13) associated with node j, $(i, j) \in \mathcal{A}$ as an explicit function of the initial states:

$$u_{[0,N-1],0}^{j,j} = \left[{}^{i}K_{11}^{j} \;\; {}^{i}K_{12}^{j} \;\; {}^{i}K_{13}^{j}\right] \bar{x}_0^{i}, \tag{5.31}$$

and thus

$$u_{k,0}^{j,j} = \underbrace{\left[{}^{i}K_{11,k}^{j} \;\; {}^{i}K_{12,k}^{j} \;\; {}^{i}K_{13,k}^{j}\right]}_{{}^{i}K_{1,k}^{j}} \bar{x}_0^{i}. \tag{5.32}$$

Note that the control input $u_{k,0}^{j,j}$ can also be expressed as a function of $\bar{x}_0^j$ (e.g. in the local stability condition associated with node j), thus the upper left index i is needed to distinguish the above controller gain matrix entries.

Using (5.30) and (5.32), we can express predicted states for any node j, $(i, j) \in \mathcal{A}$ as

$$x_{k,0}^{j,j} = \Psi_k^{j,j} \bar{x}_0^{i}, \quad x_{k,0}^{j,i} = \Psi_k^{j,i} \bar{x}_0^{i}, \tag{5.33}$$

where the matrix $\Psi_k^{j,j}$ is a function of A^i, $\{A^j | (i, j) \in \mathcal{A}\}$, B^i, $\{B^j | (i, j) \in \mathcal{A}\}$ and ${}^{i}K_1^{j}$. Similarly, matrix $\Psi_k^{j,i}$ is a function of A^i, $\{A^j | (i, j) \in \mathcal{A}\}$, B^i, $\{B^j | (i, j) \in \mathcal{A}\}$ and $K_2^{j,i}$.

Based on Equations (5.30), (5.32) and (5.33), all the terms in the stability condition (5.26) can be expressed as a quadratic form of $\bar{x}_0^i$. The terms on the left side of (5.26) can be expressed using the following two matrices:

$$\Theta_N^i = \sum_{k=1}^{N-1} \sum_{j|(i,j)\in\mathcal{A}} 2(\Psi_k^{j,j} - \Psi_k^{j,i})' Q (\Psi_k^{j,j} - \Psi_k^{j,i}), \tag{5.34a}$$

$$\Gamma_N^i = \sum_{k=1}^{N-1} \sum_{j|(i,j)\in\mathcal{A}} ({}^{i}K_{1,k}^{j} - K_{2,k}^{j,i})' R ({}^{i}K_{1,k}^{j} - K_{2,k}^{j,i}). \tag{5.34b}$$

Denoting the number of neighbors of node i by N_v^i, the following matrices are used to express terms on the right side of (5.26):

$$R_1^i = (K_{1,0}^i)' R (K_{1,0}^i), \tag{5.35a}$$

$$R_2^i = (K_{2,0}^i)' (I_{N_v^i} \otimes R)(K_{2,0}^i), \tag{5.35b}$$

$$Q_1 = \begin{bmatrix} Q & \mathbf{0} & \mathbf{0} \\ \mathbf{0} & \mathbf{0} & \mathbf{0} \\ \mathbf{0} & \mathbf{0} & \mathbf{0} \end{bmatrix}, \quad Q_2^i = \begin{bmatrix} \mathbf{0} & \mathbf{0} & \mathbf{0} \\ \mathbf{0} & I_{N_v^i} \otimes Q & \mathbf{0} \\ \mathbf{0} & \mathbf{0} & \mathbf{0} \end{bmatrix}, \tag{5.35c}$$

$$Q_3^i = D_1' \left(I_{N_v^i} \otimes Q \right) D_1, \tag{5.35d}$$

$$D_1 = \begin{bmatrix} [\mathbf{1}_{N_v^i} \;\; -I_{N_v^i}] \otimes I_n \\ \mathbf{0} \end{bmatrix}, \tag{5.35e}$$

where $\mathbf{1}_{N_v^i}$ denotes a column vector of '1'-s of size N_v^i. Using the matrices in (5.34)-(5.35), the stability condition (5.26) for node i is equivalent to testing whether

$$\Theta_N^i + \Gamma_N^i \leq Q_1 + R_1^i + Q_2^i + R_2^i + Q_3^i. \tag{5.36}$$

Stability of the aggregate system can be concluded if (5.36) holds for all $i \in \{1, \ldots, N_v\}$. For a more detailed description, the reader is referred to Keviczky *et al.* (2006a).

5.5 DISTRIBUTED DESIGN FOR IDENTICAL UNCONSTRAINED LTI SUBSYSTEMS

In this section, we build on the philosophy introduced in the previous sections, where at each node, the models of its neighbors are used to predict their behavior. We continue to explore an even simpler region of the 'complexity spectrum' and consider arbitrary interconnections of identical and unconstrained LTI subsystems in continuous time. This allows us to propose an appealingly simple distributed controller design approach and focus on a class of systems, for which existing methods are either not efficient or would not even be directly applicable. Our method applies to large-scale systems composed of finite number of identical subsystems where the interconnection structure or sparsity pattern is not required to have any special invariance properties. We show that in absence of state and input constraints, and for identical linear system dynamics, such an approach leads to an extremely powerful result: the synthesis of stabilizing distributed control laws can be obtained by using a simple local LQR design, whose size is limited by the maximum vertex degree of the interconnection graph plus one. Furthermore, the design procedure described in this section illustrates how stability of the overall large-scale system is related to the robustness of local controllers and the spectrum of a matrix representing the desired sparsity pattern. In addition, the constructed distributed controller is stabilizing independent of the tuning parameters in the local LQR cost function. This leads to a method for designing distributed controllers for a finite number of dynamically decoupled systems, where the local tuning parameters can be chosen to obtain a desirable global performance. Such result can be used to improve current stability analysis and controller synthesis in the field of distributed predictive control for dynamically decoupled systems. Although a continuous-time formulation is used in this section, similar results can be derived for the discrete-time case.

In the following exposition we will make use of some notational conventions and propositions given below.

Notation 5.5.1 *Let $\lambda_i(M)$ denote the i-th eigenvalue of $M \in \mathbb{R}^{n\times n}$, $i = 1, \ldots, n$. The spectrum of M will be denoted by $\mathcal{S}(M) = \{\lambda_1(M), \ldots, \lambda_n(M)\}$.*

Proposition 5.5.1 *Consider two matrices $A = \alpha I_n$ and $B \in \mathbb{R}^{n\times n}$. Then $\lambda_i(A + B) = \alpha + \lambda_i(B)$, $i = 1, \ldots, n$.*

Proof. Take any eigenvalue $\lambda_i(B)$ and the corresponding eigenvector $v_i \in \mathbb{C}^n$. Then $(A + B)v_i = Av_i + Bv_i = \alpha v_i + \lambda_i v_i = (\alpha + \lambda_i)v_i$.□

Proposition 5.5.2 *Given $A, C \in \mathbb{R}^{m\times m}$ and $B \in \mathbb{R}^{n\times n}$, consider two matrices $\overline{A} = I_n \otimes A$ and $\overline{C} = B \otimes C$, where $\overline{A}, \overline{C} \in \mathbb{R}^{nm\times nm}$. Then $\mathcal{S}(\overline{A} + \overline{C}) = \bigcup_{i=1}^{n} \mathcal{S}(A + \lambda_i(B)C)$, where $\lambda_i(B)$ is the i-th eigenvalue of B.*

Proof. Let $v \in \mathbb{C}^n$ be an eigenvector of B corresponding to $\lambda(B)$, and $u \in \mathbb{C}^m$ be an eigenvector of $M = (A + \lambda(B)C)$ with $\lambda(M)$ as the associated eigenvalue. Consider the vector $v \otimes u \in \mathbb{C}^{nm}$. Then $(\overline{A} + \overline{C})(v \otimes u) = v \otimes Au + Bv \otimes Cu = v \otimes Au + \lambda(B)v \otimes Cu = v \otimes (Au + \lambda(B)Cu)$. Since $(A + \lambda(B)C)u = \lambda(M)u$, we get $(\overline{A} + \overline{C})(v \otimes u) = \lambda(M)(v \otimes u)$.□

In this section we focus on *undirected* graphs, for which the adjacency matrix $\mathbf{A}(\mathcal{G})$ is symmetric. The following properties of graph Laplacians and adjacency matrices will be useful.

We define the Laplacian matrix of a graph $\mathcal{G}$ in the following way

$$L(\mathcal{G}) = \mathbf{D}(\mathcal{G}) - \mathbf{A}(\mathcal{G}), \tag{5.37}$$

where $\mathbf{D}(\mathcal{G})$ is the diagonal matrix of vertex degrees d_i (also called the valence matrix). Eigenvalues of Laplacian matrices have been widely studied by graph theorists. Their properties are strongly related to the structural properties of their associated graphs (Agaev and Chebotarev 2005; Merris 1994).

For undirected graphs, $L(\mathcal{G})$ is a symmetric, positive semidefinite matrix, which has only real eigenvalues. Let $\mathcal{S}(L(\mathcal{G})) = \{\lambda_1, \ldots, \lambda_{N_v}\}$ be the spectrum of the Laplacian matrix L associated with an undirected graph $\mathcal{G}$ arranged in nondecreasing semi-order. Then,

Property 5.5.1 $\lambda_n \leq d_{max}(\mathcal{G})$.

Property 5.5.2 *(i) $\lambda_1 = 0$ with corresponding eigenvector of all ones, and $\lambda_2 \neq 0$ iff $\mathcal{G}$ is connected. In fact, the multiplicity of 0 as an eigenvalue of $L(\mathcal{G})$ is equal to the number of connected components of $\mathcal{G}$.*

(ii) The modulus of λ_i, $i = 1, \ldots, N_v$ is less then N_v.

Property 5.5.3 $\lambda_2(L(\mathcal{G})) \geq 2\eta(\mathcal{G})(1 - \cos\frac{\pi}{N_v})$*, where $\eta(\mathcal{G})$ is the edge connectivity of the graph $\mathcal{G}$ (Bollobás 2002).*

Further relationships between the graph topology and Laplacian eigenvalue locations are discussed in Mohar (1991) for undirected graphs. Spectral characterization of Laplacian matrices for directed graphs can be found in Agaev and Chebotarev (2005).

5.5.1 LQR properties for dynamically decoupled systems

Consider a set of N_L identical, decoupled linear time-invariant dynamical systems, the i-th system being described by the continuous-time state equation:

$$\dot{x}_i = Ax_i + Bu_i, \quad x_i(0) = x_{i0}. \tag{5.38}$$

where $x_i(t) \in \mathbb{R}^n$, $u_i(t) \in \mathbb{R}^m$ are states and inputs of the i-th system at time t, respectively. Let $\tilde{x}(t) \in \mathbb{R}^{nN_L}$ and $\tilde{u}(t) \in \mathbb{R}^{mN_L}$ be the vectors which collect the states and inputs of the N_L systems at time t:

$$\begin{aligned} \dot{\tilde{x}} &= \tilde{A}\tilde{x} + \tilde{B}\tilde{u}, \\ \tilde{x}(0) &= \tilde{x}_0 \triangleq [x_{10}, \ldots, x_{N_L0}]', \end{aligned} \tag{5.39}$$

with

$$\tilde{A} = I_{N_L} \otimes A, \quad \tilde{B} = I_{N_L} \otimes B. \tag{5.40}$$

We consider an LQR control problem for the set of N_L systems where the cost function couples the dynamic behavior of individual systems:

$$\begin{aligned} J(\tilde{u}, \tilde{x}_0) = \int_0^\infty & \sum_{i=1}^{N_L} x_i(\tau)' Q_{ii} x_i(\tau) + u_i(\tau)' R_{ii} u_i(\tau) + \\ & \sum_{i=1}^{N_L} \sum_{j \neq i}^{N_L} x_i(\tau) - x_j(\tau))' Q_{ij} (x_i(\tau) - x_j(\tau) \, d\tau \end{aligned} \tag{5.41}$$

with

$$R_{ii} = R'_{ii} = R > 0, \quad Q_{ii} = Q'_{ii} = Q \geq 0 \quad \forall i, \tag{5.42a}$$

$$Q_{ij} = Q'_{ij} = Q_{ji} \geq 0 \quad \forall i \neq j. \tag{5.42b}$$

The cost function (5.41) contains terms which weigh the i-th system states and inputs, as well as the difference between the i-th and the j-th system states (similarly to the local MPC cost defined in Assumption 5.4.1 with $p = 2$), and can be rewritten using the following compact notation:

$$J(\tilde{u}(t), \tilde{x}_0) = \int_0^\infty \tilde{x}(\tau)' \tilde{Q} \tilde{x}(\tau) + \tilde{u}(\tau) \tilde{R} \tilde{u}(\tau) \, d\tau, \tag{5.43}$$

where the matrices $\tilde{Q}$ and $\tilde{R}$ have a special structure defined next. $\tilde{Q}$ and $\tilde{R}$ can be decomposed into N_L^2 blocks of dimension $n \times n$ and $m \times m$ respectively:

$$\tilde{Q} = \begin{bmatrix} \tilde{Q}_{11} & \tilde{Q}_{12} & \cdots & \tilde{Q}_{1N_L} \\ \vdots & \ddots & \vdots & \vdots \\ \tilde{Q}_{N_L1} & \cdots & \cdots & \tilde{Q}_{N_LN_L} \end{bmatrix}, \quad \tilde{R} = I_{N_L} \otimes R. \tag{5.44}$$

with

$$\tilde{Q}_{ii} = Q + \sum_{k=1,\ k\neq i}^{N_L} Q_{ik}, \quad i = 1, \ldots, N_L.$$

$$\tilde{Q}_{ij} = -Q_{ij}, \quad i, j = 1, \ldots, N_L, \quad i \neq j. \tag{5.45}$$

Let $\tilde{K}$ and $\tilde{x}_0'\tilde{P}\tilde{x}_0$ be the optimal controller and the value function corresponding to the following LQR problem:

$$\begin{aligned} \min_{\tilde{u}} \quad & J(\tilde{u}, \tilde{x}_0) \\ \text{subj. to} \quad & \dot{\tilde{x}} = \tilde{A}\tilde{x} + \tilde{B}\tilde{u} \\ & \tilde{x}(0) = \tilde{x}_0 \end{aligned} \tag{5.46}$$

Throughout this section we will assume that a stabilizing solution to the LQR problem (5.46) with finite performance index exists and is unique (see Anderson and Moore 1990, p. 52 and references therein):

Assumption 5.5.1 *System $\tilde{A}$, $\tilde{B}$ is stabilizable and system $\tilde{A}$, $\tilde{C}$ is observable, where $\tilde{C}$ is any matrix such that $\tilde{C}'\tilde{C} = \tilde{Q}$.*

We will also assume local stabilizability and observability:

Assumption 5.5.2 *System A, B is stabilizable and systems A, C are observable, where C is any matrix such that $C'C = Q$.*

It is well known that $\tilde{K} = -\tilde{R}^{-1}\tilde{B}'\tilde{P}$, where $\tilde{P}$ is the symmetric positive definite solution to the following ARE:

$$\tilde{A}'\tilde{P} + \tilde{P}\tilde{A} - \tilde{P}\tilde{B}\tilde{R}^{-1}\tilde{B}'\tilde{P} + \tilde{Q} = 0 \tag{5.47}$$

We decompose $\tilde{K}$ and $\tilde{P}$ into N_L^2 blocks of dimension $m \times n$ and $n \times n$, respectively. Denote by $\tilde{K}_{ij}$ and $\tilde{P}_{ij}$ the (i, j) block of the matrix $\tilde{K}$ and $\tilde{P}$, respectively. In the following theorems we show that $\tilde{K}_{ij}$ and $\tilde{P}_{ij}$ satisfy certain properties which will be critical for the design of stabilizing distributed controllers in Section 5.5.2. These properties stem from the special structure of the LQR problem (5.46). In the following we will use the matrix $X = BR^{-1}B'$ for notational brevity.

Theorem 5.5.3 *Let $\tilde{K}$ and $\tilde{x}_0'\tilde{P}\tilde{x}_0$ be the optimal controller and the value function solution to the LQR problem (5.46). Let $\tilde{K}_{ij} = \tilde{K}[(i-1)m : im, (j-1)n : jn]$ and $\tilde{P}_{ij} = \tilde{P}[(i-1)n : in, (j-1)n : jn]$ with $i = 1, \ldots, N_L$, $j = 1, \ldots, N_L$.*
Then,

(i) $\sum_{j=1}^{N_L} \tilde{P}_{ij} = P$ for all $i = 1, \ldots, N_L$, where P is the symmetric positive definite solution of the ARE associated with a single node local problem:

$$A'P + PA - PBR^{-1}B'P + Q = 0. \tag{5.48}$$

(ii) $\sum_{j=1}^{N_L} \tilde{K}_{ij} = K$ *for all* $i = 1, \ldots, N_L$, *where* $K = -R^{-1}B'P$.

Proof. The proof can be found in Borrelli and Keviczky (2006a). □

Theorem 5.5.4 *Assume the weighting matrices (5.45) of the LQR problem (5.46) are chosen as*

$$\begin{aligned} Q_{ii} &= Q_1 \; \forall i = 1, \ldots, N_L \\ Q_{ij} &= Q_2 \; \forall i, j = 1, \ldots, N_L, \; i \neq j. \end{aligned} \tag{5.49}$$

Let $\tilde{x}_0' \tilde{P} \tilde{x}_0$ *be the value function of the LQR problem (5.46) with weights (5.49), and the blocks of the matrix* $\tilde{P}$ *be denoted by* $\tilde{P}_{ij} = \tilde{P}[(i-1)n : in, (j-1)n : jn]$ *with* $i, j = 1, \ldots, N_L$. *Then* $\tilde{P}_{ij}$ *is a symmetric negative semidefinite matrix for all* $i \neq j$. *Furthermore, the matrix* $\tilde{P}$ *is composed of identical* $P - (N_L - 1)\tilde{P}_2$ *matrix blocks on its diagonal and identical* $\tilde{P}_2$ *matrices as off-diagonal block entries.*

Proof. The proof can be found in Borrelli and Keviczky (2006a). □

Under the hypothesis of Theorem 5.5.4, because of symmetry and equal weights Q_2 on the neighboring state differences and equal weights Q_1 on absolute states, the LQR optimal controller will have the following structure:

$$\tilde{K} = \begin{bmatrix} K_1 & K_2 & \cdots & K_2 \\ K_2 & K_1 & \cdots & K_2 \\ \vdots & \ddots & \ddots & \vdots \\ K_2 & \ldots & \ldots & K_1 \end{bmatrix}, \tag{5.50}$$

with K_1 and K_2 functions of N_L, A, B, Q_1, Q_2 and R.

The following corollaries of Theorem 5.5.4 follow from the stability and the robustness of the LQR controller $-R^{-1}B'(-N_L\tilde{P}_2)$ for system $A - XP$.

Corollary 5.5.5 $A - XP + N_L X \tilde{P}_2$ *is a Hurwitz matrix.*

From the gain margin properties (Safonov and Athans 1977) we have:

Corollary 5.5.6 $A - XP + \alpha N_L X \tilde{P}_2$ *is a Hurwitz matrix for all* $\alpha > \frac{1}{2}$, *with* $\alpha \in \mathbb{R}$.

Remark 5.5.1 $A - XP$ is a Hurwitz matrix, thus the system in Corollary 5.5.6 is stable for $\alpha = 0$ ($A - XP = A + BK$ with K being the LQR gain for system (A, B) with weights (Q_1, R)).

The following condition defines a class of systems and LQR weighting matrices which will be used in later sections to extend the set of stabilizing distributed controller structures.

Condition 5.5.1 $A - XP + \alpha N_L X \tilde{P}_2$ *is a Hurwitz matrix for all* $\alpha \in [0, \frac{1}{2}]$, *with* $\alpha \in \mathbb{R}$.

Essentially, Condition 5.5.1 characterizes systems for which the LQR gain stability margin described in Corollary 5.5.6 is extended to any positive α.

Checking the validity of Condition 5.5.1 for a given tuning of P and $\tilde{P}_2$ may be performed as a stability test for a simple affine parameter-dependent model

$$\dot{x} = \underbrace{(A_0 + \alpha A_1)}_{A(\alpha)} x, \tag{5.51}$$

where $A_0 = A - XP$, $A_1 = N_L X \tilde{P}_2$ and $0 \leq \alpha \leq \frac{1}{2}$. This test can be posed as an LMI problem (Proposition 5.9 in Scherer and Weiland (2000)) searching for quadratic parameter-dependent Lyapunov functions.

5.5.2 Distributed LQR design

We consider a set of N_v linear, identical and decoupled dynamical systems, described by the continuous-time time-invariant state equation (5.38), rewritten below

$$\dot{x}_i = Ax_i + Bu_i,$$
$$x_i(0) = x_{i0}.$$

where $x_i(t) \in \mathbb{R}^n$, $u_i(t) \in \mathbb{R}^m$ are states and inputs of the i-th system at time t, respectively. Let $\hat{x}(t) \in \mathbb{R}^{N_v n}$ and $\hat{u}(t) \in \mathbb{R}^{N_v m}$ be the vectors which collect the states and inputs of the N_v systems at time t, then

$$\dot{\hat{x}} = \hat{A}\hat{x} + \hat{B}\hat{u},$$
$$\hat{x}(0) = \hat{x}_0 \triangleq [x_{10}, \ldots, x_{N_v 0}]', \tag{5.52}$$

with

$$\hat{A} = I_{N_v} \otimes A, \ \hat{B} = I_{N_v} \otimes B.$$

Remark 5.5.2 Systems (5.52) and (5.39) differ only in the number of subsystems. We will use system (5.39) with N_L subsystems when referring to local problems, and system (5.52) with N_v subsystems when referring to the global problem. Accordingly, tilded matrices will refer to local problems and hatted matrices will refer to the global problem.

The class of $\mathcal{K}^{N_v}_{n,m}(\mathcal{G})$ matrices corresponding to a graph interconnection $\mathcal{G} = (\mathcal{V}, \mathcal{A})$ of N_v nodes is defined as follows:

Definition 5.5.7 $\mathcal{K}^{N_v}_{n,m}(\mathcal{G}) = \{M \in \mathbb{R}^{nN_v \times mN_v} | M_{ij} = \mathbf{0} \textit{ if } (i, j) \notin \mathcal{A},\ M_{ij} = M[(i-1)n : in, (j-1)m : jm],\ i, j = 1, \ldots, N_v\}$

In the following theorem, we propose a suboptimal control design procedure leading to a controller $\hat{K}$ with the following properties:

$$(1)\ \ \hat{K} \in \mathcal{K}^{N_v}_{m,n}(\mathcal{G}) \tag{5.53a}$$

(2) $\hat{A} + \hat{B}\hat{K}$ is Hurwitz. (5.53b)

(3) Simple tuning of absolute and relative state errors and control effort within $\hat{K}$. (5.53c)

Such a controller will be referred to as *distributed suboptimal controller*.

Theorem 5.5.8 *Consider the LQR problem (5.46) with $N_L = d_{max}(\mathcal{G}) + 1$ and weights chosen as in (5.49) and its solution $\tilde{P}$, $\tilde{K}$.*

Let $M \in \mathbb{R}^{N_v \times N_v}$ be a symmetric matrix with the following property:

$$\lambda_i(M) > \frac{N_L}{2}, \quad \forall \lambda_i(M) \in \mathcal{S}(M)\backslash\{0\}. \tag{5.54}$$

and construct the feedback controller:

$$\hat{K} = -I_{N_v} \otimes R^{-1}B'P + M \otimes R^{-1}B'\tilde{P}_2. \tag{5.55}$$

Then, the closed loop system

$$A_{cl} = I_{N_v} \otimes A + (I_{N_v} \otimes B)\hat{K} \tag{5.56}$$

is asymptotically stable.

Proof. Consider the eigenvalues of the closed-loop system A_{cl}:

$$\mathcal{S}(A_{cl}) = \mathcal{S}(I_{N_v} \otimes (A - XP) + M \otimes (X\tilde{P}_2))$$

By Proposition 5.5.2:

$$\mathcal{S}(I_{N_v} \otimes (A - XP) + M \otimes (X\tilde{P}_2)) = \bigcup_{i=1}^{N_v} \mathcal{S}(A - XP + \lambda_i(M)X\tilde{P}_2). \tag{5.57}$$

We will prove that $(A - XP + \lambda_i(M)X\tilde{P}_2)$ is a Hurwitz matrix $\forall i = 1, \ldots, N_v$, and thus prove the theorem. If $\lambda_i(M) = 0$ then $A - XP + \lambda_i(M)X\tilde{P}_2$ is Hurwitz based on Remark 5.5.1. If $\lambda_i(M) \neq 0$, then from Corollary 5.5.6 and from condition (5.54), we conclude that $A - XP + \lambda_i(M)X\tilde{P}_2$ is Hurwitz. □

Theorem 5.5.8 has several main consequences:

(i) If $M \in \mathcal{K}_{1,1}^{N_v}(\mathcal{G})$, then $\hat{K}$ in (5.55) is an asymptotically stable distributed controller.

(ii) We can use one local LQR controller to compose distributed stabilizing controllers for a collection of identical dynamically decoupled subsystems.

(iii) The first two consequences imply that not only can we find a stabilizing distributed controller with a desired sparsity pattern (which is in general a formidable task by itself), but it is enough to solve a low-dimensional problem (characterized by $d_{max}(\mathcal{G})$) compared to the full centralized problem size. This attractive feature of our approach relies on the specific problem structure defined in Sections 5.5.1 and 5.5.2.

(iv) The eigenvalues of the closed-loop large-scale system $\mathcal{S}(A_{cl})$ can be computed through N_v smaller eigenvalue computations as $\bigcup_{i=1}^{N_v} \mathcal{S}(A - XP + \lambda_i(M)X\tilde{P}_2)$.

(v) The result is independent from the local LQR tuning. Thus Q_1, Q_2 and R in (5.49) can be used in order to influence the compromise between minimization of absolute and relative terms, and the control effort in the global performance.

For the special class of systems defined by Condition 5.5.1, the hypothesis of Theorem 5.5.8 can be relaxed as follows:

Theorem 5.5.9 *Consider the LQR problem (5.46) with $N_L = d_{max}(\mathcal{G}) + 1$ and weights chosen as in (5.49) and its solution $\tilde{P}$, $\tilde{K}$. Assume that Condition 5.5.1 holds.*

Let $M \in \mathbb{R}^{N_v \times N_v}$ be a symmetric matrix with the following property:

$$\lambda_i(M) \geq 0, \ \forall \lambda_i \in \mathcal{S}(M). \tag{5.58}$$

Then, the closed loop system (5.56) is asymptotically stable when $\hat{K}$ is constructed as in (5.55).

Proof. Notice that if Condition 5.5.1 holds, then $\mathcal{S}(A - XP + \lambda_i X\tilde{P}_2)$ is Hurwitz for all $\lambda_i(M) \geq 0$ (from Corollary 5.5.5 and Corollary 5.5.6). Proposition 5.5.2 leads to the relationship shown in (5.57), which together with condition (5.58) proves the theorem. □

Next we show how to choose M in Theorem 5.5.8 and Theorem 5.5.9 in order to construct distributed suboptimal controllers. The matrix M will: (1) reflect the structure of the graph $\mathcal{G}$; (2) satisfy (5.54) or (5.58); and (3) be computed by using the graph adjacency matrix or the Laplacian matrix. The distributed control design is presented next for a generic graph structure. Illustrative examples for a simple finite string and for a finite square mesh interconnection can be found in Borrelli and Keviczky (2006a); Online (2006).

We consider a generic graph $\mathcal{G}$ for N_v nodes with an associated Laplacian $L(\mathcal{G})$ and maximum vertex degree d_{max}. Let $0 = \lambda_1(G) \leq \lambda_2(G) \ldots \leq \lambda_{N_v}(G)$ be the eigenvalues of the Laplacian $L(\mathcal{G})$. In the following Corollaries 5.5.10, 5.5.11 and 5.5.12 we present three ways of choosing M in (5.55) which lead to distributed suboptimal controllers.

Corollary 5.5.10 *Compute M in (5.55) as $M = aL(\mathcal{G})$. If*

$$a > \frac{N_L}{2\lambda_2(\mathcal{G})}, \tag{5.59}$$

then the closed loop system (5.56) is asymptotically stable when $\hat{K}$ is constructed as in (5.55). In addition, if Condition 5.5.1 holds, then the closed loop system (5.56) is asymptotically stable for all $a \geq 0$.

Proof. The proof is a direct consequence of Theorems 5.5.8 and 5.5.9, and Property 5.5.2 of the Laplacian matrix. □

Remark 5.5.3 By using Property 5.5.3, condition (5.59) can be linked to the edge connectivity as follows

$$a > \frac{N_L}{2\eta(\mathcal{G})1 - \cos\frac{\pi}{N_v}}. \tag{5.60}$$

Remark 5.5.4 Corollary 5.5.10 links the stability of the distributed controller to the size of the second smallest eigenvalues of the graph Laplacian. It is well known that graphs with large λ_2 (with respect to the maximal degree) have some properties which make them very useful in several applications such as computer science. Interestingly enough, this property is shown here to be crucial also for the design of distributed controllers. We refer the reader to Mohar (1991) for a more detailed discussion on the importance of the second largest eigenvalue of a Laplacian.

Corollary 5.5.11 *Compute M in (5.55) as $M = aI_{N_v} - b\mathbf{A}(\mathcal{G})$, $b \geq 0$. If $a - bd_{max} > \frac{N_L}{2}$, then the closed loop system (5.56) is asymptotically stable when $\hat{K}$ is constructed as in (5.55). In addition, if Condition 5.5.1 holds, then the closed loop system (5.56) is asymptotically stable if $a - bd_{max} \geq 0$.*

Proof. Notice that $\lambda_{min}(M) = a - b\lambda_{max}(\mathbf{A}(\mathcal{G})) \geq a - bd_{max}$. The proof is a direct consequence of Theorems 5.5.8 and 5.5.9 and Property 5.5.1 of the adjacency matrix. □

Consider a weighted adjacency matrix $\mathbf{A}^w = \mathbf{A}^w(\mathcal{G})$, defined as follows. Denote by $\mathbf{A}^w_{i,j} \in \mathbb{R}$ its i, j element, then $\mathbf{A}^w_{i,j} = 0$, if $i = j$ and $(i, j) \notin \mathcal{A}$ and $\mathbf{A}^w_{i,j} = w_{ij}$ if $(i, j) \in \mathcal{A}$, $\forall\, i, j = 1, \ldots, N_v$, $i \neq j$. Assume $w_{ij} = w_{ji} > 0$. Define w_{max} as $w_{max} = \max_i \sum_j w_{ij}$

Corollary 5.5.12 *Compute M in (5.55) as $M = aI_{N_v} - \mathbf{A}^w(\mathcal{G})$. If $a > w_{max} - \frac{N_L}{2}$, then the closed loop system (5.56) is asymptotically stable when $\hat{K}$ is constructed as in (5.55). In addition, if Condition 5.5.1 holds, then the closed loop system (5.56) is asymptotically stable if $a \geq w_{max}$.*

Proof. $\mathbf{A}^w\mathbf{1} \leq w_{max}\mathbf{1}$ and by Perron-Frobenius Theorem $\lambda_{max}(\mathbf{A}^w) \leq w_{max}$. Notice that $\lambda_{min}(M) = a - \lambda_{max}(\mathbf{A}(\mathcal{G})) \geq a - w_{max}$, then the proof is a direct consequence of Theorems 5.5.8 and 5.5.9. □

The results of Corollaries 5.5.10–5.5.12 are summarized in Table 5.1.

Corollaries 5.5.10–5.5.12 present three choices of distributed control design with increasing degrees of freedom. In fact, a, b and w_{ij} are additional parameters, which

Table 5.1 Summary of stability conditions in Corollaries 5.5.10–5.5.12 for the closed loop system (5.55)–(5.56)

Choice of M	Stability condition	Stability condition if Condition 5.5.1 holds
$aL(\mathcal{G})$	$a > \frac{N_L}{2\lambda_2(\mathcal{G})}$	$a \geq 0$
$aI_{N_v} - b\mathbf{A}(\mathcal{G}), b \geq 0$	$a - bd_{max} > \frac{N_L}{2}$	$a - bd_{max} \geq 0$
$aI_{N_v} - \mathbf{A}^w(\mathcal{G})$	$a > w_{max} - \frac{N_L}{2}$	$a \geq w_{max}$

together with Q_1, Q_2 and R, can be used to tune the closed-loop system behavior. We recall here that from Theorem 5.5.8, the eigenvalues of the closed-loop large-scale system are related to the eigenvalues of M through the simple relation (5.57). Thus as long as the stability conditions defined in Table 5.1 are satisfied, the overall system architecture can be modified arbitrarily by adding or removing subsystems and interconnection links. This leads to a very powerful modular approach for designing distributed control systems.

5.6 ENSURING FEASIBILITY

Although the distributed LQR control design method presented in the previous section provides an attractive and simple way of dealing with collections of identical, unconstrained LTI systems, we cannot avoid the fact that most real-life cooperative control problems involve coupling constraints between subsystems. However, if coupling constraints are present, ensuring feasibility in a distributed receding horizon control scheme without introducing excessively conservative assumptions is a challenging problem.

Consider for instance the problem of choosing local terminal regions for each distributed MPC problem (5.11). Even if we assume terminal point constraints (the most conservative and simplest formulation), infeasibility for the aggregate system can occur since coupling constraints may be violated due to the mismatch between neighbors' predictions and their closed loop behavior. Indeed, decoupled terminal regions do not enforce feasibility of coupling constraints, which renders them less effective than in a centralized approach. For this reason we will not focus on possible ways of computing local terminal regions based on the centralized problem. Instead, we will investigate different options for reducing the uncertainty about the behavior of neighboring systems in order to ensure feasibility in the presence of coupling constraints. Problem (5.11) needs to be modified to accomplish this objective. Specifically, we focus on using robust constraint fulfillment and review other alternative approaches that could be used to tackle the problem.

5.6.1 Robust constraint fulfillment

Consider the coupling constraints of problem $\mathcal{P}_i$ in (5.11) at step k

$$g^i(x^i_{k,t}, \tilde{x}^i_{k,t}) \le 0, \tag{5.61}$$

and by using the state update equations

$$\begin{aligned} x^i_{k+1,t} &= f^i(x^i_{k,t}, u^i_{k,t}), \quad k \ge 0, \\ x^j_{k+1,t} &= f^j(x^j_{k,t}, u^j_{k,t}), \quad (j,i) \in \mathcal{A}, \ k \ge 0, \end{aligned} \tag{5.62}$$

rewrite them as

$$g^i_k(x^i_t, \tilde{x}^i_t, u^i_{[0,\ldots,k-1]}, \tilde{u}^i_{[0,\ldots,k-1]}) \le 0, \tag{5.63}$$

where $u^i_{[0,\ldots,k-1]} \triangleq \{u^i_0, \ldots, u^i_{k-1}\}$ and $\tilde{u}^i_{[0,\ldots,k-1]} \triangleq \{\tilde{u}^i_0, \ldots, \tilde{u}^i_{k-1}\}$. In order to ensure the feasibility of the aggregate system, a possible approach is to 'robustify' the constraint (5.63) for all nodes at all time steps. In other words, we may require that the coupling constraints at each node are satisfied for *all* possible behaviors of the neighboring nodes, once their initial condition is known. Therefore, the vector $\tilde{u}^i_{[0,\ldots,k-1]}$ can be considered as a disturbance which may lead to infeasibility of constraint (5.63). There are two possible schemes: open-loop and closed-loop constraint fulfillment (Bemporad *et al.* 2003; Scokaert and Mayne 1998). An open-loop robust constraint fulfillment is formulated next.

Consider g^i_k and compute $\overline{g}^i_k : \mathbb{R}^{n^i} \times \mathbb{R}^{\tilde{n}^i} \times \mathbb{R}^{m^i} \to \mathbb{R}^{n^i}$ such that for all $x^i_t, \tilde{x}^i_t, u^i_{[0,\ldots,k-1]}$ which satisfy

$$\overline{g}^i_k(x^i_t, \tilde{x}^i_t, u^i_{[0,\ldots,k-1]}) \leq 0, \tag{5.64}$$

constraint (5.63) will be satisfied as well for all admissible[1] $\tilde{u}^i_{[0,\ldots,k-1]}$. Assuming the sets described by (5.64) for $i = 1, \ldots, N_v$ and $k = 1, \ldots, N-1$ are nonempty, problem $\mathcal{P}_i$ in (5.11) is rewritten by replacing the modified form of constraint functions g^i_k in (5.63) with $\overline{g}^i_k$. Numerical methods for computing the function $\overline{g}^i_k$ depend on its properties. For linear constraints Kerrigan (2000) provides available tools, while Rowe and Maciejowski (2003) describe an approach that can be used for piecewise linear constraint functions. Algorithms for computing such sets in case of general nonlinear constraints are treated in Mitchell and Templeton (2005).

Robust closed-loop formulation (Scokaert and Mayne 1998) is less conservative but more computationally involved. Robust constraint fulfillment applied to distributed control schemes results in a very conservative approach even for the closed-loop case. For instance, consider a formation flight application example. Assume there are only two aircraft and the goal is to design a local controller on the first aircraft using robust constraint fulfillment. The worst case scenario will include, in most cases, the collision of the two aircraft if they are not very far from each other and have the same dynamics and constraints. However, in reality, neighboring nodes collaborate between each other to fly in formation. The conservativeness introduced by robust constraint fulfillment and other worst-case approaches suggest that new methods are needed, which rely on the cooperation between neighboring nodes or exploit other degrees of freedom in the problem formulation, such as the interconnection graph structure. A brief review of different classes of methodologies are provided in the next section.

5.6.2 Review of methodologies

The methodologies listed in this section provide different ways of approaching the problem of feasibility in distributed RHC schemes by reducing the uncertainty and prediction mismatch made about neighboring solutions. They consider various modifications to the basic framework presented in Section 5.3, which rely either on some sense of cooperation between neighboring nodes, utilize special interconnection structures, employ safety controllers or robustify against realistic prediction mismatch models. Each of these methods represents a separate approach and a research topic on its own.

[1] Admissible inputs have to satisfy constraints (5.2).

5.6.2.1 *Neighboring uncertainty models for robust constraint fulfillment*

In some application examples, the mismatch between predicted and actual dynamics of neighboring systems can be modeled as an additive disturbance whose domain is time-varying. For instance, consider a constrained robust finite-time optimal control problem for a set of dynamically decoupled vehicles. In this case the domain of additive disturbance on the predicted neighboring states is increasing with the prediction horizon. Neighboring vehicles can be modeled with the following system:

$$\begin{aligned} x_{k+1}^j &= f^j(x_k^j, u_k^j) + w_k^j, \\ \mathcal{W}_k^j &\subset \mathcal{W}_{k+1}^j, \quad k = 0, \ldots, N-1, \end{aligned} \tag{5.65}$$

which is used to replace the model (5.11b) in problem (5.11). The term w_k^j, with $w_k^j \in \mathcal{W}_k^j \subseteq \mathbb{R}^{n_j}$, represents an additive disturbance on the j-th system state, whose polyhedral domain is increasing with time over the prediction horizon. Such neighboring uncertainty model has been justified by extensive simulations, and intends to mimic the fact that predictions about neighboring trajectories are typically reliable only for short prediction horizons and become worse as the horizon grows.

5.6.2.2 *Cooperation with information exchange*

A less conservative approach for ensuring feasibility of the distributed scheme takes into consideration the cooperation between local controllers, and therefore the trajectory that a node is predicting should not be extremely different from what its neighbors are executing. This idea can be formulated in several ways. For instance, one could allow the exchange of optimizers between the nodes in order to try to be as close as possible to what the neighboring system has predicted about a certain node as mentioned in Section 5.4.3. Another possibility is to tighten the coupling constraints (5.6) by a quantity which is an indirect measure of the cooperativeness of the team (Camponogara *et al.* 2002)

$$g_k^i(x_t^i, \tilde{x}_t^i, u_{[0,\ldots,k-1]}^i, \tilde{u}_{[0,\ldots,k-1]}^i) \leq \rho_k^i, \tag{5.66}$$

where $\rho_k^i \leq 0$ is a new optimization variable. Finding efficient methods to compute ρ_k^i off-line, based on a priori knowledge of the team behavior is still an open problem and the focus of current, ongoing research. Also, the idea of tightening these constraints (5.66) can be exploited in a two-stage process. In the first stage of the optimization problems (5.11), the coupling constraints are substituted with the one in (5.66). Their parametric solution (Fiacco 1983) with respect to ρ_k^i yields the optimizer function $u^{*i}(\rho_0^i, \ldots, \rho_N^i)$. In a second stage, the nodes communicate between themselves in order to agree on a set of $\overline{\rho}_k^i$ for $i = 1, \ldots, N_v$, $k = 1, \ldots, N$, which ensures feasibility of the distributed trajectories. If the agreement algorithm ends with a positive answer, each node will implement $u^{*i}(\overline{\rho}_0^i, \ldots, \overline{\rho}_N^i)$.

5.6.2.3 *Hierarchy in the interconnection graph*

Enforcing hierarchy in the interconnection links can be one way to exploit an additional degree of freedom that is available to address feasibility issues. Different approaches based on hierarchical decomposition have been proposed in Cassandras *et al.* (2001); Gokbayrak and Cassandras (1999); Stipanovic *et al.* (2002). A certain priority is assigned to each node of the graph and nodes with high priorities compute control laws for nodes with lower priorities. Gokbayrak and Cassandras (1999) decompose an optimal control problem for hybrid systems with a separable cost structure into lower-level time-driven and higher-level event-driven components. A hybrid controller is then developed, which jointly optimizes the performance of both hierarchical components. The work in Keviczky (2005); Keviczky *et al.* (2004a) describes an approach, where higher priority solutions are respected by additional state constraints determined using a feasible set projection algorithm. If the initial propagation of feasible sets through the hierarchy chain does not lead to an empty set, then feasibility is guaranteed for all subsequent time steps for the aggregate system.

5.6.2.4 *Constraint fulfillment using emergency controllers*

One way to address infeasibility in practice is by switching to an emergency controller that keeps the state trajectories within specified bounds in case the RHC subproblem becomes infeasible (Borrelli *et al.* 2004). For instance, in formation flight the emergency controller would perform a 'safety' maneuver such as flying in circles or stopping in mid-air (Schouwenaars *et al.* 2004). In order to guarantee that coupling constraints (such as collision-avoidance) are not violated, nominal constraints are chosen based on the invariant set associated with the emergency controller. This approach is discussed in detail in Keviczky (2005); Keviczky *et al.* (2007).

5.6.2.5 *Hybrid logic rules for cooperation*

It is not difficult to observe that everyday life is full of constrained distributed control problems, even if in general it is very difficult to provide formal feasibility guarantees for such problems. Although feasible solutions are not always found or even possible at all, these problems are solved day-by-day by relying on certain rules that help coordinate the single subsystem efforts. Examples range from traffic laws to behavior of individuals in a community. This suggests that it is beneficial to make use of coordination rules in some distributed engineering control problems as well, such as the MPC scheme presented in Section 5.3.

Hybrid control design techniques are able to cope with the hybrid nature of a problem governed by differential equations and logic rules. For this reason, hybrid system techniques can be beneficial in implementing coordination rules within the distributed control framework presented in Section 5.3. For instance, one could make use of logic rules to improve stability and feasibility of the distributed method by enforcing coordination (Borrelli *et al.* 2005b). The distributed control laws which respect the rules can be computed using hybrid control design. It was shown in Keviczky *et al.* (2006c) that it is possible to introduce coordinating functions, which if chosen appropriately, have the benefit of guiding towards feasible sequences of distributed solutions. When a coordinating function is used in the cost function, trajectories which respect rules can be penalized less and have a cost which is less than the cost of trajectories, which do not enforce the rules. When coordinating functions are used in the constraints, the local domain of feasibility is reduced to the domain where only trajectories respecting rules are feasible. A crucial assumption underlying these ideas is that each component has to abide by the same or at least similar set of rules.

5.7 CONCLUSION

We have presented an overview of our research on distributed predictive control (DMPC) for decoupled systems (Borrelli and Keviczky 2006b; Borrelli *et al.* 2005b; Keviczky 2005; Keviczky *et al.* 2006a, 2007). The chapter was divided intro three parts. In the first part a rigorous mathematical framework for designing DMPCs was presented and its stability investigated. We highlighted how the derived stability conditions lead to complexity reduction in stability analysis. As an example, local matrix semi-definiteness tests have been provided for the case of heterogeneous unconstrained LTI systems. Each test involves the states of as many nodes as are included in the 'one-neighbor-expansion' of the subgraph associated with each subproblem. This means that the size of these local tests are limited by the maximum size of any subgraph with diameter less than or equal to four. Thus a significant reduction in complexity can be expected when the diameter of the overall interconnection graph is large.

In the second part we confined the general framework presented in the first part to the case of identical unconstrained LTI systems. In particular, we described a simple way of constructing stabilizing distributed controllers for identical subsystems in continuous time by solving a single local LQR problem whose size is limited by the maximum vertex degree of the interconnection graph. For such class of systems, where the interconnection structure or sparsity pattern is not required to have any special invariance properties, existing methods are either not efficient or perhaps would not even be directly applicable. Our design procedure has also illustrated how stability of the overall large-scale system is related to the robustness of local controllers and the spectrum of a matrix representing the desired sparsity pattern. In addition, the constructed distributed controllers were stabilizing independent of the tuning parameters in the local LQR cost function.

In the third and last part of the chapter, different methodologies addressing the feasibility issue of DMPCs were presented.

The DMPC approach described in this work has been successfully, applied in simulation to a number of large-scale control problems. In particular, case studies for a formation flight application example and a paper machine control problem can be found in Borrelli *et al.* (2005a); Keviczky (2005); Keviczky *et al.* (2007).

REFERENCES

Agaev R and Chebotarev P 2005 On the spectra of nonsymmetric Laplacian matrices. *Linear Algebra and Its Applications* **399**(5), 157–168.

Anderson BDO and Moore JB 1990 *Optimal Control: Linear Quadratic Methods*. Prentice Hall, Englewood Cliffs, NJ.

Bemporad A, Borrelli F and Morari M 2003 Min-max control of constrained uncertain discrete-time linear systems. *IEEE Trans. Automatic Control* **48**(9), 1600–1606.

Bollobás B 2002 *Modern Graph Theory*. Springer, New York.

Borrelli F and Keviczky T 2006a Distributed LQR design for dynamically decoupled systems. Technical Report 2005-01, University of Minnesota, Minneapolis, MN. Online: http://sciweb.lib.umn.edu/repository/Repository.phtml.

Borrelli F and Keviczky T 2006b Distributed LQR design for identical dynamically decoupled systems. *IEEE Trans. Automatic Control (to appear)*.

Borrelli F, Keviczky T and Balas GJ 2004 Collision-free UAV formation flight using decentralized optimization and invariant sets. In *Proc. 43rd IEEE Conf. on Decision and Control*, pp. 1099–1104.

Borrelli F, Keviczky T and Stewart GE 2005a Decentralized constrained optimal control approach to distributed paper machine control. *44th IEEE Conf. on Decision and Control, and European Control Conf.*, Seville, Spain, pp. 3037–3042

Borrelli F, Keviczky T, Balas GJ, Stewart G, Fregene K and Godbole D 2005b Hybrid decentralized control of large scale systems. *Hybrid Systems: Computation and Control*, vol. 3414 of *Lecture Notes in Computer Science*, pp. 168–183. Springer Verlag, Berlin.

Camponogara E, Jia D, Krogh BH and Talukdar S 2002 Distributed model predictive control. *IEEE Control Systems Magazine* **22**(1), 44–52.

Cassandras CG, Pepyne DL and Wardi Y 2001 Optimal control of a class of hybrid systems. *IEEE Trans. Automatic Control* **46**(3), 398–415.

Chen H and Allgöwer F 1998 A quasi-infinite horizon nonlinear model predictive scheme with guaranteed stability. *Automatica* **14**(10), 1205–1217.

D'Andrea R and Dullerud GE 2003 Distributed control design for spatially interconnected systems. *IEEE Trans. Automatic Control* **48**(9), 1478–1495.

Dunbar WB and Murray RM 2006 Distributed receding horizon control for multi-vehicle formation stabilization. *Automatica* **42**(4), 549–558.

Fiacco AV 1983 *Introduction to Sensitivity and Stability Analysis in Nonlinear Programming*. Academic Press, London.

Gokbayrak K and Cassandras CG 1999 A hierarchical decomposition method for optimal control of hybrid systems. In *Proc. 38th IEEE Conf. on Decision and Control*, pp. 1816–1821, Phoenix, AZ.

Jadbabaie A, Lin J and Morse AS 2003 Coordination of groups of mobile autonomous agents using nearest neighbor rules. *IEEE Trans. Automatic Control* **48**(6), 988–1001.

Jia D and Krogh BH 2002 Min-max feedback model predictive control for distributed control with communication. In *Proc. American Contr. Conf.*

Kerrigan EC 2000 *Robust Constraint Satisfaction: Invariant Sets and Predictive Control*, PhD thesis, Department of Engineering, University of Cambridge, Cambridge, England.

Keviczky T 2005 Decentralized receding horizon control of large scale dynamically decoupled systems, PhD thesis, Control Science and Dynamical Systems Center, University of Minnesota, Minneapolis.

Keviczky T, Borrelli F and Balas GJ 2004a Hierarchical design of decentralized receding horizon controllers for decoupled systems. In *Proc. 43rd IEEE Conf. on Decision and Control*, pp. 1592–1597.

Keviczky T, Borrelli F and Balas GJ 2004b A study on decentralized receding horizon control for decoupled systems. In *Proc. American Contr. Conf.*, pp. 4921–4926.

Keviczky T, Borrelli F and Balas GJ 2006a Decentralized receding horizon control for large scale dynamically decoupled systems. *Automatica* **42**(12), 2105–2115.

Keviczky T, Borrelli F and Balas GJ 2006b Decentralized receding horizon control for large scale dynamically decoupled systems. Technical Report 2006–01, University of Minnesota, Minneapolis, MN. Online: http://sciweb.lib.umn.edu/repository/Repository.phtml.

Keviczky T, Vanek B, Borrelli F and Balas GJ 2006c Hybrid decentralized receding horizon control of vehicle formations. In *Proc. American Contr. Conf.*, Minneapolis, MN, pp. 3358–3363.

Keviczky T, Borrelli F, Balas GJ, Fregene K and Godbole D 2007 Decentralized receding horizon control and coordination of autonomous vehicle formations. *IEEE Trans. Control Systems Technology in press.*

Magni L, Nicolao GD, Scattolini R and Allgöwer F 2003 Robust nonlinear model predictive control for nonlinear discrete-time systems. *Int. J. Robust Nonlinear Control* **13**(3–4), 229–246.

Mayne DQ, Rawlings JB, Rao CV and Scokaert POM 2000 Constrained model predictive control: Stability and optimality. *Automatica* **36**(6), 789–814.

Merris R 1994 Laplacian matrices of graphs: A survey. *Linear Algebra and Its Applications* **197/198**, 143–176.

Mitchell IM and Templeton JA 2005 A toolbox of Hamilton-Jacobi solvers for analysis of nondeterministic continuous and hybrid systems. In *Hybrid Systems: Computation and Control*, vol. 3414 of *Lecture Notes in Computer Science*, pp. 480–494. Springer Verlag, Berlin.

Mohar B 1991 The Laplacian spectrum of graphs. *Graph Theory, Combinatorics, and Applications* **2**, 871–898.

Online 2006 http://www.cds.caltech.edu/˜tamas/simulations.html.

Richards A and How J 2004a A decentralized algorithm for robust constrained model predictive control. *Proc. American Contr. Conf.*, pp. 4261–4266.

Richards A and How J 2004b Decentralized model predictive control of cooperating UAVs. In *Proc. 43rd IEEE Conf. on Decision and Control*, pp. 4286–4291.

Rotkowitz M and Lall S 2002 Decentralized control information structures preserved under feedback. In *Proc. 41st IEEE Conf. on Decision and Control*, pp. 561–575.

Rowe C and Maciejowski J 2003 Robust constrained receding horizon control of PWA systems with norm-bounded input uncertainty. In *Proc. American Contr. Conf.*, pp. 3949–3954, Denver, Colorado.

Safonov MG and Athans M 1977 Gain and phase margin for multiloop LQG regulators. *IEEE Trans. Automatic Control* **AC-22**(2), 173–179.

Sandell NR, Varaiya P, Athans M and Safonov M 1978 Survey of decentralized control methods for large scale systems. *IEEE Trans. Automatic Control* **AC-23**(2), 195–215.

Scherer C and Weiland S 2000 *Linear Matrix Inequalities in Control.* Version 3.0.

Schouwenaars T, How J and Feron E 2004 Receding horizon path planning with implicit safety guarantees. In *Proc. American Contr. Conf.*, pp. 5576–5581.

Scokaert POM and Mayne DQ 1998 Min-max feedback model predictive control for constrained linear systems. *IEEE Trans. Automatic Control* **43**(8), 1136–1142.

Stipanovic D, Inalhan G, Teo R and Tomlin C 2002 Decentralized overlapping control of a formation of unmanned aerial vehicles. In *Proc. 41st IEEE Conf. on Decision and Control*, pp. 2829–2835.

Vadigepalli R and Doyle FJ 2003 A distributed state estimation and control algorithm for plantwide processes. *IEEE Trans. Control Systems Technology* **11**(1), 119–127.

Wang S and Davison EJ 1973 On the stabilization of decentralized control systems. *IEEE Trans. Automatic Control* **18**(5), 473–478.

6

Task assignment for mobile agents

Brandon J. Moore and Kevin M. Passino

6.1 INTRODUCTION

Within the field of cooperative control there is special interest in problems where a group of mobile agents must act in concert to accomplish a common goal. Generating the individual agent trajectories and associated actions that accomplish this goal can be viewed as the assignment of each agent to a certain subset of spatially distributed tasks. Since the agents will generally not have the same initial position or capabilities, it follows that some assignments will result in better performance than others and thus we would like to determine the optimal one subject to the problem's constraints. Given the mobile nature of the agents, the performance measure used is often some function of the distance each agent travels or the times that various tasks are completed. In this chapter, we focus on a distributed method autonomous agents may use in order to solve a particular type of assignment problem despite issues that arise when inter-agent communications are subject to significant delays.

To date, research on these types of problems has focused on either optimization algorithms or stability analysis. In the former case, the application of mixed-integer linear programming (MILP) has proved very useful when applied to the optimization of agent trajectories (Richards and How 2002) or to coordinating the efficient interaction of multiple agents in scenarios with many sequential tasks and tight timing constraints (Howlett *et al.* 2003; Schumacher *et al.* 2003). Given the potential complexity of these problems, however, the solution time of these algorithms can become prohibitive. In some cases the optimality of the solution may degrade gracefully with the time of its computation. In other cases, particularly those involving the popular nonholonomic model known as the Dubins Car (Dubins 1957), the feasibility of the solution may be destroyed if it takes too long to compute it. By their nature, such algorithms are also highly centralized; this often means that a large amount of information must be passed over the agents' communication network (a potential problem if that network is subject to imperfections such as delays, noise, or changing topology). Some of these issues can be overcome by settling for sub-optimal solutions either through the use of heuristics (such as those developed by Walker

Cooperative Control of Distributed Multi-Agent Systems edited by Jeff S. Shamma

et al. (2003)) or by decomposing the larger problem into smaller ones that can each be solved optimally and then recombined to generate a complete solution (as in Schumacher *et al.* (2002)). Through the use of Monte Carlo simulations, statistics can be generated to check if these algorithms are 'good enough', but this may fail to identify pathological situations in which they perform poorly or even fail, particularly when unrealistic assumptions are made concerning communications.

When we start to consider communication imperfections, it often makes sense to forgo optimality of the assignment in favor of stability. That is, given asynchronous communications with delays and the resulting information imbalances between agents, it is no longer trivial to assume an algorithm will accomplish its goal. Stability can be defined in many ways. In the next paragraph we discuss this concept in terms of termination (i.e. guarantees that an algorithm will eventually produce a feasible solution), but there are equally important definitions more along the lines of traditional control theory. In (Gil *et al.* 2003), a group of agents must continually process a set of reoccurring tasks, so in this context it makes sense to show that their control law guarantees an upper bound on the longest time any particular task will be ignored. In (Finke *et al.* 2003), a load balancing approach is used to divide a number of tasks between agents. Here they prove that an equilibrium set (i.e. all agents balanced within a certain range) is exponentially stable in the large. In this scenario, if the agents were capable of balancing perfectly, then this approach would also result in an optimal solution; however, the discrete nature of the load means it can only be balanced to within a fairly wide margin of error. In this work, we have attempted to find a middle ground between the two approaches of optimality and stability; the algorithm we present cannot guarantee optimality due to the unavoidable effects of communication delays, but does seek to minimize the harm done to performance and does so in a way that allows us to analytically quantify a guaranteed performance level.

The problem we focus on in this chapter is a variant of the assignment problem from the field of combinatorial optimization. This problem formulation has received much attention in cooperative control studies because it models the need to generate a suitable mapping between two finite sets subject to certain constraints, and many algorithms of polynomial complexity exist to solve it (Bertsekas 1998; Bertsekas and Castañon 1991; Bertsekas and Tsitsiklis 1997a). In (Castañon and Wu 2003), the distributed sequential shortest augmenting path is modified in a manner that allows the agents to cooperatively compute an optimal solution despite information delays between agents and the dynamic arrival of additional tasks. The existing algorithms for the assignment problem have also been used heuristically to deal with tasks that are highly coupled (i.e., the completion of certain tasks must be preceded by the completion of another). In (Schumacher *et al.* 2002), multiple assignment problems with evolving subsets of the agents and tasks are iteratively solved, thus allowing their algorithm to produce a multiple task tour solution that simulations show to be close to optimal for their scenario. That work uses a centralized redundant approach in which each agent solves the same problem independently and then implements its own portion of the resulting assignment. It is therefore highly dependent on the agents starting from the same information in order to ensure they all arrive at the same solution. This becomes problematic when communication delays are introduced; because of changes in the problem that may occur during planning (i.e., vehicle movement), the final solution may no longer be optimal or even feasible. In this work we seek to incorporate these changes in problem data directly into a modification of the distributed auction algorithm of (Bertsekas and Castañon 1991) to determine an optimal solution to the standard assignment problem. To the best of our knowledge, this is the first time an assignment method has explicitly dealt with a performance measure of a time-varying nature.

This chapter is organized as follows. In Section 6.2 we present some background on the assignment problem and the asynchronous distributed auction algorithm of (Bertsekas and Castañon 1991) in order to establish some notation for the chapter. In Section 6.3 we define the specific problem to be solved and in Section 6.4 we develop a modification of the distributed auction to solve it. We determine the conditions under which that algorithm is guaranteed to terminate and perform a worst case analysis of the resulting solution. Section 6.5 contains simulation results demonstrating the practical performance of the modified algorithm. Conclusions appear in Section 6.6.

6.2 BACKGROUND

The standard *assignment problem* is one in which members of one set must be matched to members from another distinct set on a one-to-one basis. By convention (Bertsekas 1998), the former set is usually referred to as the set of *persons* and the latter as the set of *objects*. An *assignment* is a collection of pairings such that each person and each object appears in at most one pair. An assignment is considered full if it contains every member from the smaller of the two sets (persons or objects) and partial otherwise. For each possible pairing between a person and an object, there exit's some scalar value that represents the *benefit* gained if that pair is a member of an assignment. The sum of the benefits associated with each pair in an assignment is referred to as its collective benefit.

The goal of an assignment problem is to find a full assignment that is optimal in that it maximizes the collective benefit of the individual pairings (i.e., their sum). If the sizes of the person and object sets are the same (making it so that every person and object must participate in any full assignment), we have a special case referred to as the symmetric assignment problem. In this chapter, we consider an asymmetric assignment problem where the number of objects is greater than or equal to the number of persons, thus allowing for the possibility that some objects may be excluded from a full assignment. We give a brief overview of the assignment problem in this section; for a more thorough discussion, see (Bertsekas 1998) or (Bertsekas and Tsitsiklis 1997a).

6.2.1 Primal and dual problems

Let $\mathcal{A}$ be the collection of all pairs (i, j) denoting an allowable assignment between person i and object j and let a_{ij} be the scalar benefit associated with that pair. If object j is assigned to person i, then let a decision variable x_{ij} be equal to 1 (and equal to 0 otherwise). Since we wish to consider an asymmetric assignment with more objects than persons, we introduce another decision variable x_{sj} which is equal to 1 if object j is not assigned to any person (in which case we say it is assigned to a *supersource* instead). With this notation, the asymmetric assignment problem between m persons and n objects can be stated as the constrained optimization problem,

$$\text{maximize} \sum_{(i,j)\in\mathcal{A}} a_{ij}x_{ij} \tag{6.1}$$

subject to

$$\sum_{\{j|(i,j)\in\mathcal{A}\}} x_{ij} = 1, \forall i \in \{1, \ldots, m\} \tag{6.2}$$

$$\sum_{\{i|(i,j)\in\mathcal{A}\}} x_{ij} + x_{sj} = 1, \forall j \in \{1, \ldots, n\} \tag{6.3}$$

$$\sum_{j=1}^{n} x_{sj} = n - m, \tag{6.4}$$

$$x_{ij}, x_{sj} \in \{0, 1\} \tag{6.5}$$

Where (6.1) is the collective benefit and the constraints in (6.2)–(6.5) ensure that each person is assigned to one unique object. This formulation is known as the *primal problem*. Due to the linear nature of the problem, it is possible to form the *dual problem* (Bertsekas 1998) given by

$$\text{minimize} \sum_{i=1}^{m} \pi_i + \sum_{j=1}^{n} p_j - (n-m)\lambda$$

$$\text{subject to} \begin{array}{ll} \pi_i + p_j \geq a_{ij}, & \forall (i,j) \in \mathcal{A} \\ \lambda \leq p_j, & \forall j = \{1, \ldots, n\} \end{array}$$

where π_i, p_j, and λ, are Lagrange multipliers associated with each person i, each object j, and the supersource, respectively. For convenience, let $\pi = [\pi_1, \ldots, \pi_m]$ and $p = [p_1, \ldots, p_n]$. The structure of the two problems dictates that the decision variables x_{ij} are completely dependent upon π, p, and λ. Therefore, once we have a solution to the dual problem, we can easily reconstruct the solution to the primal problem as

$$x_{ij} = \arg \min_{w_{ij}\in\{0,1\}} (a_{ij} - \pi_i - p_j)w_{ij}$$

The above result and indeed the motivation for forming the dual problem come from the existence of the complementary slackness (CS) criteria (Bertsekas 1998). This criteria allows us to determine whether a given assignment is the optimal solution, and provides a guide for constructing such an assignment. Specifically, we use a relaxed criteria, ε*−complementary slackness* (ε−CS), that allows us to generate assignments that are sub-optimal but whose collective benefit is within a proscribed bound of the optimal solution. An assignment S is said to satisfy ε−CS for the asymmetric assignment problem if

$$\pi_i + p_j \geq a_{ij} - \varepsilon, \ \forall (i,j) \in \mathcal{A} \tag{6.6}$$

$$\pi_i + p_j = a_{ij}, \quad \forall (i,j) \in S \tag{6.7}$$

$$p_j \leq \min_{k \text{ assigned under } S} p_k, \ \forall j \text{ unassigned under } S \tag{6.8}$$

where (6.6) and (6.7) are often combined to yield

$$a_{ij} - p_j \geq \max_{k \in A(i)} \{a_{ik} - p_k\} - \varepsilon \quad \forall\ (i, j) \in S \tag{6.9}$$

with $A(i) = \{j|(i, j) \in \mathcal{A}\}$.

Theorem 6.2.1 *(Bertsekas 1998) If a feasible full assignment S along with a specific π and p satisfies ε–CS, then the collective benefit of S is within $m\varepsilon$ of the optimal solution to the asymmetric assignment problem.*

6.2.2 Auction algorithm

The auction algorithm (Bertsekas 1998; Bertsekas and Tsitsiklis 1997a) is a method of solving the assignment problem using elements from both the primal and dual formulations that is motivated by an economic interpretation of the ε–CS condition. The variables p and π are thought of as the *prices* and *profits* of the objects and persons, respectively. The ε–CS conditions are then interpreted as stating that if person i is assigned to object j, then his profit is equal to the benefit he receives from that object minus its price (6.7), and that object j provides person i more profit (within some tolerance ε) than any other object (6.9).

The auction algorithm starts with an empty assignment S (which trivially satisfies ε–CS) and iteratively modifies it by either adding new assignment pairs or replacing current ones while adjusting prices and profits in a manner that maintains ε–CS. Thus, the size of S is strictly non-decreasing as the algorithm progresses, and when a full assignment is reached it has the optimality guarantees of Theorem 6.2.1. The modification of the working assignment takes place by having unassigned persons bid against each other for objects (or objects for persons), thereby raising their prices (profits) much in the same way as a real auction. Bidding decisions are based on *values*, which are the differences between the benefits a bidder associates with objects (persons) and their prices (profits).

Although many variations of the auction algorithm exist, this chapter will be concerned exclusively with the forward auction algorithm in which persons bid for objects and prices are the only variable of concern (π and λ are not directly altered and can easily be calculated from other information). Under mild restrictions this algorithm can be used even in a distributed asynchronous manner to solve the asymmetric assignment problem in a finite period of time (Bertsekas and Castañon 1991). The algorithm consists of two main operations, bidding and assignment, which are described below.

Bidding: *Let U be the (nonempty) set of persons unassigned under a partial assignment S and $R \subseteq U$ be those persons with a pending bid. Let p^i_j represent person i's most recent knowledge about the true price of object j, p_j. To submit a bid, person $i \in U - R$ takes the following steps:*

1. Finds its preferred object j_i (the one that returns the best possible value v_i)

$$j_i = \arg\max_{j \in A(i)} \{a_{ij} - p^i_j\}$$

$$v_i = \max_{j \in A(i)} \{a_{ij} - p^i_j\}$$

as well as the value of the next best object w_i

$$w_i = \max_{\substack{k \in A(i) \\ k \neq j_i}} \{a_{ik} - p^i_k\}$$

(if j_i *is the only object in* $A(i)$*, let* $w_i = -\infty$*).*

2. *Calculates its bid for object* j_i *as*

$$b_{ij_i} = p^i_{j_i} + v_i - w_i + \varepsilon = a_{ij_i} - w_i + \varepsilon$$

and communicates this information to object j_i.

Assignment: *Let* $P(j)$ *be the set of persons with bids pending for object* j *since the last time it was reassigned (if ever). If* $P(j)$ *is not empty, then object* j *takes the following steps to resolve the bidding:*

1. *Selects the person* i_j *from* $P(j)$ *with the highest bid* $b_{i_j j}$

$$i_j = \arg \max_{i \in P(j)} b_{ij}$$

$$b_{i_j j} = \max_{i \in P(j)} b_{ij}$$

2. *If* $b_{i_j j} \geq p_j + \varepsilon$ *then it sets* $p_j := b_{i_j j}$*, updates the partial assignment S by replacing the current person assigned to it (if any) with person,* i_j*, and communicates this new information to all persons.*
3. *If* $b_{i_j j} < p_j + \varepsilon$ *then all bids for object* j *are cleared, no reassignment or price change is made, and it communicates this information to all persons in* $P(j)$.

From the description above, we can see that prices are strictly increasing and that once an object becomes assigned, it remains assigned for the duration of the algorithm. Thus, if the initial prices of the objects are identically zero, the ε–CS condition of (6.8) will be automatically satisfied because the price of any assigned object must be at least ε.

It should be noted that 'person i bids ... ' and 'object j assigns ... ' each refer to calculations done by a specific processor. Which processor is responsible for performing the calculations for a specific person or object at a given moment in time will depend on what makes the most sense considering the application. While there is a great deal of flexibility, the need to ensure the integrity of the algorithm places some limitations on the methods we may use. It is imperative that the previously mentioned restrictions are met, and, in a distributed processing environment, the reliability of the communication network and of the processors themselves may be of concern. Therefore, there must be protocols in place that ensure price information is (eventually) transmitted correctly to every processor. Also, if one processor fails, then there must be some form of redundancy built into the system to ensure that another processor will assume its responsibilities. However, adding such redundancy can create problems of its own. If two processors with different knowledge about the current price of an object simultaneously make two different assignments for that object, then there will be conflicting information in the system that must be resolved somehow. Clearly, it is preferable to avoid this hazard by implementing a method ensuring that one and only one processor may take action for a particular person or object at a time. That processor must also be guaranteed to have correct information concerning that person or object (i.e., a person's assignment or bidding status and an object's true price).

6.3 PROBLEM STATEMENT

In this section, we will consider a very specific case of the general task assignment problem in the context of cooperative control. We assume that there are m mobile agents of possibly different type that are initially distributed in the environment. There are n tasks ($n \geq m$) at fixed locations, also of possibly differing type. An agent is considered to have completed a task when it arrives there, where arrival can consist of more or less restrictive conditions than mere collocation (e.g., a particular orientation or earliest arrival time are common specifications). Agents are non-renewable resources in that they can perform one and only one task (e.g., a 'kamikaze' attack by an autonomous munition). Whether or not an agent has the capability to perform a specific task depends on both its type and the type of that task. Upon completion of a task, each agent receives a benefit that depends on both its type and the target's type and is decremented by a linear function of the time of completion. Optimization for this problem will mean maximizing the collective benefit received by the agents upon completion of m of the tasks, with zero benefit received from the $n - m$ unfinished tasks.

Given the one-to-one nature of the above problem, if the location of initial positions of the agents and tasks are known prior to deployment (i.e., the start of the scenario), then it clearly can be modeled as an asymmetric assignment problem (see Section 6.2) and solved accordingly. Consider, however, the case where the agents must calculate an assignment 'on-the-fly' because they lack complete knowledge of the situation prior to deployment. If communications between the agents are subject to substantial delays and/or the agents' available computing power is limited, it could easily take a nontrivial length of time to arrive at an assignment, during which the benefits each agent associates with each task have most likely changed.

6.3.1 Feasible and optimal vehicle trajectories

Because the benefits for the assignment problem above are based on the time an agent arrives at a task, in order to understand how those benefits change as the agents move, we must first consider the vehicle dynamics involved. If we define $s_i(t)$ as a state vector containing all the information needed to represent an agent i at a particular point in time (i.e., physical coordinates plus possibly other pertinent data such as orientation), then

$s_i(t), t \geq 0$ denotes the state space trajectory an agent follows as a function of time. A trajectory $s_i(t), t \geq 0$ is *feasible* so long as its components satisfy the vehicle dynamics of the agent, which we will usually state in the form of a system of differential equations and associated constraints.

We define the constant d_j as a vector associated with task j and representing the pertinent information describing its location and arrival criteria. If the agents' only concern is to reach the tasks (i.e. collocation), then there is no need for this arrival criteria. In general, however, we allow for its inclusion in order to provide flexibility in modeling more complex situations. For example, an agent might have to arrive at a task along a certain heading. Alternatively, the agent may be allowed to perform the task from any of a number of different points defined in relation to the nominal location of the task. Unless stated otherwise, we will assume that if agent i can ever reach task j in a finite length of time, then it can always do so (i.e., the set $\mathcal{A}$ of pairs (i, j) denoting agent i is capable of completing task j in the network flow model of Section 6.2 is time invariant). Under this assumption, for each agent i that can reach task j, there exists a time optimal trajectory from any point $s_i(t)$ to that task. Let us denote that trajectory by $\sigma[s_i(t), d_j]$ and its travel time by the metric $|\sigma[s_i(t), d_j]|$. Since the tasks are stationary, by the definition of optimality the travel time of the optimal trajectory to any task from a point $s_i(t + \Delta t)$, can be no less than the difference between the travel time of the optimal trajectory to the same task and the elapsed time interval $\Delta t > 0$. Using our notation,

$$|\sigma[s_i(t + \Delta t), d_j]| \geq |\sigma[s_i(t), d_j]| - \Delta t \quad \forall t \geq 0, \ \forall 0 \leq \Delta t \leq |\sigma[s_i(t), d_j]| \quad (6.10)$$

with equality if and only if $\sigma[s_i(t + \Delta t), d_j]$ is a subpath of $\sigma[s_i(t), d_j]$. When that is the case, we will say that agent i is *tracking* task j from t to $t + \Delta t$.

For this work we will require that for any feasible trajectory agent i might follow, the increase in the travel time of the optimal trajectory to a given task j can be upper bounded as

$$|\sigma[s_i(t + \Delta t), d_j]| \leq |\sigma[s_i(t), d_j]| + W + V\Delta t \qquad \forall t, \ \forall \Delta t \geq 0 \qquad (6.11)$$

where the constants $W \geq 0$ and $V \geq -1$ are identical for all i and j. An inequality of this form should be satisfied for many situations with realistic vehicle dynamics (i.e., anywhere the difference $|\sigma[s_i(t + \Delta t), d_j]| - |\sigma[s_i(t), d_j]|$ has a bounded derivative and only limited discontinuities). Consider the following examples:

Pivoting Robot: An agent in a planar environment that can either move forward along its current heading or pivot in place has a kinematic model given by

$$\begin{aligned} \dot{x} &= v\delta[u]\cos\theta \\ \dot{y} &= v\delta[u]\sin\theta \\ \dot{\theta} &= \omega_{max}u \end{aligned} \qquad \text{subject to:} \qquad \begin{aligned} 0 \leq\ & v \leq v_{max} \\ -1 \leq\ & u \leq 1 \end{aligned}$$

where v_{max} and ω_{max} are the maximum forward and radial velocities respectively, and $\delta[u]$ is the Kronecker delta function ($\delta[u]$ equals 1 if $u = 0$ and 0 otherwise). Optimal trajectories for this type of vehicle consist of a pivot (of up to half a revolution) to face the robot towards its target destination followed by forward travel at its maximum velocity until it gets there. Therefore, the length of the optimal trajectory satisfies an inequality in the form of (6.11) with $V = 1$ and $W = \frac{\omega_{max}}{\pi}$. □

Dubins Car: Uninhabited air vehicles (UAVs) are often modeled as agents in a planar environment that are forced to move at a constant forward velocity and possess limited steering ability. This can be represented by the kinematic model known as the Dubins Car (Dubins 1957; Sussman and Tang 1991) given by

$$\begin{aligned} \dot{x} &= v_c \cos\theta \\ \dot{y} &= v_c \sin\theta \quad \text{subject to:} \; -1 \leq u \leq 1 \\ \dot{\theta} &= \omega_{max} u \end{aligned} \tag{6.12}$$

where $v_c > 0$ is the fixed forward velocity and ω_{max} is the maximum radial velocity of which the vehicle is capable while traveling at v_c. Optimal trajectories for this type of vehicle consist of movement along paths made up of straight line segments or arcs of radius $R_{min} = \frac{v_c}{\omega_{max}}$. If the arrival criteria for the tasks prescribe a specific orientation, then the length of the optimal trajectory satisfies an inequality in the form of (6.11) with $V = 1$ and $W = \frac{(2+3\pi)R_{min}}{v_c}$. While somewhat secondary to our main focus, that derivation demonstrates the applicability of this work to an important class of vehicles and implies that the assumption of (6.11) is not unreasonable. □

We note that different approaches to determining the pair (V, W) may, in some cases, produce values that are not trivially different (i.e., one approach may produce a smaller V than another, but at the expense of a larger W). In these instances, we have the freedom to choose the pair that gives the best results for the application. For example, if we are concerned with large time intervals the $V\Delta t$ term will dominate (6.11), and we should select a pair with a smaller V if possible. If, on the other hand, we know the time interval is small, we may want to choose a pair with small W. Of course, we can also combine the inequalities of the different pairs provided the resulting piecewise linear bound is of some use.

6.3.2 Benefit functions

Now that we have a method to characterize how the movement of agents affects their ability to reach the tasks, we propose a motion-dependent formulation of the benefit received by agent i from task j,

$$a_{ij}(t) = C_{ij} - B(t + |\sigma[s_i(t), d_j]|) \tag{6.13}$$

where C_{ij} is a constant particular to the pairing (i, j) (i.e., a function of the types of both the agent and the task), B is constant across all such pairs and defines the relative importance of completing tasks quickly, and the quantity $t + |\sigma[s_i(t), d_j]|$ is the time of arrival of agent i if it tracks task j from time t onward. (A more general formulation replaces the constant B with values B_{ij} for each agent/task pair and is considered in Section 6.4.3.) We assume that our algorithm starts at $t = 0$, so this benefit function is equivalent to saying that each task has its nominal benefit C_{ij} discounted proportional to the time it takes for an agent to complete it. If the algorithm starts at some time other than $t = 0$, each benefit function $a_{ij}(t)$ can be scaled accordingly, but this is unnecessary since only the relative benefit between tasks matters when finding the optimal assignment. If we want to take into account that tasks have been waiting to be completed for varying amounts of time before $t = 0$ then we can incorporate relative shifts into the appropriate values of C_{ij}. Note that this seems to be the first time that solving an assignment problem with time-varying benefits has been addressed.

With the benefit function defined above, using the inequalities in (6.10) and (6.11) we can place the following bounds on a_{ij},

$$a_{ij}(t + \Delta t) \geq a_{ij}(t) - B(W + (V + 1)\Delta t) \quad \forall \quad t, \Delta t \geq 0 \tag{6.14}$$

$$a_{ij}(t + \Delta t) \leq a_{ij}(t) \quad \forall \quad t \geq 0, \quad \forall \quad 0 \leq \Delta t \leq \left|\sigma[s_i(t), d_j]\right| \tag{6.15}$$

where equality is reached in (6.15) if and only if agent i is tracking task j from time t to time $t + \Delta t$. The qualification in (6.15) stems from the fact that once an agent reaches a task, the benefit will start to decrease again if it does not complete that task. If the agent continues to track the task and is able to remain at the task location, $a_{ij}(t)$ will decrease at a rate of $B\Delta t$. If it cannot stop and wait, the agent must follow a loop back to the task, and so $a_{ij}(t)$ will drop by the length of that loop and then remains steady until the next time the agent reaches the task.

At this point let us denote t_a as the time a full assignment is reached and t_f^i as the time agent i completes its task. Since each agent will track its assigned task from t_a until t_f^i, it is clear from (6.15) that the collective benefit of the assignment at t_a is the same as what the agents will receive when they complete their tasks. Equally obvious from these two inequalities is that during the time the agents take to calculate and communicate in order to reach their eventual assignment, the collective benefit of that assignment is degrading (except for the unlikely case where every agent somehow happens to always track the task to which they will eventually be assigned). Given that fact, we are motivated to develop an algorithm that controls the motion of the agents during the time interval $[0, t_a]$ in a manner that seeks to reduce the degradation of the collective benefit. The modification of the traditional auction algorithm proposed in the next section attempts to do just this.

6.4 ASSIGNMENT ALGORITHM AND RESULTS

6.4.1 Assumptions

In this section we will consider two sources of delay: one which arises from limited computing power and another which comes from the communication network. For computational delays we assume that there exists a maximum bound D_{calc} on the time it takes to complete all the calculations associated with an iteration of the algorithm (i.e., the time from when an agent receives new information until the time it is ready to transmit the results based on that information). If desired, this delay could be described in terms that were proportional to the size of the problem (i.e., the number of tasks). When considering communication delays, the specific network topology and protocols used are unimportant so long as information transmitted by an agent is guaranteed to reach every other agent without error and within a known maximum delay D_{comm}. This delay could also be broken down into the sum of a value representing the maximum transmission delay and a term proportional to the amount of data being sent. Given the low computational complexity of the individual operations used in the auction algorithm and a desire to focus on situations involving significant communication delays, we assume from this point on that $D_{calc} \ll D_{comm}$ and just use $D = D_{calc} + D_{comm}$ as the maximum length of time it takes one agent to receive updated information from another.

We will also assume that every agent knows d_j or an acceptable approximation thereof for all the tasks prior to the start of the algorithm. In addition, it is important that the numbering scheme used to identify the targets has been agreed upon as well, or that there exists a 'natural' way of distinguishing targets (e.g., by their coordinates if they lie far enough apart to avoid confusion). While this last assumption is not trivial, it is not the emphasis of this work. Under these assumptions, we need only concern ourselves with information directly related to the operation of the auction algorithm (i.e., bids, prices, and assignment updates). For prices in particular, the maximum information delay D and the fact that prices never decrease in a forward auction allow us to put the following bounds on agent i's knowledge of the price of task j,

$$p^i_j(t) \leq p_j(t) \tag{6.16}$$

$$p^i_j(t) \geq p_j(t - D) \tag{6.17}$$

with the implication that if $p_j(t) = p_j(t - D)$, then each agent is guaranteed to have perfect knowledge of task j's price.

6.4.2 Motion control for a distributed auction

In this section we propose a partially asynchronous algorithm that can be used to solve an assignment problem with motion-dependent benefits. This algorithm is a modification of the distributed asynchronous auction presented in (Bertsekas and Castañon 1991) that we summarized in Section 6.2:

- Starting with an empty partial assignment S we add or replace agent/task pairs (i, j) based on competing bids from the individual agents that are communicated across the network. This process terminates when every agent has a task assignment.
- Each unassigned agent calculates its own preferred task and transmits its associated bid using a procedure identical to that of Section 6.2 but with the exception that they will have to recalculate the benefits $a_{ij}(t)$ each time. We require that each agent performs this process immediately upon learning that it is eligible to bid (i.e., after becoming unassigned, having its last bid rejected, or learning about an assignment update that makes its pending bid obsolete).
- Some agents will be designated to perform the assignment calculations for the tasks, with one and only one agent responsible for each task at a time. We require that each agent performs this task immediately upon receiving a bid, and that it immediately transmits the results (i.e., new price and assignment and/or bid rejection messages).
- The motion of the agents is controlled by two simple rules throughout the duration of the algorithm. If an agent/task pair (i, j) belongs to the partial assignment S, then agent i tracks task j. If an unassigned agent i has a bid pending for task j, then agent i also tracks task j.

- In order to guarantee termination of the algorithm, we will require that the bidding increment used exceeds a certain threshold specified in terms of the problem's parameters.

6.4.3 Assignment algorithm termination

Given this description of the algorithm, we will first show that it maintains an arbitrary $\varepsilon-$CS condition for the partial assignment S, and then prove that the algorithm terminates in finite time when the bidding increment meets the stated criteria. All the proofs presented in this section are based on those found in (Bertsekas and Castañon 1991) and (Bertsekas 1998) but have been augmented to handle the time-varying benefit function in (6.13).

Lemma 6.4.1 *For an assignment S that satisfies $\varepsilon-$CS at time t, if every agent in S follows the time optimal trajectory to its assigned target then S will satisfy $\varepsilon-$CS at time $t+\Delta t \geq t$ when S is constant from t to $t+\Delta t$, benefits are defined as in (6.13), and no agent reaches its assignment before $t+\Delta t$.*

Proof. If a partial assignment S satisfies $\varepsilon-$CS (see Section 6.2) at time t, then for each assignment pair we have

$$a_{ij}(t) - p_j(t) \geq \max_{k \in A(i)} \{a_{ik}(t) - p_k(t)\} - \varepsilon \quad \forall\ (i, j) \in S$$

Consider the quantity $\max_{k \in A(i)} \{a_{ik}(t+\Delta t) - p_k(t+\Delta t)\} - \varepsilon$,

$$\begin{aligned}
\max_{k \in A(i)} \{a_{ik}(t+\Delta t) - p_k(t+\Delta t)\} - \varepsilon &\leq \max_{k \in A(i)} \{a_{ik}(t) - p_k(t+\Delta t)\} - \varepsilon \\
&\leq \max_{k \in A(i)} \{a_{ik}(t) - p_k(t)\} - \varepsilon \\
&\leq a_{ij}(t) - p_j(t) \\
&= a_{ij}(t+\Delta t) - p_j(t+\Delta t)
\end{aligned}$$

where the first line makes use of the fact that $a_{ij}(t+\Delta t)$ is upper bounded by $a_{ij}(t)$, the second uses the fact that prices are strictly increasing (i.e., $p_k(t) \leq p_k(t+\Delta t)$), the third uses the definition of the $\varepsilon-$CS condition, and the last line follows from the stipulation that every agent in S tracks its assigned task from time t to time $t+\Delta t$ and the fact that the price of a task cannot change unless the assignment does. The resulting inequality shows that $\varepsilon-$CS still holds at time $t+\Delta t$ for every agent-task pair continuously in the assignment from t to $t+\Delta t$. □

Lemma 6.4.2 *Providing that an agent tracks the minimum-time path to the task it is currently bidding on, the proposed algorithm maintains $\varepsilon-$CS for a partial assignment S for all $t \geq 0$ when benefits are defined as in (6.13).*

Proof. Consider agent i which has just submitted a bid b_{ij_i} for his preferred task j at time t. Since agent i is working with possibly outdated price information, its preferred target satisfies

$$a_{ij}(t) - p^i_j(t) \geq \max_{k \in A(i)} \{a_{ik}(t) - p^i_k(t)\} - \varepsilon \quad \forall \ (i, j) \in A(i)$$

where b_{ij_i} is calculated to satisfy the above equation if $p^i_j(t) = b_{ij_i}$. Since agent i tracks task j while its bid is pending, then by logic analogous to that used in Lemma 6.4.1 and the fact that the auctioneer for task j knows its true price, the ε–CS condition is guaranteed for pairs entering the partial assignment and thus the overall ε–CS of S is maintained. □

Theorem 6.4.3 *For a feasible asymmetric assignment problem where there exists at least one full assignment S such that every pair (i, j) in S agent i is capable of completing task j, if the benefits $a_{ij}(t)$ are defined as in (6.13) and information transmission delays are bounded by D, the proposed algorithm terminates in finite time t_a provided that $\varepsilon > 2DB(m-1)(V+1)$ and no agent reaches its preferred task before t_a. Furthermore, the final assignment has a corresponding benefit that is within $m\varepsilon$ of the optimal assignment given the configuration of the agents at time t_a.*

Proof. We first assume that the algorithm does not terminate in a finite amount of time t_a and then prove the theorem by contradiction. We do this in a manner similar to the termination proofs for the standard auction algorithm that appear in (Bertsekas and Tsitsiklis 1997a) and (Bertsekas and Castañon 1991). To ensure that enough tasks receive bids to create a full assignment, it was sufficient in (Bertsekas and Tsitsiklis 1997a) and (Bertsekas and Castañon 1991) to show that the value of each agent's preferred task eventually fell below the benefit of some unassigned task. We must, however, show that the value of the agent's preferred task falls fast enough in order to overcome the effects of potentially decreasing benefits.

If the algorithm does not terminate, it must be the case that at least one (but not necessarily the same) agent is unassigned at all times. Therefore, it is also the case that some agents submit (and thus some tasks receive) an infinite number of bids. This is ensured by the stipulation that unassigned agents must bid immediately if eligible to do so and are guaranteed to receive confirmation or rejection within at most $2D$ time units, after which time they must bid again unless they were accepted into the assignment. If accepted into the assignment, they either displace another agent (who must then bid) or add another agent/task pair to the assignment (which can only happen a finite number of times if the algorithm does not terminate since the size of assignment is strictly non-decreasing).

Let us denote the subset of agents that bid indefinitely as I^∞ and the subset of tasks that receive an infinite number of bids as J^∞. There must then be a sufficiently large time index t_∞ such that for all time $t > t_\infty$ bidding is confined to I^∞ and J^∞.

Consider an agent $i \in I^\infty$ and the set of tasks $A(i)$ to which it may be assigned. Let t_0 be the time a task from $A(i)$ gets reassigned. If no other tasks from $A(i)$ get reassigned before $t_0 + D$, then agent i will have perfect knowledge of the prices for those tasks. With such knowledge, agent i can make a bid for some task in $A(i)$ that will be accepted no later $t_0 + 2D$. Therefore, at least one task from $A(i)$ must receive a successful bid every $2D$ time units.

Given that information, examine the maximum value agent i associates with tasks that are in both $A(i)$ and J^{∞},

$$v_i(t) = \max_{j \in A(i) \bigcap J^{\infty}} \{a_{ij}(t) - p_j(t)\} \tag{6.18}$$

and note that after t_{∞}, the price of the task that achieves this value decreases by no less than the bidding increment ε every $2D$ time units. Since the upper bound on the benefits in (6.15) tells us that $a_{ij}(t)$ cannot increase, and the algorithm dictates that prices do not decrease, this means that v_i either decreases by at least ε every $2D$ time units or by some lesser amount if the value of the second best task is within ε of the value of the best task. The latter can happen at most $N \leq m - 1$ times (the number of tasks in $A(i) \bigcap J^{\infty}$) before the quantity $a_{ij}(t) - p_j(t)$ for each task j must be at least ε less than the original value of $v_i(t)$, so we can over bound $v_i(t)$ with

$$v_i(t) \leq \max_{j \in A(i) \bigcap J^{\infty}} \{a_{ij}(t_{\infty}) - p_j(t_{\infty})\} - \varepsilon \left\lfloor \frac{t - t_{\infty}}{2DN} \right\rfloor$$

where $\lfloor \cdot \rfloor$ is the floor operator (which takes the value of the largest integer less than its operand). Since $-x\lfloor \frac{t}{y} \rfloor$ is a descending staircase function bounded from above by $x - \frac{x}{y}t$, we have

$$v_i(t) \leq \max_{j \in A(i) \bigcap J^{\infty}} \{a_{ij}(t_{\infty}) - p_j(t_{\infty})\} + \varepsilon - \frac{\varepsilon}{2DN}(t - t_{\infty})$$

Substituting $\varepsilon = 2DB(m-1)(V+1)\delta$ where $\delta > 1$ (so that ε satisfies the assumed bound) and noting that N can be no greater than $m - 1$ or else at least as many tasks as there are agents would receive an infinite number of bids (in which case the algorithm would have terminated),

$$\begin{aligned} v_i(t) \leq & \max_{j \in A(i) \bigcap J^{\infty}} \{a_{ij}(t_{\infty}) - p_j(t_{\infty})\} + 2DB(m-1)(V+1)\delta \\ & - \frac{2DB(m-1)(V+1)\delta}{2D(m-1)}(t - t_{\infty}) \end{aligned}$$

Cancelling like terms and collecting the constants on the right gives us

$$\begin{aligned} v_i(t) \leq & -\delta B(V+1)(t - t_{\infty}) + \max_{j \in A(i) \bigcap J^{\infty}} \{a_{ij}(t_{\infty}) - p_j(t_{\infty})\} \\ & + 2DB(m-1)(V+1)\delta \end{aligned} \tag{6.19}$$

which tells us that the value of the best object from the set $A(i) \bigcap J^{\infty}$ can be bounded by a decreasing affine function of time after t_{∞}.

Now, since the prices of tasks $j \notin J^{\infty}$ stop changing after time t_{∞}, using the lower bound for the change in a benefit from (6.14), the value $\overline{v}_i$ agent i associates with the best of those tasks is lower bounded by

$$\overline{v}_i(t) = \max_{j \notin J^{\infty}} \{a_{ij}(t) - p^i_j(t)\}$$

$$\geq \max_j\{a_{ij}(t_\infty) - B(W + (V+1)(t - t_\infty)) - p_j(t_\infty)\}$$

$$= -B(V+1)(t - t_\infty) + \left[\max_j\{a_{ij}(t_\infty) - p_j(t_\infty)\} - BW\right] \tag{6.20}$$

which is also a decreasing affine function of time after t_0. Since the slope of the line described in (6.19) is less than the slope of the one described in (6.20), some task $j \notin J^\infty$ must eventually become the best value for each agent $i \in I^\infty$ at some time $t > t_\infty$. Since these tasks never receive a bid after t_∞ it must be the case that none of these tasks are in $A(i)$ for any agent in I^∞, implying that $A(i) \subset J^\infty \ \forall\, i \in I^\infty$ and $\bigcup_{i \in I^\infty} A(i) = J^\infty$. In a forward auction, once tasks are assigned they remain assigned, so after a finite length of time, every task in J^∞ will be assigned to some agent from I^∞. Since there will still be some agent from I^∞ bidding, there must be more agents in I^∞ than tasks in J^∞, contradicting the assumption that a feasible solution exists. Since the algorithm terminates, the final assignment has a collective benefit within $m\varepsilon$ of the maximum possible (with respect to the configuration of the agents at the time of termination) given the results of Lemmas 6.4.1 and 6.4.2 and Theorem 6.2.1. □

Here we note a few important things concerning the previous theorem. First, it should be clear that in the more general formulation with values B_{ij} for each agent-task pair instead of a single common value B, the validity of the proof is maintained by simply replacing B with $\max_{i,j} B_{ij}$. Secondly, the assumption that no agent reaches a task it is tracking prior to the termination of the algorithm is virtually impossible to guarantee for cases in which the communication delay between agents is large in relation to their speed. Thirdly, the value required for the bidding increment ε tends to be extremely large, and in practice (see Section 6.5) forces the auction to conclude quickly and with poor results (particularly for a large number of agents since the sub-optimality bound $m\varepsilon$ increases quadratically with m). The reason ε must be so large is that Theorem 6.4.3 requires an analysis of the worst case scenario. In this scenario, the agents are all bidding (and moving towards) a group of $m - 1$ tasks that lie directly in front of them while moving directly away from the remaining tasks. The benefits of the first group only decrease when their prices rise, while the benefits of the latter continually decrease as the agents move away from them at the fastest possible rate.

For these reasons we would like to guarantee the termination of algorithm under less restrictive conditions. The inequalities of (6.14) and (6.15) were based on the assumption that an agent's trajectory could lead anywhere so long as it was feasible. Considering instead a case where the agent's movement is restricted to a subset of the environment that includes all the tasks. Then for most situations the length of the optimal trajectory from an agent to a task should satisfy

$$0 \leq \left|\sigma[s_i(t), d_j]\right| \leq X \qquad \forall\, t > 0,\ \forall\, i, j \tag{6.21}$$

for some positive scalar X. Thus the associated benefit of that pair satisfies

$$C_{ij} - BX - Bt \leq a_{ij}(t) \leq C_{ij} - Bt \qquad \forall\, t > 0,\ \forall\, i, j \tag{6.22}$$

Theorem 6.4.4 *If the motion of the agents is restricted in such a way that the inequality in (6.22) holds, then the auction algorithm terminates for any* $\varepsilon > 0$.

Proof. Without loss of generality, assume that no tasks are completed before a full assignment is reached. If they are handled correctly (see the next section), then an early task completion simply changes the assignment problem to another with one less agent and task.

In Theorem 6.4.3 we have already established that the price of every task in a partial assignment decreases by at least ε every $2D(m-1)$ time units. Then for some agent i, the value it associates with its preferred task from the current partial assignment $S(t)$

$$v_i(t) = \max_{j \in S(t)} \{a_{ij}(t) - p_j(t)\}$$

is upper bounded by

$$v_i(t) \leq \max_{j \in S(t)} \{C_{ij} - Bt - p_j(t)\} \leq -Bt - \min_{j \in S(t)} p_j(t) + \max_j C_{ij}$$

whereas the value $\overline{v}_i(t)$ that agent i associates with an arbitrary task not in $S(t)$ can be bounded from below by

$$\overline{v}_i(t) \geq -Bt + \min_j C_{ij} - BX$$

Since the price term $\min_{j \in S(t)} p_j(t) \to \infty$ as $t \to \infty$ if a full assignment is not reached, it is clear that after some finite time period, $v_i(t)$ will fall below $\overline{v}_i(t)$ and some task will be added to the assignment. This process must repeat until the assignment is full, therefore causing the algorithm to terminate in finite time. □

Note that the condition stated in (6.21) can be satisfied by a number of simple motion control schemes, including those laid out in Section 6.4.2. In this case, however, we lose the optimality bounds guaranteed from maintaining an ε−CS condition since Lemmas 6.4.1 and 6.4.2 do not hold when an agent can reach a task prior to termination. As we will see in Section 6.5, the choice of the bidding increment ε will have a large effect on the practical performance of the algorithm as it will determine the trade-off between the advantage of acting quickly (to prevent too much degradation in collective benefit) and the advantage of optimizing the assignment with respect to the benefits at the time of termination.

6.4.4 Optimality bounds

In this section we derive worst case bounds for the collective benefit received from the assignment reached at time t_a. Since we know that the collective benefit of any assignment will degrade as a function of the time it takes to achieve a full assignment, we start by considering a worst case bound on t_a. For the algorithm outlined in Section 6.4.2, we can use the intersection of the lines described in (6.19) and (6.20) as a guide for determining this value. Consider the mth task to enter the assignment and let t_a be the time that one of the agents bids for it (since all agents will be tracking their final assignment from that time forward). The value $\overline{v}(t)$ of the final task to any agent is lower bounded by

$$\overline{v}(t) \geq \min_{i,j} a_{ij}(0) - B(W + (V+1)t)$$

and the value any of the agents places on any object already in the assignment must be upper bounded by

$$v(t) \leq \max_{i,j} a_{ij}(0) - \delta B(V+1)t + \varepsilon$$

with t_a given as the intersection of the two lines defined by the right-hand sides of the previous two equations because after that point, some unassigned task must be the preferred task for all agents (one of which must enter a bid). For the algorithm with motion control, the termination time t_a therefore can be bounded from above as

$$t_a \leq 2D(m-1)\frac{\max a_{ij}(0) - \min a_{ij}(0) + BW + \varepsilon}{\varepsilon - 2DB(V+1)(m-1)} \tag{6.23}$$

For comparison, we state a similar bound for the termination of a distributed auction with fixed benefits (the benefits 'frozen' at their initial values) adapted from the results of (Bertsekas n.d.) and (Bertsekas and Tsitsiklis 1997b). In this case the termination time is bounded from above by

$$t_a \leq 2D(m-1)\left(\frac{\max a_{ij}(0) - \min a_{ij}(0)}{\varepsilon} + 1\right) \tag{6.24}$$

where it is apparent that the worst case termination time for an auction performed with static benefits is always better than that of one in which the current values of benefits are used (given the same ε). That said, a worst case collective benefit based solely on termination time and the related possible change in benefit given by (6.14) will always favor a fixed-benefit distributed auction (Bertsekas and Castañon 1991). We can, however, take another approach to give us an interesting result for the algorithm with motion control.

Let j_i denote the final task assignment of agent i and let $i^\star$ denote the agent who made the final bid and $j^\star$ the task receiving that bid. We are interested in finding a worst case lower bound for the collective benefit of that assignment, i.e.,

$$\sum_{i=1}^{m} a_{ij_i}(t_a) = a_{i^\star j^\star}(t_a) + \sum_{i=1, i\neq i^\star}^{m} a_{ij_i}(t_a) \tag{6.25}$$

where we have separated the benefit of the last assignment for the purpose of analysis. The lower bound for the benefit of this pair is given by the inequalities in (6.15) and (6.14) as

$$a_{i^\star j^\star}(t_a) \geq a_{i^\star j^\star}(0) - BW - B(V+1)t_a \tag{6.26}$$

Before returning to the other half of (6.25) we note two facts concerning the task prices at the moment of termination. First, the price of $j^\star$ is still equal to zero (it has received a bid, but that does not become its real price until the auctioneer responsible for $j^\star$ makes the assignment). Second, since we are guaranteed to have a price increase of at least ε every $2D$ time units, the sum of the prices of the other tasks must satisfy

$$\sum_{i=1, i\neq i^\star}^{m} p_{j_i}(t_a) \geq \varepsilon\left\lceil \frac{t_a}{2D} \right\rceil \geq \varepsilon\frac{t_a}{2D} \tag{6.27}$$

where $\lceil \cdot \rceil$ is the ceiling operator. Now consider the benefit terms under the sum on the right-hand side of (6.25). Since the algorithm maintains an ε-CS condition for all assignments, we know that for each $i \neq i^\star$ the benefit of an agent's assigned task and the price of that task satisfy the following inequality

$$\begin{aligned} a_{ij_i}(t_a) - p_{j_i}(t_a) &\geq \max_k \{a_{ik}(t_a) - p_k(t_a)\} - \varepsilon \\ &\geq a_{ij^\star}(t_a) - \varepsilon \\ &\geq a_{ij^\star}(0) - BW - B(V+1)t_a - \varepsilon \end{aligned} \tag{6.28}$$

where we have used the fact that the $p_{j^\star}(t_a) = 0$ and the lower bound on the possible change in benefit. Returning to the sum on the right-hand side of (6.25)

$$\begin{aligned} \sum_{i=1, i\neq i^\star}^{m} a_{ij_i}(t_a) &= \sum_{i=1, i\neq i^\star}^{m} \Big(a_{ij_i}(t_a) - p_{j_i}(t_a) + p_{j_i}(t_a)\Big) \\ &= \sum_{i=1, i\neq i^\star}^{m} \Big(a_{ij_i}(t_a) - p_{j_i}(t_a)\Big) + \sum_{i=1, i\neq i^\star}^{m} p_{j_i}(t_a) \\ &\geq \sum_{i=1, i\neq i^\star}^{m} \Big(a_{ij_i}(0) - [BW + B(V+1)t_a + \varepsilon]\Big) \\ &\quad + \sum_{i=1, i\neq i^\star}^{m} p_{j_i}(t_a) \end{aligned} \tag{6.29}$$

which allows us to start to determine a bound on the collective benefit obtained through the algorithm

$$\begin{aligned} \sum_{i=1}^{m} a_{ij_i}(t_a) \geq\ & a_{i^\star j^\star}(0) - BW - B(V+1)t_a + \sum_{i=1, i\neq i^\star}^{m} p_{j_i}(t_a) \\ & + \sum_{i=1, i\neq i^\star}^{m} \Big(a_{ij_i}(0) - [BW + B(V+1)t_a + \varepsilon]\Big) \end{aligned}$$

For convenience we let $\underline{a} = \min_{i,j} a_{ij}(0)$ and $\overline{a} = \max_{i,j} a_{ij}(0)$, and let A^* denote the optimal collective benefit at time $t = 0$. Taking the minimum value for the benefit terms, the previous equation becomes

$$\begin{aligned} \sum_{i=1}^{m} a_{ij_i}(t_a) \geq\ & m\underline{a} - (m-1)\varepsilon - mBW - B(V+1)t_a \\ & - B(V+1)(m-1)t_a + \sum_{i=1, i\neq i^\star}^{m} p_{j_i}(t_a) \end{aligned}$$

Using the lower bound for the sum of prices in (6.27) and the fact that $A^* \leq m\overline{a}$

$$\sum_{i=1}^{m} a_{ij_i}(t_a) \geq A^* - m(\overline{a} - \underline{a}) - (m-1)\varepsilon - mBW$$
$$-B(V+1)t_a - B(V+1)(m-1)t_a + \varepsilon\frac{t_a}{2D}$$

and rewriting ε in the last term as the product $\delta 2DB(V+1)(m-1)$, where $\delta > 1$ if the termination criteria of Theorem 6.4.3 is met, and $0 < \delta \leq 1$ otherwise.

$$\begin{aligned}\sum_{i=1}^{m} a_{ij_i}(t_a) \geq\; & A^* - m(\overline{a} - \underline{a}) - (m-1)\varepsilon - mBW - B(V+1)t_a \\ & -B(V+1)(m-1)t_a + \delta B(V+1)(m-1)t_a \\ =\; & A^* - m(\overline{a} - \underline{a}) - (m-1)\varepsilon - mBW \\ & -[1-(\delta-1)(m-1)]B(V+1)t_a\end{aligned} \quad (6.30)$$

The bound in (6.30) is linear in terms of the termination time t_a which allows us to easily find its minimum on the interval $[0, \max\{t_a\}]$ by substituting the value for $\max\{t_a\}$ from (6.23) if $(\delta - 1)(m-1) < 1$ or simply taking the value of the first four terms if otherwise. Since the bound provided in the latter case is extremely poor, we note that since the final assignment arrived at by the algorithm is guaranteed to be within $m\varepsilon$ of the optimal solution at t_a combined with the maximum possible degradation of the optimal solution given the bounds on the change in benefits, we have a second lower bound given by

$$\sum_{i=1}^{m} a_{ij_i}(t_a) \geq A^* - m(BW + B(V+1)t_a) - m\varepsilon \quad (6.31)$$

then the worst case benefit in the case where $(\delta - 1)(m-1) \geq 1$ is given by the value of (6.30) and (6.31) at their intersection. Note that this value can be optimized by varying ε but, as simulations will show, the practical performance of the algorithm in most cases is significantly better than its worst case performance, so it will make little sense to optimize in this fashion.

As an example, we compare the worst case performance of the algorithm and a fixed-benefit distributed auction for the Dubins car model. For a generic situation involving this model, the best worst case bound for the latter algorithm is achieved by having the agents circle (loiter) until the auction is complete and is given by

$$\sum_{i=1}^{m} a_{ij_i}(t_a) \geq A^* - mB\left(\frac{2\pi R_{min}}{v_c} + t_a\right) - m\varepsilon \quad (6.32)$$

For a given set of problem parameters ($m = 10$, $\overline{a} - \underline{a} = 100$, $B = 1$, $R_{min}/v_c = 0.1$) and with δ set to $1 + \frac{1}{m-1}$ (see preceding discussion) we plot the worst case bound as

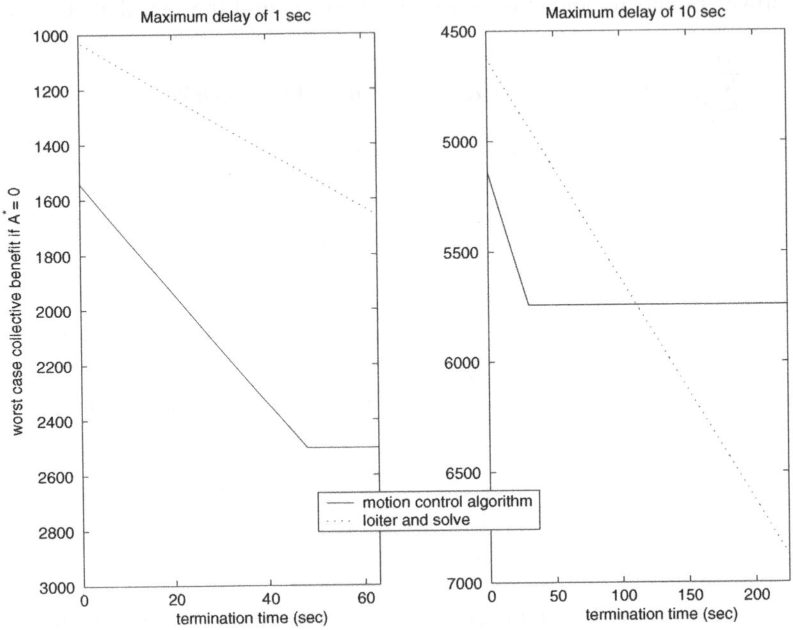

Figure 6.1 Worst case comparison between algorithms.

a function of termination time for both the algorithm with motion control and the fixed-benefit distributed auction over the interval $t = 0$ to $t = 2D(m-1)\frac{\overline{a}-\underline{a}}{\varepsilon}$ (the upper bound of the termination time of the fixed-benefit distributed auction). The left plot in Figure 6.1, for a maximum delay of $D = 1$, clearly shows that the worst case performance of the fixed-benefit distributed auction (given by the minimum of the dashed line) is better than that of the algorithm with motion control. The right plot shows the same case for a maximum delay of $D = 10$; the worst case bound of both algorithms has decreased, but that of the fixed-benefit distributed auction has now fallen below that of the one with motion control.

6.4.5 Early task completion

The preceding analysis was all based on the assumption in Theorem 6.4.3 that no agent reaches a task prior to the termination of the algorithm. Since this is difficult (if not impossible) to guarantee in many situations, we are curious about how much the maximum possible benefit can degrade if we let agents complete a task prior to the termination of the algorithm. Intuition tells us that for cases where the benefits are highly dependent on the time of task completion (i.e., B large relative to $\max_{i,j} C_{ij} - \min_{i,j} C_{ij}$) there is more to gain from completing a task early than waiting for the algorithm to arrive at a full assignment. For a potentially large class of vehicle models, this turns out to be the case.

Theorem 6.4.5 *Consider the configuration of agents and tasks at a specific instant in time. If*

1. *The vectors describing the position of the agents (s_i, $i \in \{1, \ldots, m\}$) and those describing the position of the tasks (d_j, $j = \{1, \ldots, n\}$) all belong to the same state space M,*
2. *The optimal trajectory distance function $|\sigma| : M \times M \to \mathbb{R}_{\geq 0}$ satisfies $\left|\sigma[s_i, d_j]\right| = \left|\sigma[s_{i'}, d_{j'}]\right|$ whenever $s_i = s_{i'}$ and $d_j = d_{j'}$, and*
3. *There exists an agent $i^\star$ and a task $j^\star$ that are collocated (i.e., $s_{i^\star} = d_{j^\star}$),*

then the minimum possible total travel time of a one-to-one assignment between agents and tasks is achieved by an assignment containing the pair $(i^\star, j^\star)$.

Proof. Let A be the minimum possible total travel time for a configuration of agents and tasks where there exists agent $i^\star$ and task $j^\star$ such that $s_{i^\star} = d_{j^\star}$. Let S be an assignment that achieves that minimum. Likewise, let $A^\star$ be the minimum possible total travel time for the same configuration and $S^\star$ an assignment that achieves it under the constraint that $(i^\star, j^\star) \in S^\star$. Since S is a less restricted assignment than $S^\star$, it follows that $A \leq A^\star$.

Assume $(i^\star, j^\star) \notin S$ (if $(i^\star, j^\star) \in S$, the hypothesis is trivial). Let i' be the agent assigned to $j^\star$ and j' the task assigned to $i^\star$ under S. For convenience, define the subset of S that contains these assignment pairs as $S' = \{(i^\star, j'), (i', j^\star)\}$. Now, using the optimal trajectory distance function we can state the value A as

$$
\begin{aligned}
A &= \sum_{(i,j)\in S} \left|\sigma[s_i, d_j]\right| \\
&= \left|\sigma[s_{i'}, d_{j^\star}]\right| + \left|\sigma[s_{i^\star}, d_{j'}]\right| + \sum_{(i,j)\in S-S'} \left|\sigma[s_i, d_j]\right| \\
&= \left|\sigma[s_{i'}, d_{j^\star}]\right| + \left|\sigma[d_{j^\star}, d_{j'}]\right| + \sum_{(i,j)\in S-S'} \left|\sigma[s_i, d_j]\right| \\
&\geq \left|\sigma[s_{i'}, d_{j'}]\right| + \sum_{(i,j)\in S-S'} \left|\sigma[s_i, d_j]\right| \\
&= \left|\sigma[s_{i^\star}, d_{j^\star}]\right| + \left|\sigma[s_{i'}, d_{j'}]\right| + \sum_{(i,j)\in S-S'} \left|\sigma[s_i, d_j]\right| \\
&\geq A^\star
\end{aligned}
$$

where we use the assumption that $s_{i^\star} = d_{j^\star}$ and the fact that the optimal trajectory distance function must satisfy both the triangle inequality and $|\sigma[x, x]| = 0$. The final inequality follows from the fact that the prior sum is the total travel time of an assignment that includes the pair $(s_{i^\star}, d_{j^\star})$, which, by definition, is lower bounded by $A^\star$. Thus $A \leq A^\star \leq A \Rightarrow A^\star = A$, proving that the constrained assignment achieves the optimal value. □

The above theorem demonstrates that letting an agent complete a task early cannot degrade the maximum achievable benefit in problems where $C_{ij} = 0 \; \forall \; i, j$ (or when $C_{ij} = C_{kj} \; \forall \; i, j, k$ in symmetric assignment problems) provided that agents and tasks are defined as points in the same space and the optimal trajectory distance function has the stated properties. Note that if the speed and turning radius of each agent are identical, the

Dubins car model satisfies these criteria. The pivoting robot model, on the other hand, does not meet the first assumption of Theorem 6.4.5 except in the case where $\omega_{max} = \infty$.

Note that in the more general situation where the fixed portion of the benefit terms C_{ij} varies widely among agents, the early completion of a task $j^{\star}$ by agent $i^{\star}$ has the potential to be much more detrimental to the maximum possible benefit. Also, if agents are allowed to complete tasks early, it is important that no other agents waste their resources on the same task. Unless the agents have the ability to sense the status of a task before committing themselves, they must have a mechanism to avoid possible duplication. One manner of accomplishing this is to have auctioneer responsibility for a task fall on the last agent to which that task was assigned (with responsibility for an unassigned task remaining with the agent at which it started).

6.5 SIMULATIONS

In this section we present results from simulations using the modified auction algorithm from Section 6.4 and agents with vehicle dynamics described by a Dubins car model. We do not attempt to characterize the algorithm's behavior in terms of all the salient parameters since there are a large number of them, many of which are interdependent (i.e., vehicle dynamics, communication methodology, number of agent/tasks, initial configurations, benefit range, weighting of benefit terms, etc.). Instead, we have selected a few scenarios that best illustrate the more important aspects of the problem. All simulations with the exception of the last one used 10 agents, since simulation time quickly became prohibitive with more, and 10 tasks since assignment problems are generally more difficult to solve when agents do not have the option of ignoring some tasks. In the cases where we compare the algorithm of Section 6.4 to the fixed-benefit distributed auction, we sought to use an implementation of the latter that was proper for the scenario involved.

6.5.1 Effects of delays

As previously discussed, the presence of communication delays will generally lengthen the time it takes the agents to reach an assignment regardless of the algorithm employed. The associated delay in getting the agents to the tasks will therefore have a detrimental impact on the collective benefit that is achieved. In order to assess the effects of communication delays, we simulated a situation in which the tasks are randomly distributed over one area and the agents over another area some distance away from the first (with agents initially headed towards the general area of the targets). See Figure 6.2 for an example of this configuration. For this scenario, the agents were given a speed of 100 m/s and a turning radius of 1000 m. The static benefit of the task ranged from 0 to 100 and the time-dependent term entered at a rate of 1/sec which put the static and the initial time-dependent terms of the benefit equations on roughly the same scale for this set-up. A bidding increment of 25 was chosen as a value that ensured the algorithms terminated prior to an agent reaching a task most of the time. Early completion of tasks was allowed under the method of trading auctioneer duty discussed in Section 6.4.5. Random but bounded communication delays were simulated using a uniform distribution over a specified range. The maximum communication delay was varied from 0.5 sec to 10 sec for one case in

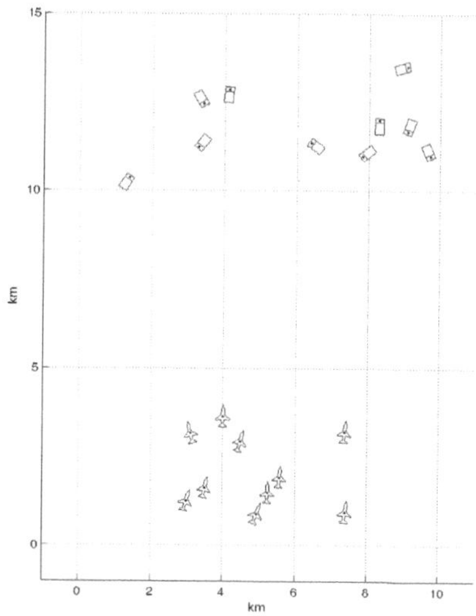

Figure 6.2 Example of scenario for investigating effects of delays. Rectangles and planes denote tasks and agents respectively.

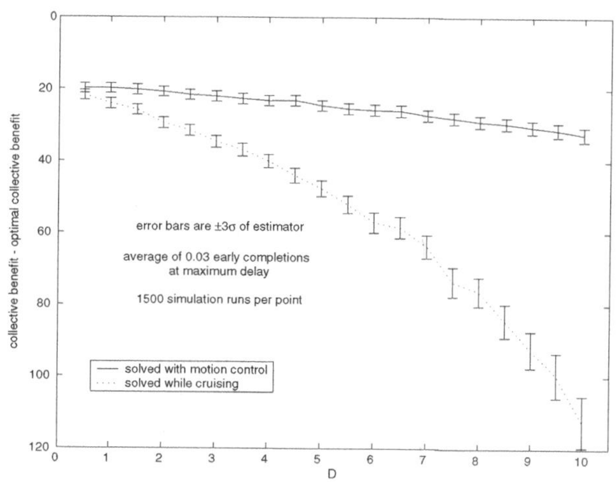

Figure 6.3 Comparison of algorithms when delay is uniform on [0, D].

which the minimum delay was zero and another in which the minimum delay was 80% of the maximum.

Figures 6.3 and 6.4 clearly show how the average benefit decreases with increased delay for both the algorithm with motion control and a fixed-benefit distributed auction in which the agents move along their initial heading towards the tasks until a full assignment is reached (labeled 'solved while cruising' in the figures). They also demonstrate that the motion control algorithm tends to achieve a better collective benefit in comparison to the fixed-benefit distributed auction. This is particularly the case when the average delay is high, as shown in Figure 6.4. With large delay, the average benefit achieved with the

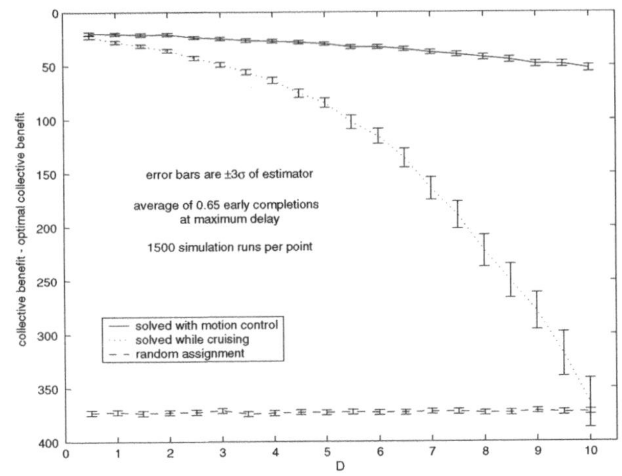

Figure 6.4 Comparison of algorithms when delay is uniform on $[0.8D, D]$.

fixed-benefit auction falls below even that of the average random assignment. We note here that the error bars on all the figures in this section are based on the standard deviation of the estimate of the mean (i.e., $\sigma = S/\sqrt{k}$ where S^2 is the sample variance and k the number of simulation runs).

6.5.2 Effects of bidding increment

As discussed in Section 6.4.3, in most situations it is virtually impossible to guarantee the assumption of Theorem 6.4.3 that no agent will reach a task before termination of the algorithm. On the other hand, the bidding increment specified in that theorem is often so large as to effectively eliminate the optimization process of the auction (i.e., the first bid a task receives will raise its price so high that it is unlikely to receive another bid). In order to study the effects of varying the bid increment, a scenario similar to that in Figure 6.2 was used. The area covered was approximately twice as large and the agents' speed was increased to 300 m/s while its turning radius decreased to 500 m. These changes reduced the variance of the simulations, allowing us to highlight the effect of small changes in the bidding increment without having to run an excessive number of trials.

As Figure 6.5 shows, the best average performance of the algorithm for a given delay lies at a certain value of the bidding increment. As the bidding increment decreases from that value, the average benefit falls off steeply since it takes more time to reach an assignment. As the bidding increment increases from that value, the average benefit decreases as the algorithm is not attempting to achieve as much optimization. As is apparent from Figure 6.5, the optimal bidding increment also tends to decrease as the communication delay decreases. The balance between the competing goals of acting quickly and optimizing the assignment according to the current benefits is determined by the size of the bidding increment; a large value emphasizes the former and a small value the latter. When the communication delay decreases, the algorithm runs faster with respect to the changes in benefits, thereby tilting that balance in favor of performing a better optimization.

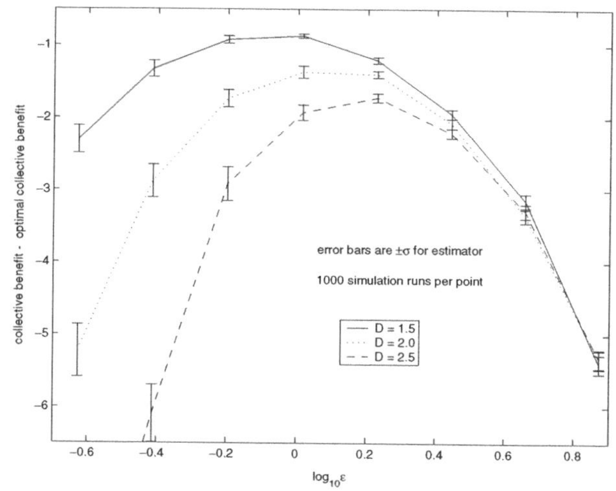

Figure 6.5 Average performance as a function of the bidding increment (delays uniform on $[0.8D, D]$).

6.5.3 Early task completions

In Theorem 6.4.5 we saw that letting an agent complete a task early cannot decrease the achievable collective benefit in a total path minimization problem when a few simple properties hold for the length of the optimal trajectory between two points. To illustrate this we simulated a total path minimization scenario for a configuration of agents and tasks in which both are randomly distributed on a disc with radius twice the minimum turn radius of the vehicles (see Figure 6.6). For this simulation the speed of the agents was 100 m/s and their turning radius 1000 m. The bidding increment was kept constant at 25 and the maximum communication delay again varied from 0.5 sec to 10 sec. One simulation was run in which agents where allowed to complete tasks early, and another in which they were forced to circle back to the task until the algorithm terminated (or

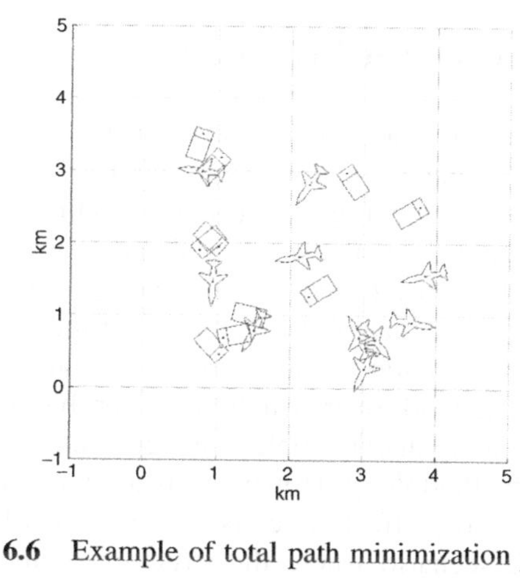

Figure 6.6 Example of total path minimization problem.

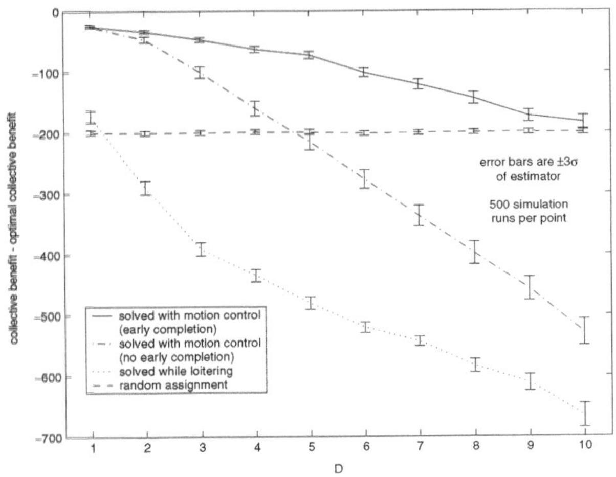

Figure 6.7 Comparison of algorithms in total path minimization problem (delays uniform on $[0.8D, D]$).

their preferred task changed). Both cases were compared to a fixed-benefit distributed auction based on initial benefits in which the agents moved on a circle passing through their initial position (they 'loiter') until the auction was complete.

The results appear in Figure 6.7. As expected, the motion control algorithm in which agents were allowed to complete tasks early performed much better than both of the other algorithms and is the only algorithm whose performance stays above that of a random assignment for the given values of D. The performance of the two motion control algorithms starts close together, but quickly diverges as communication delays increase. This is expected since the close proximity of agents and tasks will result in more and more agents having to pass tasks as the time it takes to reach an assignment increases. In fact, as the delays increase, the performance of the motion control algorithm without early task completions starts to resemble that of the fixed-benefit distributed auction.

6.5.4 Distributed vs. centralized computation

Given that the number of agents and tasks involved in many of these problems is relatively small, solving the associated assignment problem on a single processor can take considerably less time than implementing a distributed auction that is prone to protracted 'price wars' (i.e., agents making small incremental bids for a few tasks and incurring a communication delay with each bid and assignment). In order for the motion control algorithm to be practical, it would have to be the case that the benefits of its parallel computation exceeded the communication penalty associated with the bidding process. If we are given a communication network over which all the problem data could be efficiently sent to every agent and the assignment problem solved redundantly, it would be hard to justify using a distributed methodology. However, as an example of when the algorithm of Section 6.4.2 may prove more effective, consider the case where communications are limited by bandwidth. In a scenario where the structure of the benefits tends to result in the modified auction taking only a few iterations, then that algorithm might tend to terminate faster (and produce a higher collective benefit) than it takes for the agents to transmit all the problem data to each other.

The total path minimization problem is just such a scenario; if the agents and tasks are distributed in the same manner across a wide area, most agents will wind up assigned to the task to which they are closest initially. For this simulation, we used a time division multiplex (TDM) scheme for the communication network, with the time slot allocated to each agent the size of a distributed auction message, denoted as T. To approximate the time it would take to transmit all the problem data to the agents, we roughly estimate the time required for a single field to be $\frac{1}{4}T$ since an auction message has four fields (i.e., task, agent, bid/price, and a time stamp). Then in order to synchronize their data each agent would have to send a message of $2 + n$ fields (time stamp, agent, and its benefit for each of n tasks) if the tasks can be placed in a known order or $2 + 2n$ fields otherwise. Then transmitting the entire problem data for every agent would require $\frac{1}{4}Tm(2 + n)$ in the first case (scheme 1) and $\frac{1}{4}Tm(2 + 2n)$ in the second (scheme 2).

For this simulation we distributed the agents and tasks randomly over a 20 km square, with the speed and turning radius of the agents again at 100 m/s and 1000 m respectively. In both 'synchronized schemes', the agents move towards their closest initial task while they wait to collect all the benefit data. Once that is complete, they proceeded towards their assignment from the optimal solution. The results appear in Figure 6.8 and clearly show that there are crossover points at which the average performance of the distributed auction with motion control becomes greater than that of the two synchronized schemes as the bandwidth of the communication network decreases. Hence, we can conclude that with a poor quality communication network the distributed auction with motion control can outperform a centralized algorithm. We note that the performance of the synchronized schemes appears to degrade linearly with decreasing bandwidth (with slopes roughly proportional to their synchronization time), while the performance of the motion control algorithm becomes decidedly nonlinear for low bandwidth values in such a way that cannot be attributed to the variance of the simulation data. While we do not have a proper

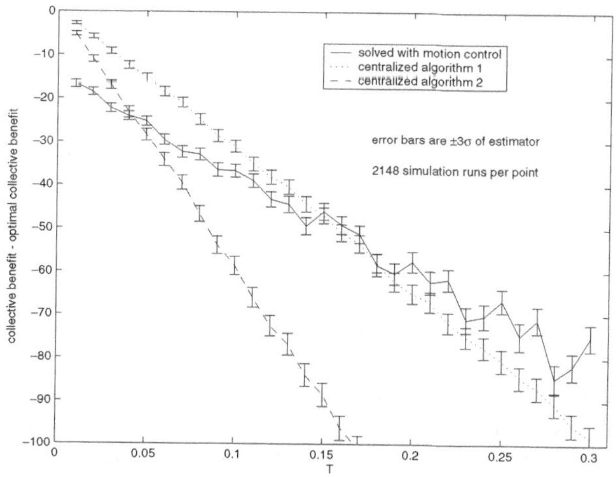

Figure 6.8 Comparison of algorithms in bandwidth constrained scenario (10 agents/10 tasks).

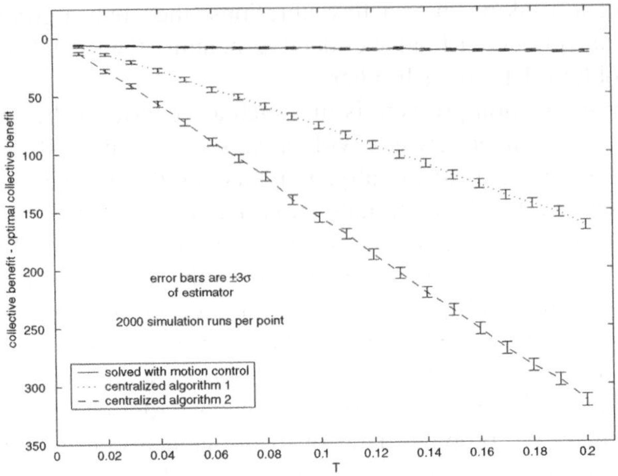

Figure 6.9 Comparison of algorithms in bandwidth constrained scenario (10 agents/20 tasks).

explanation for this phenomenon, we assume it is the effect of some interaction between the switching behavior of the motion control algorithm, the synchronicity imposed by TDM communication, and the geometry of the task scenario. The practical advantages of the motion control algorithm in this scenario are even more evident (and without when the number of tasks is large relative to the number of agents (see Figure 6.9). The additional tasks do not generally make these small assignment problems more difficult to solve (since we can still assume calculation times are negligible), but the communication requirements of the synchronized schemes quickly cause their performance to degrade as the available bandwidth shrinks.

6.6 CONCLUSIONS

In this chapter we have sought to address an assignment problem between mobile agents and stationary tasks where the benefits (and hence the optimal assignment quality) have the potential to decrease during the time used to calculate a solution. We presented a modification of the distributed auction algorithm of (Bertsekas and Castañon 1991) that controls the motion of the agents during the algorithm's progress in an attempt to minimize that loss of benefit. We showed that this algorithm is guaranteed to terminate in finite time at an assignment that is within a known bound of the optimal solution under one set of assumptions, and simply guaranteed to terminate under less restrictive conditions. Simulations demonstrated that the motion-controlled algorithm has an average performance superior to that of a fixed-benefit distributed auction in many cases, and may even outperform a centralized approach in certain situations.

There are two key extensions of the studied problem that deserve further research. One of these is a multi-assignment version in which agents are capable of completing more than one task. In this case, an agent's trajectory to a task (and the benefit it will receive from it) are directly related to the other tasks that agent must visit. Another case that presents similar problems is a version in which tasks are coupled to each other (i.e., certain tasks cannot be completed before others). Here, the benefit an agent will receive from a task may depend on what other agents are doing and when. In both these cases the non-linear coupling between tasks and benefits rules out algorithms, such as the auction, that are designed for linear network flow models. Given that these problems often take a longer time to solve, the concept of directing agents' motion during that process would seem to be of even more importance.

ACKNOWLEDGEMENTS

This work was supported by the AFRL/VA and AFOSR Collaborative Center for Control Science (Grant F33615-01-2-3154). In addition, the work was supported by a DAGSI Fellowship to Brandon Moore.

REFERENCES

Bertsekas DP 1998 *Network Optimization: Continuous and Discrete Models.* Athena Scientific, Belmont, MA.

Bertsekas DP n.d. Auction algorithms for network flow problems: A tutorial introduction. Technical Report LIDS-P-2108, Laboratory for Information and Decision Systems Report, M.I.T., Cambridge, MA.

Bertsekas DP and Castañon DA 1991 Parallel synchronous and asynchronous implementations of the auction algorithm. *Parallel Computing* **17**, 707–732.

Bertsekas DP and Tsitsiklis JN 1997a *Introduction to Linear Optimization.* Athena Scientific, Belmont, MA.

Bertsekas DP and Tsitsiklis JN 1997b *Parallel and Distributed Computation: Numerical Methods.* Athena Scientific, Belmont, MA.

Castañon DA and Wu C 2003 Distributed algorithms for dynamic reassignment. *42nd IEEE Conference on Decision and Control,* pp. 13–18, Maui, Hawaii.

Dubins L 1957 On curves of minimal length with a constraint on average curvature, and with prescribed initial and terminal position. *American Journal of Math* **79**, 497–516.

Finke J, Passino KM and Sparks A 2003 Cooperative control via task load balancing for networked uninhabited autonomous vehicles. *42nd IEEE Conference on Decision and Control,* pp. 31–36, Maui, Hawaii.

Gil AE, Passino KM, Ganapathy S and Sparks A 2003 Cooperative scheduling of tasks for networked uninhabited autonomous vehicles. *42nd IEEE Conference on Decision and Control,* pp. 522–527, Maui, Hawaii.

Howlett JK, McClain TW and Goodrich MA 2003 Learning Real-Time A* path planner for sensing closely-spaced targets from an aircraft. *AIAA Guidance, Navigation, and Control Conference,* number 2003–5338, Austin, Texas.

Richards A and How JP 2002 Aircraft trajectory planning with collision avoidance using mixed integer linear programming. *American Control Conference,* pp. 1936–1941, Anchorage, Alaska.

Schumacher C, Chandler PR and Rasmussen SJ 2002 Task allocation for wide area search munitions via iterative network flow. *AIAA Guidance, Navigation, and Control Conference,* number 2002–4586, Monterey, CA.

Schumacher C, Chandler PR, Pachter M and Pachter L 2003 UAV task assignment with timing constraints. *AIAA Guidance, Navigation, and Control Conference,* number 2003–5664, Austin, Texas.

Sussman HJ and Tang G 1991 Shortest paths for the Reeds-Shepp car: a worked out example of the use of geometric techniques in nonlinear optimal control. Technical Report SYCON-91-10, Rutgers University, New Brunswick, NJ.

Walker DH, McClain TW and Howlett JK 2003 Coordinated UAV target assignment using distributed tour calculations. *4th Annual Conference on Cooperative Control and Optimization*, Gainesville, Florida.

7

On the value of information in dynamic multiple-vehicle routing problems

Alessandro Arsie, John J. Enright and Emilio Frazzoli

7.1 INTRODUCTION

A very active research area today addresses cooperative control of distributed multi-agent systems: groups of autonomous robots and large-scale mobile networks are being considered for a broad class of applications, ranging from environmental monitoring, to search and rescue operations, and national security.

An area of particular interest is concerned with the generation of efficient cooperative strategies for several mobile agents to move through a certain number of given target points, possibly avoiding obstacles or threats (Beard *et al.* 2002; Earl and D'Andrea 2005; Li and Cassandras 2006; Richards *et al.* 2002; Schumacher *et al.* 2003). Trajectory efficiency in these cases is understood in terms of cost for the agents: in other words, efficient trajectories minimize the total path length, the time needed to complete the task, or the fuel/energy expenditure. A related problem has been investigated as the Weapon-Target Assignment (WTA) problem, in which mobile agents are allowed to team up in order to enhance the probability of a favorable outcome in a target engagement (Arslan and Shamma 2006; Murphey 1999). In this set-up, targets locations are known and an assignment strategy is sought that maximizes the global success rate. In a biological setting, the closest parallel to many of these problems is the development of foraging strategies, and of territorial vs. gregarious behaviors (Tanemura and Hasegawa 1980), in which individuals choose to identify and possibly defend a hunting ground. Ideally, distributed multi-agent systems are desired to being able to perform complex tasks, comparable to the complexity achieved in animal populations or even in interacting human groups. For instance it would be extremely interesting, both on a theoretical ground and from a practical perspective, to have distributed multi-agent systems which are able to engage efficiently teams of hostile intelligent beings. Just considering what happens in a biological setting, it is evident

Cooperative Control of Distributed Multi-Agent Systems edited by Jeff S. Shamma

that information is extremely valuable in achieving highly complex task coordination or more generally collective behaviors. Our analysis, in this contribution will be much more modest in scope and breadth, so we are not going even to touch the issue of information value for complex collective behaviors that can be found in a biological set-up.

Indeed, we are going to focus our attention on the value of information just for a class of cooperative motion coordination problems, which we can call *dynamic vehicle routing*. In these problems service requests are not known a priori, but are dynamically generated over time by a stochastic process in a geographic region of interest. Each service request is associated to a target point in the plane, and is fulfilled when one of a team of mobile agents visits that point. For example, service requests can be thought of as threats to be investigated in a surveillance application, events to be measured in an environmental monitoring scenario, and as information packets to be picked up and delivered to a user in a wireless sensor network. It is desired to design a control strategy for the mobile agents that provably minimizes the expected waiting time between the issuance of a service request and its fulfillment. In other words, our focus is on the quality of service as perceived by the 'end user,' rather than, for example, fuel economies achieved by the mobile agents. Similar problems were also considered by Bertsimas and van Ryzin (1991) and Psaraftis (1988), and decentralized strategies were presented by Frazzoli and Bullo (2004). This problem has connections to the Persistent Area Denial (PAD) and area coverage problems discussed, e.g., in (Cortés *et al.* 2004; Liu *et al.* 2004; Moore and Passino 2005; Schumacher *et al.* 2003).

In particular, we will focus our attention on the effects of different communication and information-sharing protocols on the system performance. Despite the fact that the modeling we are studying is a rough simplification compared to the actual complex coordination behaviors of biological populations, it will present some surprising characteristics. Clearly, the ability to access more information at each single agent cannot decrease the performance level; hence, it is commonly believed that by providing better communication among agents will provide a (strict) improvement of the system's performance. Remarkably, this is not always the case. Indeed, we present a class of certain dynamic vehicle routing problems which can, in fact, be solved (almost) optimally without any explicit communication between agents; in other words, the no-communication constraint in such cases is not binding, and does not limit the steady-state performance, while it obviously limits the speed with which the steady-state performance is reached.

In this peculiar class of dynamic vehicle routing problems it is assumed that the agents have unlimited sensing capabilities, or, more precisely, that they are able to detect the presence of service requests in any bounded subset of the plane. We extend our analysis also to the case in which the agents do not communicate explicitly with one another (or communicate just among closest neighbors), but also have limited sensor capabilities. In particular we will get analytical results in two asymptotic cases, corresponding to target generation rate going to infinity or to the effective area of sensing shrinking to zero. Also in this set-up, quite surprisingly, the absence of explicit communication among the agents and the limited sensing range have no effect on the collective efficiency of the system in the limit in which the target generation rate diverges.

The chapter is structured as follows: in Section 7.2 we set up our contribution. In Section 7.3 we introduce the proposed solution algorithms, and discuss some of their characteristics. Section 7.4 is the technical part in which we prove the convergence of the performance provided by some of the proposed algorithms to a critical point (either a local minimum or a saddle point) of the global performance function. Moreover, we show that any optimal configuration corresponds to a class of tessellations of the plane that we call Median Voronoi Tessellations. Section 7.5 is devoted to an analogous analysis for another class of control policies which are characterized by the fact that the agents have limited sensing capabilities. In Section 7.6 we present some numerical results, while Section 7.7 is dedicated to final remarks and further perspective along this line of research.

7.2 PROBLEM FORMULATION

Let $\Omega \subset \mathbb{R}^2$ be a convex domain on the plane, with non-empty interior; we will refer to Ω as the *workspace*. A stochastic process generates *service requests* over time, which are associated to points in Ω; these points are also called *targets*. The process generating service requests is modeled as a spatio-temporal Poisson point process, with temporal intensity $\lambda > 0$, and an absolutely continuous spatial distribution described by the density function $\varphi : \Omega \to \mathbb{R}_+$, with bounded and convex support within Ω (i.e., $\varphi(q) > 0 \Leftrightarrow q \in \mathcal{Q} \subseteq \Omega$, with $\mathcal{Q}$ bounded and convex). The spatial density function φ is normalized in such a way that $\int_\Omega \varphi(q)\, dq = 1$. Both λ and φ are not necessarily known.

A spatio-temporal Poisson point process is a collection of functions $\{\mathcal{P} : \overline{\mathbb{R}}_+ \to 2^\Omega\}$ such that, for any $t > 0$, $\mathcal{P}(t)$ is a random collection of points in Ω, representing the service requests generated in the time interval $[0, t)$, and such that

- The total numbers of events generated in two disjoint time-space regions are *independent* random variables;
- The total number of events occurring in an interval $[s, s+t)$ in a measurable set $\mathcal{S} \subseteq \Omega$ satisfies

$$\Pr\left[\text{card}\left((\mathcal{P}(s+t) - \mathcal{P}(s)) \cap \mathcal{S}\right) = k\right] = \frac{\exp(-\lambda t \cdot \varphi(\mathcal{S}))(\lambda t \cdot \varphi(\mathcal{S}))^k}{k!}, \qquad \forall k \in \mathbb{N},$$

  where $\varphi(\mathcal{S})$ is a shorthand for $\int_{\mathcal{S}} \varphi(q)\, dq$.

Each particular function $\mathcal{P}$ is a realization, or trajectory, of the Poisson point process. A consequence of the properties defining Poisson processes is that the *expected* number of targets generated in a measurable region $\mathcal{S} \subseteq \Omega$ during a time interval of length Δt is given by:

$$\text{E}[\text{card}\left((\mathcal{P}(t + \Delta t) - \mathcal{P}(t)) \cap \mathcal{S}\right)] = \lambda \Delta t \cdot \varphi(\mathcal{S}).$$

Without loss of generality, we will identify service requests with targets points, and label them in order of generation; in other words, given two targets $e_i, e_j \in \mathcal{P}(t)$, with $i < j$, the service request associated with these targets have been issued at times $t_i \leq t_j \leq t$ (since events are almost never generated concurrently, the inequalities are in fact strict almost surely).

A service request is fulfilled when one of m mobile agents, modeled as point masses, moves to the target point associated with it; m is a possibly large, but finite number. Let $p(t) = (p_1(t), p_2(t), \ldots, p_m(t)) \in \Omega^m$ be a vector describing the positions of the agents at time t. (We will tacitly use a similar notation throughout this contribution.) The agents are free to move, with bounded speed, within the workspace Ω; without loss of generality, we will assume that the maximum speed is unitary. In other words, the dynamics of the agents are described by differential equations of the form

$$\frac{d\ p_i(t)}{dt} = u_i(t), \qquad \text{with } \|u_i(t)\| \leq v, \quad \forall t \geq 0, i \in \{1, \ldots, m\}, \tag{7.1}$$

where v is a strictly positive real constant, namely the maximum speed. The agents are identical, and have unlimited range and target-servicing capability. Let $\mathcal{B}_i(t) \subset \Omega$ indicate the set of targets serviced by the i-th agent up to time t. (By convention, $\mathcal{B}_i(0) = \emptyset$, $i = 1, \ldots, m$.) We will assume that $\mathcal{B}_i \cap \mathcal{B}_j = \emptyset$ if $i \neq j$, i.e., that service requests are fulfilled by at most one agent. (In the unlikely event that two or more agents visit a target at the same time, the target is arbitrarily assigned to one of them.)

Let $\mathcal{D} : t \to 2^\Omega$ indicate (a realization of) the stochastic process obtained combining the service request generation process $\mathcal{P}$ and the removal process caused by the agents servicing outstanding requests; in other words,

$$\mathcal{P}(t) = \mathcal{D}(t) \cup \mathcal{B}_1(t) \cup \ldots \cup \mathcal{B}_m(t), \qquad \mathcal{D}(t) \cap \mathcal{B}_i(t) = \emptyset, \ \forall i \in \{1, \ldots, m\}.$$

The random set $\mathcal{D}(t) \subset \Omega$ represents the *demand*, i.e., the service requests outstanding at time t; let $n(t) = \text{card}(\mathcal{D}(t))$. Let us remark that in order for an agent to know the set $\mathcal{D}(t)$, it is not necessary to exchange directly any information with other agents. Indeed, the set $\mathcal{D}(t)$ can be known because of large sensing capabilities of each single agent compared to the dimension of the work-space Ω, or because there is a satellite or an airplane identifying the outstanding targets and broadcasting their position to the agents (in this case, there is no communication complexity, since there is no dedicated communication among agents and satellite/airplane).

It is interesting also to consider the case in which the sensing range of each agent is limited and there is no device broadcasting globally the positions of outstanding targets. In this case, the set $\mathcal{D}(t)$ is generally not known to all agents. In the case of limited sensing range, for the sake of simplicity, we will model the sensing region of an agent as a circle of radius σ centered at the agent's position; indicate with

$$\mathcal{S}(p) = \{q \in \mathbb{R}^2 : \|p - q\| \leq \sigma\}$$

such sensing region for an agent at position p. Other shapes of the sensor footprint can be considered with minor modifications to our analysis, and affect the results at most by a constant. We will assume that the sensor footprint is small enough that it is contained within $\mathcal{Q}$ for at least one position of the agent; moreover, we will be interested in analyzing the effect of a limited sensor range as $\sigma \to 0$. Define the set $\mathcal{D}_i(t) \subseteq \mathcal{D}(t)$, $i \in \{1, \ldots, m\}$ as the set of outstanding targets that have been detected by the i-th agent, i.e., outstanding targets that have been approached by the i-th agent to within a distance σ at some instant prior to t.

In this contribution we will present some motion coordination strategies that allow the mobile agents to fulfill service requests efficiently (we will make this more precise in the following), focusing on the value of information. In particular, we will concentrate on motion coordination strategies of the following three forms:

$$\pi_i : (p_i, \mathcal{B}_i, \mathcal{D}) \mapsto u_i, \qquad i \in \{1, \ldots, m\}, \tag{7.2}$$

$$\pi_i : (p_i, \mathcal{D}_i) \mapsto u_i, \qquad i \in \{1, \ldots, m\}, \tag{7.3}$$

and

$$\pi_i : (p_1, \ldots, p_m, \mathcal{B}_i, \mathcal{D}) \mapsto u_i, \qquad i \in \{1, \ldots, m\}. \tag{7.4}$$

An agent executing a control policy of the form (7.2) relies on the knowledge of its own current position, on a record of targets it has previously visited, and on the current demand, while an agent executing a control policy of the form (7.3) relies on the knowledge of its own current position and the part of the current demand which is under its sensing capabilities. In other words, such control policies do not need any explicit information exchange between agents; as such, we will refer to them as *no communication* (nc) policies. Such policies are trivially decentralized. On the other hand, an agent executing a control policy of the form (7.4) can sense the current position of other agents, but still has information only on the targets itself visited in the past (i.e., does not know what, if any, targets have been visited by other agents). We call these *sensor-based* (sb) policies, to signify the fact that only factual information is exchanged between agents – as opposed to information related to intent and past history. Note that just control policies (7.4) and (7.2) rely, in principle, on the knowledge of the locations of all outstanding targets. A policy $\pi = (\pi_1, \pi_2, \ldots, \pi_m)$ is said to be *stabilizing* if, under its effect, the expected number of outstanding targets does not diverge over time, i.e., if

$$\overline{n}_\pi = \lim_{t\to\infty} \mathrm{E}[n(t) \| \dot{p}_i(t) = \pi_i(p(t), \mathcal{B}_i(t), \mathcal{D}(t)), i \in \{1, \ldots, m\}] < \infty. \tag{7.5}$$

Intuitively, a policy is stabilizing if the mobile agents are able to visit targets at a rate that is – on average – at least as fast as the rate at which new service requests are generated.

Let T_j be the time elapsed between the issuance of the j-th service request, and the time it is fulfilled. If the system is stable, then the following balance equation (also known as Little's formula (Little 1961)) holds:

$$\overline{n}_\pi = \lambda \overline{T}_\pi, \tag{7.6}$$

where $\overline{T}_\pi := \lim_{j\to\infty} \mathrm{E}[T_j]$ is the system time under policy π, i.e., the expected time a service request must wait before being fulfilled, given that the mobile agents follow the strategy defined by π. Note that the system time $\overline{T}_\pi$ can be thought of as a measure of the quality of service, as perceived by the user issuing the service requests. At this point we can finally state our problem: we wish to devise a policy that is (1) stabilizing, and (2) yields a quality of service (i.e., system time) achieving, or approximating, the theoretical optimal performance given by

$$\overline{T}_{\mathrm{opt}} = \inf_{\pi \text{ stabilizing}} \overline{T}_\pi \tag{7.7}$$

Centralized and decentralized strategies are known that optimize or approximate (7.7) in a variety of cases of interest (Bertsimas and van Ryzin 1991, 1993a, b; Frazzoli and Bullo 2004). However, all such strategies rely either on a central authority with the ability to communicate to all agents, or on the exchange of certain information about each agent's strategy with other neighboring agents. In addition, these policies require the knowledge of the spatial distribution φ; decentralized versions of these implement versions of Lloyd's algorithm for vector quantization (Lloyd 1982).

In the remainder of this contribution, we will investigate how the additional constraints posed on the exchange of information between agents by the models (7.2), (7.3) and

(7.4) impact the achievable performance and quality of service. Remarkably, some of the policies we will present do not rely on the knowledge of the spatial distribution φ, and are a generalized version of MacQueen's clustering algorithm (MacQueen 1967).

7.3 CONTROL POLICY DESCRIPTION

In this section, we introduce three control policies of the forms, respectively, (7.2), (7.3) and (7.4). An illustration of two of the control policies is given in Figure 7.1.

7.3.1 A control policy requiring no explicit communication: the unlimited sensing capabilities case

Let us begin with an informal description of a policy π_{nc} requiring no explicit information exchange between agents. At any given time t, each agent computes its own control input according to the following rule:

1. If $\mathcal{D}(t)$ is not empty, move towards the nearest outstanding target.
2. If $\mathcal{D}(t)$ is empty, move towards the point minimizing the average distance to targets *serviced in the past* by each agent. If there is no unique minimizer, then move to the nearest one.

In other words, we set

$$\pi_{\text{nc}}(p_i(t), \mathcal{B}_i(t), \mathcal{D}(t)) = \text{vers}(F_{\text{nc}}(p_i(t), \mathcal{B}_i(t), \mathcal{D}(t)) - p_i(t)), \tag{7.8}$$

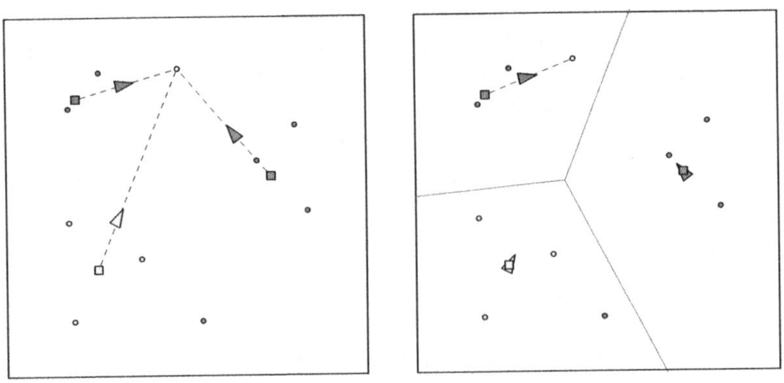

Figure 7.1 An illustration of two of the control policies proposed in Section 7.3. While no targets are outstanding, vehicles wait at the point that minimizes the average distance to targets they have visited in the past; such points are depicted as squares, while targets are circles and vehicles triangles. In the no-communication policy, at the appearance of a new target, all vehicles pursue it (left). In the sensor-based policy, only the vehicle that is closest to the target will pursue it (right).

where

$$F_{\text{nc}}(p_i, \mathcal{B}_i, \mathcal{D}) = \begin{cases} \arg\min_{q \in \mathcal{D}} \|p_i - q\|, & \text{if } \mathcal{D} \neq \emptyset, \\ \arg\min_{q \in \Omega} \sum_{e \in \mathcal{B}_i} \|e - q\|, & \text{otherwise}, \end{cases} \tag{7.9}$$

$\|\cdot\|$ is the Euclidean norm, and

$$\text{vers}(v) = \begin{cases} v/\|v\|, & \text{if } v \neq 0, \\ 0 & \text{otherwise}. \end{cases}$$

The convex function $W : q \mapsto \sum_{e \in \mathcal{B}} \|q - e\|$, often called the (discrete) Weber function in the facility location literature (Agarwal and Sharir 1998; Drezner 1995) (modulo normalization by card($\mathcal{B}$)), is not strictly convex only when the point set $\mathcal{B}$ is empty – in which case we set $W(\cdot) = 0$ by convention – or contains an even number of collinear points. In such cases, the minimizer nearest to p_i in (7.9) is chosen. We will call the point $p_i^*(t) = F_{\text{nc}}(\cdot, \mathcal{B}_i(t), \emptyset)$ the *reference point* for the i-th agent at time t. In the π_{nc} policy, whenever one or more service requests are outstanding, all agents will be pursuing a target; in particular, when only one service request is outstanding, all agents will move towards it. When the demand queue is empty, agents will either (1) stop at the current location, if they have visited no targets yet, or (2) move to their reference point, as determined by the set of targets previously visited.

7.3.2 A control policy requiring communication among closest neighbors: the limited sensing capabilities case

To present this control policy, let us focus first of all on the single-vehicle case. Moreover since the algorithm is based on solution or on an approximation of the Euclidean Traveling Salesperson Problem (ETSP), let us review some asymptotic properties related to this. Given a set of points $\mathcal{D} \subset \mathbb{R}^n$, the ETSP is the problem of finding the shortest closed path (tour) through all points in $\mathcal{D}$; let ETSP($\mathcal{D}$) be the length of such tour. Even though the exact optimal solutions of a large TSP can be very hard to compute, several exact and heuristic algorithms and software tools are available for the numerical solution of Euclidean TSPs.

The most advanced TSP solver to date is arguably `concorde` (Applegate *et al.* 1998). Heuristic polynomial-time algorithms are available for constant-factor approximations of TSP solutions, among which we mention Christofides' work (Christofides 1972). On a more theoretical note, Arora proved the existence of polynomial-time approximation schemes, providing a $(1 + \varepsilon)$ constant-factor approximation for any $\varepsilon > 0$ (Arora 1997).

A modified version of the Lin-Kernighan heuristic (Lin and Kernighan 1973) is implemented in `linkern`; this powerful solver yields approximations in the order of 5% of the optimal tour cost very quickly for many instances. For example, in our numerical experiments on a 2.4 GHz Pentium machine, approximations of random TSPs with 1,000 points typically required about two seconds of CPU time.[1]

[1] Both `concorde` and `linkern` are written in ANSI C and are freely available for academic research use at `http://www.math.princeton.edu/tsp/concorde.html`.

Furthermore, studying the performance of this class of policies we will make use of the following remarkable results:

Theorem 7.3.1 (Beardwood, Halton, and Hammersley (Beardwood *et al.* 1959)) *Let $\mathcal{D}_n \subset \mathcal{Q}$ be a set of n points independently sampled from an absolutely continuous distribution with spatial density φ with compact support $\mathcal{Q} \subset \mathbb{R}^2$. There exists a constant $\beta \in \mathbb{R}$ such that the length of the Euclidean Traveling Salesperson tour through all points in $\mathcal{D}_n$ satisfies the following limit, almost surely:*

$$\lim_{n\to+\infty} \frac{\mathrm{ETSP}(\mathcal{D}_n)}{\sqrt{n}} = \beta \int_{\mathcal{Q}} \sqrt{\varphi(q)}\, dq, \qquad \text{a.s.} \tag{7.10}$$

The current best estimate of the constant is $\beta = 0.7120 \pm 0.0002$.(Johnson *et al.* 1996; Percus and Martin 1996) Interestingly, the asymptotic cost of the ETSP for uniform point distributions is an upper bound on the asymptotic cost for general point distributions, as can be proven by applying Jensen's inequality to (7.10). In other words,

$$\lim_{n\to+\infty} \frac{\mathrm{ETSP}(\mathcal{D}_n)}{\sqrt{n}} \leq \beta\sqrt{A}, \qquad \text{a.s.,}$$

where $A = \mathrm{Area}(\mathcal{Q})$.

There is also a deterministic version of the above result, which we can state as

Theorem 7.3.2 (e.g., Steele (Steele 1990)) *Let $\mathcal{D}_n \subset \mathcal{Q}$ be a set of n points arbitrarily chosen from a set $\mathcal{Q} \subset \mathbb{R}^2$. There exists a constant $\overline{\beta}_{\mathcal{Q}} \in \mathbb{R}$, depending only on the shape of $\mathcal{Q}$, such that the length of the Euclidean Traveling Salesperson tour through all points in $\mathcal{D}_n$ satisfies the following inequality:*

$$\frac{\mathrm{ETSP}(\mathcal{D}_n)}{\sqrt{n}} \leq \overline{\beta}_{\mathcal{Q}}\sqrt{\mathrm{Area}(\mathcal{Q})}, \qquad \forall n \in \mathbb{N}. \tag{7.11}$$

The type of policy we are going to present requires on-line solutions of large TSPs. Practical implementations of the algorithms will rely on heuristics, such as Lin-Kernighan's or Christofides'. If a constant-factor approximation algorithm is used, the effect on the asymptotic performance guarantees of our algorithms can simply be modeled as a scaling of the constant β. We consider first the case of a single-vehicle with the single-vehicle tiling policy (sTP).

The tiling policy is based on a tiling of the plane (or of the region of interest) with tiles small enough to be contained in the sensor's footprint, e.g., squares of side length $\eta\sigma\sqrt{2}$, where $\eta \in (0, 1]$ is a design parameter. Let C be the set of centers of the squares that have a non-empty intersection with the environment $\mathcal{Q}$. Construct a tour of the points in C, e.g., as an ordered set $C_{\mathrm{tour}}(c_1, c_2, \ldots)$. The sTP can be described as Algorithm 7.1

Actually, in the Step 4 of the algorithm one has to construct a solution to the ETSP inside the current tile. If we want to have a fast implementation of the algorithm, it is necessary to use approximations to the optimal solutions. In order to extend the single-vehicle policy to the multiple-vehicle case we introduce certain points, that we call *virtual generators*, associated to each agent. These are not physical points, but are rather logical

Algorithm 7.1 The single-vehicle Tiling Policy

Require: A tiling of the plane, and an ordered list $\mathcal{C}_{\text{tour}} = (c_1, c_2, \ldots)$ of tile centers.
1: Let $i \leftarrow 1$.
2: **loop**
3: Move to the point c_i.
4: Visit all targets currently in the tile centered at c_i, in minimum time. {I}gnore targets generated after arrival of the agent to the center of the tile.
5: Increment the counter $i \leftarrow i + 1$, modulo the cardinality of $\mathcal{C}_{\text{tour}}$.

variables that are used to identify regions of dominance for each vehicle within $\mathcal{Q}$. Let these virtual generators be given by $g = (g_1, g_2, \ldots, g_m) \in \mathcal{Q}^m$.

Let $\mathcal{V}(g) = (\mathcal{V}_1(g), \mathcal{V}_2(g), \ldots, \mathcal{V}_m(g))$ be the Voronoi partition of $\mathbb{R}^2$ generated by the points in g, i.e.,

$$\mathcal{V}_i(g) = \{q \in \mathbb{R}^2 : \|q - g_i\| \leq \|q - g_j\|, \forall j \in \{1, \ldots, m\}\}, \qquad i \in \{1, \ldots, m\}.$$

Let each agent execute the sTP policy within the intersection of its Voronoi region and the environment, for fixed generators. Let us call this policy multiple-vehicle Tiling Policy, with fixed generators (mTP/FG)

Let $A_i(g) = \text{Area}[\mathcal{V}_i(g) \cap \mathcal{Q}]$, for all $i \in \{1, \ldots, m\}$. It is possible to consider a generalization of this policy, enabling the virtual generators to move, at least in the case in which the spatial density distribution of the process generating the targets is uniform. Indeed, we will see that in this case, the upper-bounds for the system performance are minimized when the areas of all Voronoi regions are equal, i.e. when $A_i(g) = A_j(g)$, for all $i, j \in \{1, \ldots, m\}$. One way to accomplish this is to update the virtual generators according to a gradient descent law. Remarkably, such gradient descent law can be computed in a decentralized way. In fact, by the definition of Voronoi regions, we have that

$$\frac{\partial A_i(g)}{\partial g_j} = \begin{cases} \dfrac{1}{2} L_{ij} \dfrac{g_j - g_i}{\|g_j - g_i\|} & \text{if } i \neq j \\ \dfrac{1}{2} \displaystyle\sum_{l \neq j} L_{lj} \dfrac{g_l - g_j}{\|g_l - g_j\|} & \text{if } i = j, \end{cases} \tag{7.12}$$

where L_{ij} is the length of the Voronoi region boundary shared by agents i and j within $\mathcal{Q}$. ($L_{ij} = 0$ if the two agents' regions have an empty intersection.)

Using the above formula, one obtains that, for any exponent d,

$$\frac{\partial \sum_{i=1}^m A_i(g)^d}{\partial g_j} = d \sum_{i \neq j} (A_i(g)^{d-1} - A_j(g)^{d-1}) L_{ij} \frac{g_j - g_i}{\|g_j - g_i\|} \tag{7.13}$$

The multiple-vehicle Tiling Policy can hence be stated as in Algorithm 7.2

Let us observe that the mTP policy requires a minimum amount of information exchange among the agents, since it is required to know at any instant of time (or at any time step), the position of the virtual generators of Delaunay neighbors. Communication among closest neighbors is thus required.

Algorithm 7.2 The multiple-vehicle Tiling Policy

Require: A tiling of the plane, and an ordered list $\mathcal{C}_{\text{tour}} = (c_1, c_2, \ldots)$ of tile centers.
Require: The position of the own virtual generator and of the virtual generators of Delaunay neighbors.
Let $i \leftarrow 1$.
Let j be the identifier of the vehicle.
loop {Thread 1, in discrete time}
  while $c_i \notin \mathcal{V}_j(g)$ **do**
    $i \leftarrow i + 1$, modulo the cardinality of $\mathcal{C}_{\text{tour}}$
  Move to the point c_i.
  Visit all targets currently in the tile centered at c_i, in minimum time. {I}gnore targets generated after arrival of the agent to the center of the tile.
  Increment the counter $i \leftarrow i + 1$, modulo the cardinality of $\mathcal{C}_{\text{tour}}$.
loop {Thread 2, in continuous time}
  For a given gain $k > 0$, and $d > 1$, update own generator as follows: $\dot{g}_j = -k \dfrac{\partial \sum_{i=1}^{m} A_i(g)^d}{\partial g_j}$.

7.3.3 A sensor-based control policy

The control strategy in Section 7.3.1 can be modified to include information on the current position of other agents, if available (e.g., through on-board sensors). In order to present the new policy, indicate with $\mathcal{V}(p) = \{\mathcal{V}_1(p), \mathcal{V}_2(p), \ldots, \mathcal{V}_m(p)\}$ the Voronoi partition of the workspace Ω, defined as:

$$\mathcal{V}_i(p) = \{q \in \Omega : \|q - p_i\| \leq \|q - p_j\|, \forall j = 1 \ldots m\}. \tag{7.14}$$

As long as an agent has never visited any target, i.e., as long as $\mathcal{B}_i(t) = \emptyset$, it executes the π_{nc} policy. Once an agent has visited at least one target, it computes its own control input according to the following rule:

1. If $\mathcal{D}(t) \cap \mathcal{V}_i(t)$ is not empty, move towards the nearest outstanding target in the agent's own Voronoi region.
2. If $\mathcal{D}(t) \cap \mathcal{V}_i(t)$ is empty, move towards the point minimizing the average distance to targets in $\mathcal{B}_i(t)$. If there is no unique minimizer, then move to the nearest one.

In other words, we set

$$\pi_{\text{sb}}(p(t), \mathcal{B}_i(t), \mathcal{D}(t)) = \text{vers}(F_{\text{sb}}(p(t), \mathcal{B}_i(t), \mathcal{D}(t)) - p_i(t)), \tag{7.15}$$

where

$$F_{\text{sb}}(p, \mathcal{B}_i, \mathcal{D}) = \begin{cases} \arg\min_{q \in \mathcal{D}} \|p_i - q\|, & \text{if } \mathcal{D} \cap \mathcal{V}_i \neq \emptyset, \text{ and } \mathcal{B}_i = \emptyset \\ \arg\min_{q \in \mathcal{D} \cap \mathcal{V}_i(p)} \|p_i - q\|, & \text{if } \mathcal{D} \cap \mathcal{V}_i \neq \emptyset, \text{ and } \mathcal{B}_i \neq \emptyset \\ \arg\min_{q \in \Omega} \sum_{e \in \mathcal{B}_i} \|e - q\|, & \text{otherwise.} \end{cases} \tag{7.16}$$

In the π_{sb} policy, at most one agent will be pursuing a given target, at any time after an initial transient that terminates when all agents have visited at least one target each. The agents' behavior when no outstanding targets are available in their Voronoi region is similar to that determined by the π_{nc} policy previously discussed, i.e., they move to their reference point, determined by previously visited targets.

While we introduced Voronoi partitions in the definition of the control policy, the explicit computation of each agent's Voronoi region is not necessary. In fact, each agent only needs to check whether it is the closest agent to a given target or not. In order to check whether a target point q is in the Voronoi region of the i-th agent, it is necessary to know the current position only of agents within a circle or radius $\|p_i - q\|$ centered at q (see Figure 7.2). For example, if such a circle is empty, then q is certainly in $\mathcal{V}_i$; if the circle is not empty, distances of the agents within it to the target must be compared. This provides a degree of spatial decentralization – with respect to other agents – that is even stronger than that provided by restricting communications to agents sharing a boundary in a Voronoi partition (i.e., neighboring agents in the Delaunay graph, dual to the partition (7.14)).

The sensor-based policy is more efficient than the no-communication policy (with unlimited sensing capabilities) in terms of the length of the path traveled by each agent, since there is no duplication of effort as several agents pursue the same target. However, in terms of 'quality of service,' we will show that there is no difference between the two policies, for low target generation rates. Numerical results show that the sensor-based policy is more efficient in a broader range of target generation rates, and in fact provides almost optimal performance both in light and heavy load conditions. Finally notice that the sensor-based policy and the no-communication policy (with unlimited sensing capabilities) are *dynamical* policies, in the sense that they are able to adapt in principle to a spatial distribution of the target generation process which can (slowly) vary

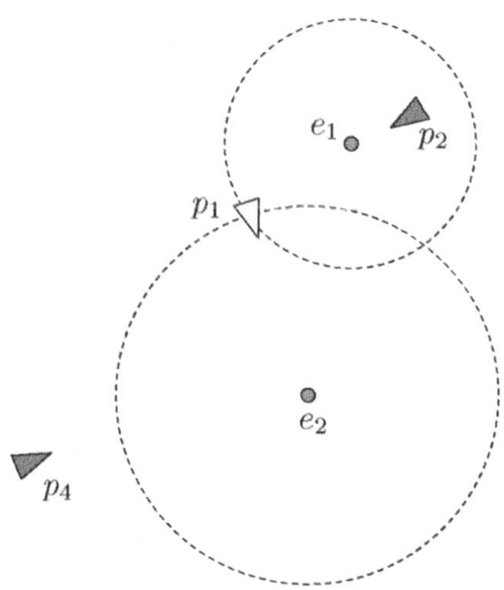

Figure 7.2 Implicit computation of Voronoi regions: Even though target e_1 is the nearest target to p_1, it is not in the Voronoi region of the 1st agent. In fact, the circle of radius $\|e_1 - p_1\|$ centered at e_1 contains p_2, and the 2nd agent is closer to e_1. However, the circle of radius $\|e_2 - p_1\|$ centered at e_2 does not contain any other agent, ensuring that e_2 is in the Voronoi region generated by p_1.

in time. Compared to these, the sTP, the mTP/FG and even the mTP policies do not have this adaptive characteristics. Anyway, in Section 7.5 we will analyze the performance also of the policies sTP, mTP/FG and mTP in some asymptotic cases.

7.4 PERFORMANCE ANALYSIS IN LIGHT LOAD

In this section we analyze the performance of the control policy no communication (with unlimited sensing capabilities) and sensor-based control policy in the light load case, in which the target generation rate is very small, i.e., as $\lambda \to 0^+$. This will allow us to prove analytically certain interesting and perhaps surprising characteristics of the proposed policies. The performance analysis in the general case is more difficult; we will discuss the results of numerical investigation in Section 7.6, but no analytical results are available at this time. A performance analysis for the policies sTP, mTP/FG and mTP will be presented in Section 7.5.

7.4.1 Overview of the system behavior in the light load regime

Before starting a formal analysis, let us summarize the key characteristics of the agents' behavior in light load, i.e., for small values of λ.

1. At the initial time the m agents are assumed to be deployed in general position in Ω, and the demand queue is empty, $\mathcal{D}(0) = \emptyset$.
2. The agents do not move until the first service request appears. At that time, if the policy π_{nc} is used, all agents will start moving towards the first target. If the sensor-based policy π_{sb} is used, only the closest agent will move towards the target.
3. As soon as one agent reaches the target, all agents start moving towards their current reference point, and the process continues.

For small λ, with high probability (i.e., with probability approaching 1 as $\lambda \to 0$) at most one service request is outstanding at any given time. In other words, new service requests are generated so rarely that most of the time agents will be able to reach a target and return to their reference point before a new service request is issued.

Consider the j-th service request, generated at time t_j. Assuming that at t_j all agents are at their reference position, the expected system time T_j can be computed as

$$T_j = \int_\Omega \min_{i=1,\dots,m} \|p_i^*(t_j) - q\| \, \varphi(q)dq.$$

Assume for now that the sequences $\{p_i^*(t_j) : j \in \mathbb{N}\}$ converge, and let

$$\lim_{j\to\infty} p_i^*(t_j) = \hat{p}_i^*.$$

Note that $\hat{p}_i^*$ is a random variable, the value of which depends in general on the particular realization of the target generation process. If all service requests are generated with the agents at their reference position, the average service time (for small λ) can be evaluated as

$$\overline{T}_{\mathrm{nc}} = \overline{T}_{\mathrm{sb}} = \int_{\Omega} \min_{i=1,\ldots,m} \|\hat{p}_i^* - q\| \, \varphi(q)dq = \sum_{i=1}^{m} \int_{\mathcal{V}_i(p^*)} \|\hat{p}_i^* - q\| \, \varphi(q)dq. \tag{7.17}$$

Since the system time depends on the random variable $\hat{p}^* = (\hat{p}_1^*, \ldots, \hat{p}_m^*)$, it is itself a random variable. The function appearing on the right-hand side of the above equation, relating the system time to the asymptotic location of reference points, is called the continuous multi-median function (Drezner 1995). This function admits a global minimum (in general not unique) for all non-singular density functions φ, and in fact it is known (Bertsimas and van Ryzin 1991) that the optimal performance in terms of system time is given by

$$\overline{T}_{\mathrm{opt}} = \min_{p \in \Omega^m} \sum_{i=1}^{m} \int_{\mathcal{V}_i(p)} \|p_i - q\| \, \varphi(q)dq. \tag{7.18}$$

In the following, we will investigate the convergence of the reference points as new targets are generated, in order to draw conclusions about the average system time $\overline{T}$ in light load. In particular, we will prove not only that the reference points converge with high probability (as $\lambda \to 0$) to a local critical point (more precisely, either local minima or saddle points) for the average system time, but also that the limiting reference points $\hat{p}^*$ are *generalized medians* of their respective Voronoi regions, where

Definition 7.4.1 (Generalized median) *The generalized median of a set $\mathcal{S} \subset \mathbb{R}^n$ with respect to a density function $\varphi : \mathcal{S} \to \overline{\mathbb{R}}_+$ is defined as*

$$\overline{p} := \arg\min_{p \in \mathbb{R}^n} \int_{\mathcal{S}} \|p - q\| \varphi(q) \, dq.$$

We call the resulting Voronoi tessellation *Median Voronoi Tessellation* (MVT for short), analogy with what is done with Centroidal Voronoi Tessellations. A formal definition is as follows:

Definition 7.4.2 (Median Voronoi Tessellation) *A Voronoi tessellation $\mathcal{V}(p) = \{\mathcal{V}_1(p), \ldots, \mathcal{V}_m(p)\}$ of a set $\mathcal{S} \subset \mathbb{R}^n$ is said to be a Median Voronoi Tessellation of $\mathcal{S}$ with respect to the density function φ if the ordered set of generators p is equal to the ordered set of generalized medians of the sets in $\mathcal{V}(p)$ with respect to φ, i.e., if*

$$p_i = \arg\min_{s \in \mathbb{R}^n} \int_{\mathcal{V}_i(p)} \|s - q\| \varphi(q) \, dq, \qquad \forall i \in \{1, \ldots, m\}.$$

Since the proof builds on a number of intermediate results, we provide an outline of the argument as a convenience to the reader.

1. First we prove that the reference point of any agent that visits an unbounded number of targets over time converges almost surely.

2. Second, we prove that, if $m \geq 1$ agents visit an unbounded number of targets over time, their reference points will converge to the generators of a MVT almost surely, as long as agents are able to return to their reference point infinitely often.
3. Third, we prove that all agents will visit an unbounded number of targets (this corresponds to a property of distributed algorithms that is often called *fairness* in computer science).
4. Finally, we prove that agents are able to return to their reference point infinitely often with high probability as $\lambda \to 0^+$.

Combining these steps, together with (7.7), will allow us to state that the reference points converge to a local critical point of the system time, with high probability as $\lambda \to 0^+$.

7.4.2 Convergence of reference points

Let us consider an agent i, such that

$$\lim_{t\to\infty} \mathrm{card}(\mathcal{B}_i(t)) = \infty,$$

i.e., an agent that services an unbounded number of requests over time. Since the number of agents m is finite, and the expected number of targets generated over a time interval $[0, t)$ is proportional to t, at least one such agent will always exist. In the remainder of this section, we will drop the subscript i, since we will consider only this agent, effectively ignoring all others for the time being.

For any finite t, the set $\mathcal{B}(t)$ will contain a finite number of points. Assuming that $\mathcal{B}(t)$ contains at least three non-collinear points, the discrete Weber function $p \mapsto \sum_{q\in\mathcal{B}(t)} \|p - q\|$ is strictly convex, and has a unique optimizer $p^*(t) = \arg\min_{p\in\Omega} \sum_{q\in\mathcal{B}(t)} \|p - q\|$. The optimal point $p^*(t)$ is called the Fermat-Torricelli (FT) point – or the Weber point in the location optimization literature – associated with the set $\mathcal{B}(t)$; (see Agarwal and Sharir 1998; Chandrasekaran and Tamir 1990; Wesolowsky 1993) for a historical review of the problem and for solution algorithms. It is known that the FT point is unique and algebraic for any set of non-collinear points. While there are no general analytic solutions for the location of the FT point associated to more than 4 points, numerical solutions can easily be constructed relying on the convexity of the Weber function, and on the fact that is differentiable for all points not in $\mathcal{B}$. Polynomial-time approximation algorithms are also available (see, e.g., Carmi *et al.* 2005; Fekete *et al.* 2000). Remarkably, a simple mechanical device can be constructed to solve the problem, based on the so-called Varignon frame, as follows. Holes are drilled on a horizontal board, at locations corresponding to the points in $\mathcal{B}$. A string attached to a unit mass is passed through each of these holes, and all strings are tied together at one end. The point reached by the knot at equilibrium is a FT point for $\mathcal{B}$.

Some useful properties of FT points are summarized below. If there is a $q_0 \in \mathcal{B}$ is such that

$$\left\| \sum_{q\in\mathcal{B}\setminus q_0} \mathrm{vers}(q_0 - q) \right\| \leq 1 \tag{7.19}$$

then $p^* = q_0$ is a FT point for $\mathcal{B}$. If no point in $\mathcal{B}$ satisfies such condition, then the FT point p^* can be found as a solution of the following equation:

$$\sum_{q \in \mathcal{B}} \text{vers}(p^* - q) = 0. \tag{7.20}$$

In other words, p^* is simply the point in the plane at which the sum of unit vectors starting from it and directed to each of the points in $\mathcal{B}$ is equal to zero; this point is unique if $\mathcal{B}$ contains non-collinear points. Clearly, the FT point is in the convex hull of $\mathcal{B}$.

Note that the points in $\mathcal{B}(t)$ are randomly sampled from an unknown absolutely continuous distribution, described by a spatial density function $\tilde{\varphi}$ – which is not necessarily the same as φ, and in general is time-varying, depending on the past actions of all agents in the system. Even though $\tilde{\varphi}$ is not known, it can be expressed as

$$\varphi(q, t) = \begin{cases} \dfrac{\varphi(q)}{\int_{\mathcal{I}(t)} \varphi(q)\, dq} & \text{if } q \in \mathcal{I}(t) \\ 0 & \text{otherwise,} \end{cases}$$

for some convex set $\mathcal{I}(t)$ containing $p(t)$. (In practical terms, such a set will be the Voronoi region generated by $p(t)$.)

The function $t \mapsto p^*(t)$ is piecewise constant, i.e., it changes value at the discrete time instants $\{t_j : j \in \mathbb{N}\}$ at which the agent visits new targets. As a consequence, we can concentrate on the sequence $\{p^*(t_j) : j \in \mathbb{N}\}$, and study its convergence.

Definition 7.4.3 *For any $t > 0$, let the* solution set *$C(t)$ be defined as*

$$C(t) := \left\{ p \in \Omega : \left\| \sum_{q \in \mathcal{B}(t)} \text{vers}(p - q) \right\| \leq 1 \right\}.$$

An example of such set is shown in Figure 7.3. The reason for introducing such solution sets is that they have quite remarkable properties as shown by the following

Proposition 7.4.4 *For any $j \in \mathbb{N}$, $p^*(t_{j+1}) \in C(t_j)$. More specifically, if $e_{j+1} \notin C(t_j)$ (i.e., the target point associated to the j-th service request is outside the solution set) then the FT point $p^*(t_{j+1})$ is on the boundary of C_i. If $e_{j+1} \in C(t_j)$, then $p^*(t_{j+1}) = e_{j+1}$.*

Proof. If e_{j+1} lies outside $C(t_j)$, we search for $p^*(t_{j+1})$ as the solution of the equation

$$\sum_{q \in \mathcal{B}(t_j)} \text{vers}(p - q) + \text{vers}(p - e_{j+1}) = 0,$$

from which it turns out immediately

$$\left\| \sum_{q \in \mathcal{B}(t_j)} \text{vers}(p - q) \right\| = \left\| -\text{vers}(p - e_{j+1}) \right\| = 1,$$

thus $p^*(t_{j+1}) \in \partial C(t_j)$. Notice that it is not true in general that the solution $p^*(t_{j+1})$ will lie on the line connecting $p^*(t_j)$ with the new target e_{j+1}. In the other case, if e_{j+1} lies in $C(t_j)$, then it satisfies condition (7.19), and is the new FT point.

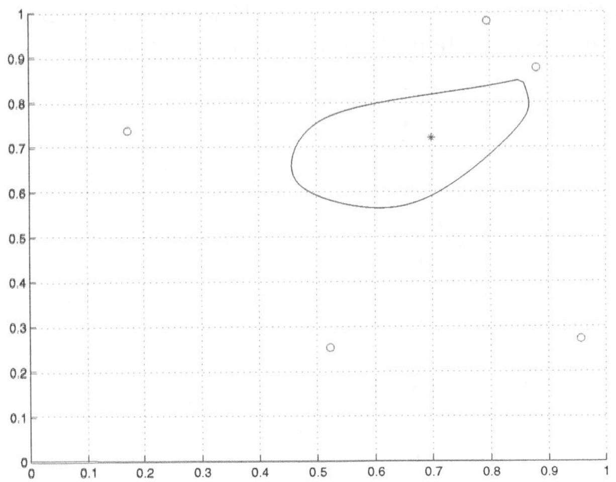

Figure 7.3 Example of a Fermat-Torricelli point (star) and solution set corresponding to five target points (circles). Upon the addition of an arbitrarily chosen sixth target point, the Fermat-Torricelli is guaranteed to remain within the region bounded by the curve.

Now, in order to prove that the $\{p^*(t_j)\}_{j\in\mathbb{N}}$ converges to a point $\hat{p}^*$, we will prove that the diameter of the solution set $C(t_j)$ vanishes almost surely as j tends to infinity. First we need the following result.

Proposition 7.4.5 *If $\mathcal{Q} = \mathrm{Supp}(\varphi)$ is convex with non-empty interior, then $p^*(t) \in \mathrm{int}(\mathcal{Q})$ almost surely, for all t such that $\mathcal{B}(t) \neq \emptyset$.*

Proof. For any non-empty target set $\mathcal{B}(t)$, the FT point lies within the convex hull of $\mathcal{B}(t)$. All points in the set $\mathcal{B}(t)$ are contained within the interior of $\mathcal{Q}$ with probability one, since the boundary of $\mathcal{Q}$ is a set of measure zero. Since $\mathrm{int}(\mathcal{Q})$ is convex, and $\mathcal{B}(t) \subset \mathrm{int}(\mathcal{Q})$ almost surely, $p^*(t) \in \mathrm{co}(\mathcal{B}(t)) \subset \mathrm{int}(\mathcal{Q})$, almost surely.

Proposition 7.4.6 *If the support of φ is convex and bounded,*

$$\lim_{j\to\infty} \mathrm{diam}(C(t_j)) = 0, \qquad \text{a.s.}$$

Proof. Consider a generic point $p \in C(t_j)$, and let $\delta = p - p^*(t_j)$, and consider $\alpha_q = \arccos\left[\mathrm{vers}(p - p^*(t_j)) \cdot \mathrm{vers}(q - p^*(t_j))\right]$, and $\alpha'_q = \arccos[\mathrm{vers}(p - p^*(t_j)) \cdot \mathrm{vers}(q - p)]$, see Figure 7.4.

Since $p \in C(t_j)$, the magnitude of the sum of unit vectors $\sum_{q\in\mathcal{B}(t_j)} \mathrm{vers}(p - q)$ is no more than one, and the following inequality is true:

$$\left|\left(\sum_{q\in\mathcal{B}(t_j)} \mathrm{vers}(p - q)\right) \cdot \mathrm{vers}(p - p^*(t_j))\right|$$

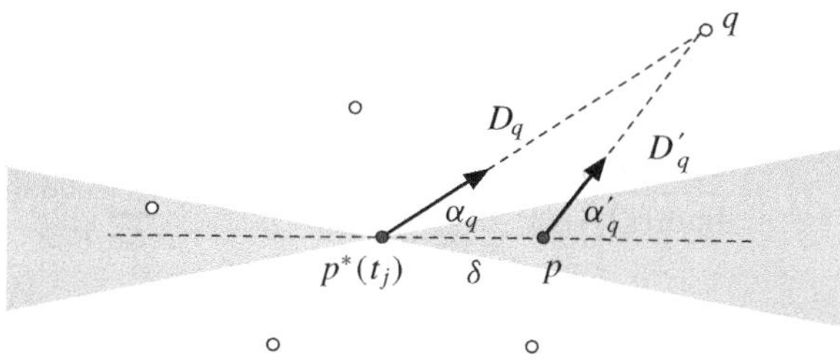

Figure 7.4 Geometric constructions in the proof of Proposition 7.4.6.

$$= \left| \sum_{q \in \mathcal{B}(t_j)} \left(\text{vers}(p - q) \cdot \text{vers}(p - p^*(t_j)) \right) \right|$$

$$= \left| \sum_{q \in \mathcal{B}(t_j)} \cos(\alpha'_q) \right| \leq 1. \quad (7.21)$$

Using elementary planar geometry, we obtain that

$$\alpha'_q - \alpha_q \geq \sin(\alpha'_q - \alpha_q) = \frac{\delta \sin(\alpha_q)}{\|q - p\|}.$$

Pick a small angle $0 < \alpha_{\min} < \pi/2$, and let

$$\mathcal{B}_{\alpha_{\min}}(t_j) = \{q \in \mathrm{B}(t_j) : \sin(\alpha_q) \geq \sin(\alpha_{\min})\}.$$

(In other words, $\mathcal{B}_{\alpha_{\min}}(t)$ contains all points in $\mathcal{B}$ that are not in a conical region of half-width $\alpha_{\min}$, as shown in Figure 7.4). For all $q \in \mathcal{B}(t_j)$, $\cos(\alpha'_q) \leq \cos(\alpha_q)$; moreover, for all $q \in \mathcal{B}_{\alpha_{\min}}$,

$$\cos(\alpha'_q) \leq \cos(\alpha_q) - \sin(\alpha_{\min})(\alpha'_q - \alpha_q) \leq \cos(\alpha_q) - \frac{\delta \sin(\alpha_{\min})^2}{\|q - p\|} \leq$$

$$\leq \cos(\alpha_q) - \frac{\delta \sin(\alpha_{\min})^2}{\text{diam}(\mathcal{Q}) + \delta}.$$

Hence, summing over all $q \in \mathcal{B}(t_j)$, we get:

$$\sum_{q \in \mathcal{B}(t_j)} \cos(\alpha'_q) \leq \sum_{q \in \mathcal{B}(t_j)} \cos(\alpha_q) - \sum_{q \in \mathcal{B}_{\alpha_{\min}}(t_j)} \frac{\delta \sin(\alpha_{\min})^2}{\text{diam}(\mathcal{Q}) + \delta} \quad (7.22)$$

Observe now that in any case

$$\left| \sum_{q \in \mathcal{B}(t_j)} \cos(\alpha_q) \right| \leq 1,$$

(it is zero in case $p^*(t_j) \notin \mathcal{B}(t_j)$, and bounded in absolute value by one if $p^*(t_j) \in \mathcal{B}(t_j)$). Therefore, rearranging equation (7.22):

$$\mathrm{card}(\mathcal{B}_{\alpha_{\min}}(t_j))\frac{\delta \sin(\alpha_{\min})^2}{\mathrm{diam}(\mathcal{Q}) + \delta} \leq \sum_{q\in\mathcal{B}(t_j)} \cos(\alpha_q) - \sum_{q\in\mathcal{B}(t_j)} \cos(\alpha'_q) \leq 2$$

Solving this inequality with respect to δ we get:

$$\delta \leq \frac{2\,\mathrm{diam}(\mathcal{Q})}{\mathrm{card}(\mathcal{B}_{\alpha_{\min}}(t_j))\sin(\alpha_{\min})^2 - 2}. \tag{7.23}$$

Since (i) $\alpha_{\min}$ is a positive constant, (ii) $p^*(t_j)$ is in the interior of $\mathcal{Q}$, and moreover we have that (iii) $\lim_{j\to\infty} \mathrm{card}(\mathcal{B}(t_j)) = +\infty$, the right-hand side of (7.23) converges to zero with probability one. Since the bound holds for all points $p \in C(t_j)$, for all $j \in \mathbb{N}$, the claim follows.

In the previous proposition, we have proven that $\|p^*(t_{j+1}) - p^*(t_j)\|$ tends to zero a.s. as $j \to \infty$, under some natural assumptions on the distribution φ and its support. Unfortunately this is not sufficient to prove that the sequence $\{p^*(t_j)\}_{t_j\in\mathbb{N}}$ is Cauchy; convergence of the sequence is, however, ensured by the following

Proposition 7.4.7 *The sequence $\{p^*(t_j)\}_{j\in\mathbb{N}}$ converges almost surely.*

Proof. See (Arsie and Frazzoli 2006).

7.4.3 Convergence to the generalized median

From the discussion in the previous section, we know that the reference points of all agents that visit an unbounded number of targets converge to a well-defined limit. So do, trivially, the reference points of all agents that visit a bounded set of targets. Hence, we know that the sequence of reference points $p_i^*(t_j)$ converges to a limit $\hat{p}_i^*$, almost surely for all $i \in \{1, \ldots, m\}$. Let us denote by $\mathcal{V}_i(p^*(t_j))$ the Voronoi region associated to the generator $p_i^*(t_j)$, and by $\mathcal{V}_i(\hat{p}^*)$ the Voronoi region corresponding to the limit point $\hat{p}_i^*$.

Proposition 7.4.8 *If the limit reference points $\hat{p}^* = (\hat{p}_1^*, \ldots, \hat{p}_2^*)$ are distinct, then the sequence of Voronoi partitions $\{\mathcal{V}(p^*(t_j))\}_{j\in\mathbb{N}}$ converges to the Voronoi partition generated by the limit of reference points, i.e.,*

$$\lim_{j\to\infty} \mathcal{V}_i(p^*(t_j)) = \mathcal{V}_i(\hat{p}^*), \qquad \text{a.s.}$$

Proof. The boundaries of regions in a Voronoi partition are algebraic curves that depend continuously on the generators, as long as these are distinct. Hence, under this assumption, almost sure convergence of the generators implies the almost sure convergence of the Voronoi regions.

As a next step, we wish to understand what is the relation between the asymptotic reference positions and their associated Voronoi regions. More precisely, let $\mathcal{A} \subset \{1, \ldots, m\}$ be the subset of indices of agents that visit an unbounded number of targets; we want to prove that $\hat{p}_i^*$ is indeed the generalized median $\overline{p}_i$ associated to agent i, with respect to the limiting set $\mathcal{V}_j(\hat{p}^*)$ and distribution $\varphi(x)$, $\forall i \in \mathcal{A}$. First, we need the following technical result.

Lemma 7.4.9 *Let $\{f_i\}_{i \in \mathbb{N}} : \mathcal{Q} \to \mathbb{R}$ be a sequence of strictly convex continuous functions, defined on a common compact subset $\mathcal{Q} \subset \mathbb{R}^n$. Assume that each f_i has a unique $x_i := \arg\min_x f_i$ belonging to the interior of Ω for any i and that this sequence of function converges uniformly to a continuous strictly convex function f admitting a unique minimum point $\overline{x}$ belonging also to the interior of Ω. Then $\lim_{i \to \infty} x_i = \overline{x}$.*

Proof. See (Arsie and Frazzoli 2006).

We conclude this section with the following:

Proposition 7.4.10 *Assume that all agents in $\mathcal{A}$ are able to return infinitely often to their reference point between visiting two targets. Then, the limit reference points of such agents coincide, almost surely, with the generalized medians of their limit Voronoi regions, i.e.,*

$$\hat{p}_i^* = \arg\min_{p \in \Omega} \int_{\mathcal{V}_i(\hat{p}^*)} \varphi(q) dq, \qquad \text{a.s.,} \qquad \forall i \in \mathcal{A}.$$

Proof. See (Arsie and Frazzoli 2006).

7.4.4 Fairness and efficiency

In this section, we show that, as long as φ is strictly positive over a convex set, the non-communication policy and the sensor-based policy introduced in Section 7.3 are fair. Moreover, we show that the system time provided by either one of these two policies converges to a critical point (either a saddle point or a local minimum) with high probability as $\lambda \to 0$.

Proposition 7.4.11 (Fairness) *If $\mathcal{Q} = \text{Supp}(\varphi)$ is convex, all agents eventually visit an unbounded number of targets, almost surely, i.e.,*

$$\lim_{t \to +\infty} \text{card}(\mathcal{B}_i(t)) = +\infty, \qquad \text{a.s.,} \qquad \forall i \in \{1, \ldots, m\},$$

under either the non-communication policy or the sensor-based policy.

Proof. See (Arsie and Frazzoli 2006).

We have seen that – as long as each agent is able to return to its reference point between servicing two targets, infinitely often – the reference points $p_i^*(t_j)$ converge to points $\hat{p}_i^*$, which generate a MVT. In such case, we know that the average time of service will converge to

$$\overline{T}_\pi = \int_\Omega \min_{i=1,\ldots,m} \|\hat{p}_i^* - q\| \; \varphi(q) dq = \sum_{i=1}^{m} \int_{\mathcal{V}_i(\hat{p}^*)} \|\hat{p}_i^* - q\| \; \varphi(q) dq. \tag{7.24}$$

Consider now functions $\mathcal{H}_m$ of the form:

$$\mathcal{H}_m(p_1, \ldots, p_m) = \int_\Omega \min_{i=1,\ldots,m} \|p_i - q\| \; \varphi(q)dq = \sum_{i=1}^{m} \int_{\mathcal{V}_i(p)} \|p_i - q\| \; \varphi(q)dq. \quad (7.25)$$

Observe that $\overline{T}_\pi$ belongs to the class of functions of the form $\mathcal{H}_m$ where each point p_i is constrained to be the generalized median of the corresponding Voronoi region (i.e., $\mathcal{V}(p)$ is a MVT).

We want to prove that $\overline{T}_\pi$ is a critical point of $\mathcal{H}_m$. To do so, we consider an extension of $\mathcal{H}_m$, i.e. a functional $\mathcal{K}_m$ defined as follows:

$$\mathcal{K}_m(p_1, \ldots, p_m, \mathcal{V}_1, \ldots, \mathcal{V}_m) := \sum_{i=1}^{m} \int_{y \in \mathcal{V}_i} \|y - p_i\| \; \varphi(y)dy.$$

Observe that in this case the regions $\{\mathcal{V}_i\}_{i=1,\ldots,m}$ are not restricted to form a MVT with respect to the generators $\{x_i\}_{i=1,\ldots,m}$. Thus we can view the functional $\mathcal{H}_m$ we are interested in as a constrained form of the unconstrained functional $\mathcal{K}_m$. It turns out therefore that critical points of $\mathcal{K}_m$ are also critical points of $\mathcal{H}_m$. With respect to critical points of $\mathcal{K}_m$ we have the following result:

Proposition 7.4.12 *Let* $\{p_i\}_{i=1,\ldots,m}$ *denote* any set *of m points belonging to* Supp(φ) *and let* $\{\mathcal{V}_i\}_{i=1,\ldots,m}$ *denote any tessellation of* Supp(φ) *into m regions. Moreover, let us define* $\mathcal{K}_m$ *as above. Then a sufficient condition for* $\{p_1, \ldots, p_m, \mathcal{V}_1, \ldots, \mathcal{V}_m\}$ *to be a critical point (either a saddle point or a local minimum), is that the* $\mathcal{V}_i$*s are the Voronoi regions corresponding to the* p_i*s, and,* simultaneously, *the* p_i*s are the generalized median of the corresponding* $\mathcal{V}_i$*s.*

Proof. Consider first the variation of $\mathcal{K}_m$ with respect to a single point, say p_i. Now let v be a vector in $\mathbb{R}^2$, such that $p_i + \varepsilon v \in \Omega$. Then we have

$$\mathcal{K}_m(p_i + \varepsilon v) - \mathcal{K}_m(p_i) = \int_{y \in \mathcal{V}_i} \{\|y - p_i - \varepsilon v\| - \|y - p_i\|\} \; \varphi(y)dy,$$

where we have not listed the other variables on which $\mathcal{K}_m$ depends since they remain constant in this variation. By the very form of this variation, it is clear that if the point p_i is the generalized median for the *fixed* region $\mathcal{V}_i$, we will have that $\mathcal{K}_m(p_i + \varepsilon v) - \mathcal{K}_m(p_i) > 0$, for any v. Now consider the points $\{p_i\}_{i=1,\ldots,m}$ fixed and consider a tessellation $\{\mathcal{U}_i\}_{i=1,\ldots,m}$ different from the Voronoi regions $\{\mathcal{V}_i\}_{i=1,\ldots,m}$ generated by the points p_is. We compare the value of $\mathcal{K}_m(p_1, \ldots, p_m, \mathcal{V}_1, \ldots, \mathcal{V}_m)$, with the value of $\mathcal{K}_m(p_1, \ldots, p_m, \mathcal{U}_1, \ldots, \mathcal{U}_m)$. Consider those y which belong to the Voronoi region $\mathcal{V}_j$ generated by p_j, and possibly not to the Voronoi region of another p_i. Anyway, since $\mathcal{U}_i$ is not a Voronoi tessellation, it can happen that in any case these y belong to $\mathcal{U}_i$. Thus for these particular ys we have $\varphi(y)\|y - p_j\| \leq \varphi(y)\|y - p_i\|$. Moreover, since $\{\mathcal{U}_i\}_{i=1,\ldots,m}$ are not the Voronoi tessellation associated to the p_is, the last inequality must be strict over some set of positive measure. Thus we have that $\mathcal{K}_m(p_1, \ldots, p_m, \mathcal{V}_1, \ldots, \mathcal{V}_m) < \mathcal{K}_m(p_1, \ldots, p_m, \mathcal{U}_1, \ldots, \mathcal{U}_m)$, and therefore $\mathcal{K}_m$ is minimized, keeping fixed the p_is exactly when the subset $\mathcal{V}_i$s are chosen to be the Voronoi regions associated with the point p_is.

By the previous proposition and by the fact that critical points of the unconstrained functional $\mathcal{K}_m$ are also critical points of the constrained functional $\mathcal{H}_m$, we have that the MVT are always critical points for the functional $\mathcal{H}_m$, and in particular $\overline{T}$ is either a saddle point or a local minimum for the functional $\mathcal{H}_m$.

Before we conclude, we need one last intermediate result.

Proposition 7.4.13 *Each agent will be able to return to its reference point before the generation of a new service request infinitely often with high probability as* $\lambda \to 0$.

Proof. See (Arsie and Frazzoli 2006).

We can now conclude with following:

Theorem 7.4.14 (Efficiency) *The system time provided by the no-communication policy* π_{nc} *and by the sensor-based policy* π_{sb} *converges to a critical point (either a saddle point or a local minimum) with high probability as* $\lambda \to 0$.

Proof. Combining results in Propositions 7.4.7 and 7.4.10 we conclude that the reference points of all agents that visit an unbounded number of targets converge to a MVT, almost surely – provided agents can return to the reference point between visiting targets. Moreover, the fairness result in Proposition 7.4.11 shows that in fact all agents do visit an unbounded number of targets almost surely; as a consequence, Proposition 7.4.12 the limit configuration is indeed a critical point for the system time. Since agents return infinitely often to their reference positions with high probability as $\lambda \to 0$, the claim is proven.

Thus we have proved that the suggested algorithm enables the agents to realize a coordinated task, such that 'minimizing' the cost function without explicit communication, or with mutual position knowledge only. Let us underline that, in general, the achieved critical point strictly depends on the initial positions of the agents inside the environment Ω. It is known that the function $\mathcal{H}_m$ admits (not unique, in general) *global minima*, but the problem to find them is *NP-hard.*

We cannot exclude that the algorithm so designed will converge indeed to a saddle point instead of a local minimum. This is due to the fact that the algorithm provides a sort of implementation of the *steepest descent* method, where, unfortunately we are not following the steepest direction of the gradient of the function $\mathcal{H}_m$, but just the gradient with respect to one of the variables. For a broader point of view of steepest descent in this framework, see for instance (Okabe *et al.* 2000).

On the other hand, since the algorithm is based on a sequence of targets and at each phase we are trying to minimize a different cost function, it can be proved that the critical points reached by this algorithm are *no worse* than the critical points reached knowing a priori the distribution φ. This is a remarkable result proved in a different context by Sabin and Gray (1986), where also is presented an example in which the use of a sample sequence provides a better result (with probability one) than the a priori knowledge of φ. In that specific example the algorithm with the sample sequence does converge to a global minimum, while the algorithm based on the a priori knowledge of the distribution φ gets stuck in a saddle point.

7.4.5 A comparison with algorithms for vector quantization and centroidal Voronoi tessellations

The use of Voronoi tessellations is ubiquitous in many fields of science, ranging from operative research, animal ethology (territorial behaviour of animals), computer science (design of algorithms), to numerical analysis (construction of adaptive grids for PDEs and general quadrature rules), and algebraic geometry (moduli spaces of abelian varieties). For a detailed account of possible applications, see for instance (Okabe *et al.* 2000). In the available literature, most of the analysis is devoted to applications of centroidal Voronoi tessellations, i.e., Voronoi tessellation such that

$$p_i = \arg\min_{s\in\mathbb{R}^n} \int_{\mathcal{V}_i(p_1,p_2,\ldots,p_m)} \|s-q\|^2 \varphi(q)\, dq, \qquad \forall i \in \{1,\ldots,m\}.$$

A popular algorithm due to Lloyd (Lloyd 1982) is based on the iterative computation of centroidal Voronoi tessellations. The algorithm can be summarized as follows. Pick m generator points, and consider a large number n of samples from a certain distribution. At each step of the algorithm generators are moved towards the centroid of the samples inside their respective Voronoi region. The algorithm terminates when each generator is within a given tolerance from the centroid of samples in its region, thus obtaining a centroidal Voronoi tessellation weighted by the sample distribution. There is also a continuous version of the algorithm, which requires the a priori knowledge of a spatial density function, and computation of the gradient of the polar moments of the Voronoi regions with respect to the positions of the generators. An application to coverage problems in robotics and sensor networks of Lloyd's algorithm is available in (Cortés *et al.* 2004).

The presented algorithms are more closely related to an algorithm due to MacQueen (MacQueen 1967), originally designed as a simple online adaptive algorithm to solve clustering problems, and later used as the method of choice in several vector quantization problems where little information about the underlying geometric structure is available. MacQueen's algorithm can be summarized as follows. Pick m generator points. Then iteratively sample points according to the probability density function φ. Assign the sampled point to the nearest generator, and update the latter by moving it in the direction of the sample. In other words, let q_j be te j-th sample, let $i^*(j)$ be the index of the nearest generator, and let $c = (c_1, \ldots, c_m)$ be a counter-vector, initialized to a vector of ones. The update rule takes the form

$$p_{i^*(j)} \leftarrow \frac{c_{i^*(j)} q_j + p_{i^*(j)}}{c_{i^*(j)} + 1},$$

$$c_{i^*(j)} \leftarrow c_{i^*(j)} + 1.$$

The process is iterated until some termination criterion is reached. Compared to Lloyd's algorithm, MacQueen's algorithm has the advantage of being a learning adaptive algorithm, not requiring the a priori knowledge of the distribution of the objects, but rather allowing the online generation of samples. It is recognized that the update rule in MacQueen's algorithm corresponds to moving the generator points to the centroids of the samples assigned to them. The algorithm we propose is very similar in spirit to MacQueen's algorithm, however, there is a remarkable difference. MacQueen's algorithm deals with centroidal Voronoi tessellations, thus with the computation of k-means. Our algorithm instead is based on MVT, and on the computation of k-medians. In general, very little is known about Voronoi diagrams generated using simply the Euclidean distance instead of the square of the Euclidean distance. For instance, the medians of a sequence of points can exhibit a quite peculiar behavior if compared to the one of the means. Consider the following example. Given a sequence of points $\{q_i\}_{i\in\mathbb{N}}$ in a compact set $K \subset \mathbb{R}^2$, we can construct the induced sequence of means:

$$m_N := \frac{1}{N}\sum_{i=1}^{N} q_j$$

and analogously the induce sequence of FT points we considered in the previous sections. Call FT_N the FT point corresponding to the first N points of the sequence $\{q_i\}_{i\in\mathbb{N}}$. We want to point out that induced sequence $\{m_j\}$ and $\{FT_j\}$ have a very different behaviour. Indeed, the induced sequence of means will always converge as long as the points q_js belong to a compact set. To see this, just observe that if $\text{diam}(K) \leq L$, then $\|m_j\| \leq L$. Moreover, it is clear to see that $\|m_{N+1} - m_N\| \leq \frac{2L}{N}$. Then one can conclude using the same argument of Theorem (7.4.7). On the other hand, one can construct a special sequence of points q_js in a compact set K for which the induced sequence of FT points does not converge. This is essentially due to the fact that while the contribution of each single point q_j in moving the position of the mean decreases as j increases, this could not happen in the case of the median. To give a simple example, start with the following configuration of points in $\mathbb{R}^2$: $q_1 = (1, 0), q_2 = (-1, 0), q_3 = (0, 1)$ and $q_4 = (0, -1)$. Then the sequence of points q_js continues in the following way: $q_k = (0, 1)$ if $k > 4$, and k odd, $q_k = (0, -1)$ if $k > 4$ and k is even. Using the characterization of FT points, it is clear to see that $FT_k = (0, 0)$ for $k > 4$ and k even, while $FT_k = (0, \tan(\pi/6))$ for $k > 4$ and k odd, so the induced sequence cannot converge. This phenomenon cannot happen to the sequence of means, which is instead always convergent. Therefore, it should be clear that the use of MVT instead of centroidal Voronoi tessellations makes it much more difficult to deal with the technical aspects of the algorithm such as its convergence.

7.5 A PERFORMANCE ANALYSIS FOR sTP, mTP/FG AND mTP POLICIES

We first present some estimates of the achievable system time and then we compare the system time provided by sTP, mTP/FG and mTP algorithms to these estimates to assess the performance of these policies. Throughout all this section, we will assume that the spatial density function φ is *uniform* (and constant in time), namely we assume that the targets are generated uniformly in $\mathcal{Q}$. As a further reference for the material presented here, see (Enright and Frazzoli 2006).

7.5.1 The case of sTP policy

Let us first state a lower bound on the system time that always holds, even though it is particularly useful when the target generation rate is very small.

Theorem 7.5.1 *If there exists a point $p \in \mathcal{Q}$ such that $\mathcal{S}(p) \subseteq \mathcal{Q}$, then the system time satisfies*

$$\overline{T}_{\text{opt}} \geq \frac{A}{4\sigma v} + \left(\frac{\pi}{4} - \frac{1}{3}\right)\frac{\pi\sigma^3}{Av} + \left(1 - \frac{\pi}{2}\right)\frac{\sigma}{v}.$$

Proof. Consider the generic target point $e_j \in \mathcal{Q}$, and indicate with p the position of the agent at the time e_j is generated. We can distinguish the following two mutually exclusive cases.

- If the target is generated with the agent's sensor footprint, the minimum time needed by the agent to visit it is proportional to the distance between the agent and the target, i.e., $T_j = \|e_j - p\|/v$.
- If the target is generated outside the agent's sensor footprint, it is not detected until the time at which the agent moves in such a way that it detects the target. The time to visit the target in this case will be the sum of the time spent by the agent searching for it, plus a term that is no smaller than σ/v.

The probability of the first case occurring is proportional to the ratio of the area of the intersection of the agent's sensing area with the environment. Let $A_s(p) = \text{Area}(\mathcal{S}(p) \cap \mathcal{Q})$. A lower bound on the expected time to visit a target appearing within $\mathcal{S}(p) \cap \mathcal{Q}$ is given by the expected time to visit a point sampled (from a uniform distribution) within a circle of area $A_s(p)$, i.e.,

$$\text{E}[T_j | e_j \in \mathcal{S}(p)] = \int_{\mathcal{S}(p)\cap\mathcal{Q}} \frac{\|q - p\|}{A_s(p)v}\, dq \geq \frac{2}{3}\sqrt{\frac{A_s(p)}{\pi v^2}}.$$

If the target is generated outside of the sensing region $\mathcal{S}(p)$, the agent must first search for it. The area of the region within $\mathcal{Q}$ uncovered per unit time as the agent moves at speed v is at most $2\sigma v$. In order to search $\mathcal{Q} \setminus \mathcal{S}(p)$, the agent will hence need at least a time interval equal to $(A - A_s(p))/(2\sigma v)$. Since targets are generated uniformly on $\mathcal{Q}$, the expected search time will be one half of the time needed to search the whole environment. In other words,

$$\text{E}[T_j | e_j \notin \mathcal{S}(p)] = \frac{A - A_s(p)}{4\sigma v} + \frac{\sigma}{v}.$$

Summarizing, we get that

$$\text{E}[T_j] \geq \frac{2}{3}\sqrt{\frac{A_s(p)^3}{\pi A^2 v^2}} + \left(1 - \frac{A_s(p)}{A}\right)\left(\frac{A - A_s(p)}{4\sigma v} + \frac{\sigma}{v}\right). \tag{7.26}$$

The right-hand side of Equation (7.26) is a continuous function of $A_s(p)$ over the compact interval $[0, A]$, and therefore attains the minimum value; furthermore, it has only one critical point in the interval, which corresponds to a maximum. The minimum value is attained either at 0 or at the maximum value attainable by $A_s(p)$, i.e., at $A_s(p) = \pi\sigma^2$. (Recall that we assumed that there exists p such that $\mathcal{S}(p) \subseteq \mathcal{Q}$.) Direct substitution reveals that the minimum is in fact attained at $A_s(p) = \pi\sigma^2$. Substituting, rearranging terms, and noting that the bound holds uniformly for all targets, we obtain the claimed result.

Note that in the case in which σ is very small, the search term dominates and we get

$$\lim_{\sigma\to 0^+} \sigma \overline{T}_{\mathrm{opt}} \geq \frac{A}{4v} \tag{7.27}$$

When the target generation rate is very high, the bound in Theorem 7.5.1 is very optimistic. A tighter lower bound applicable in the heavy-load case ($\lambda \to +\infty$), and all values of σ, is given below.

Theorem 7.5.2 (Bertsimas and van Ryzin (Bertsimas and van Ryzin 1991)) *The system time in heavy load satisfies:*

$$\lim_{\lambda\to+\infty} \frac{\overline{T}_{\mathrm{opt}}}{\lambda} \geq \gamma^2 \frac{A}{v^2}, \tag{7.28}$$

where $\gamma = \frac{2}{3\sqrt{\pi}}$.

We are going to use these bounds to evaluate the performance of given policies in some asymptotic cases of interest, namely the heavy load and the small sensor range.

In order to assess the system time under the sTP policy we need the following two Propositions:

Proposition 7.5.3 *The number of tiles with a non-empty intersection with* $\mathcal{Q}$ *satisfies*

$$2\eta^2\sigma^2 \mathrm{card}(\mathcal{C}) = A + O\left(f(\eta^2\sigma^2)\right), \qquad (\eta\sigma \to 0^+), \tag{7.29}$$

where $f(x)$ *is a vanishing function for* $x \to 0^+$.

Proof. Obviously we have $2\eta^2\sigma^2 \mathrm{card}(\mathcal{C}) \geq A$ and moreover

$$\lim_{\eta\sigma\to 0^+} 2\eta^2\sigma^2 \mathrm{card}(\mathcal{C}) = A,$$

since the set $\mathcal{Q}$ is a convex compact domain of the plane with non-empty interior and so its measure can be computed via the Riemann measure.

Furthermore,

Proposition 7.5.4 *The length of a tour through the points in* $\mathcal{C}$ *satisfies:*

$$\eta\sigma\, \mathrm{ETSP}(\mathcal{C}) \leq \sqrt{2}A + O\left(h(\eta\sigma)\right), \qquad (\eta\sigma \to 0^+), \tag{7.30}$$

where $h(x)$ *is a vanishing function for* $x \to 0^+$.

Proof. Since all tiles corresponding to $\mathcal{C}$ are adjacent, and hence share an edge with at least another tile in $\mathcal{C}$, a minimum spanning tree of $\mathcal{C}$ will have length no greater than $\sqrt{2}\eta\sigma \cdot \mathrm{card}(\mathcal{C})$. A (non-minimal) tour of $\mathcal{C}$ can be built by doubling all edges in the spanning tree, hence

$$\mathrm{ETSP}(\mathcal{C}) \leq 2\,\mathrm{MST}(\mathcal{C}) \leq 2\sqrt{2}\eta\sigma \cdot \mathrm{card}(\mathcal{C}),$$

therefore, multiplying the previous chain of inequalities by $\eta\sigma$ and using (7.29) we get the result.

Then, the sTP performance in the limit in which the sensing radius is very small is a constant factor approximation of the optimum performance, as shown by the following:

Proposition 7.5.5 *The system time under the sTP policy, satisfies*

$$\lim_{\sigma\to 0^+} \frac{\overline{T}_{\mathrm{sTP}}}{\overline{T}_{\mathrm{opt}}} \leq \frac{2\sqrt{2}}{\eta} \tag{7.31}$$

Proof. We want to find out a bound for $\lim_{\sigma\to 0^+} \sigma\overline{T}_{\mathrm{sTP}}$ in order then to compare it with the bound for $\lim_{\sigma\to 0^+} \sigma\overline{T}_{\mathrm{opt}}$ already determined and get (7.31). Define a *phase* of the sTP algorithm as the interval of time between two consecutive arrivals of the agent at the point c_1. The duration of the j-th phase is given by

$$T_j^{\mathrm{phase}} = \frac{\mathrm{ETSP}(\mathcal{C})}{v} + \sum_{c\in\mathcal{C}} T_{c,j}^{\mathrm{tile}}, \tag{7.32}$$

where $T_{c,j}^{\mathrm{tile}}$ is the cost of visiting all targets in the tile centered at c at the j-th phase.

The expected cost of visiting all targets in the first tile at the $(j+1)$-th phase satisfies

$$\mathrm{E}[T_{1,j+1}^{\mathrm{tile}} | T_j^{\mathrm{phase}}] \leq \frac{\overline{\beta}_Q\sqrt{2}\eta\sigma}{v}\mathrm{E}[\sqrt{n_{1,j+1}}] \leq$$

$$\leq \frac{\overline{\beta}_Q\sqrt{2}\eta\sigma}{v}\sqrt{\frac{2\lambda\eta^2\sigma^2}{A}T_j^{\mathrm{phase}}} = 2\frac{\overline{\beta}_Q\eta^2\sigma^2}{v}\sqrt{\frac{\lambda T_j^{\mathrm{phase}}}{A}},$$

where $n_{1,j+1}$ is the number of targets in the first tile at the beginning of the $(j+1)$-th phase, and we applied Jensen's inequality. Since the choice of the 'first' tile is arbitrary, the bound on the expected time to visit all targets in a tile is in fact uniform over all tiles. Summing over all tiles in $\mathcal{C}$, we get:

$$\mathrm{E}[T_{j+1}^{\mathrm{phase}} | T_j^{\mathrm{phase}}] \leq \frac{\mathrm{ETSP}(\mathcal{C})}{v} + \mathrm{card}(\mathcal{C})\frac{2\overline{\beta}_Q\eta^2\sigma^2}{v}\sqrt{\frac{\lambda T_j^{\mathrm{phase}}}{A}},$$

that is, multiplying the previous inequality by σ, using (7.29), (7.30) and neglecting higher order terms as $\sigma \to 0^+$,

$$\sigma\mathrm{E}[T_{j+1}^{\mathrm{phase}} | T_j^{\mathrm{phase}}] \leq \sqrt{2}\frac{A}{\eta v} + \overline{\beta}_Q\sigma\frac{\sqrt{A}}{v}\sqrt{\lambda T_j^{\mathrm{phase}}}.$$

We want to find under which condition on T_j^{phase} we have

$$\sigma \mathrm{E}[T_{j+1}^{\text{phase}} | T_j^{\text{phase}}] \leq \sigma T_j^{\text{phase}}. \tag{7.33}$$

A sufficient condition is clearly given by:

$$\sqrt{2}\frac{A}{\eta v} + \overline{\beta}_Q \sigma \frac{\sqrt{A}}{v}\sqrt{\lambda T_j^{\text{phase}}} \leq \sigma T_j^{\text{phase}},$$

or equivalently

$$\sqrt{2}\frac{A}{\eta v} + \overline{\beta}_Q \frac{\sqrt{\lambda A \sigma}}{v}\sqrt{y} \leq y, \tag{7.34}$$

where $y = \sigma T_j^{\text{phase}}$. Inequality (7.34) is satisfied for

$$y \geq \frac{\sqrt{2}A}{\eta v}, \tag{7.35}$$

in the limit $\sigma \to 0^+$. Moreover, whenever (7.35) is strict, then also (7.33) is strict, always in the limit $\sigma \to 0^+$. Applying the Law of Total Expectation to (7.33), we get

$$\lim_{\sigma \to 0^+} \sigma \mathrm{E}[T_{j+1}^{\text{phase}}] = \lim_{\sigma \to 0^+} \sigma \mathrm{E}[\mathrm{E}[T_{j+1}^{\text{phase}} | T_j^{\text{phase}}]] \leq \lim_{\sigma \to 0^+} \sigma \mathrm{E}[T_j^{\text{phase}}] \tag{7.36}$$

Moreover, the inequality in (7.36) is strict as long as inequality (7.35) is strictly satisfied. It is therefore easy to see that it must be

$$\lim_{j \to +\infty} \lim_{\sigma \to 0^+} \sigma \mathrm{E}[T_j^{\text{phase}}] \leq \frac{\sqrt{2}A}{\eta v} \tag{7.37}$$

Indeed, assume that

$$\lim_{j \to +\infty} \lim_{\sigma \to 0^+} \sigma \mathrm{E}[T_j^{\text{phase}}] =: L > \frac{\sqrt{2}A}{\eta v},$$

then it means that inequality (7.35) is strictly satisfied eventually in j. Therefore inequality (7.36) is also strictly satisfied eventually in j, so

$$\lim_{\sigma \to 0^+} \sigma \mathrm{E}[T_{j+1}^{\text{phase}}] < \lim_{\sigma \to 0^+} \sigma \mathrm{E}[T_j^{\text{phase}}],$$

and taking the limit for $j \to \infty$ we get $L < L$. Contradiction.

On average, each target will have to wait one half of a phase before being visited, and we get the following estimate for the system time

$$\lim_{\sigma \to 0^+} \sigma \overline{T}_{\text{sTP}} \leq \frac{\sqrt{2}A}{2\eta v}, \tag{7.38}$$

from which we get the desired result, using equation (7.27).

Let us turn to the case in which the target generation rate is very high.

Proposition 7.5.6 *The system time under the sTP policy, with* $\eta = \Theta(\frac{1}{\lambda^\alpha})$, $0 < \alpha < 1$, *satisfies*

$$\lim_{\lambda \to +\infty} \frac{\overline{T}_{\text{sTP}}}{\overline{T}_{\text{opt}}} \leq \frac{\beta^2}{2\gamma^2}$$

Proof. Proceeding as in the previous case, define a *phase* of the sTP algorithm as the interval of time between the consecutive arrivals of the agent at the point c_1. The duration of the j-th phase is given by

$$T_j^{\text{phase}} = \frac{\text{ETSP}(C)}{v} + \sum_{c \in C} T_{c,j}^{\text{tile}}. \tag{7.39}$$

The expected cost of visiting all targets in the first tile at the $(j+1)$-th phase satisfies (note that when $\lambda \to +\infty$ we can apply Theorem 7.3.1)

$$\text{E}[T_{1,j+1}^{\text{tile}} | T_j^{\text{phase}}] \leq \frac{\beta\sqrt{2}\eta\sigma}{v} \text{E}[\sqrt{n_{1,j+1}}] \leq \frac{\beta\sqrt{2}\eta\sigma}{v}\sqrt{\frac{2\lambda\eta^2\sigma^2}{A} T_j^{\text{phase}}} = 2\frac{\beta\eta^2\sigma^2}{v}\sqrt{\frac{\lambda T_j^{\text{phase}}}{A}}.$$

Summing over all tiles in C, we get:

$$\text{E}[T_{j+1}^{\text{phase}} | T_j^{\text{phase}}] \leq \frac{\text{ETSP}(C)}{v} + \text{card}(C)\frac{2\beta\eta^2\sigma^2}{v}\sqrt{\frac{\lambda T_j^{\text{phase}}}{A}}.$$

Using the bounds (7.29), (7.30) we the fact that $\eta = \Theta(\frac{1}{\lambda^\alpha})$ (so that $\eta = c/(\lambda)^\alpha$) for a suitable constant c, we get

$$\text{E}[T_{j+1}^{\text{phase}} | T_j^{\text{phase}}] \leq \frac{\sqrt{2}A\lambda^\alpha}{c\sigma v} + \beta\frac{\sqrt{A}}{v}\sqrt{\lambda T_j^{\text{phase}}}. \tag{7.40}$$

The following inequality

$$\frac{\sqrt{2}A\lambda^\alpha}{c\sigma v} + \beta\frac{\sqrt{A}}{v}\sqrt{\lambda T_j^{\text{phase}}} \leq T_j^{\text{phase}}, \tag{7.41}$$

is satisfied (in the limit $\lambda \to +\infty$), whenever

$$\lim_{\lambda \to +\infty} \frac{T_j^{\text{phase}}}{\lambda} \geq \frac{\beta^2 A}{v^2}, \tag{7.42}$$

as it is immediate to verify. Moreover, (7.41) is strict whenever (7.42) is strict. Therefore, using the same technique used in the proof of the previous Proposition, we can claim that

$$\lim_{j \to +\infty} \lim_{\lambda \to +\infty} \frac{\text{E}[T_j^{\text{phase}}]}{\lambda} \leq \frac{\beta^2 A}{v^2}. \tag{7.43}$$

On average, each target will have to wait one half of a phase before being visited, and we get the following estimate for the system time

$$\lim_{\lambda \to +\infty} \frac{\overline{T}_{\mathrm{sTP}}}{\lambda} \leq \frac{\beta^2}{2} \frac{A}{v^2}, \tag{7.44}$$

from which we get the desired result, using equation (7.28).

7.5.2 The case of mTP/FG and mTP policies

We know that the mTP/FG policy is a direct extension of the sTP policy, in which any agent basically behaves according to a sTP policy in its own region of dominance. So, after having recalled that $A_i(g) = \mathrm{Area}[\mathcal{V}_i(g) \cap \mathcal{Q}]$, for all $i \in \{1, \ldots, m\}$, it is not hard to prove the following:

Proposition 7.5.7 *The system time for the mTP/FG policy satisfies the following inequalities:*

$$\lim_{\sigma \to 0+} \frac{\overline{T}_{\mathrm{mTP/FG}}}{\overline{T}_{\mathrm{opt}}} \leq \frac{2\sqrt{2}}{\eta} \frac{\sum_{i=1}^{m} A_i(g)^2}{A^2/m}, \tag{7.45}$$

and

$$\lim_{\lambda \to +\infty} \frac{\overline{T}_{\mathrm{mTP/FG}}}{\overline{T}_{\mathrm{opt}}} \leq \frac{\beta^2}{2\gamma^2} \frac{\sum_{i=1}^{m} A_i(g)^2}{A^2/m} \quad \textit{when } \eta = \Theta\left(\frac{1}{\lambda^\alpha}\right), \quad \alpha < 1 \tag{7.46}$$

Proof. The $\overline{T}_{\mathrm{mTP/FG}}$ is defined as

$$\overline{T}_{\mathrm{mTP/FG}} = \sum_{i=1}^{m} \overline{T}_{\mathrm{sTP(i)}} \cdot \frac{A_i(g)}{A},$$

where $\overline{T}_{\mathrm{sTP(i)}}$ is the corresponding system time in the region serviced by agent i. Using the bound (7.38) applied at each agent we get:

$$\lim_{\sigma \to 0+} \sigma \overline{T}_{\mathrm{mTP/FG}} \leq \frac{\sqrt{2}}{2\eta v} \sum_{i=1}^{m} \frac{A_i(g)^2}{A}. \tag{7.47}$$

In this situation, $\overline{T}_{\mathrm{opt}}$ is just the $\overline{T}_{\mathrm{opt}}$ computed in the case of a single agent, divided by the number of the agents. Therefore, using bound (7.27), we get:

$$\lim_{\sigma \to 0+} \sigma \overline{T}_{\mathrm{opt}} \geq \frac{A}{4vm} \tag{7.48}$$

Combining (7.47) and (7.48) we immediately get (7.45). To get the second part of the Proposition we reason analogously. Namely, since as before

$$\overline{T}_{\text{mTP/FG}} = \sum_{i=1}^{m} \overline{T}_{\text{sTP(i)}} \cdot \frac{A_i(g)}{A},$$

using bound (7.44) applied to each single agent (this bound is valid provided that $\eta = \Theta(1/\lambda^\alpha)$), we get

$$\lim_{\lambda\to+\infty} \frac{\overline{T}_{\text{mTP/FG}}}{\lambda} \leq \frac{\beta^2}{2v^2}\sum_{i=1}^{m}\frac{A_i(g)^2}{A}, \qquad \text{when } \eta = \Theta\left(\frac{1}{\lambda^\alpha}\right). \tag{7.49}$$

Analogously, from (7.28) the optimal system time satisfies the following bound:

$$\lim_{\lambda\to+\infty} \frac{\overline{T}_{\text{opt}}}{\lambda} \geq \gamma^2 \frac{A}{mv^2}. \tag{7.50}$$

Again, combining (7.49) and (7.50), we get (7.46).

Since $\sum_{i=1}^{m} A_i(g) = A$, it is easy to show that the upper bounds in Proposition 7.5.7 are minimized when the areas of all Voronoi regions are equal, i.e., when $A_i(g) = A_j(g)$, for all $i, j \in \{1, \ldots, m\}$. As we have already observed, one way to accomplish $A_i(g) = A_j(g)$, for all $i, j \in \{1, \ldots, m\}$ is to update the virtual generators according to a gradient descent law. Remarkably, such gradient descent law can be computed in a decentralized way, that is to say, each agent must be aware only of the position of the virtual generators for its neighbors. In fact, by the definition of Voronoi regions, we have that

$$\frac{\partial A_i(g)}{\partial g_j} = \begin{cases} \frac{1}{2} L_{ij} \frac{g_j - g_i}{\|g_j - g_i\|} & \text{if } i \neq j \\ \frac{1}{2} \sum_{l \neq j} L_{lj} \frac{g_l - g_j}{\|g_l - g_j\|} & \text{if } i = j, \end{cases} \tag{7.51}$$

where L_{ij} is the length of the Voronoi region boundary shared by agents i and j within $\mathcal{Q}$. ($L_{ij} = 0$ if the two agents' regions have an empty intersection.) Therefore the mTP policy requires a certain amount of information exchange at least among neighboring agents in order to be implemented. Generally speaking, the mTP policy provides a better performance compared to the mTP/FG policy as we are going to show in the next few lines. It turns out, then, that in this case the effective decentralized information exchange is essential in order to get an improvement in the overall performance of system time. This should be contrasted with the policy analyzed in Section 7.4, where instead an optimal performance can be reached without any explicit communication among the agents.

It is easy to find out asymptotic estimates for the mTP policy. Indeed, these simply are consequences of the gradient descent law applied to a non-negative function, and of the properties of the single-vehicle policies:

Proposition 7.5.8 *The bounds for the system time in the mTP policy converge to a critical point of the bounds in Proposition 7.5.7. In particular, if the generators are close enough to the optimal configuration, the system time will locally converge to*

$$\lim_{\sigma \to 0^+} \frac{\overline{T}_{\mathrm{mTP}}}{\overline{T}_{\mathrm{opt}}} \le \frac{2\sqrt{2}}{\eta},$$

and

$$\lim_{\lambda \to +\infty} \frac{\overline{T}_{\mathrm{mTP/FG}}}{\overline{T}_{\mathrm{opt}}} \le \frac{\beta^2}{2\gamma^2}.$$

Proof. This is simply the result of applying the gradient descent law to the bounds of Proposition 7.5.7.

7.6 SOME NUMERICAL RESULTS

In this section, we present simulation results showing the performance of the proposed policies for various scenarios.

7.6.1 Uniform distribution, light load

In the numerical experiments, we first consider $m = 9$, choose $\mathcal{Q}$ as a unit square, and set $\varphi = 1$ (i.e., we consider a spatially uniform target-generation process). This choice allows us to easily determine the optimal placement of reference points, at the centers of a tesselation of $\mathcal{Q}$ into nine equal squares, and compute analytically the optimal system time. In fact, it is known that the expected distance of a point q randomly sampled from a uniform distribution within a square of side L from the center of the square c is

$$\mathrm{E}[\|q - c\|] = \frac{\sqrt{2} + \log(1 + \sqrt{2})}{6} L \approx 0.3826L.$$

The results for a small value of λ, i.e., $\lambda = 0.5$, are presented in Figure 7.5. The average service time converges to a value that is very close to the theoretical limit computed above, taking $L = 1/\sqrt{m} = 1/3$. In both cases, the reference points converge – albeit very slowly – to the generators of a MVT, while the average system time quickly approaches the optimal value.

7.6.2 Non-uniform distribution, light load

We also present in Figure 7.6 results of similar numerical experiments with a non-uniform distribution, namely an isotropic normal distribution centered at (0.25, 0.25), with standard deviation equal to 0.25.

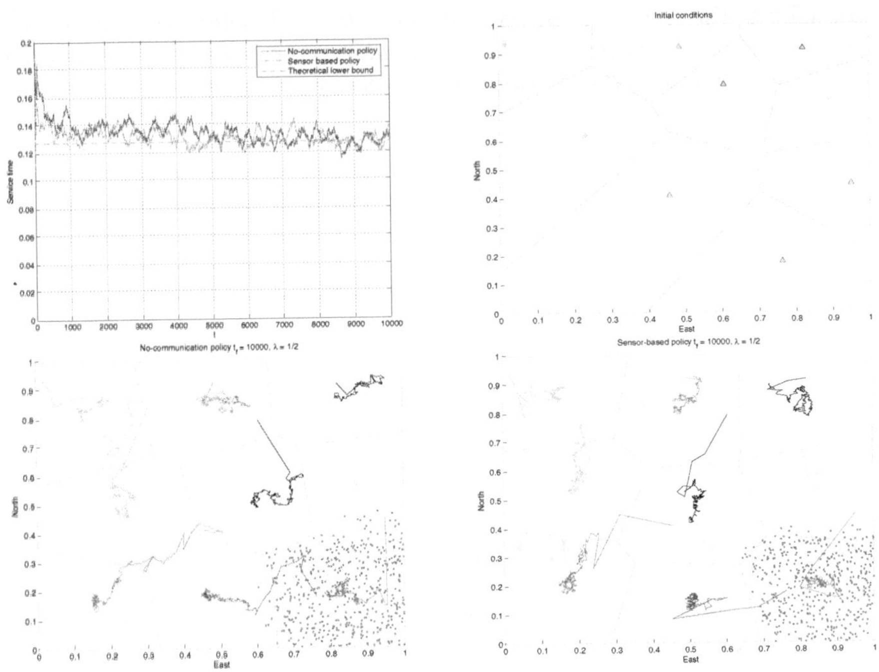

Figure 7.5 Numerical simulation in the light-load case, for a uniform spatial distribution. Top left: the actual service times as a function of time, for the two policies, compared with the optimal system time. Top right: the initial configuration of the nine agents. Bottom left and right: paths followed by the reference points up to $t = 10^4$ (corresponding to approximately 5,000 targets), using the two policies. The locations of all targets visited by one of the agents are also shown.

7.6.3 Uniform distribution, dependency on the target generation rate

An interesting set of numerical experiments evaluates the performance of the proposed policies over a large range of values of the target generation rate λ. In Section 7.4, we proved the convergence of the system's behavior to an efficient steady state, with high probability as $\lambda \to 0$, as confirmed by the simulations discussed above. For large values of λ, however, the assumption that vehicles are able to return to their reference point breaks down, and the convergence result is no longer valid. In Figure 7.7 we report results from numerical experiments on scenarios involving $m = 3$ agents, and values of λ ranging from 1/2 to 32. In the figure, we also report the known (asymptotic) lower bounds on the system time (with 3 agents), as derived in (Bertsimas and van Ryzin 1991), and the system time obtained with the proposed policies in a single-agent scenario.

The performance of both proposed policies is close to optimal for small λ, as expected. The sensor-based policy behaves well over a large range of target generation rates; in fact, the numerical results suggest that the policy provides a system time that is a constant-factor approximation of the optimum, by a factor of approximately 1.6.

However, as λ increases, the performance of the no-communication policy degrades significantly, almost approaching the performance of a single-vehicle system over an intermediate range of values of λ. Our intuition in this phenomenon is the following. As

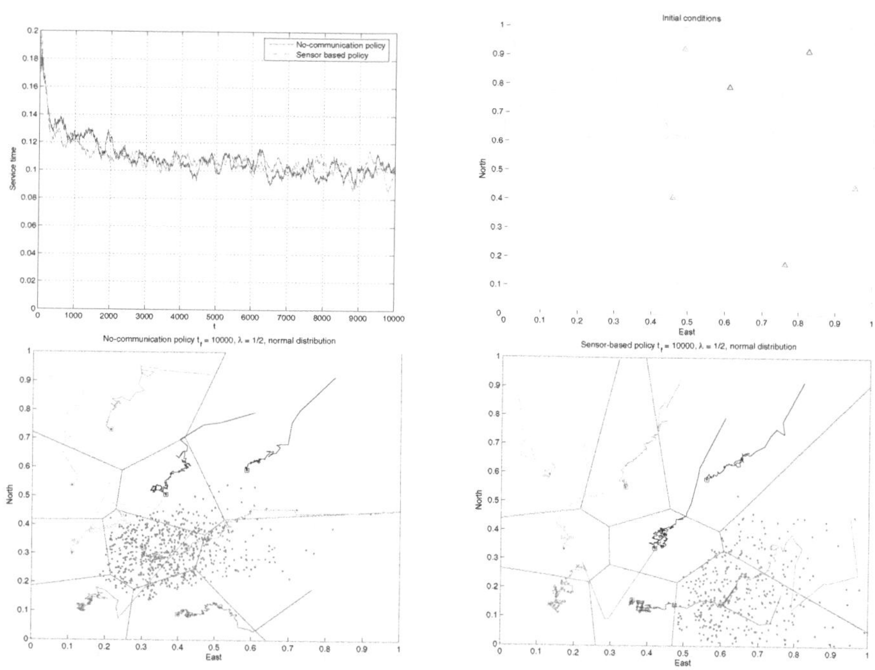

Figure 7.6 Numerical simulation in the light-load case, for a normal spatial distribution. Top left: the actual service times as a function of time, for the two policies. Top right: the initial configuration of the nine agents. Bottom left and right: paths followed by the reference points up to $t = 10^4$ (corresponding to approximately 5,000 targets), using the two policies. The locations of all targets visited by one of the agents are also shown.

agents do not return to their own reference points between visiting successive targets, their efficiency decreases since they are no longer able to effectively separate regions of responsibility. In practice – unless they communicate and concentrate on their Voronoi region, as in the sensor-based policy – agents are likely to duplicate efforts as they pursue the same target, and effectively behave as a single-vehicle system. Interestingly, this efficiency loss seems to decrease for large λ, and the numerical results suggest that the no-communication policy recovers a similar performance as the sensor-based policy in the heavy load limit. Unfortunately, we are not able at this time to provide a rigorous analysis of these policies for general values of the target generation rate.

7.6.4 The sTP policy

In this section we present the results of numerical experiments of the sTP policy. The sTP algorithm was implemented in Matlab 7.0, with external calls to `linkern` for approximate ETSP solutions when needed. In Figure 7.8 the results of our experiments in the heavy load are summarized and compared to the theoretical prediction of the asymptotic upper bound. While T_j^{phase} asymptotically goes to $\beta^2 A\lambda/v^2$, the system time

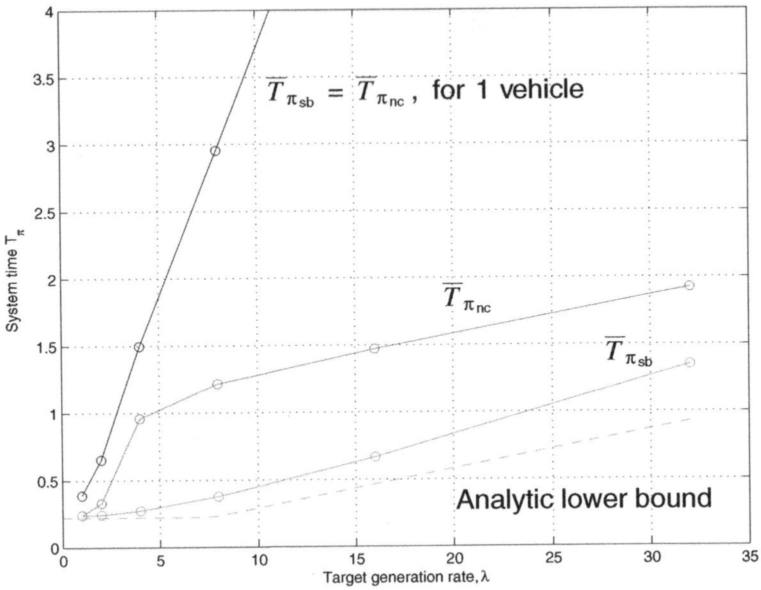

Figure 7.7 System time provided by some of the policies proposed in this chapter, as a function of the target generation rate λ. The system is composed of three vehicles, and the target points are generated uniformly in the unit square.

depends on the number of tiles, k, by the following relationship:

$$\overline{T}_{\text{sTP}} = \frac{1}{2}\frac{k+1}{k}T_j^{\text{phase}}.$$

The experimental results match well with this asymptotic prediction of the upper bound. Also note that for any finite λ, there exists a number of tiles k for which the assumptions used in the proof of the asymptotic upper bound do not hold. That is, there exists a k such that the local target generation process, with time-intensity λ/k does not generate a large number of targets in the tile during a phase. This effect is visible in the experimental results: for each λ, the system time decreases as k increases from 1 until some critical value of $k_{\text{opt}}(\lambda)$ is reached, whereupon the system time grows thereafter. Upon inspection of the data in Figure 7.8 it is apparent that $k_{\text{opt}}(\lambda)$ increases with λ. This is further evidence that η must be chosen in such a way that $\eta = \Theta(1/\lambda^{\alpha})$ for the heavy load upper bound to hold. The results of our experiments with small sensing radius are summarized in Figure 7.9. Note that the numerical results stay within the theoretically predicted lower and upper bounds.

7.7 CONCLUSIONS

In this chapter we considered some very simple strategies for multiple vehicle routing in the presence of dynamically-generated targets, and analyzed their performance in some

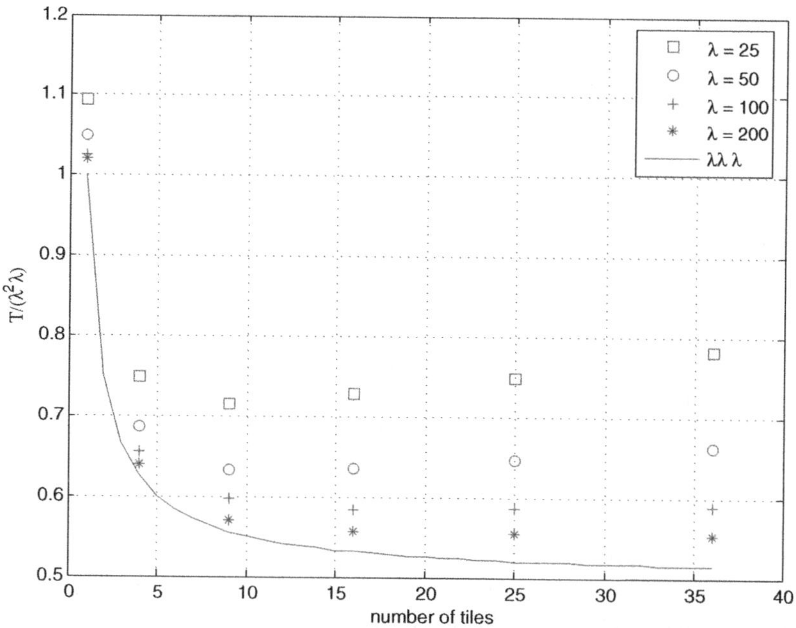

Figure 7.8 Numerical experiment results for the single-vehicle Tiling Policy as $\lambda \to +\infty$.

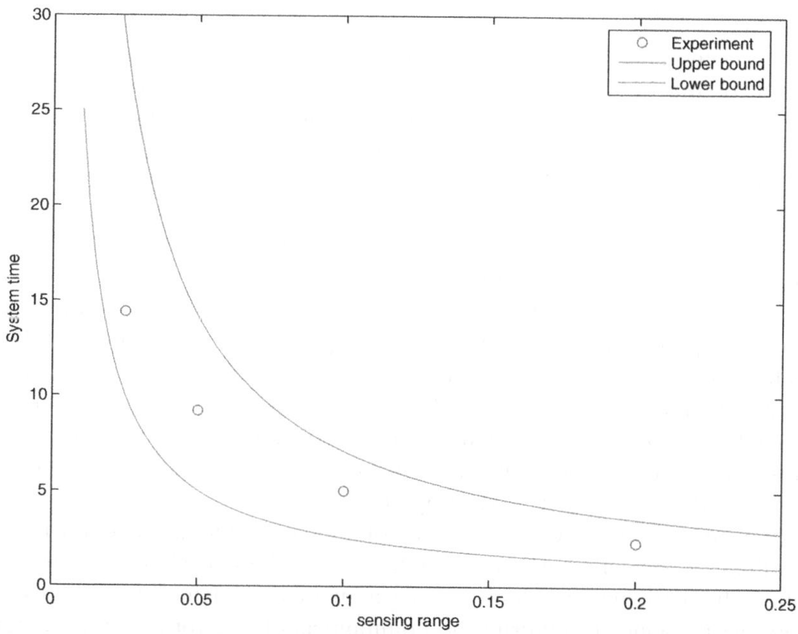

Figure 7.9 Numerical experiment results for the single-vehicle Tiling Policy as $\sigma \to 0^+$.

asymptotic conditions, ranging to light load, heavy load and to vanishing sensing range. Generally speaking, the strategies we addressed are based on minimal assumptions on the ability of the agents to exchange information: in one case they do not explicitly communicate at all, in one other case, agents are only aware of other agents' current location and in the mTP policy agents are require to exchange information with their Delaunay neighbors. In any case, all the proposed strategies are decentralized. A possibly unexpected and striking result of our analysis is the following: the collective performance of the agents using such minimal or no-communication strategies is (locally) optimal, and is in fact as good as that achieved by the best-known decentralized strategies. Moreover, two proposed strategies do not rely on the knowledge of the target generation process, and make minimal assumptions on the target spatial distribution; in fact, the convexity and boundedness assumptions on the support of the spatial distribution can be relaxed, as long as path connectivity of the support, and absolute continuity of the distribution are ensured. Also, the distribution needs not be constant: Indeed, the algorithm for the implementation of the non-communication strategy will provide a good approximation to a local optimum for the cost function as long as the characteristic time it takes for the target generation process to vary significantly is much greater than the relaxation time of the algorithm. In summary, the first two strategies can be seen as a learning mechanism in which the agents learn the target generation process, and the ensuing target spatial distribution, and adapt their own behavior to it.

Moreover, we have investigated the effects of limited sensor range in dynamic vehicle routing problems, in which a number of mobile agents must visit stochastically-generated targets in a time-efficient manner. We introduced a new single-vehicle control policy, based on a tiling of the plane into regions that can be covered instantaneously by the on-board sensor, and analyzed its performance, proving that it provides a constant-factor approximation of the optimal performance when the sensing range is very small, and recovers the performance of the best available full-information policies when the target generation rate is very large – thus demonstrating that in such a case limited sensing range has no effect on the collective efficiency of the system. We then extended the policies to the multiple-vehicle case, providing spatially-decentralized control laws that locally converge to the same performance levels of their single-vehicle counterpart. This has been presented in the case of uniform-density.

The proposed strategies are very simple to implement, as they only require storage of the coordinates of points visited in the past and simple algebraic calculations; the 'sensor-based' strategy also requires a device to estimate the position of other agents, while the effective implementation of the mTP policy requires information exchange among Delaunay neighbors. On the other hand, it must be observed that simple implementation and the absence of active communication make the proposed strategies attractive, for example, in embedded systems and stealthy applications.

While we were able to prove that the non-communication strategy performs efficiently for small values of the target generation rate, little is known about its performance in other regimes. In particular, we have shown numerical evidence that suggests that the first strategy we introduced, requiring no communication, performs poorly when targets are generated very frequently, whereas the performance of the sensor-based strategy is in fact comparable to that of the best-known strategies for the heavy load case.

Extensions of this work will include the analysis and design of efficient strategies for general values of the target generation rate, for different vehicle dynamics models (e.g., including differential constraints on the motion of the agents), and heterogeneous systems in which both service requests and agents can belong to several different classes with different characteristics and abilities. It would be interesting to address also some other issues related to the effects of other models of sensing and communication.

REFERENCES

Agarwal PK and Sharir M 1998 Efficient algorithms for geometric optimization. *ACM Computing Surveys* **30**(4), 412–458.

Applegate D, Bixby R, Chvátal V and Cook W 1998 On the solution of traveling salesman problems. *Documenta Mathematica, Journal der Deutschen Mathematiker-Vereinigung*, pp. 645–656, Berlin, Germany. Proceedings of the International Congress of Mathematicians, Extra Volume ICM III.

Arora S 1997 Nearly linear time approximation scheme for Euclidean TSP and other geometric problems. In *Proc. 38th IEEE Annual Symposium on Foundations of Computer Science*, pp. 554–563, Miami Beach, FL.

Arsie A and Frazzoli E 2006 Efficient routing with no communications. *International Journal of Robust and Nonlinear Control.* Submitted, downloadable at http://www.arxiv.org/math.OC/45940504.

Arslan G and Shamma J 2006 Autonomous vehicle-target assignment: a game theoretic formulation. *Submitted to the IEEE Trans. on Automatic Control.*

Beard RW, McLain TW, Goodrich MA and Anderson EP 2002 Coordinated target assignment and intercept for unmanned air vehicles. *IEEE Trans. on Robotics and Automation* **18**(6), 911–922.

Beardwood J, Halton J and Hammersley J 1959 The shortest path through many points. In *Proceedings of the Cambridge Philosophy Society* **55**, 299–327.

Bertsimas DJ and van Ryzin GJ 1991 A stochastic and dynamic vehicle routing problem in the Euclidean plane. *Operations Research* **39**, 601–615.

Bertsimas DJ and van Ryzin GJ 1993a Stochastic and dynamic vehicle routing in the Euclidean plane with multiple capacitated vehicles. *Operations Research* **41**(1), 60–76.

Bertsimas DJ and van Ryzin GJ 1993b Stochastic and dynamic vehicle routing with general interarrival and service time distributions. *Advances in Applied Probability* **25**, 947–978.

Carmi P, Har-Peled S and Katz M 2005 On the Fermat-Weber center of a convex object. *Computational Geometry* **32**(3), 188–195.

Chandrasekaran R and Tamir A 1990 Algebraic optimization: the Fermat-Weber location problem. *Mathematical Programming* **46**, 219–224.

Christofides N 1972 Bounds for the travelling-salesman problem. *Operations Research* **20**, 1044–1056.

Cortés J, Martínez S, Karatas T and Bullo F 2004 Coverage control for mobile sensing networks. *IEEE Transactions on Robotics and Automation* **20**(2), 243–255.

Drezner Z (ed.) 1995 *Facility Location: A Survey of Applications and Methods*. Springer Series in Operations Research. Springer Verlag, New York.

Earl M and D'Andrea R 2005 Iterative MILP methods for vehicle control problems. *IEEE Trans. on Robotics* **21**, 1158–1167.

Enright JJ and Frazzoli E 2006 Cooperative UAV routing with limited sensor range. *AIAA Conf. on Guidance, Navigation, and Control*, Keystone, CO, AIAA Paper 2006-2008.

Fekete SP, Mitchell JSB and Weinbrecht K 2000 On the continuous Weber and k-median problems. In *Proceedings of the Sixteenth Annual Symposium on Computational Geometry (Hong Kong, 2000)*, pp. 70–79. ACM, New York.

Frazzoli E and Bullo F 2004 Decentralized algorithms for vehicle routing in a stochastic time-varying environment. In *Proc. 43rd IEEE Conf. on Decision and Control*, Paradise Island, Bahamas, pp. 3357–3363.

Johnson DS, McGeoch LA and Rothberg EE 1996 Asymptotic experimental analysis for the held-karp traveling salesman bound. In *Proc. 7th Annual ACM-SIAM Symposium on Discrete Algorithms*, pp. 341–350.

Li W and Cassandras C 2006 A cooperative receding horizon controller for multivehicle uncertain environments. *IEEE Trans. on Automatic Control* **51**(2), 242–257.

Lin S and Kernighan BW 1973 An effective heuristic algorithm for the traveling-salesman problem. *Operations Research* **21**, 498–516.

Little J 1961 A proof of the queueing formula $l = \lambda w$. *Operations Research* **9**, 383–387.

Liu Y, Cruz J and Sparks AG 2004 Coordinated networked uninhabited AER vehicles for persistent area denial. *IEEE Conf. on Decision and Control*, pp. 3351–3356, Paradise Island, Bahamas.

Lloyd SP 1982 Least squares quantization in PCM. *IEEE Trans. on Information Theory* **28**(2), 129–137.

MacQueen J 1967 Some methods for the classification and analysis of multivariate observations. In *Proceedings of the Fifth Berkeley Symposium on Math. Stat. and Prob.* (ed. LeCam LM and Neyman J), pp. 281–297. University of California Press, Berkeley, CA.

Moore B and Passino K 2005 Distributed balancing of AAVs for uniform surveillance coverage. *44th IEEE Conference on Decision and Control*, pp. 7060–7065.

Murphey R 1999 Target-based weapon target assignment problems. In *Nonlinear Assignment Problems: Algorithms and Applications* (ed. Pardalos P and Pitsoulis L). pp. 39–53, Kluwer Academic Publisher.

Okabe A, Boots B, Sugihara K and Chiu S 2000 *Spatial Tessellations: Concepts and Applications of Voronoi Diagrams* Wiley Series in Probability and Statistics second edn. John Wiley & Sons, Chichester, UK.

Percus G and Martin OC 1996 Finite size and dimensional dependence of the Euclidean traveling salesman problem. *Physical Review Letters* **76**(8), 1188–1191.

Psaraftis H 1988 Dynamic vehicle routing problems. In *Vehicle Routing: Methods and Studies* (ed. Golden B and Assad A) Studies in Management Science and Systems, Elsevier.

Richards A, Bellingham J, Tillerson M and How J 2002 Coordination and control of multiple UAVs. In *Proc. of the AIAA Conf. on Guidance, Navigation, and Control*, Monterey, CA, AIAA Paper 2002–4588.

Sabin M and Gray R 1986 Global convergence and empirical consistency of the generalized Lloyd algorithm. *IEEE Trans. on Information Theory*, **32**(2), 148–155.

Schumacher C, Chandler PR, Rasmussen SJ and Walker D 2003 Task allocation for wide area search munitions with variable path length. In *Proc. of the American Control Conference*, pp. 3472–3477, Denver, CO.

Steele JM 1990 Probabilistic and worst case analyses of classical problems of combinatorial optimization in Euclidean space. *Mathematics of Operations Research* **15**(4), 749.

Tanemura M and Hasegawa H 1980 Geometrical models of territory I: Models for synchronous and asynchronous settlement of territories. *Journal of Theoretical Biology* **82**, 477–496.

Wesolowsky G 1993 The Weber problem: History and perspectives. *Location Science* **1**, 5–23.

8

Optimal agent cooperation with local information

Eric Feron and Jan DeMot

8.1 INTRODUCTION

In recent years, multi-agent navigation problems have become of major interest to many researchers (see Arai *et al.* (2002) for an overview). The main idea is that a group of multiple, possibly cheap and diverse agents executes tasks faster, more reliably, and more efficiently than a single agent. Applications range from coordinated terrain exploration, search and rescue operations (Haley and Stone 1980), to mine detection (Cassinis 2000; Cassinis *et al.* 1999), coordinated thermal search and deceptive reconnaissance missions. In the last example application, a group of unmanned autonomous agents is required to gather information on a specific environment area in a deceptive manner, so that to a group of observers on the ground, the location of the area of interest remains unknown (Hespanha *et al.* 2000). The intuition is that a group of agents deceive far more efficiently than single agents. The coordinated thermal search problem focuses on finding cooperative strategies for multiple sailplanes so that the group traverses a thermal-ridden environment faster. A sailplane only detects a thermal upon experiencing lift. Practical examples in soaring competitions demonstrate that multiple information-sharing sailplanes reach higher average velocities than single sailplanes traversing the same environment (Fukada 2000; McCready 1954; Reichmann 1978).

In general, the main challenge multi-agent systems pose is the problem of computational complexity. DasGupta *et al.* (2004b) show NP-hardness of the problem of searching for a hidden target in a bounded region where a single robot evolves using only local sensory information. Therefore, many researchers focus on decentralized strategies, which are computationally simple, but in general lack optimality. Examples include the motion of swarms (Jadbabaie *et al.* 2003; Olfati-Saber and Murray 2004) where local single-agent navigation rules lead to agent flocking. Similarly, in the area of formation flight, researchers focus on computing adequate decentralized controllers to maintain a certain fixed agent formation (Ogren *et al.* 2002). In (Ribichini and Frazzoli 2003), a group of

Cooperative Control of Distributed Multi-Agent Systems edited by Jeff S. Shamma

agents exploits the energy efficiency of agents in formation to coordinate the traversal of a set of agents as energy efficiently as possible, even when the agents do not share the same destination. The two-agent case is solved exactly; for larger problems, approximate solutions are needed. Other approaches include the casting of the multi-agent problem in the pursuit-evader framework in (Hespanha *et al.* 1999), where an agent cluster pursues a single evader. In (Hespanha *et al.* 1999), approximate solutions are presented. Decentralized solutions to the problem where a set of agents intends to minimize the expected waiting time to service randomly generated targets are described by Frazzoli and Bullo (2004).

An essential property of the problem we intend to solve is the limited available information on the environment and the agent capability to improve environment information in a local area around its current position by observation. In the literature, there are examples of similar problems exhibiting agents with local observation capabilities. Searching for unknown targets with local sensory information and a single autonomous agent is treated in (DasGupta *et al.* 2004a), where the problem is transformed into a tractable discrete version and leads to suboptimal behavior. In (Burgard *et al.* 2000; Simmons *et al.* 2000), the authors implemented a heuristic map building and localization algorithm that prevents multiple agents from observing the same area by sequentially assigning paths to agents and discounting the utility of the area expected to be covered. Experimentally, the authors verify the faster coverage of a map with an increasing number of agents. Beard and McLain (2003) consider a multi-agent navigation problem whereby only lateral agent distances are controlled. Constraints imposing upper and lower bounds on the agent separation reduce the computational complexity, but oversimplify the problem, in our opinion. In (Cortes *et al.* 2004), agents with a limited sensing and communication radius are required to optimally cover a particular area. Distributed algorithms are shown to converge.

In this chapter, we focus on obtaining the optimal strategy for a two-agent graph traversal problem, since there is much to learn from the exact problem solution. To each edge is associated a cost. *A priori*, only the edge cost statistics are given, but on a particular set of edges around the current agent position (the *local observation zone*), the exact edge cost is observed, and communicated to the other agent. We intend to compute the optimal agent strategies and no approximations are involved. In our chapter, we mathematically characterize the following underlying trade-off. First, the agents (named A and B) tend to spread, increasing the size of the union of the local observation zones ($\mathcal{O} = \mathcal{O}^A \cup \mathcal{O}^B$). Indeed, agents close to each other have overlapping observation zones and not as many edge costs are observed. More information yields enhanced efficiency, hence the spreading tendency. On the other hand, the agents tend to converge so that agent A's reachable set of edges ($\mathcal{R}^A$) intersects with $\mathcal{O}^B$ and vice versa. Only then can each agent take advantage of potentially cheap edges the other agent observes and exploit the benefit multiple agents have over single agents. Hence, our problem differs significantly from a problem where agents are merely required to explore.

In past work (De Mot and Feron 2003), we presented a method to compute optimal two-agent policies whereby the observation zone structure (*tunnel vision*, i.e. each agent only observes the cost of edges straight ahead) is such that the agent separation is upper bounded. This fact allows for the computation of the optimal policy by artificially upper bounding the agent separation by this upper bound, reducing the problem size significantly. In (De Mot *et al.* 2002), we solved an even simpler version of this problem by considering a cylindrical navigation terrain, artificially upper bounding the agent separation but maintaining spatial invariance properties. In this chapter, we study an observation zone for which the agent separation under an optimal policy is unbounded, which presents a very significant extension to our past work, and in fact allows us to extend our work to clusters of larger sizes.

The chapter is organized as follows. In Section 8.2 we introduce the notation and formulate the problem. In Section 8.3, we present a mathematical model. First, a *Dynamic Programming* (DP) formulation is given, then, we present an equivalent *Linear Program* (LP). Section 8.4 provides intuition, gives an overview of the algorithm, and presents the LP decomposition into a function whose fixed point corresponds to the optimal value function. Section 8.5 details the fixed point computation. In Section 8.6, we study the two-agent optimal policy for different environment parameters. Finally, we conclude our chapter in Section 8.7.

8.2 NOTATION AND PROBLEM FORMULATION

In this section, we introduce the notation and formulate the two-agent navigation problem as a graph traversal problem with partial edge cost information. In particular, two agents are required to traverse a graph while minimizing the sum of the discounted costs of the traversed edges. To each agent is associated a set of edges whose costs the agents observe and share, which is the incentive for cooperation.

We grid the navigation terrain into sectors and associate a vertex $v \in V$ to each sector. Edges $e \in E$ connect pairs of vertices, which reduces the navigation problem into a graph traversal problem on the graph $G(V, E)$. We refer the reader to the literature for descriptions of algorithms to convert generic navigation terrains into a graph (see, for example, Latombe 1991) and references therein). One typical example is the Voronoi graph for obstacle cluttered environments, where edges consist of points at equal distance from its closest obstacles (see, for example, Choset 1996).

The structured transition graph we consider (see Figure 8.1) consists of an infinite number of vertical lines of vertices, having an infinite number of vertices each. A horizontal vertex array is referred to as a *lane*, while a *stage* refers to a vertical line of vertices. Vertex v_{ij} denotes the vertex at stage i and lane j where the first stage and lane are chosen arbitrarily. Three edges e_{ij}^k, $k \in \{-1, 0, 1\}$, connect each vertex v_{ij} to the set of three vertices $v_{i+1,j+k}$ ($k \in \{-1, 0, 1\}$).

To each edge e_{ij}^k is associated a cost c_{ij}^k reflecting the edge traversal cost and the reward an agent collects upon reaching $v_{i+1,j+k}$. For example, if *a priori* information suggests that the environment sector associated with vertex $v_{i+1,j+k}$ is a particularly unsafe sector, then $c_i(j, k)$ is relatively high. The traversal cost in practical applications relates, for instance, to the energy an agent requires to traverse an edge. In this work, we assume that the edge costs are independent identically distributed (i.i.d.) random variables picked from a finite set $\mathcal{L} = \{0, 1\}$. We denote p as the probability of encountering a zero edge cost. *A priori*, only the edge cost statistics are available and the edge costs are constant.

In this work, we consider the graph traversal problem for a set of two agents, referred to as agents A and B. The graph traversal problem is modeled as a discrete time decision-making problem where at time zero A and B are positioned at the vertices on lanes l_0^A and l_0^B ($\in \mathbf{N}$ and with $l_0^A \leq l_0^B$) at stage zero, indicated by the subscript. At each time step, each agent chooses to traverse one of the three available links to the next stage. Therefore, we let the time index coincide with the stage index and at time i the agents reach stage i at lanes l_i^A and l_i^B (where the agent are labeled such that $l_i^A \leq l_i^B$, $i \geq 0$). Since at each time step agents A and B are positioned at the same stage, we refer to this as *spatially synchronous* motion. In a more general problem version, the spatially synchronous motion requirement can be dropped. However, the work presented here illustrates the richness of the less general problem. Furthermore, we believe spatial asynchrony is an extension the algorithm developed in this work can handle.

At each stage i, we associate a local observation zone to both agents, namely $\mathcal{O}_i^A$ and $\mathcal{O}_i^B$, a set of edges whose cost is observed and known. In this work, we define the local observation zone for agent A positioned at vertex l_i^A as $\mathcal{O}_i^A = \{e_{ij}^{-1}, e_{ij}^0, e_{ij}^1, e_{i+1,j}^0\}$ (see Figure 8.1). Local observation zone $\mathcal{O}_i^B$ is defined similarly. Of unobserved edges, only the *a priori* available information, i.e. the edge cost statistics, is known. The term *local* stems from the fact that only edges in the neighborhood of the current agent position are included in $\mathcal{O}_i^{(\cdot)}$. At stage i, the set of agents incurs the costs associated with the edges traversed to reach stage $i+1$. Upon arrival at a vertex, each agent observes the costs of until then unknown edges and communicates these to the other agent. This provides an incentive for cooperation since observations by agent A possibly are of use to agent B and vice versa.

The consideration of an infinitely wide and long graph involves an idealization with respect to practical path planning problems which have bounded environments. However, this idealization is analogous to the approach taken in control of dynamical systems, where a long but finite time horizon is typically modeled with infinite horizon tools. Limiting the graph width and length imposes boundary conditions on the problem which often do not lead to elegant and insightful solutions.

We formulate the following main problem.

Main Problem: *Let a set of two agents be located at positions l_0^A and l_0^B of graph $G(V, E)$. The edge costs are time invariant and i.i.d. random variables over $\mathcal{L}$ with probability p for a zero edge cost. The agents navigate synchronously through the graph in the*

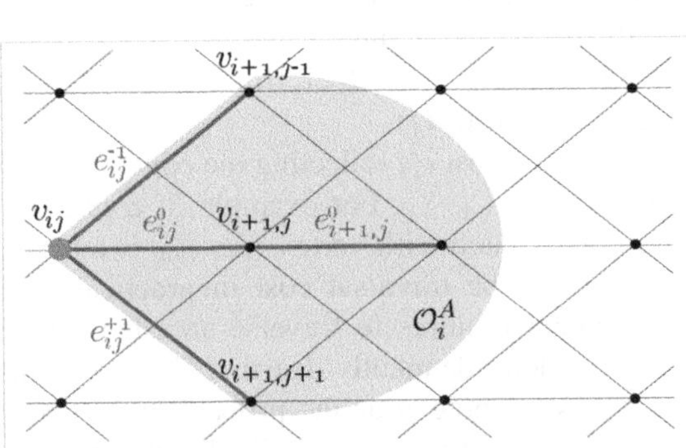

Figure 8.1 Notation related to graph G. Black dots represent vertices, thin lines are edges. The gray area conceptually represents the local observation zone of an agent A located at vertex v_{ij}, formally, $\mathcal{O}_i^A$ is the set of highlighted edges.

direction of increasing stage indices, infinitely far. Furthermore, the agents share the costs of the edges in their respective local observation zones perfectly and instantaneously upon reaching a vertex.

Then, find the navigation strategy for both agents so that the expected discounted sum of costs incurred at each stage is minimized. The expected value is taken over all initially unobserved edge costs; costs incurred at stage i are discounted by factor α^i, where $0 < \alpha < 1$.

8.3 MATHEMATICAL PROBLEM FORMULATION

In Section 8.3.1, we present a *Dynamic Programming* (DP) formulation of the graph traversal problem. Then, in Section 8.3.2, we transform the DP into an equivalent *Linear Program* (LP), whose structure we analyze in subsequent sections.

8.3.1 DP formulation

We cast the two-agent navigation problem as a discounted cost, infinite horizon DP problem as follows. Since graph G exhibits spatial invariance properties in all directions, and since we have spatially synchronous motion, we choose the system state x, an element of set $\mathcal{S}$, to be independent from the current stage and from the absolute positions of the agent pair. In particular, let

$$\mathcal{C}^A = \left[\, a_{-1}^A \;\; a_0^A \;\; a_1^A \;\; b_0^A \,\right]^T \tag{8.1}$$

denote the vector with as entries the costs of the edges in $\mathcal{O}^A$ at the current position of A, where $a_k^A = c_i(l_i^A, k)$, for all $k \in \{-1, 0, 1\}$, where $b_0^A = c_{i+1}(l_i^A, 0)$, and where i is the current stage. Vector $\mathcal{C}^B$ is defined similarly. Then, $x = (s, \mathcal{C}^A, \mathcal{C}^B)$, contains the agent *separation* $s = l^B - l^A \in \mathbf{N}^+$ and the costs of the edges belonging to the local observation zones of A and B. Let $\mathcal{S}(s) \subset \mathcal{S}$ denote the set of states associated with separation s. Let $\mathbf{u} = (u^A, u^B) \in U$ denote the decision vector where $u^{(\cdot)} \in \{-1, 0, 1\}$ represents the three possible decisions available to each agent (*i.e.* traversing edges with cost $a_k^{(\cdot)}$, with $k \in \{-1, 0, 1\}$). The set $U = \{-1, 0, 1\}^2$ comprises the nine possible decision combinations for the agents and is invariant. Given x_i and $\mathbf{u}_i$ at time i, the agent cluster moves into the new state

$$x_{i+1} = f(x_i, \mathbf{u}_i), \tag{8.2}$$

where $f : \mathcal{S} \times U \to \mathcal{S}$ is the state transition function. The cost incurred in the transition from state x_i to x_{i+1} is $g(x_i, \mathbf{u}_i) = a_{u_A}^A + a_{u_B}^B$, which is the sum of the edge costs of the edges agents A and B choose to traverse.

Let policy $\mu : \mathcal{S} \to U$ be a particular agent policy. Then, the expected discounted cost $J_\mu(x_0)$ the agents incur in advancing for an infinite number of time steps, given initial state x_0 and under policy μ is

$$J_\mu(x_0) = \lim_{N\to\infty} E\left[\sum_{i=0}^{N} \alpha^i g(x_i, \mu(x_i))\right],$$

subject to the state transition function (8.2), and where $0 < \alpha < 1$ is the discount factor. The cost function $J_\mu(x)$, for all $x \in \mathcal{S}$, satisfies

$$J_\mu(x) = E\left[g(x, \mu(x)) + \alpha J_\mu(f(x, \mu(x)))\right], \tag{8.3}$$

where we drop the time index i since we have time invariance and an infinite horizon. The discount factor α ensures that J_μ is well defined since with a bounded local cost g, the total cost incurred is bounded. Using the principle of optimality, the optimal value function $J^*(x)$ solves the corresponding Bellman equation:

$$J^*(x) = \min_{\mathbf{u} \in U} E\left[g(x, \mathbf{u}) + \alpha J^*(f(x, \mathbf{u}))\right]. \tag{8.4}$$

A policy that minimizes the right-hand side (RHS) of Equation (8.4) is referred to as an optimal policy μ^*.

8.3.2 LP formulation

We reformulate the DP defined in Section 8.3.1 as an equivalent LP. For any function $J : \mathcal{S} \to \mathbf{R}$, we denote the function obtained after applying the DP mapping as

$$(TJ)(x) = \min_{\mathbf{u} \in U} E[g(x, \mathbf{u}) + \alpha J(f(x, \mathbf{u}))], \qquad x \in \mathcal{S}.$$

Value iteration creates a sequence of functions $(T^k J) : \mathcal{S} \to \mathbf{R}$, for $k = 0, 1, \ldots$ and $x \in \mathcal{S}$, such that

$$\lim_{k \to \infty} (T^k J)(x) = J^*(x), \qquad x \in \mathcal{S}, \tag{8.5}$$

where T^k indicates the application of operator T with itself, k times. If J is chosen such that $J(x) \leq (TJ)(x)$, for all $x \in \mathcal{S}$, then, using the monotonicity property of DP (see, for example, Bertsekas (2001)), we have that $(T^k J)(x) \leq (T^{k+1} J)(x)$, for $x \in \mathcal{S}$, and for $k = 0, 1, \ldots$. With the convergence property of value iteration [see Equation (8.5)], it follows that $J(x) \leq J^*(x)$, for all $x \in \mathcal{S}$. Since we have a monotonously increasing sequence of functions $T^k J \leq J^*$ and since $J^* = TJ^*$, we have that J^* is the 'largest' function J that satisfies $J \leq TJ$, which is equivalent to the set of constraints

$$J(x) \leq E\left[g(x, \mathbf{u}) + \alpha J(f(x, \mathbf{u}))\right], \qquad x \in \mathcal{S}, \mathbf{u} \in U. \tag{8.6}$$

In particular, J^* is the 'upper right corner' of the polyhedron formed by the set of equality constraints (8.6). With the purpose of formulating the LP cost function and of rewriting the previous set of constraints in typical LP form, we define new variables. Let the vector $\mathbf{x}^0 \in \mathbf{R}^{n_0}$ contain the value function $J(x)$ for $x \in \mathcal{S}(0)$ as its entries and, similarly, let the vectors $\mathbf{x}^s \in \mathbf{R}^n$, for all $s \geq 1$, contain $J(x)$ for $x \in \mathcal{S}(s)$ as entries. The dimensions n_0 and n represent the number of different sets of local observation zones the agents can encounter for separations $s = 0$ and $s \geq 1$, respectively. For clarity of the exposition, keeping the respective DP and LP notation conventions, we write the previous in short as

$$J(x) \sim \mathbf{x}^s, \qquad x \in \mathcal{S}(s). \tag{8.7}$$

Let $\mathbf{e}$ denote a vector whose entries are all ones. Let $\mathbf{T}$ denote a linear operator so that, when applied to a vector $\mathbf{x}$ of size n, the vector $\mathbf{Tx}$, of size $n|U|$, has as its first $|U|$ entries the first entry of $\mathbf{x}$, the next $|U|$ entries are the second entry of $\mathbf{x}$, etc. In what follows, we assume the dimensions of $\mathbf{T}$ to be clear from the context. Then, we have the following equivalent LP:

$$\begin{aligned} &\text{maximize} \quad \mathbf{e}^T \mathbf{x}^{\tilde{s}} \\ &\text{subject to} \quad \mathbf{T}\mathbf{x}^s \leq \sum_{\sigma=\max\{-2,-s\}}^{2} \mathbf{A}_s^\sigma \mathbf{x}^{s+\sigma} + \mathbf{b}^s, \quad s = 0, 1, \ldots, \end{aligned} \tag{8.8}$$

where $(\cdot)^T$ denotes the transpose of a vector and for any $0 \leq \tilde{s} < \infty$. Since the optimal value function is the maximum element of the lattice (Davey and Priestley 1990) formed by the constraints in Equation (8.8), the objective function can be any linear combination with positive coefficients of any set of entries of the vectors $\mathbf{x}^s$, $s = 0, 1, \ldots$, where the sum of the coefficients is finite. In the remainder of this chapter, we adopt different versions according to our needs. We denote this LP by $\mathcal{LP}_g$. The matrices $\mathbf{A}_s^\sigma$, for $s = 0, 1, \ldots$, reflect the dependence of the constraint set associated with $\mathbf{x}^s$ [Equation (8.8)] on $\mathbf{x}^{s+\sigma}$ and are deducted from Equation (8.6) in a strenuous but straightforward manner. For $s \geq 2$, we have that $\mathbf{A}^\sigma = \mathbf{A}_s^\sigma$, for $\sigma = -2, \ldots, 2$.

In order to illustrate the LP problem structure adequately, we define the vectors

$$\begin{aligned} \mathbf{X}_g &= \left[\, (\mathbf{x}^0)^T \ \ (\mathbf{x}^1)^T \ \ \ldots \,\right]^T, \\ \mathbf{B}_g &= \left[\, (\mathbf{b}^0)^T \ \ (\mathbf{b}^1)^T \ \ \ldots \,\right]^T, \end{aligned}$$

and the matrix $\mathbf{A}_g$ such that Equation (8.8) can be rewritten in typical LP-format as $(\mathbf{T} - \mathbf{A}_g)\mathbf{X}_g \leq \mathbf{B}_g$. Note that $\mathbf{X}_g$, $\mathbf{B}_g$, and $\mathbf{A}_g$ have infinite dimensions. Figure 8.2 shows the block structure of $\mathbf{A}_g$. Empty squares are zero block matrices, the gray squares denote the relevant matrices that appear in Equation (8.8). Since $\mathbf{T}$ is block-diagonal, the matrix

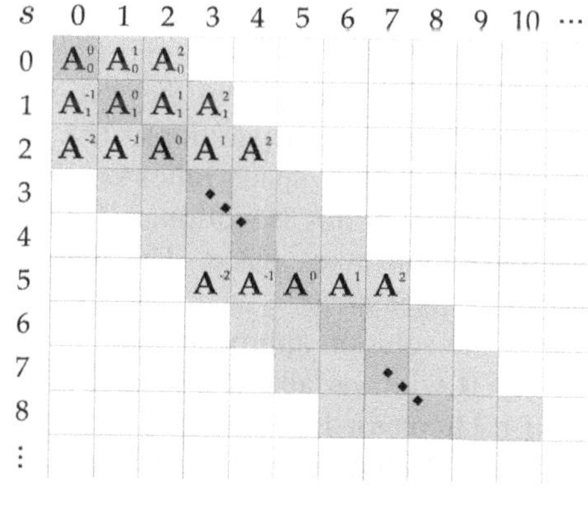

Figure 8.2 Matrix $\mathbf{A}_g$. Empty squares are zero-matrices, the gray squares denote the relevant matrices in Equation (8.8).

$\mathbf{T} - \mathbf{A}_g$ has the same structure as $\mathbf{A}_g$. The LP has a general block multi-diagonal structure, but is not of the *staircase* type since non-zero blocks appear above and under the main diagonal. For staircase LPs, efficient distributed algorithms based on nested decomposition have been developed (see (Ho and Loute 1980; Ho and Sundarraj 1997) and references therein). In our problem, the optimal value function at separation $s \geq 2$ depends on the optimal value function at lower *and* greater separations (namely, at separations $s + \sigma$, for $\sigma = -2, \ldots, 2$), which renders the problem significantly different from staircase LPs where, in equivalent terms, there is dependence uniquely on the optimal value function at lower separations.

8.4 ALGORITHM OVERVIEW AND LP DECOMPOSITION

We give intuition and an overview of the solution approach in Section 8.4.1 and present the essential decomposition of $\mathcal{LP}_g$ into an infinite set of small LPs in Section 8.4.2. This leads to the definition of a function F, whose unique fixed point is the optimal value function for the Main Problem 8.2.

8.4.1 Intuition and algorithm overview

We provide intuition for the agent behavior in the Main Problem 8.2. A trade-off exists between two competing trends. First, since more environment information yields an enhanced efficiency, the agents tend to spread, increasing the size of the union of the local observation zones ($\mathcal{O} = \mathcal{O}^A \cup \mathcal{O}^B$). For the local observation zone at hand, at $s = 0$ four edges are observed, while at $s = 1$, we have eight observed edges. Note that for $s \geq 1$, the size of $\mathcal{O}$ is constant, for the particular set-up considered in this work. On the other hand, the agents tend to converge so that agent A's reachable set of edges ($\mathcal{R}^A$) intersects with $\mathcal{O}^B$ and vice versa. Only then can each agent take advantage of opportunities in the form of potentially cheap edges the other agent observes and exploit the benefit multiple agents have over single agents.

For large s, it is intuitively clear that the optimal agent policy converges to a 'stationary' policy in the sense that as s increases, the optimal agent decision at each state is only dependent on the set of local observation zones, $\mathcal{O}^A$ and $\mathcal{O}^B$. For small s, the optimal policy is expected to be significantly different from the stationary policy valid at large s, since opportunities are exploited at small separations. The algorithm to compute the optimal value function over the whole state space relies on the two mentioned areas in the state space: for large s, the optimal policy is stationary (independent of separation s), and at small s, the actual cooperation takes place. The algorithm ensures the correct *connection* between the two state space areas.

We now give an overview of the solution approach. As a first step, we define a family $\mathcal{Q}$ of LPs, one LP for each $s \geq 0$ with variable $\mathbf{x}^s$, so that $\mathcal{Q}$ is equivalent to $\mathcal{LP}_g$. In particular, we establish that if each vector in a set $\{\mathbf{x}^s\}$, $s \geq 0$, solves the corresponding LP in $\mathcal{Q}$, then $\{\mathbf{x}^s\}$, $s \geq 0$, solves $\mathcal{LP}_g$ and therefore represents the optimal value function (Section 8.4.2).

The problem we now face is finding the solution of an infinite set of small LPs, the subject of Section 8.5. To this end, we devise from $\mathcal{Q}$ a causal *linear time invariant* (LTI) system with state $\mathbf{x}^s$, and with s as the equivalent of 'time', for $s \geq \bar{s}$, with $\bar{s}$ a particular small separation. We carefully choose the output so that its positivity indicates optimality of $\mathbf{x}^s$ in the corresponding LP of $\mathcal{Q}$, for $s \geq \bar{s}$ (Sections 8.5.2, 8.5.3 and 8.5.4).

To determine the initial conditions for this LTI system, we formulate another LP (called $\mathcal{LP}_{in}$) with variables $\mathbf{x}^s$ for $s < \bar{s}$ and with an extra set of equality constraints so that the solution to $\mathcal{LP}_{in}$ is the particular value function that excites specific modes in the LTI system for $s \geq \bar{s}$. The added equality constraints are the *connection* between the system behavior at small and large separations. We establish that this provides the optimal value function (Section 8.5.5).

8.4.2 LP decomposition

In this section we define a set of LPs $\mathcal{Q} = \{\mathcal{LP}(s)\}$, $s = 0, 1, \ldots$, where $\mathcal{LP}(s)$ refers to the LP associated with separation s. We show that $\mathcal{Q}$ is equivalent to $\mathcal{LP}_g$.

We have the following definition of $\mathcal{LP}(s)$ (for all $s \geq 0$).

Definition 8.4.1 *Let $\hat{\mathbf{x}}^{s-2}$, $\hat{\mathbf{x}}^{s-1}$, $\hat{\mathbf{x}}^{s+1}$ and $\hat{\mathbf{x}}^{s+2}$ be given vectors of appropriate dimensions. Then, we define $\mathcal{LP}(s)$ for $s = 0, 1, \ldots$ as*

$$\begin{aligned} &\textit{maximize} \quad \mathbf{e}^T\mathbf{x}^s \\ &\textit{subject to} \quad (\mathbf{T} - \mathbf{A}_s^0)\mathbf{x}^s \leq \sum_{\sigma \in S^s} \mathbf{A}_s^\sigma \hat{\mathbf{x}}^{s+\sigma} + \mathbf{b}^s, \end{aligned} \tag{8.9}$$

where $S^0 = \{1, 2\}$, $S^1 = \{-1, 1, 2\}$ and $S^s = \{-2, -1, 1, 2\}$, for $s \geq 2$. □

The matrices $\mathbf{e}$, $\mathbf{T}$, $\mathbf{A}_s^\sigma$ and $\mathbf{b}^s$ of the appropriate dimensions are defined as in Section 8.3.2. Note that the RHS of the constraint set (8.9) is known.

In the rest of this section, we establish the equivalence between $\mathcal{LP}_g$ and the set $\mathcal{Q}$ of LPs. In particular, we show that the vectors $\mathbf{x}^s$, $s = 0, 1, \ldots$ of suitable dimension solve $\mathcal{LP}_g$ if and only if $\mathbf{x}^s$ solves $\mathcal{LP}(s)$ where $\hat{\mathbf{x}}^{\sigma+s}$ in Equation (8.9) is equal to $\mathbf{x}^{\sigma+s}$ for $\sigma \in S^s$ and for all $s \geq 0$.

We first introduce some notation. Let $\mathcal{K} = \mathbf{R}^\mathbf{N}$. Define the vector $\mathbf{X} \in \mathcal{K}$ as the concatenation of the set of vectors $\{\mathbf{x}^s\}$ ($s = 0, 1, \ldots$). In particular,

$$\mathbf{X}^T \doteq \left[\, (\mathbf{x}^0)^T \;\; (\mathbf{x}^1)^T \;\; \ldots \,\right]^T, \tag{8.10}$$

where $\mathbf{x}^0 \in \mathbf{R}^{n_0}$, and $\mathbf{x}^s \in \mathbf{R}^{n_1}$ ($s \geq 1$). The vector $\mathbf{X}^* \in \mathcal{K}$ is defined similarly from the set $\{\mathbf{x}^{s,*}\}$ (for $s = 0, 1, \ldots$), *i.e.*, the solution of $\mathcal{LP}_g$. The following function definition is key to the development in this section.

Definition 8.4.2 *We define the function $F : \mathcal{K} \to \mathcal{K} : F(\mathbf{X}) = \mathbf{X}'$, where $\mathbf{X}$ and $\mathbf{X}'$ are the concatenation of respective sets $\{\mathbf{x}^s\}$ and $\{(\mathbf{x}^s)'\}$ ($s = 0, 1, \ldots$) [see Equation (8.10)], and where $(\mathbf{x}^s)'$ solves $\mathcal{LP}(s)$ with constraint parameters [see Equation (8.9)] $\hat{\mathbf{x}}^{s+\sigma} = \mathbf{x}^{s+\sigma}$, for $\sigma \in S^s$.* □

We show that $\mathcal{Q}$ is equivalent to $\mathcal{LP}_g$ by establishing that $\mathbf{X}^*$ is the unique fixed point of F, presented in the remaining part of this section. We essentially prove that F is a contraction.

We need the following set of auxiliary problems $\mathcal{P}(s)$, one for each separation $s \geq 0$. Let the set $\mathcal{M}_s = \bigcup_{\sigma \in S^s} \mathcal{S}(s+\sigma)$. Then, each auxiliary problem is an optimal stopping version of the Main Problem 8.2, as follows:

Problem 8.4.1 *[Auxiliary] Let two agents be positioned on an arbitrary stage of the graph $G(V, E)$ at separation s. The edge costs are time invariant and i.i.d. random variables over $\mathcal{L}$ with probability p for a zero edge cost. The agents navigate synchronously through the graph in the direction of increasing stage indices. Furthermore, the agents share the costs of the edges in their respective local observation zones perfectly and instantaneously upon reaching a vertex. Let $g(x, \mathbf{u})$ be defined as before as the sum of the edge costs the two agents incur at state x and choosing decision $\mathbf{u}$, and let $f(x, \mathbf{u})$ be the state transition function. Let $\hat{J}_s : \mathcal{M}_s \to \mathbf{R}$ be any function. Two situations occur:*

- *The current agent separation equals s. Then, the local cost incurred in advancing to the next stage equals $g(x, \mathbf{u})$ if $f(x, \mathbf{u}) \in \mathcal{S}(s)$ and $g(x, \mathbf{u}) + \alpha \hat{J}(f(x, \mathbf{u}))$ otherwise.*
- *The current agent separation differs from s. Then, the process terminates and the agents do not incur further costs.*

Then, find the navigation strategy for both agents so that the expected discounted sum of costs incurred at each stage is minimized (discount factor $0 < \alpha < 1$). □

In other words, in Problem 8.4.1, the agents stick to separation s incurring the associated aggregate edge costs, or move to a separation differing from s incurring a one-time cost and stop.

Before we relate the solution of problem $\mathcal{P}(s)$ to the solution of $\mathcal{LP}(s)$, we define operator $T_{s,\hat{J}_s}$ as follows.

Definition 8.4.3 *Given an arbitrary function $J_s : \mathcal{S}(s) \to \mathbf{R}$, and an associated function $\hat{J}_s : \mathcal{M}_s \to \mathbf{R}$ then, for all $x \in \mathcal{S}(s)$,*

$$(T_{s,\hat{J}_s} J_s)(x) = \min_{\mathbf{u} \in U} E\left[v(x, \mathbf{u}) + \alpha V_s(f(x, \mathbf{u}))\right], \tag{8.11}$$

where

$$v(x, \mathbf{u}) = \begin{cases} g(x, \mathbf{u}), & f(x, \mathbf{u}) \in \mathcal{S}(s) \\ g(x, \mathbf{u}) + \alpha \hat{J}_s(f(x, \mathbf{u})), & f(x, \mathbf{u}) \in \mathcal{M}_s, \end{cases} \tag{8.12}$$

and where

$$V_s(x) = \begin{cases} J_s(x), & x \in \mathcal{S}(s) \\ 0, & x \in \mathcal{M}_s. \end{cases} \tag{8.13}$$

□

In words, $T_{s,\hat{J}_s}$ executes one value iteration step for the DP associated with Problem 8.4.1, given $\hat{J}_s$ and J_s. From the traditional value iteration results, we have that for an arbitrary bounded function $J_s : \mathcal{S}(s) \to \mathbf{R}$ and a given associated function $\hat{J}_s : \mathcal{M}_s \to \mathbf{R}$

$$\tilde{J}_s(x) = \lim_{k \to \infty} \left(T^k_{s,\hat{J}_s} J_s\right)(x), \qquad x \in \mathcal{S}(s), \tag{8.14}$$

where $\tilde{J}_s(x)$ $(x \in \mathcal{S}(s))$ denotes the optimal value function for Auxiliary Problem 8.4.1, where in the notation we do not explicitly mention the dependence of $\tilde{J}_s$ on the problem parameter $\hat{J}_s$.

The following lemma relates $\tilde{J}_s$ to the solution of $\mathcal{LP}(s)$.

Lemma 8.4.4 *Let $\tilde{\mathbf{x}}^s \sim \tilde{J}_s(x)$, for $x \in \mathcal{S}(s)$, where $\tilde{J}_s$ is computed as in Equation (8.14), given any function $\hat{J}_s : \mathcal{M}_s \to \mathbf{R}$. Then $\tilde{\mathbf{x}}^s$ solves $\mathcal{LP}(s)$, with $\hat{\mathbf{x}}^{s+\sigma} \sim \hat{J}_s(x)$, for $x \in \mathcal{S}(s+\sigma)$, and $\sigma \in S^s$.* □

Proof. With the LP formulation of the DP associated with the Auxiliary Problem 8.4.1 (see Section 8.3.2 for details), we have that $\tilde{J}_s$ is the solution to the following LP:

$$\begin{aligned} &\text{maximize} \quad \sum_{x \in \mathcal{S}(s)} J_s(x) \\ &\text{subject to} \quad J_s(x) \leq E[v(x, \mathbf{u}) + \alpha V_s(f(x, \mathbf{u}))], \qquad x \in \mathcal{S}(s), \mathbf{u} \in U, \end{aligned}$$

with $v(x, \mathbf{u})$ and $V_s(x, \mathbf{u})$ as defined in Equations (8.12,8.13). This LP is equivalent to a LP with the same cost function but with constraints $J_s(x) \leq E[g(x, \mathbf{u}) + \alpha \hat{V}_s(f(x, \mathbf{u}))]$, for all $x \in \mathcal{S}(s)$, and $\mathbf{u} \in U$, and where

$$\hat{V}_s(x) = \begin{cases} J_s(x), & x \in \mathcal{S}(s) \\ \hat{J}_s(x), & x \in \mathcal{M}_s. \end{cases}$$

With $\mathbf{x}^s \sim J_s(x)$, for $x \in \mathcal{S}(s)$, and for $\hat{\mathbf{x}}^{s+\sigma} \sim \hat{J}_s(x)$, for $x \in \mathcal{S}(s)$, $\sigma \in S^s$, the equivalence of the last LP with $\mathcal{LP}(s)$ follows easily. Hence, $\tilde{\mathbf{x}}^s \sim \tilde{J}_s$, for $x \in \mathcal{S}(s)$ is the solution to $\mathcal{LP}(s)$ and the proof is complete. ■

The following lemma establishes an essential monotonicity property of $T_{s,\hat{J}_s}$.

Lemma 8.4.5 *For any two functions $J_s : \mathcal{S}(s) \to \mathbf{R}$ and $J'_s : \mathcal{S}(s) \to \mathbf{R}$ and for any two functions $\hat{J}_s : \mathcal{M}_s \to \mathbf{R}$ and $\hat{J}'_s : \mathcal{M}_s \to \mathbf{R}$ such that*

$$\begin{aligned} J_s(x) &\leq J'_s(x), \qquad x \in \mathcal{S}(s), \\ \hat{J}_s(x) &\leq \hat{J}'_s(x), \qquad x \in \mathcal{M}_s, \end{aligned} \tag{8.15}$$

we have

$$(T^k_{s,\hat{J}_s} J_s)(x) \leq (T^k_{s,\hat{J}'_s} J'_s)(x), \qquad x \in \mathcal{S}(s),\ k = 1, 2, \ldots \tag{8.16}$$

□

Proof. We use induction. For $k = 0$, Equation (8.16) holds trivially by Equation (8.15). For the induction hypothesis, assume

$$(T^k_{s,\hat{J}_s} J_s)(x) \leq (T^k_{s,\hat{J}'_s} J'_s)(x), \qquad x \in \mathcal{S}(s), \tag{8.17}$$

for some k, then we have, for $x \in \mathcal{S}(s)$,

$$(T^{k+1}_{s,\hat{J}_s} J_s)(x) = \min_{\mathbf{u} \in U} E\left[v(x, \mathbf{u}) + \alpha(V^k_s)(f(x, \mathbf{u}))\right]$$

$$\leq \min_{\mathbf{u} \in U} E\left[v'(x, \mathbf{u}) + \alpha(V_s'^k)(f(x, \mathbf{u}))\right] \tag{8.18}$$
$$= (T^{k+1}_{s,\hat{J}_s'} J_s')(x).$$

Here, $v(x, \mathbf{u})$ and $v'(x, \mathbf{u})$ are as defined in Equation (8.12), with $\hat{J}_s(x)$ and $\hat{J}_s'(x)$, respectively. Further,

$$V_s^k(x) = \begin{cases} (T^k_{s,\hat{J}_s} J_s)(x), & x \in \mathcal{S}(s), \\ 0, & x \in \mathcal{M}_s. \end{cases}$$

The function $V_s'^k$ is defined similarly. In Equation (8.18), we use induction hypothesis (8.17) for the inequality relation and twice the definition of $T_{s,\hat{J}_s}$ [Equation (8.11)]. This concludes the induction and hence the proof. ■

Let $e : \mathcal{S} \to \mathbf{R}$ denote the unity function such that $e(x) = 1$, for all $x \in \mathcal{S}$. Then, we have the following property of $T_{s,\hat{J}_s}$.

Lemma 8.4.6 *For $k = 1, 2, \ldots$, for any function $J_s : \mathcal{S}(s) \to \mathbf{R}$, for any function $\hat{J}_s : \mathcal{M}_s \to \mathbf{R}$ and for any scalar $r \geq 0$, we have*

$$(T^k_{s,\hat{J}_s+re}(J_s + re))(x) \leq (T^k_{s,\hat{J}_s} J_s)(x) + \alpha r, \quad x \in \mathcal{S}(s), \tag{8.19}$$
$$(T^k_{s,\hat{J}_s-re}(J_s - re))(x) \geq (T^k_{s,\hat{J}_s} J_s)(x) - \alpha r, \quad x \in \mathcal{S}(s). \tag{8.20}$$

□

Proof. We give the proof of Equation (8.19) by induction (the proof of Equation (8.20) is similar). From the definition of $T_{s,\hat{J}_s}$ [Equation (8.11)], we have

$$(T_{s,\hat{J}_s+re}(J_s \pm re))(x) = (T_{s,\hat{J}_s} J_s)(x) \pm \alpha r, \quad x \in \mathcal{S}(s). \tag{8.21}$$

For $k = 1$, Equation (8.19) holds, by Equation (8.21). Assume, as induction hypothesis that

$$(T^k_{s,\hat{J}_s+re}(J_s + re))(x) \leq (T^k_{s,\hat{J}_s} J_s)(x) + \alpha r, \quad x \in \mathcal{S}(s).$$

Then, applying operator $T_{s,\hat{J}_s+re}$ to both sides of this equation and using Lemma 8.4.5, yields

$$\left[T^{k+1}_{s,\hat{J}_s+re}(J_s + re)\right](x) \leq \left[T_{s,\hat{J}_s+re}\left((T^k_{s,\hat{J}_s} J_s) + \alpha r e\right)\right](x), \quad x \in \mathcal{S}(s). \tag{8.22}$$

Two cases occur:

- First, consider the set of pairs of $x \in \mathcal{S}(s)$ and $\mathbf{u} \in U$ such that $f(x, \mathbf{u}) \in \mathcal{S}(s)$. Then, for the RHS of Equation (8.22), we have the following inequality:

$$\left[T_{s,\hat{J}_s+re}\left((T^k_{s,\hat{J}_s} J_s) + \alpha r e\right)\right](x) \leq \min_{\mathbf{u}} E\left[g(x, \mathbf{u}) + \alpha\left((T^k_{s,\hat{J}_s} J_s)(f(x, \mathbf{u})) + \alpha r\right)\right],$$

$$= \min_{\mathbf{u}} E\left[g(x,\mathbf{u}) + \alpha(T^k_{s,\hat{J}_s} J_s)(f(x,\mathbf{u}))\right] + \alpha^2 r,$$

$$\leq \min_{\mathbf{u}} E\left[g(x,\mathbf{u}) + \alpha(T^k_{s,\hat{J}_s} J_s)(f(x,\mathbf{u}))\right] + \alpha r, \quad (8.23)$$

where, for all $x \in \mathcal{S}(s)$, the minimization is over all $\mathbf{u} \in U$ such that $f(x,\mathbf{u}) \in \mathcal{S}(s)$.

- Second, consider the set of pairs of $x \in \mathcal{S}(s)$ and $\mathbf{u} \in U$ such that $f(x,\mathbf{u}) \in \mathcal{M}_s$. Then, for the RHS of Equation (8.22), we have the following inequality:

$$\left[T_{s,\hat{J}_s+re}\left((T^k_{s,\hat{J}_s} J_s) + \alpha re\right)\right](x) \leq \min_{\mathbf{u}}\left[g(x,\mathbf{u}) + \alpha(\hat{J}_s(f(x,\mathbf{u})) + r)\right],$$

$$= \min_{\mathbf{u}}\left[g(x,\mathbf{u}) + \alpha \hat{J}_s(f(x,\mathbf{u}))\right] + \alpha r, \quad (8.24)$$

where, for all $x \in \mathcal{S}(s)$, the minimization is over all $\mathbf{u} \in U$ such that $f(x,\mathbf{u}) \in \mathcal{M}_s$.

Combining Equations (8.23) and (8.24), yields

$$\left[T_{s,\hat{J}_s+re}\left((T^k_{s,\hat{J}_s} J_s) + \alpha re\right)\right](x) \leq \left[T^{k+1}_{s,\hat{J}_s} J_s\right](x) + \alpha r, \qquad x \in \mathcal{S}(s),$$

and with Equation (8.22), we have

$$\left[T^{k+1}_{s,\hat{J}_s+re}(J_s + re)\right](x) \leq \left[T^{k+1}_{s,\hat{J}_s} J_s\right](x) + \alpha r, \qquad x \in \mathcal{S}(s),$$

which concludes the induction and the proof. ■

We now establish two properties of F, based on which we show that F is a contraction. We first introduce some notation. For an arbitrary $\mathbf{X}^0 \in \mathcal{K}$, define $\mathbf{X}^k \in \mathcal{K}$, for $k = 1, 2, \ldots$ as

$$\mathbf{X}^k = F(\mathbf{X}^{k-1}), \quad (8.25)$$
$$= \left[\ (\mathbf{x}^{0,k})^T \ \ (\mathbf{x}^{1,k})^T \ \ \ldots\ \right]^T.$$

Vector $(\mathbf{X}')^k \in \mathcal{K}$ is defined similarly. We establish the first property of F, monotonicity.

Lemma 8.4.7 *For any two vectors* $\mathbf{X}^0$, $(\mathbf{X}')^0 \in \mathcal{K}$ *such that*

$$\mathbf{X}^0 \leq (\mathbf{X}')^0, \quad (8.26)$$

we have

$$F^k(\mathbf{X}^0) \leq F^k((\mathbf{X}')^0), \qquad k = 1, 2, \ldots \quad (8.27)$$

□

Proof. We use induction. For $k = 0$, Equation (8.27) holds trivially by Equation (8.26). For the induction hypothesis, assume that $F^k(\mathbf{X}^0) \leq F^k((\mathbf{X}')^0)$, or using the notation in Equation (8.25), we have that

$$\mathbf{X}^k \leq (\mathbf{X}')^k, \tag{8.28}$$

for some k. From Equation (8.25), we have that $F(\mathbf{X}^k) = \mathbf{X}^{k+1}$, and with Lemma 8.4.4, this yields for all $s = 0, 1, \ldots,$

$$\mathbf{x}^{s,k+1} \sim \left(\lim_{l \to \infty} T^l_{s,\hat{J}^k_s} J^k_s \right)(x), \qquad x \in \mathcal{S}(s),$$

where $J^k_s(x) \sim \mathbf{x}^{s,k}$, for $x \in \mathcal{S}(s)$ and where $\hat{J}^k_s(x) \sim \mathbf{x}^{s+\sigma,k}$, for $x \in \mathcal{S}(s+\sigma)$, for all $\sigma \in S^s$. A similar statement holds for $(\mathbf{x}')^{s,k+1}$. From Lemma 8.4.5, and with the induction hypothesis (8.28), we have that for $s = 0, 1, \ldots$

$$\left(T^l_{s,\hat{J}^k_s} J^k_s \right)(x) \leq \left(T^l_{s,(\hat{J}'_s)^k} (J'_s)^k \right)(x), \qquad x \in \mathcal{S}(s), \tag{8.29}$$

for $l = 0, 1, \ldots$. Therefore, taking the limit for $l \to \infty$ in Equation (8.29), we have that $\mathbf{x}^{s,k+1} \leq (\mathbf{x}')^{s,k+1}$, for $s = 0, 1, \ldots,$ from which it follows that $\mathbf{X}^{k+1} \leq (\mathbf{X}')^{k+1}$, concluding the induction and hence the proof. ■

The following lemma provides the second property of F required to show that F is a contraction.

Lemma 8.4.8 *For $k = 1, 2, \ldots,$ for any $\mathbf{X} \in \mathcal{K}$, and for any scalar $r \geq 0$, we have*

$$F^k(\mathbf{X} + r\mathbf{e}) \leq F^k(\mathbf{X}) + \alpha^k r\mathbf{e}, \tag{8.30}$$

$$F^k(\mathbf{X} - r\mathbf{e}) \geq F^k(\mathbf{X}) - \alpha^k r\mathbf{e}, \tag{8.31}$$

where $\mathbf{e} \in \mathcal{K}$ contains ones as its entries. □

Proof. We prove Equation (8.30). The proof of Equation (8.31) is similar. We define the functions $J_s : \mathcal{S}(s) \to \mathbf{R}$, for $s = 0, 1, \ldots,$ such that

$$J_s(x) \sim \mathbf{x}^s, \qquad x \in \mathcal{S}(s), \tag{8.32}$$

with

$$\mathbf{X} = \left[\, (\mathbf{x}^0)^T \;\; (\mathbf{x}^1)^T \;\; \ldots \, \right]^T.$$

Similarly, we define the functions $\hat{J}_s : \mathcal{M}_s \to \mathbf{R}$ such that

$$\hat{J}_s(x) \sim \{\mathbf{x}^{s+\sigma}\}, \qquad x \in \mathcal{M}_s, \sigma \in S^s. \tag{8.33}$$

From Lemma 8.4.6, we have

$$\left(T^l_{s,\hat{J}_s + re}(J_s + re) \right)(x) \leq (T^l_{s,\hat{J}_s} J_s)(x) + \alpha r, \qquad x \in \mathcal{S}(s),$$

and taking the limit for $l \to \infty$, with Equations (8.32, 8.33) and with Lemma 8.4.4, this yields

$$F(\mathbf{X} + r\mathbf{e}) \le F(\mathbf{X}) + \alpha r\mathbf{e}. \tag{8.34}$$

We use induction. For $k = 1$, Equation (8.30) holds by Equation (8.34). For the induction hypothesis, suppose $F^k(\mathbf{X} + r\mathbf{e}) \le F^k(\mathbf{X}) + \alpha^k r\mathbf{e}$. Applying F to both sides of the previous equation, and using Lemma 8.4.7, yields

$$\begin{aligned} F^{k+1}(\mathbf{X} + r\mathbf{e}) &\le F(F^k(\mathbf{X}) + \alpha^k r\mathbf{e}), \\ &\le F^{k+1}(\mathbf{X}) + \alpha(\alpha^k r\mathbf{e}), \\ &= F^{k+1}(\mathbf{X}) + \alpha^{k+1} r\mathbf{e}, \end{aligned}$$

where the second inequality follows from Equation (8.34). This concludes the induction, and hence the proof. ■

The following lemma provides the basis for main result of this section.

Lemma 8.4.9 *The function F is a contraction mapping. In particular, given any two bounded vectors $\mathbf{X}, \mathbf{X}' \in \mathcal{K}$, we have that*

$$\max |F^k(\mathbf{X}) - F^k(\mathbf{X}')| \le \alpha^k \max |\mathbf{X} - \mathbf{X}'|,$$

where the $\max(\cdot)$ *is taken over all components of its argument.* □

Proof. The proof, based on Lemmas 8.4.7 and 8.4.8, is identical to the proof of the fact that operator T is a contraction as can be found in (Bertsekas 2001). We give it here for completeness. Let $m = \max |\mathbf{X} - \mathbf{X}'|$. It follows that

$$\mathbf{X} - m\mathbf{e} \le \mathbf{X}' \le \mathbf{X} + m\mathbf{e}. \tag{8.35}$$

We apply F^k to both sides of Equation (8.35) and use Lemmas 8.4.7 and 8.4.8 to obtain

$$F^k(\mathbf{X}) - \alpha^k m\mathbf{e} \le F^k(\mathbf{X}') \le F^k(\mathbf{X}) + \alpha^k m\mathbf{e},$$

which yields

$$\max |F^k(\mathbf{X}) - F^k(\mathbf{X}')| \le \alpha^k m,$$

which concludes the proof. ■

The following lemma provides an essential property of contraction mappings (see Bertsekas 2001).

Lemma 8.4.10 *If the function $F : \mathcal{K} \to \mathcal{K}$ is a contraction mapping, then there exists a unique fixed point of F. In particular, there exists a unique vector $\mathbf{X}^* \in \mathcal{K}$ such that $F(\mathbf{X}^*) = \mathbf{X}^*$. Furthermore, if $\mathbf{X}$ is any vector in $\mathcal{K}$, then*

$$\lim_{k \to \infty} \|F^k(\mathbf{X}) - \mathbf{X}^*\| = 0.$$

□

Proof. For a proof, see Liusternik and Sobolev (1961) and Luenberger (1969). □

We have now left to show that the solution of $\mathcal{LP}_g$ is the unique fixed point of F. The following theorem provides the main result.

Theorem 8.4.11 *The solution* $\mathbf{X}^*$ *of* $\mathcal{LP}_g$ *is the unique fixed point of* F. □

Proof. From Lemma 8.4.9, we have that F is a contraction and from Lemma 8.4.10, F has a unique fixed point. Therefore, we only need to show that the solution of $\mathcal{LP}_g$, i.e.

$$\mathbf{X}^* = \left[\, (\mathbf{x}^{0,*})^T \;\; (\mathbf{x}^{1,*})^T \;\; \ldots \right]^T,$$

is a fixed point of F. For a all $s \geq 0$, let

$$\hat{\mathbf{x}}^{s+\sigma} = \mathbf{x}^{s+\sigma,*}, \qquad \sigma \in S^s. \tag{8.36}$$

Furthermore, let the solution of $\mathcal{LP}(s)$, with $\hat{\mathbf{x}}^{s+\sigma}$ ($\sigma \in S^s$) as in Equation (8.36), be denoted as $\tilde{\mathbf{x}}^s$. Then, for $\mathbf{X}^*$ to be a fixed point of F, we need to show that for all $s = 0, 1, 2, \ldots$ we have that

$$\tilde{\mathbf{x}}^s = \mathbf{x}^{s,*}. \tag{8.37}$$

Since $\mathbf{X}^*$ solves $\mathcal{LP}_g$, $\mathbf{x}^{s,*}$ is also a feasible solution of $\mathcal{LP}(s)$, with $\hat{\mathbf{x}}^{s+\sigma}$ as in Equation (8.36). It follows that $\tilde{\mathbf{x}}^s \geq \mathbf{x}^{s,*}$. To establish Equation (8.37), we show that $\tilde{\mathbf{x}}^s \leq \mathbf{x}^{s,*}$, by contradiction. Assume that

$$\tilde{\mathbf{x}}^s \geq \mathbf{x}^{s,*}, \tag{8.38}$$

with for at least one i

$$\tilde{x}_i^s > x_i^{s,*}, \tag{8.39}$$

where x_i^s denotes the ith component of $\mathbf{x}^s$. We show that

$$\tilde{\mathbf{X}}^s = \left[\, (\mathbf{x}^{0,*})^T \;\; (\mathbf{x}^{1,*})^T \;\; \ldots \;\; (\mathbf{x}^{s-1,*})^T \;\; (\tilde{\mathbf{x}}^s)^T \;\; (\mathbf{x}^{s+1,*})^T \;\; \ldots \right]^T,$$

is a feasible solution for $\mathcal{LP}_g$, as follows. The constraints associated with separation s [see Equation (8.8)] are satisfied, since $\tilde{\mathbf{x}}^s$ satisfies the constraints of $\mathcal{LP}(s)$ with $\hat{\mathbf{x}}^{s+\sigma}$ as in Equation (8.36). The constraints associated with all other separations also hold since $\tilde{\mathbf{x}}^s$ appears in the RHS with nonnegative coefficient matrices $\mathbf{A}_s^{(\cdot)}$. From assumption (8.38) it follows that $\tilde{\mathbf{X}}^s \geq \mathbf{X}^*$, where, under assumption (8.39) and for at least one $l \geq 1$ we have that $\tilde{X}_l^s > X_l^*$, where X_l^s denotes the lth element of $\mathbf{X}^s$. Since $\mathbf{X}^*$ is optimal, we reached a contradiction and it follows that Equation (8.37) holds for all $s \geq 0$. This concludes the proof. ■

We have shown that it suffices to compute the unique fixed point of F to find the optimal problem value function.

8.5 FIXED POINT COMPUTATION

In this section, we provide an algorithm to compute $\mathbf{X}^*$, the fixed point of F. In Section 8.5.1, we study the single agent problem, and compute the optimal single agent policies. Then, in Section 8.5.2, we assume that there exists a particular $\bar{s} \geq 0$ such that for all $s \geq \bar{s}$ a particular policy based on set of optimal single agent policies, is optimal. We derive a *Linear Time Invariant* (LTI) system with as state, the optimal value function at one separation, where 'separation' is the equivalent of 'time' in a traditional LTI system, in Sections 8.5.3 and 8.5.4. Finally in Section 8.5.5, we formulate a LP whose solution is the optimal value function at separations $s < \bar{s}$, satisfying the fixed point condition. Furthermore, the initial condition for the LTI system derived from this LP solution, is such that the simulation of the LTI system yields a sequence of value functions for $s \geq \bar{s}$ that also satisfy the fixed point condition.

8.5.1 Single agent problem

We formulate the single agent problem and find the optimal single agent policies, a building block for the two-agent problem.

Problem 8.5.1 ***[Single Agent]*** *Let one agent be positioned on an arbitrary stage of graph $G(V, E)$. The edge costs are time invariant and i.i.d. random variables over $\mathcal{L}$ with probability p for a zero edge cost. The agent observes the costs of the links in its local observation zone and navigates through the graph in the direction of increasing stage indices. Then, find the navigation strategy so that the expected discounted sum of costs incurred at each stage is minimized (discount factor $0 < \alpha < 1$).* □

The state $x = C$ is the set of costs of the edges in the local observation zone associated with the agent, summarized in vector C [see Equation (8.1)]. We distinguish two types of edge costs: a controllable edge cost is determined at least in part by decision $u \in U_1 = \{-1, 0, 1\}$, whereas an uncontrollable edge cost is independent of u. Edge cost a_0 is controllable, as opposed to a_{-1}, a_1 and b_0, which are uncontrollable. In particular, if at the current state the agent chooses the edge straight ahead ($u = 0$), then $a_0' = b_0$, where $(\cdot)'$ denotes the value of $(\cdot)$ at the next time step. On the other hand, for any $u \in U_1$, the edge costs a_{-1}', a_1' and b_0' are only known probabilistically, according to the *a priori* information available on unobserved edge costs, and do not depend on the agent decision. Hence, a_0 is (partially) controllable. Since a_0 is the only controllable edge cost, from Bellman's equation, it suffices to compute the expected value of the optimal value function for $a_0 = 0$ and for $a_0 = 1$ to determine the optimal value function at all states.

The optimal policy is summarized with the following two rules:

1. Pick an edge with lowest cost.
2. Break ties using the obvious relation

$$E[J_1^*(x)|a_0 = 0] \leq E[J_1^*(x)] \leq E[J_1^*(x)|a_0 = 1],$$

where the expected value is taken over the unknown edge costs and where J_1^* represents the single agent optimal value function.

Then, we have that

$$E[J_1^*(x)|a_0 = 0] = \frac{(1-p)^4\left(1+p-p^2\right)\alpha}{(1-\alpha)\left(1+p\alpha-5p^2\alpha+6p^3\alpha-2p^4\alpha\right)}, \tag{8.40}$$

$$E[J_1^*(x)|a_0 = 1] = \frac{(1-p)^2\left(1-p\alpha-2p^2\alpha+3p^3\alpha-p^4\alpha\right)}{(1-\alpha)\left(1+p\alpha-5p^2\alpha+6p^3\alpha-2p^4\alpha\right)}, \tag{8.41}$$

where $0 \le p \le 1$ and $0 < \alpha < 1$. For optimality of the described policy two conditions verifying the first optimal policy rule need to hold:

$$\alpha E[J_1^*(x)|a_0 = 1] \le 1 + \alpha E[J_1^*(x)],$$
$$\alpha E[J_1^*(x)] \le 1 + \alpha E[J_1^*(x)|a_0 = 0].$$

With Equations (8.40,8.41) this yields the conditions

$$0 \le \frac{1-3\,p^2\,\alpha+5\,p^3\,\alpha-2\,p^4\,\alpha}{1+p\,\alpha-5\,p^2\,\alpha+6\,p^3\,\alpha-2\,p^4\,\alpha}, \tag{8.42}$$

$$0 \le \frac{1-\alpha+4\,p\,\alpha-8\,p^2\,\alpha+7\,p^3\,\alpha-2\,p^4\,\alpha}{1+p\,\alpha-5\,p^2\,\alpha+6\,p^3\,\alpha-2\,p^4\,\alpha}, \tag{8.43}$$

for $0 \le p \le 1$ and $0 < \alpha < 1$. In the following lemma we prove that conditions (8.42) and (8.43) hold for $0 \le p \le 1$ and $0 \le \alpha \le 1$. Hence, the policy is optimal for any relevant value of p and α.

Lemma 8.5.1 *Equation (8.42) and Equation (8.43) hold for all* $0 \le p \le 1$ *and* $0 < \alpha < 1$.
□

We omit the proof, but mention that it involves proving positivity of polynomials either by hand or using SOSTOOLS (Prajna *et al.* 2004), a software that uses semi-definite programming to that end (Parrilo 2003). Hence, for any relevant p and α ($0 \le p \le 1$, $0 < \alpha < 1$), the same set of single agent policies is optimal.

8.5.2 Mixed forward-backward recursion

In this section, we extract from $\mathcal{LP}(s)$ a recursion writing the optimal value function at separation s as a function of the optimal value function at separations $s+\sigma$, for $\sigma = \max\{-2, -s\}, \ldots, 2$. This recursion is neither forward, nor backward since at separation s, the RHS includes terms at lower and greater separations. Let $\mu_\infty^s : \mathcal{S}(s) \to U$ denote the policy, whereby at separation s each agent follows the single agent optimal policy and where ties are broken by choosing the pair of decisions that minimizes the resulting agent separation. Then, we make the following assumption:

Assumption 8.5.1 For any $0 \leq p \leq 1$ and $0 < \alpha < 1$, we look for an optimal policy $\mu^* : \mathcal{S} \rightarrow U$ in the set of policies $\{\mu \mid \mu(x) = \mu_\infty^s(x)$ for $x \in \mathcal{S}(s), s \geq \bar{s}\}$ for a particular $\bar{s} \geq 0$.

This assumption is motivated in part by Section 8.5.1, where we show that for the set of relevant p and α, the same set of policies is optimal. Note that under Assumption 8.5.1, μ_∞^s is independent of s for $s \geq \bar{s}$ and therefore, for $s \geq \bar{s}$ we omit the superscript s. One sanity check is that the optimal two-agent value function should converge to twice the optimal single-agent value function for s going to infinity, since for infinitely large s the two-agent problem is equivalent to the combination of two single-agent problems. Assumption 8.5.1 satisfies this check, since for infinitely large s, each agent indeed uses a single agent optimal policy.

We write $\mathcal{LP}(s)$ in standard form as follows:

$$\text{maximize} \quad \mathbf{c}^T \begin{bmatrix} \mathbf{x}^s \\ \mathbf{y}^s \end{bmatrix} \tag{8.44}$$

$$\text{subject to} \quad (\mathbf{T} - \mathbf{A}^0)\mathbf{x}^s + \mathbf{y}^s = \sum_{\sigma \in S^s} \mathbf{A}^\sigma \hat{\mathbf{x}}^{s+\sigma,*} + \mathbf{b}^s, \tag{8.45}$$

$$\mathbf{x}^s = \mathbf{x}_+^s - \mathbf{x}_-^s \tag{8.46}$$

$$\mathbf{x}_+^s, \mathbf{x}_-^s, \mathbf{y}^s \geq 0,$$

denoted $\mathcal{LP}_f(s)$. Here, $\mathbf{c}^T = \begin{bmatrix} \mathbf{e}^T & \mathbf{0}^T \end{bmatrix}^T$ with the vectors $\mathbf{e}$ and $\mathbf{0}$ of the same dimension as $\mathbf{x}^s$ and $\mathbf{y}^s$, respectively. Writing $\mathbf{x}^s = \mathbf{x}_+^s - \mathbf{x}_-^s$, with $0 \leq \mathbf{x}_+^s, \mathbf{x}_-^s \in \mathbf{R}^{n'}$, where $n' = n_0$ for $s = 0$, and $n' = n$ for $s \geq 1$, is a common procedure for converting a free variable to the standard form format. Similarly, introducing slack variables $0 \leq \mathbf{y}^s \in \mathbf{R}^{n'|U|}$ is a standard method to transform inequalities into equalities. We refer the reader to Bertsimas and Tsitsiklis (1997) for details on converting a general LP into standard form. The version of $\mathcal{LP}_f(s)$ where $\hat{\mathbf{x}}^{s+\sigma} = \mathbf{x}^{s+\sigma,*}$, for $\sigma \in S^s$, is denoted $\mathcal{LP}_f^*(s)$. Since $\mathbf{x}^{s+\sigma,*} \geq 0$, for any $s \geq 0$ and $\sigma \in S^s$, we can set $\mathbf{x}_-^s = 0$ in $\mathcal{LP}_f^*(s)$, a helpful simplification.

We now derive a matrix equation relating $\mathbf{x}^{s,*}$ to $\mathbf{x}^{s+\sigma,*}$ $(\sigma \in S^s)$ for $s \geq 0$. To this end, we show a property of the optimal basic feasible solution, which holds for all $s \geq 0$. In particular, consider the optimal basic feasible solution pair $(\mathbf{x}^{s,*}, \mathbf{y}^{s,*})$ of $\mathcal{LP}_f^*(s)$ for any $s \geq 0$. At $(\mathbf{x}^{s,*}, \mathbf{y}^{s,*})$, the $n|U|$ linear equality constraints (8.45) are satisfied, where n needs to be replaced by n_0 for $s = 0$ (from here on, we omit this fact). Furthermore, from the definition of a basic solution, we have that, out of all the constraints that are active, $n + n|U|$ (i.e. the number of decision variables) are linearly independent. Therefore, at least n of the inequality constraints $\mathbf{y}^s \geq 0$ are active. To each particular $x_i^{s,*}$ $(0 \leq i \leq n)$, i.e. at separation s and for the pair of observation zones corresponding to index i, is associated a set of $|U|$ equality constraints (see the definition of $\mathbf{T}$ in Section 8.3.2). Let $\mathbf{y}_i^s \in \mathbf{R}^{|U|}$ contain the slack variables associated with the equality constraint set related to x_i^s. We have the following property of $\mathbf{y}_i^{s,*}$.

Lemma 8.5.2 *At least one of the components of* $\mathbf{y}_i^{s,*}$ *equals zero, for all* $i = 1, \ldots, n$, *and for all* $s = 0, 1, \ldots$ □

Proof. One way to see this is by contradiction. Suppose the opposite is true for some i, *i.e.* $\mathbf{y}_i^{s,*} > 0$, then we can increase $x_i^{s,*}$, adapting the elements of $\mathbf{y}_i^{s,*}$ accordingly so that the set of $|U|$ equality constraints (8.45) associated with x_i^s remains satisfied, until some element of $\mathbf{y}_i^s$ equals zero. Increasing x_i^s does not jeopardize equality constraints related to other components of $\mathbf{x}^s$, since x_i^s appears in the RHS of those constraints with non-negative coefficients.

This procedure yields a feasible solution $(\overline{\mathbf{x}}^s, \overline{\mathbf{y}}^s)$ such that $\overline{\mathbf{x}}^s \geq \mathbf{x}^{s,*}$, where strict inequality holds for at least one component. Since $(\mathbf{x}^{s,*}, \mathbf{y}^{s,*})$ is assumed optimal, this is a contradiction and the proof is complete. ∎

We now derive a linear equation in $\mathbf{x}^{s,*}$, and $\mathbf{x}^{s+\sigma,*}$ for $\sigma \in S^s$ for all $s \geq 0$. Since for all $i = 1, \ldots, n$, at least one element of $\mathbf{y}_i^s$ equals zero (see Lemma 8.5.2), we can construct a set of n equations with x_i^s, *i.e.* the components of $\mathbf{x}^s$, as the n unknowns. Specifically, for each $i = 1, \ldots, n$ we can pick an index j_i^s $(1 \leq j_i^s \leq |U|)$, such that

$$(\mathbf{y}_i^{s,*})_{j_i^s} = 0, \tag{8.47}$$

where the LHS denotes the (j_i^s)th component of $\mathbf{y}_i^{s,*}$. Then, for each $i = 1, \ldots, n$, and out of the $|U|$ equality constraints associated with x_i^s, take the equality constraint with index j_i^s as row i of matrix $\mathbf{I} - \mathbf{D}_{s,0}$, *i.e.*, $\mathbf{I}$ and $\mathbf{D}_{s,0}$ contain the relevant rows of $\mathbf{T}$ and $\mathbf{A}_0$, respectively. Then, for all $s \geq 0$, we have that

$$(\mathbf{I} - \mathbf{D}_{s,0})\mathbf{x}^{s,*} = \sum_{\sigma \in S^s} \mathbf{D}_{s,\sigma}\mathbf{x}^{s+\sigma,*} + \mathbf{d}_s, \tag{8.48}$$

where $\mathbf{I} - \mathbf{D}_{s,0}, \mathbf{D}_{s,\sigma} \in \mathbf{R}^{n \times n}$ (for all $\sigma \in S^s$), where $\mathbf{D}_{s,\sigma}$ and $\mathbf{d}_s \in \mathbf{R}^n$ contain the relevant rows of $\mathbf{A}_s^\sigma$ (for all $\sigma \in S^s$) and $\mathbf{b}^s$, respectively. Note that in general the optimal basic feasible solution need not be the same for all s, hence the addition of subscript s to the composing matrices of Equation (8.48).

We now restrict our attention to separations $s \geq \overline{s}$, where under Assumption 8.5.1, the same basic feasible solution is optimal, and therefore we can write $\mathbf{D}_0 \doteq \mathbf{D}_{s',0}$, for all $s' \geq \overline{s}$. The matrices $\mathbf{D}_\sigma$, for $\sigma \in S^s$ and the vector $\mathbf{d}$ are defined similarly. We have then, for $s \geq \overline{s}$, that

$$(\mathbf{I} - \mathbf{D}_0)\mathbf{x}^{s,*} = \sum_{\sigma \in S^s} \mathbf{D}_\sigma \mathbf{x}^{s+\sigma,*} + \mathbf{d}. \tag{8.49}$$

The structure of Main Problem 8.2 allows us to reduce the dimension of the matrices and vectors in Equation (8.49). Recall that in this problem, state $x = (s, \mathcal{C}^A, \mathcal{C}^B)$ is a triple containing the agent separation s and the costs of the edges in the local observation zones associated with both agents, summarized in the vectors $\mathcal{C}^A$ and $\mathcal{C}^B$ [see Equation (8.1)]. The edge costs a_0^A and a_0^B are controllable (see Section 8.5.1 for the definition of a controllable edge cost); a_1^A, a_{-1}^A, b_0^A, a_1^B, a_{-1}^B, and b_0^B, are uncontrollable.

We introduce some notation. Let

$$E\mathbf{x}_{qr}^{s,*} = E[J^*(x)|a_0^A = q, a_0^B = r], \tag{8.50}$$

where $q, r \in \mathcal{L} \cup \{\times\}$, and where the expected value is taken over the uncontrollable edge costs. Here, $q = \times$ (resp. $r = \times$) indicates that a_0^A (resp. a_0^B) is unobserved. From symmetry, we have that $E\mathbf{x}_{qr}^{s,*} = E\mathbf{x}_{rq}^{s,*}$, for $q, r \in \mathcal{L} \cup \{\times\}$. Furthermore, we have the relation

$$E\mathbf{x}_{q\times}^{s,*} = pE\mathbf{x}_{q0}^{s,*} + (1-p)E\mathbf{x}_{q1}^{s,*}, \qquad \text{for } q \in \mathcal{L} \cup \{\times\}, \tag{8.51}$$

which follows immediately from the definition (8.50) of $E\mathbf{x}_{qr}^{s,*}$.

We now use $E\mathbf{x}_{qr}^{s,*}$ and the fact that only a_0^A and a_0^B are controllable to simplify Equation (8.49), for $s \geq \bar{s}$. We investigate two cases. First, if at a particular separation $s \geq \bar{s}$, the optimal two agent decision $\mathbf{u}_i \in U$ associated with some i $(0 \leq i \leq n)$ is such that $s' = s+2$ or $s' = s-2$, then $x_i^{s,*} = d_i + \alpha E\mathbf{x}_{\times\times}^{s',*}$. Therefore, in Equation (8.49), we have that the ith row of the matrices $\mathbf{D}_0$, $\mathbf{D}_1$ and $\mathbf{D}_{-1}$ equals zero. Furthermore, $\mathbf{D}_2$ and $\mathbf{D}_{-2}$ have at most rank one. In particular, we have

$$\mathbf{D}_2\mathbf{x}^{s+2,*} = \mathbf{D}_2' E\mathbf{x}_{\times\times}^{s+2,*} \tag{8.52}$$

where $\mathbf{D}_2' \in \mathbf{R}^{n\times 1}$. A similar equation holds in the case $s' = s-2$.

Second, for $\mathbf{u}_i$ such that $s' = s+1$ or $s' = s-1$, we have that $x_i^{s,*} = d_i + \alpha E\mathbf{x}_{q\times}^{s',*}$, with $q = b_0^A$ if $u_i^A = 0$ and $q = b_0^B$ if $u_i^B = 0$. Hence, the matrices $\mathbf{D}_1$ and $\mathbf{D}_{-1}$ are at most of rank two since at most two linearly independent linear combinations of $\mathbf{x}^{s+1,*}$ (resp. $\mathbf{x}^{s-1,*}$), namely $E\mathbf{x}_{0\times}^{s+1,*}$ and $E\mathbf{x}_{1\times}^{s+1,*}$ (resp. $E\mathbf{x}_{0\times}^{s-1,*}$ and $E\mathbf{x}_{1\times}^{s-1,*}$), suffice in Equation (8.49). We have

$$\mathbf{D}_1\mathbf{x}^{s+1,*} = \mathbf{D}_1' \begin{bmatrix} E\mathbf{x}_{0\times}^{s+1,*} & E\mathbf{x}_{1\times}^{s+1,*} \end{bmatrix}^T, \tag{8.53}$$

where $\mathbf{D}_1' \in \mathbf{R}^{n\times 2}$. A similar equation holds in the case $s' = s-1$. For policy μ_∞, we have $\text{rank}(\mathbf{D}_2') = \text{rank}(\mathbf{D}_{-2}') = 1$ and $\text{rank}(\mathbf{D}_1') = \text{rank}(\mathbf{D}_{-1}') = 2$.

Finally, if $\mathbf{u}_i$ is such that $s' = s$, we have that, in similar fashion, matrix $\mathbf{D}_0$ is of rank three and hence

$$\mathbf{D}_0\mathbf{x}^{s,*} = \mathbf{D}_0' \begin{bmatrix} E\mathbf{x}_{00}^{s,*} & E\mathbf{x}_{01}^{s,*} & E\mathbf{x}_{11}^{s,*} \end{bmatrix}^T, \tag{8.54}$$

where $\mathbf{D}_0' \in \mathbf{R}^{n\times 3}$, with the appropriate components. With Equations (8.52–8.54), we write Equation (8.49) as

$$(\mathbf{I} - \mathbf{D}_0'') \begin{bmatrix} E\mathbf{x}_{00}^{s,*} & E\mathbf{x}_{01}^{s,*} & E\mathbf{x}_{11}^{s,*} \end{bmatrix}^T = \ldots$$

$$\mathbf{D}_{-2}'' E\mathbf{x}_{\times\times}^{s-2,*} + \mathbf{D}_{-1}'' \begin{bmatrix} E\mathbf{x}_{0\times}^{s-1,*} \\ E\mathbf{x}_{1\times}^{s-1,*} \end{bmatrix} + \mathbf{D}_1'' \begin{bmatrix} E\mathbf{x}_{0\times}^{s+1,*} \\ E\mathbf{x}_{1\times}^{s+1,*} \end{bmatrix} + \mathbf{D}_2'' E\mathbf{x}_{\times\times}^{s+2,*} + \mathbf{d}'', \tag{8.55}$$

where $\mathbf{D}_\sigma'' = \mathbf{K}\mathbf{D}_\sigma'$, for $\sigma \in S^s$, and where $\mathbf{d}'' = \mathbf{K}\mathbf{d}$. Here, the matrix $\mathbf{K} \in \mathbf{R}^{3\times n}$ is such that

$$\begin{bmatrix} E\mathbf{x}_{00}^{s} & E\mathbf{x}_{01}^{s} & E\mathbf{x}_{11}^{s} \end{bmatrix}^T = \mathbf{K}\mathbf{x}^s, \tag{8.56}$$

where the entries of $\mathbf{K}$ are known polynomials in p. In words, linear operator $\mathbf{K}$ takes the appropriate linear combination of the elements of $\mathbf{x}^s$ to compute the expected values in the LHS of Equation (8.56). This recursion is neither forward nor backward in the sense that the optimal value function at separation s depends (linearly) on the optimal value function at separations both lower and greater than s. In the following section, we convert Equation (8.55) to a forward equation, so that given the optimal value function up till separation s, we can compute the optimal value function at larger separations.

8.5.3 Forward recursion

In this section, we transform Equation (8.55) into a forward recursion which allows us, under Assumption 8.5.1, to compute the optimal value function for $s \geq \bar{s}$, given the appropriate initial condition, involving $E\mathbf{x}_{\times\times}^{\bar{s}+1,*}$, $E\mathbf{x}_{q\times}^{\bar{s},*}$ $(q = 1, 2)$ and all information on the optimal value function at separations $s < \bar{s}$.

We first perform a change of coordinates in Equation (8.55) and rearrange the terms in an appropriate way. In particular, we express the expected values of the optimal value function at a particular separation s as a function of $E\mathbf{x}_{\times\times}^{s,*}$, $E\mathbf{x}_{1\times}^{s,*}$ and $E\mathbf{x}_{11}^{s,*}$. Specifically, we define $\mathbf{P}_1 \in \mathbf{R}^{3\times 3}$ and $\mathbf{P}_2 \in \mathbf{R}^{2\times 2}$ such that

$$\left[\, E\mathbf{x}_{\times\times}^{s,*} \;\; E\mathbf{x}_{1\times}^{s,*} \;\; E\mathbf{x}_{11}^{s,*} \,\right]^T = \mathbf{P}_1 \left[\, E\mathbf{x}_{00}^{s,*} \;\; E\mathbf{x}_{01}^{s,*} \;\; E\mathbf{x}_{11}^{s,*} \,\right]^T, \tag{8.57}$$

$$\left[\, E\mathbf{x}_{\times\times}^{s,*} \;\; E\mathbf{x}_{1\times}^{s,*} \,\right]^T = \mathbf{P}_2 \left[\, E\mathbf{x}_{0\times}^{s,*} \;\; E\mathbf{x}_{1\times}^{s,*} \,\right]^T. \tag{8.58}$$

Both $\mathbf{P}_1$ and $\mathbf{P}_2$ are invertible for $0 < p < 1$. The cases where $p = 0$ or $p = 1$, yield trivial navigation problems. Using Equations (8.57,8.58), Equation (8.55) becomes

$$\begin{aligned} \mathbf{R}_0 \begin{bmatrix} E\mathbf{x}_{\times\times}^{s,*} \\ E\mathbf{x}_{1\times}^{s,*} \\ E\mathbf{x}_{11}^{s,*} \end{bmatrix} &= \mathbf{R}_{-2} E\mathbf{x}_{\times\times}^{s-2,*} + \mathbf{R}_{-1} \begin{bmatrix} E\mathbf{x}_{\times\times}^{s-1,*} \\ E\mathbf{x}_{1\times}^{s-1,*} \end{bmatrix} + \mathbf{R}_1 \begin{bmatrix} E\mathbf{x}_{\times\times}^{s+1,*} \\ E\mathbf{x}_{1\times}^{s+1,*} \end{bmatrix} \\ &\quad + \mathbf{R}_2 E\mathbf{x}_{\times\times}^{s+2,*} + \mathbf{r}, \end{aligned} \tag{8.59}$$

where $\mathbf{R}_0 = (\mathbf{I} - \mathbf{D}_0'')\mathbf{P}_1^{-1}$, where $\mathbf{R}_\sigma = \mathbf{D}_\sigma''$, for $\sigma \in \{-2, 2\}$, where $\mathbf{R}_\sigma = \mathbf{D}_\sigma''\mathbf{P}_2^{-1}$, for $\sigma \in \{-1, 1\}$, and where $\mathbf{r} = \mathbf{d}''$. We regroup the terms in Equation (8.59) to obtain the forward recursion

$$\mathbf{R}_2' \begin{bmatrix} E\mathbf{x}_{\times\times}^{s+2,*} \\ E\mathbf{x}_{1\times}^{s+1,*} \\ E\mathbf{x}_{11}^{s,*} \end{bmatrix} = \sum_{\sigma=-2}^{1} \mathbf{R}_\sigma' \begin{bmatrix} E\mathbf{x}_{\times\times}^{s+\sigma,*} \\ E\mathbf{x}_{1\times}^{s+\sigma-1,*} \\ E\mathbf{x}_{11}^{s+\sigma-2,*} \end{bmatrix} + \mathbf{r}, \tag{8.60}$$

where

$$\begin{aligned} \mathbf{R}_2' &= \left[\, -\mathbf{R}_2 \;\; -\mathbf{R}_{1,2} \;\; \mathbf{R}_{0,3} \,\right], \\ \mathbf{R}_1' &= \left[\, \mathbf{R}_{1,1} \;\; -\mathbf{R}_{0,2} \;\; \mathbf{0} \,\right], \\ \mathbf{R}_0' &= \left[\, -\mathbf{R}_{0,1} \;\; \mathbf{R}_{-1,2} \;\; \mathbf{0} \,\right], \\ \mathbf{R}_{-1}' &= \left[\, \mathbf{R}_{-1,1} \;\; \mathbf{0} \;\; \mathbf{0} \,\right], \\ \mathbf{R}_{-2}' &= \left[\, \mathbf{R}_{-2} \;\; \mathbf{0} \;\; \mathbf{0} \,\right], \end{aligned}$$

where $\mathbf{R}_{k,l}$ denotes the lth column of $\mathbf{R}_k$. Equation (8.60) is only then a well-defined forward recursion if $\mathbf{R}_2'$ is invertible. We have the following lemma.

Lemma 8.5.3 *The matrix* $\mathbf{R}_2' \in \mathbf{R}^{3\times 3}$ *is invertible for all* p *and* α *such that* $0 < \alpha < 1$ *and* $0 < p < 1$, $p \neq p'$, *where*

$$p' = \frac{1}{3}\left(4 - \left(\frac{2}{25 - 3\sqrt{69}}\right)^{\frac{1}{3}} - \left(\frac{25 - 3\sqrt{69}}{2}\right)^{\frac{1}{3}}\right) \simeq 0.245.$$

□

We omit the proof of Lemma 8.5.3, but indicate that it involves proving positivity of polynomials in p and α using SOSTOOLS (Parrilo 2003; Prajna *et al.* 2004). Lemma 8.5.3 indicates that for all relevant $p \neq p'$ and α, Equation (8.60) is a well-defined forward recursion. For $p = p'$, rearranging terms in Equation (8.60) yields a different, well-defined forward recursion, for which the algorithm presented in the remainder of this chapter is valid. We omit the details regarding this special case and assume from here on that $p \neq p'$.

8.5.4 LTI system

In this section, we analyze the recursion in Equation (8.60) with control-theoretic tools. Rearranging the terms in Equation (8.60) and adding appropriate variables yields

$$E\mathbf{x}^{s+1,*} = \mathbf{A}(E\mathbf{x}^{s,*}) + \mathbf{B}, \tag{8.61}$$

where

$$E\mathbf{x}^{s,*} = \Big[\, E\mathbf{x}_{\times\times}^{s+2,*} \; E\mathbf{x}_{1\times}^{s+1,*} \; E\mathbf{x}_{11}^{s,*} \; E\mathbf{x}_{\times\times}^{s-2,*} \; E\mathbf{x}_{\times\times}^{s-1,*} \; E\mathbf{x}_{1\times}^{s-1,*} \; \ldots \; E\mathbf{x}_{\times\times}^{s,*} \; E\mathbf{x}_{1\times}^{s,*} \; E\mathbf{x}_{\times\times}^{s+1,*} \,\Big]^T. \tag{8.62}$$

Equation (8.61) is a necessary condition on the optimal value function, for $s \geq \bar{s}$. The superscript s of the state vector indicates that at 'time' s, the optimal value function is completely known at separations up until and including s. Note that the entries $E\mathbf{x}_{\times\times}^{s-2,*}$ and $E\mathbf{x}_{1\times}^{s-1,*}$ of $E\mathbf{x}^{s,*}$ are unnecessary in the state transition equation, but are required in the state vector to define the system output adequately.

We now consider any state sequence $E\mathbf{x}_c^s$, for $s = \bar{s}, \bar{s}+1, \ldots$, and formulate the state transition equation of an LTI system with system matrix $\mathbf{A}$ as follows:

$$E\mathbf{x}_c^{s+1} = \mathbf{A}(E\mathbf{x}_c^s) + \mathbf{B}, \tag{8.63}$$

where $E\mathbf{x}_c^s$ is defined as in Equation (8.62), but omitting the asterisk; the subscript c indicates that the value function associated with $E\mathbf{x}_c^s$, for $s = \bar{s}, \bar{s}+1, \ldots$, is considered a candidate optimal value function. We determine the output $\mathbf{z}^s \in \mathbf{R}^{n|U|}$ (for $s \geq \bar{s}$) so that non-negativity of $\mathbf{z}^s$ verifies that $\mathbf{x}_c^s$ is indeed the optimal solution of $\mathcal{LP}_f(s)$, where

$\hat{\mathbf{x}}^{s+\sigma} = \mathbf{x}_c^{s+\sigma}$, for $\sigma \in S^s$, an LP which we denote as $\mathcal{LP}_f^c(s)$. Here, $\mathbf{x}_c^s$ and $\mathbf{x}_c^{s+\sigma}$ (for $s \geq \overline{s}$ and $\sigma \in S^s$) are computed from the state vectors $E\mathbf{x}_c^s$ and $E\mathbf{x}_c^{s+\sigma}$ using Equation (8.3) with μ_∞ as policy [recall the notation convention in Equation (8.7)]. To this end, we rewrite the constraints of $\mathcal{LP}_f^c(s)$ [see Equation (8.45)] as

$$\mathbf{A}_{LP}\begin{bmatrix}\mathbf{x}^s \\ \mathbf{y}^s\end{bmatrix} = \mathbf{b}_{LP}^s,$$

with

$$\mathbf{A}_{LP} = \begin{bmatrix}\mathbf{T} - \mathbf{A}^0 & \mathbf{I}\end{bmatrix},$$
$$\mathbf{b}_{LP}^s = \sum_{\sigma \in S^s} \mathbf{A}^\sigma \mathbf{x}_c^{s+\sigma} + \mathbf{b}^s,$$

where $\mathbf{I}$ has the appropriate dimensions. Note that we only consider non-negative candidate optimal value functions $\mathbf{x}_c^s$, and therefore, in $\mathcal{LP}_f^c(s)$, we can set $\mathbf{x}_-^s = 0$ [see Equation (8.46)]. We need the following property of matrix $\mathbf{A}_{LP}$.

Lemma 8.5.4 *The first n columns of the matrix* $\mathbf{A}_{LP}$ *are linearly independent.* □

Proof. Equivalently, we need to establish that the matrix $\mathbf{T} - \mathbf{A}^0$ is full rank. Recall that there are $|U|$ rows of $\mathbf{T} - \mathbf{A}^0$ associated with $\mathbf{x}_i^s$ $(1 \leq i \leq n)$, one row for each agent decision pair (choose any $s \geq \overline{s}$). For any i $(1 \leq i \leq n)$, the row of $\mathbf{A}^0$ associated with the decision $\mathbf{u}$ where the two agents diverge, has all zero entries, since $s' = s + 2$, where s' is the separation at the next stage. Therefore, for each i $(1 \leq i \leq n)$, there is an r $(1 \leq r \leq n|U|)$ such that $\mathbf{T}_{ri} - \mathbf{A}_{ri}^0 = 1$ and $\mathbf{T}_{ri'} - \mathbf{A}_{ri'}^0 = 0$ for $i' \neq i$, where $(\cdot)_{ri}$ is the entry on the rth row and ith column of $(\cdot)$. Therefore, the first n columns of A_{LP} are linearly independent, which concludes the proof. ■

Let $\mu^s : \mathcal{S}(s) \to U$ be any two-agent policy for the Auxiliary Problem $\mathcal{P}(s)$, where $\hat{J}_s(x) \sim \mathbf{x}_c^{s+\sigma}$, for $\sigma \in S^s$, $s \in \mathcal{M}_s$. Let $\mathbf{x}_{\mu^s}^s \sim J_{\mu^s}(x)$, for $x \in \mathcal{S}(s)$, denote the value function associated with stationary policy μ^s, *i.e.* the cost when the agents apply policy μ^s at all stages. We show a property of $\mathbf{x}_{\mu^s}^s$ in the following lemma.

Lemma 8.5.5 *For any* μ^s, *there exists a vector* $\mathbf{y} \in \mathbf{R}^{n|U|}$ *such that* $(\mathbf{x}_{\mu^s}^s, \mathbf{y})$ *is a basic solution of* $\mathcal{LP}_f^c(s)$. □

Proof. At the state associated with index i $(1 \leq i \leq n)$, the agents take decision $\mathbf{u}_i \in U$ under policy μ^s. For each i, let index j_i $(1 \leq j_i^s \leq |U|)$ correspond to the decision $\mathbf{u}_i$. We choose

$$(\mathbf{y}_i)_{j_i} = 0 \qquad \text{for all } i = 1, \ldots, n. \tag{8.64}$$

We now define the matrix $\mathbf{B}_{LP}^{\mu^s} \in \mathbf{R}^{n|U| \times n|U|}$ as consisting of a copy of the matrix $\mathbf{A}_{LP}$, where the columns associated with $(\mathbf{y}_i)_{j_i}$ are removed. We show that $\mathbf{B}_{LP}^{\mu^s}$ is full rank. From Lemma 8.5.4, we have that the first n columns of $\mathbf{B}_{LP}^{\mu^s}$ are linearly independent. The last $(|U| - 1)n$ columns of $\mathbf{B}_{LP}^{\mu^s}$ are linearly independent since they consist of different columns of $\mathbf{I}$. Lastly, each of the first n columns is linearly independent of the last $(|U|-1)n$ columns, since the ith column of $\mathbf{B}_{LP}^{\mu^s}$ $(1 \leq i \leq n)$, has a nonzero element on the row corresponding to $(\mathbf{y}_i)_{j_i}$, and that the same row only contains zeros on its last $(|U|-1)n$ columns. Therefore, the columns of $\mathbf{B}_{LP}^{\mu^s}$ are linearly independent and $\mathbf{B}_{LP}^{\mu^s}$ is an invertible matrix.

Let $\mathbf{w} \in \mathbf{R}^{(|U|-1)n}$ contain all entries of $\mathbf{y}$ except for the entries $(\mathbf{y}_i)_{j_i}$. With $\mathbf{B}_{LP}^{\mu^s}$, we have

$$\begin{bmatrix} \mathbf{x}_{\mu^s} \\ \mathbf{w} \end{bmatrix} = (\mathbf{B}_{LP}^{\mu^s})^{-1}\mathbf{b}_{LP}^{s}. \tag{8.65}$$

The LHS of Equation (8.65), with Equation (8.64), satisfies the equality constraints of $\mathcal{LP}_f^c(s)$. Since $\mathbf{A}_{LP}$ has linearly independent rows, and given the definition of $\mathbf{B}_{LP}^{\mu^s}$, we apply Theorem 2.4 in Bertsimas and Tsitsiklis (1997) to conclude that $\mathbf{x}_{\mu^s}$ is a basic solution of $\mathcal{LP}_f^c(s)$, with corresponding matrix $\mathbf{B}_{LP}^{\mu^s}$. ∎

From Lemma 8.5.5, it follows that to policy μ_∞ is associated a basic solution of $\mathcal{LP}_f^c(s)$ with basis matrix $\mathbf{B}_{LP} \in \mathbf{R}^{n|U|\times n|U|}$. Two conditions need to hold for matrix $\mathbf{B}_{LP}$ to be optimal (see Bertsimas and Tsitsiklis (1997) for a detailed treatment). First, the basic solution associated with $\mathbf{B}_{LP}$ is required to be dual feasible. In particular, we have that

$$\bar{\mathbf{c}}^T = \mathbf{c}^T - \mathbf{c}_B^T\mathbf{B}_{LP}^{-1}\mathbf{A}_{LP} \geq 0, \tag{8.66}$$

where $\bar{\mathbf{c}}$ is the *reduced cost* vector, $\mathbf{c}$ is the cost vector of $\mathcal{LP}_f^c(s)$ and where $\mathbf{c}_B$ is the cost vector corresponding to the basis $\mathbf{B}_{LP}$. This condition is independent of $\mathbf{b}_{LP}^s$ and thus needs to be checked only once for each specific α and p, and not for all separations $s \geq \bar{s}$. We have the following lemma.

Lemma 8.5.6 *The vector* $\bar{\mathbf{c}}^T = \mathbf{c}^T - \mathbf{c}_B^T\mathbf{B}_{LP}^{-1}\mathbf{A}_{LP} \geq 0$, *for all* p *and* α *such that* $0 < p < 1$ *and* $0 < \alpha < 1$. □

Proof. The basis matrix $\mathbf{B}_{LP}$ corresponds to the policy μ_∞, a two-agent policy that is a combination of two optimal single-agent policies (see Section 8.5.2). Further, in Lemma 8.5.1, we establish that for any $0 < \alpha < 1$ and $0 \leq p \leq 1$, the same set of single agent policies is optimal. Therefore, in the limit for $s \to \infty$, the policy μ_∞ is an optimal two-agent policy for all $0 < \alpha < 1$ and $0 \leq p \leq 1$.

The vector of reduced costs $\bar{\mathbf{c}}$ is independent of separation s, for $s \geq \bar{s}$. Hence, since in the limit for $s \to \infty$ the basis matrix $\mathbf{B}_{LP}$ is optimal, we have that $\bar{\mathbf{c}}^T \geq 0$. □

Lemma 8.5.6 allows us to conclude that dual feasibility is satisfied for $s \geq \bar{s}$, under Assumption 8.5.1.

The second condition is primal feasibility. In particular, we need to check that

$$\mathbf{B}_{LP}^{-1}\mathbf{b}_{LP}^{s} \geq 0, \tag{8.67}$$

for all $s \geq \bar{s}$. Therefore, we define the output $\mathbf{z}^s$ of the LTI system as

$$\mathbf{z}^s = \mathbf{B}_{LP}^{-1}\left(\mathbf{M}_1 E\mathbf{x}_c^s + \mathbf{M}_2\right), \tag{8.68}$$

where $\mathbf{M}_1 \in \mathbf{R}^{n|U|\times 9}$ and $\mathbf{M}_2 \in \mathbf{R}^{n|U|}$ are such that $\mathbf{b}^s_{LP} = \mathbf{M}_1 E\mathbf{x}^s_c + \mathbf{M}_2$. If dual feasibility holds for a particular p and α, then $\mathbf{z}^s \geq 0$ for all $s \geq \bar{s}$, indicates optimality of $\mathbf{B}_{LP}$ for all $s \geq \bar{s}$.

Therefore, given a state and output sequence of linear system (8.63, 8.68) such that the output is non-negative at all separations, then the state sequence represents the optimal value function at separations $s \geq \bar{s}$. The next section deals with the only remaining issue of computing the adequate system initial condition $E\mathbf{x}^{\bar{s}-1}_c$.

8.5.5 Computation of the optimal value function at small separations

We first establish a necessary condition on the initial state of system (8.63). Then, we formulate a LP whose solution is the optimal value function for separations $s < \bar{s}$. The necessary condition on the initial system state is one of the LP constraints, linking the optimal value function at small separations to the optimal value function for $s \to \infty$.

Under Assumption 8.5.1, $s = \bar{s}$ is the smallest separation for which the complete optimal value function can be computed with forward recursion (8.60). Therefore, $E\mathbf{x}^{s'}_c$, for any $s' \geq \bar{s} - 1$, is a valid initial state of system (8.63). We choose $E\mathbf{x}^{\bar{s}-1}_c$. Since the optimal value function for the two-agent navigation problem is upper bounded by twice the optimal value function for the single-agent navigation problem, $E\mathbf{x}^{\bar{s}-1}_c$ can only excite the stable modes of system (8.63), leading to a necessary condition on $E\mathbf{x}^{\bar{s}-1}_c$ as follows.

First we have the following lemma, regarding the spectrum $\lambda(\mathbf{A})$ of the system matrix $\mathbf{A}$ [see Equation (8.63)].

Lemma 8.5.7 *The following holds for all α and p such that $0 < \alpha < 1$ and $0 < p < 1$, $p \neq p'$. The spectrum $\lambda(\mathbf{A})$ of matrix $\mathbf{A}$ contains exactly three eigenvalues $\lambda_{u,i} \in \lambda(\mathbf{A})$, for $i = 1, 2, 3$, such that $|\lambda_{u,i}| > 1$. Furthermore, for all other eigenvalues $\lambda_{s,j} \in \lambda(\mathbf{A})$, for $j = 1, \ldots, 6$, we have that $|\lambda_{s,j}| < 1$.* □

We omit the proof of Lemma 8.5.7, which involves proving positivity of polynomials in p and α using SOSTOOLS (Parrilo 2003; Prajna *et al.* 2004). The subscripts u and i indicate eigenvalues corresponding to unstable and stable system modes, respectively. Note that from Lemma 8.5.3, we have that for $p = p'$ and any $0 < \alpha < 1$, the forward recursion in Equation (8.60) is not well defined. Hence, $\mathbf{A}$ is not well defined at those parameter values.

With

$$\tilde{E}\mathbf{x}^s_c = (\mathbf{I} - \mathbf{A})^{-1}\mathbf{B} - E\mathbf{x}^s_c, \tag{8.69}$$

where $\mathbf{I} - \mathbf{A}$ is invertible from Lemma 8.5.7, the state transition equation (8.63) becomes

$$\tilde{E}\mathbf{x}^{s+1}_c = \mathbf{A}\tilde{E}\mathbf{x}^s_c. \tag{8.70}$$

We use Schur's unitary triangularization theorem (Horn and Johnson 1985), to write Equation (8.70) as

$$\mathbf{V}(\tilde{E}\mathbf{x}^{s+1}_c) = \mathbf{U}\mathbf{V}(\tilde{E}\mathbf{x}^s_c), \tag{8.71}$$

where $\mathbf{V}$ is a unitary matrix and where $\mathbf{U} = \text{diag}(\mathbf{U}_s, \mathbf{U}_u)$ with $\mathbf{U}_s \in \mathbf{R}^{6\times 6}$ and $\mathbf{U}_u \in \mathbf{R}^{3\times 3}$ upper triangular matrices containing as diagonal elements $\lambda_{s,j}$ for $j = 1, \ldots, 6$ and $\lambda_{u,i}$ for $i = 1, 2, 3$, respectively (see Lemma 8.5.7), and where $\mathbf{VA} = \mathbf{UV}$. The initial state $E\mathbf{x}_c^{\bar{s}-1}$ excites only the stable modes if and only if $\mathbf{V}_u(\tilde{E}\mathbf{x}_c^{\bar{s}-1}) = \mathbf{0}$, where $\mathbf{V}_u$ consists of the rows of $\mathbf{V}$ associated with the unstable system modes. With Equation (8.69), this yields $\mathbf{V}_u\left((\mathbf{I} - \mathbf{A})^{-1}\mathbf{B} - E\mathbf{x}_c^{\bar{s}-1}\right) = \mathbf{0}$, or

$$\mathbf{V}_u E\mathbf{x}_c^{\bar{s}-1} = \mathbf{V}_u(\mathbf{I} - \mathbf{A})^{-1}\mathbf{B}. \tag{8.72}$$

Let $\mathbf{V}_1 \in \mathbf{R}^{3\times 3}$ contain the first two columns and the last column of $\mathbf{V}_u$, *i.e.* the columns corresponding to the entries $E\mathbf{x}_{\times\times,c}^{s+2}$, $E\mathbf{x}_{1\times,c}^{s+1}$ and $E\mathbf{x}_{\times\times,c}^{s+1}$.

Equation (8.72) allows us to formulate linear program $\mathcal{LP}_{in}$, whose solution is the optimal value function for separations $s < \bar{s}$. In particular, the structure of $\mathcal{LP}_{in}$ is similar to the structure of $\mathcal{LP}_g$ for separations $s < \bar{s}$, with minor changes. To compute the optimal value function at separation $\bar{s} - 1$, information is required on the optimal value function at separations $s = \bar{s}, \bar{s} + 1$. This information is provided implicitly by adding Equation (8.72), the necessary condition on $E\mathbf{x}_c^{\bar{s}-1}$, as one of the $\mathcal{LP}_{in}$ constraints. This yields the following LP:

$$\text{maximize} \quad \sum_{s=0}^{\bar{s}-1} \mathbf{e}^T\mathbf{x}^s$$

$$\text{subject to} \quad (\mathbf{T} - \mathbf{A}_s^0)\mathbf{x}^s \leq \sum_{\sigma\in S^s} \mathbf{A}_s^\sigma \mathbf{x}^{s+\sigma} + \mathbf{b}^s, \qquad s = 0, 1, \ldots, \bar{s} - 3, \tag{8.73}$$

$$(\mathbf{T} - \mathbf{A}_{\bar{s}-2}^0)\mathbf{x}^{\bar{s}-2} \leq \sum_{\sigma\in\{-2,-1,1\}} \mathbf{A}_{\bar{s}-2}^\sigma \mathbf{x}^{\bar{s}-2+\sigma} + \overline{\mathbf{A}}_{\bar{s}-2}^2 E\mathbf{x}_{\times\times}^{\bar{s}} + \mathbf{b}^{\bar{s}-2}, \tag{8.74}$$

$$(\mathbf{T} - \mathbf{A}_{\bar{s}-1}^0)\mathbf{x}^{\bar{s}-1} \leq \sum_{\sigma\in\{-2,-1\}} \mathbf{A}_{\bar{s}-1}^\sigma \mathbf{x}^{\bar{s}-1+\sigma} + \overline{\mathbf{A}}_{\bar{s}-1}^1 \begin{bmatrix} E\mathbf{x}_{1\times}^{\bar{s}} \\ E\mathbf{x}_{\times\times}^{\bar{s}} \end{bmatrix} \tag{8.75}$$

$$+ \overline{\mathbf{A}}_{\bar{s}-1}^2 E\mathbf{x}_{\times\times}^{\bar{s}+1} + \mathbf{b}^{\bar{s}-1}, \tag{8.76}$$

$$\mathbf{V}_u(E\mathbf{x}^{\bar{s}-1}) = \mathbf{V}_u(\mathbf{I} - \mathbf{A})^{-1}\mathbf{B}, \tag{8.77}$$

where $\overline{\mathbf{A}}_s^1$ and $\overline{\mathbf{A}}_s^2$ are such that

$$\overline{\mathbf{A}}_s^1 \begin{bmatrix} E\mathbf{x}_{1\times}^{s+1} \\ E\mathbf{x}_{\times\times}^{s+1} \end{bmatrix} = \mathbf{A}^1\mathbf{x}^{s+1}, \tag{8.78}$$

$$\overline{\mathbf{A}}_s^2 E\mathbf{x}_{\times\times}^{s+2} = \mathbf{A}^2\mathbf{x}^{s+2}. \tag{8.79}$$

A reasoning identical to the one that led to the derivation of Equations (8.52) and (8.53) leads to Equations (8.78) and (8.79), where we use Equation (8.51) with q taken to be $\times$ to yield Equation (8.78).

The following theorem links the solution of $\mathcal{LP}_{in}$ and the stable state trajectory of system (8.63) to the fixed point of F, and therefore to the optimal value function, under some technical conditions. Let $\mathbf{x}_c^s$, for $s = 0, 1, \ldots, \bar{s} - 1$ and $E\mathbf{x}_{1\times,c}^{\bar{s}}$, $E\mathbf{x}_{\times\times,c}^{\bar{s}}$ and $E\mathbf{x}_{\times\times,c}^{\bar{s}+1}$ denote the solution of $\mathcal{LP}_{in}$. Let the vector $E\mathbf{x}_c^{\bar{s}-1}$ [see Equation (8.62)] be determined, based on $\mathbf{x}_c^s$, for $s = \bar{s} - 3, \bar{s} - 2, \bar{s} - 1$ and on $E\mathbf{x}_{1\times,c}^{\bar{s}}$, $E\mathbf{x}_{\times\times,c}^{\bar{s}}$ and $E\mathbf{x}_{\times\times,c}^{\bar{s}+1}$. Lastly, let $E\mathbf{x}_c^s$, for $s = \bar{s}, \bar{s} + 1, \ldots$ be the state of system (8.63) at separation s, with $E\mathbf{x}_c^{\bar{s}-1}$ as initial condition. For $s \geq \bar{s}$, let $\mathbf{x}_c^s$ be defined as

$$\mathbf{x}_c^s \sim J_c(x) = g(x, \mu_\infty(x)) + \alpha E[J_c(f(x, \mu_\infty(x)))], \qquad x \in \mathcal{S}(s)$$

where the expected value is taken over the unknown edge costs. Note that the required expected values of $J_c(x)$, for $x \in \mathcal{S}(s)$ and $s \geq \bar{s}$, are determined from $E\mathbf{x}_c^s$ with Equation (8.50).

Theorem 8.5.8 *For any α and p such that $0 < \alpha < 1$ and $0 < p < 1$, $p \neq p'$, the vectors $\mathbf{x}_c^s$ for $s = 0, 1, \ldots$, represent the Main Problem 8.2 optimal value function for some $\bar{s} \geq 0$ if $\mathbf{V}_1$ is rank three and if for the output of system (8.63), we have that*

$$\mathbf{z}^s \geq 0, \qquad s = \bar{s}, \bar{s} + 1, \ldots. \tag{8.80}$$

Furthermore, we have that the optimal two-agent policy for separation $s \geq \bar{s}$ is μ_∞. □

Proof. We show that

$$\mathbf{X}_c = \left[\, (\mathbf{x}_c^0)^T \;\; (\mathbf{x}_c^1)^T \;\; (\mathbf{x}_c^2)^T \cdots \,\right]^T,$$

is the unique fixed point of the function F, so that with Theorem 8.4.11, we have that $\mathbf{X}_c$ represents the optimal value function and hence is equal to $\mathbf{X}^*$.

We can see that the constraints in Equations (8.73–8.76) of $\mathcal{LP}_{in}$ are identical to the constraints associated with the separations $s = 0, \ldots, \bar{s} - 1$ in $\mathcal{LP}_g$. Without any other constraint, the presence of $E\mathbf{x}_{\times\times}^{\bar{s}}$, $E\mathbf{x}_{1\times}^{\bar{s}}$ and $E\mathbf{x}_{\times\times}^{\bar{s}+1}$ in the RHS of Equations (8.74–8.76) yields an unbounded $\mathcal{LP}_{in}$ solution. However, with the necessary condition in Equation (8.77), where the matrix $\mathbf{V}_1$ is of rank three, we can express $E\mathbf{x}_{1\times}^{\bar{s}}$, $E\mathbf{x}_{\times\times}^{\bar{s}}$ and $E\mathbf{x}_{\times\times}^{\bar{s}+1}$ as unique affine functions of $\mathbf{x}^s$, for $s = 0, \ldots, \bar{s}$. Since, $\mathbf{x}_c^s$ for $s = 0, \ldots, \bar{s} - 1$, solve $\mathcal{LP}_{in}$, we have that $\mathbf{x}_c^s$ is the solution of $\mathcal{LP}(s)$ with $\hat{\mathbf{x}}^{s+\sigma} = \mathbf{x}_c^{s+\sigma}$ ($\sigma \in S^s$). The latter follows from the definition of $\mathcal{LP}(s)$.

For the given parameter ranges, the matrices $\mathbf{R}_2'$ and $\mathbf{I} - \mathbf{A}$ are invertible (see Lemmas 8.5.3 and 8.5.7), and therefore the LTI system (8.63) and the constraint in Equation (8.77) are well defined. Then, from the latter constraint, we have that $E\mathbf{x}_c^{\bar{s}-1}$ as initial state for system (8.63) is such that only the stable modes are excited. Furthermore, the vector $\mathbf{x}_c^s$ solves $\mathcal{LP}(s)$ with $\hat{\mathbf{x}}^{s+\sigma} = \mathbf{x}_c^{s+\sigma}$ ($\sigma \in S^s$), for $s = \bar{s}, \bar{s} + 1, \ldots$ since both the primal [see Equation (8.80)] and dual [see Lemma 8.5.6] feasibility conditions are satisfied for all $s \geq \bar{s}$.

Hence, $\mathbf{x}_c^s$ solves $\mathcal{LP}(s)$ with $\hat{\mathbf{x}}_c^{s+\sigma} = \mathbf{x}_c^{s+\sigma}$ ($\sigma \in S^s$), for $s = 0, 1, \ldots$ and therefore $\mathbf{X}_c$ is the unique fixed point of F and represents the optimal value function. Lastly, from Assumption 8.5.1, we have that for $s \geq \bar{s}$, policy μ_∞ is optimal, which concludes the proof. ■

We now have a computationally simple algorithm that computes the optimal value function for $s < \bar{s}$ as the solution of $\mathcal{LP}_{in}$, which is relatively small-sized given that $\bar{s}$

is not too large. The optimal value function for $s \geq \bar{s}$ is implicitly described via a LTI system. In Section 8.6, we give $\bar{s}$ for different probabilities p; for most p, separation $\bar{s} = 3$. Furthermore, numerical experiments indicate that the technical condition on the rank of $\mathbf{V}_1$ holds in all cases computed.

8.6 DISCUSSION AND EXAMPLES

Theorem 8.5.8 allows us to solve for the two-agent optimal value function in two simple steps. First, solve $\mathcal{LP}_{in}$ to obtain the optimal value function at separations $s \leq \bar{s} - 1$. Then, use its solution to compute the initial condition for the LTI system in Equation (8.63) which is simulated to obtain the optimal value function for any $s \geq \bar{s}$. For $p = 0.7$ and $\alpha = 0.9$, Figure 8.3 shows $E\mathbf{x}_{00}^{s,*}$, $E\mathbf{x}_{01}^{s,*}$ and $E\mathbf{x}_{11}^{s,*}$ as a function of s. As expected, as $s \to \infty$, the two-agent optimal value function converges to the dashed lines which represent the two-agent value function in case both agents adopt a single-agent optimal policy at all separations $s \geq 0$. Figure 8.4 shows a sample optimal two-agent trajectory set. After the initial transition phase where, in expected sense, the agent separation decreases, the steady state is reached, where with high probability the agent separation remains small.

We now discuss the size of $\bar{s}$ for different parameter sets. Numerical experiments indicate that for $\alpha = 0.9$ and for $p = 0.1, 0.2, \ldots, 0.9$, we have that $\bar{s} = 3$. Figure 8.5 shows $\bar{s}$ for $p = 0.01, 0.02, \ldots, 0.09$. The separation $\bar{s}$ increases as p decreases. Hence, for the considered probabilities $p \geq 0.08$, the method presented to compute the optimal two-agent value function is computationally cheap, since $\mathcal{LP}_{in}$ is small. The size of $\mathcal{LP}_{in}$ increases for $p < 0.08$, but it remains computationally much more efficient than computing the optimal value function by artificially imposing a reasonable separation upper bound.

With a method similar to the ones used to compute the steady-state behavior of a Birth-Death Processes (Hillier and Lieberman 1974), we compute the probability distribution of

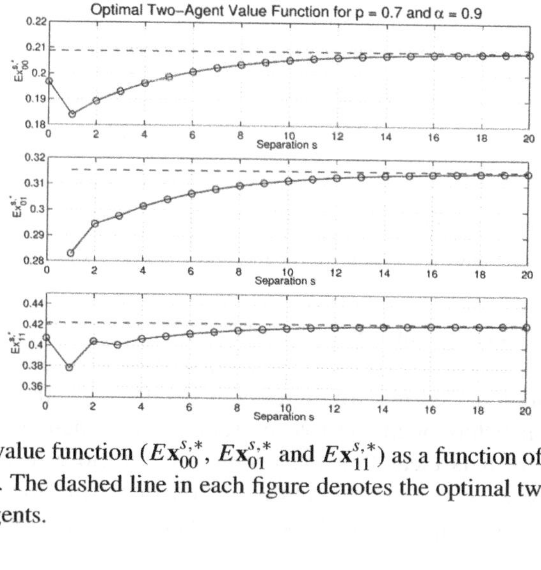

Figure 8.3 Optimal value function ($E\mathbf{x}_{00}^{s,*}$, $E\mathbf{x}_{01}^{s,*}$ and $E\mathbf{x}_{11}^{s,*}$) as a function of the agent separation s ($p = 0.7$ and $\alpha = 0.9$). The dashed line in each figure denotes the optimal two-agent value function for non-cooperating agents.

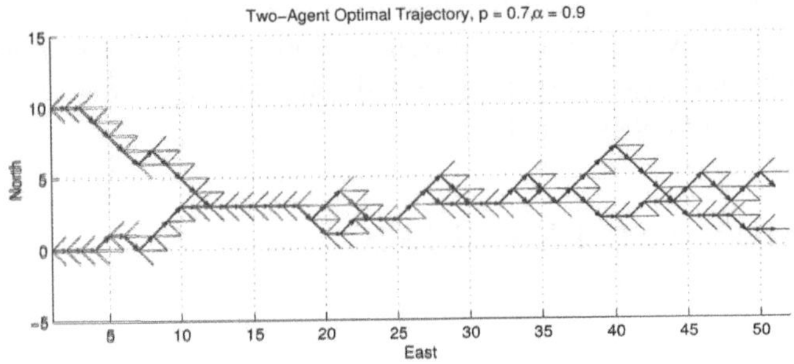

Figure 8.4 Example set of trajectories. Black and grey edges denote an edge cost of zero and one, respectively ($p = 0.7$ and $\alpha = 0.9$).

p	0.01	0.02	0.03	0.04	0.05	0.06	0.07	0.08	0.09
$\bar{s}$	11	9	7	6	5	5	4	3	3

Figure 8.5 The value of $\bar{s}$ for the problems where $p = 0.01, 0.02, \ldots, 0.09$, where p is the probability of encountering a zero edge cost. For $p \geq 0.1$, we have that $\bar{s} = 3$.

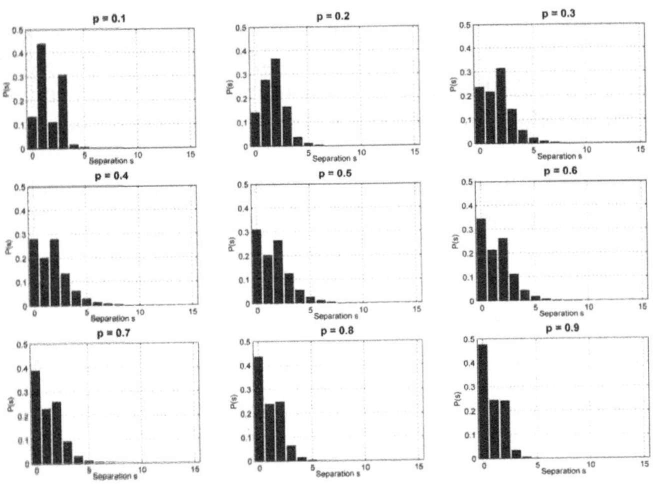

Figure 8.6 Steady state agent separation probability distributions for $p = 0.1, 0.2, \ldots, 0.9$, where p is the probability of encountering a zero edge cost. The agents navigate with an optimal policy.

the agent separation in steady state. We refer the reader to (De Mot and Feron 2005) for further details. Figure 8.6 shows the probability distribution of the two-agent separation in steady state and under an optimal policy for $p = 0.1, 0.2, \ldots, 0.9$. Later in this section, we focus on the agent behavior in the two extreme cases where p is close to zero and close to one. For all p, it is clear that the agents remain close to each other with high probability. In fact, as $s \to \infty$, the probability of encountering the agents at separation s decreases exponentially. Figure 8.7 shows the expected value of agent separation in

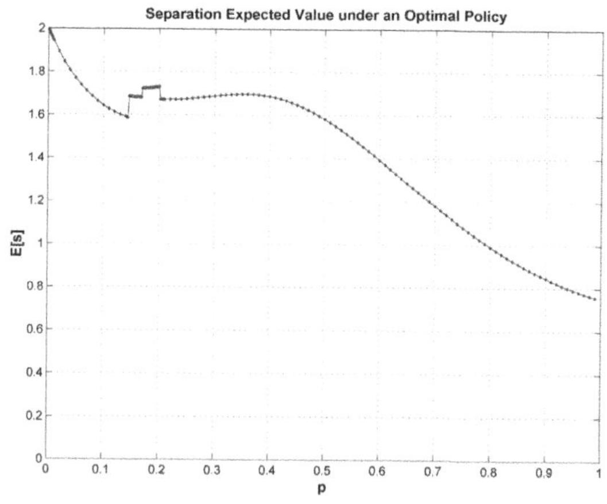

Figure 8.7 The expected value of the agent separation under an optimal policy for several values of $0 < p < 1$ (dots on the curve), where p is the probability of encountering a zero edge cost.

steady state. For p close to one, the expected agent separation is the smallest and equals approximately 0.75. On the other hand, for p close to zero, we have approximately the largest expected agent separation, equal to 2. This is further support for the intuition that as the environment is more hostile, the agents spread out to increase the size of the region where exact edge cost is observed, and thus increasing the probability of encountering a zero edge cost. The apparent discontinuous changes in the curve reflect changes in the optimal two-agent policy at the separations $s = 0, 1, 2$.

Figures 8.8 and 8.9 show two extreme but instructive cases, for $p = 0.06$ and $p = 0.94$, respectively; in each figure, the top depicts the probability distribution of the separation in steady state, while the bottom depicts a set of sample trajectories. For $p = 0.06$, i.e. the case where ones are abundant, the separations $s = 1$ and $s = 3$ are most probable. The sample trajectories indicate the mechanics of cooperation. In particular, the agents tend to the most favorable separation $s = 1$, where one agent can leverage opportunities the other agent observes. However, with high probability, only ones are observed, driving the agents apart to the separation $s = 3$, where a set of eight previously unobserved edges is observed, thus maximizing the probability of encountering a zero. With high probability, only ones are in sight, and the agents converge again to the favorable separation $s = 1$, where eight previously unobserved edges enter the observation zone.

For $p = 0.94$, the case where zeros are abundant, the separation $s = 0$ is most likely, followed by $s = 1$ and $s = 2$, both equally likely. Again, the sample trajectories indicate the mechanics of cooperation. In particular, let the agents start at $s = 1$, the most favorable separation. Most likely, only zeros are observed, and agents continue straight ahead. With probability $p_1 = 2p(1 - p)$ $(= 0.11.$ for $p = 0.94)$, an edge of cost one enters an observation zone one stage ahead (see, for example, stage 118 in Figure 8.9). As a consequence, the agents converge to $s = 0$, which is maintained till again a one appears one stage further, which happens with probability $p_2 = 1 - p$ $(= 0.06$ for $p = 0.94)$. The agents split to $s = 2$, maximizing the number newly observed edges. Again, one

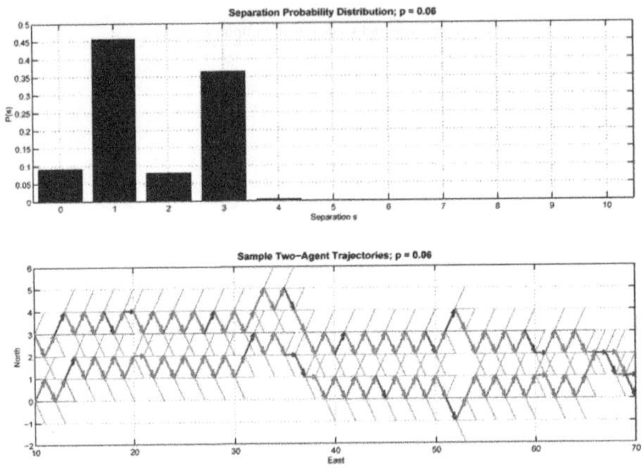

Figure 8.8 Top: Steady state agent separation probability distribution for a p close to zero ($p = 0.06$). Probability p is the probability of encountering a zero edge cost; the agents navigate with an optimal policy. Bottom: a set of sample trajectories, for $p = 0.06$. Grey (black) arrows and lines indicate traversed and observed edges of cost one (zero), respectively.

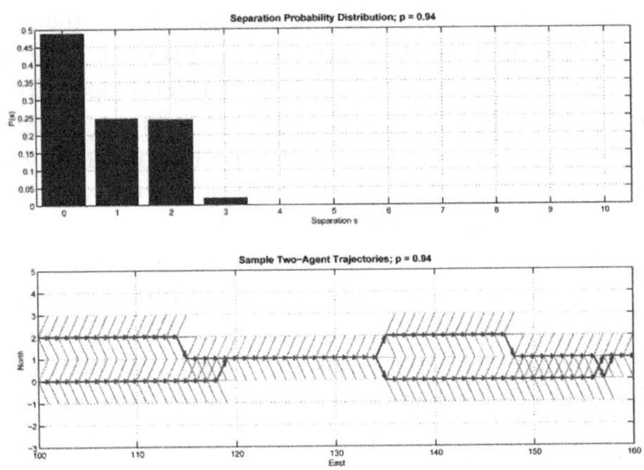

Figure 8.9 Top: Steady state agent separation probability distribution for a p close to one ($p = 0.94$). Probability p is the probability of encountering a zero edge cost; the agents navigate with an optimal policy. Bottom: a set of sample trajectories, for $p = 0.94$. Grey (black) arrows and lines indicate traversed and observed edges of cost one (zero), respectively.

edge cost of one enters an observation zone one stage ahead with probability p_1, causing the agents to converge back to $s = 1$. The difference in magnitude of p_1 and p_2 clarifies the difference of the probabilities with which agents are at separation $s = 0$ and at the separations $s = 1$ and $s = 2$.

For $s \geq \bar{s}$, policy μ_∞ is optimal allowing for a quasi-decentralized strategy for $s \geq \bar{s}$. In particular, each agent chooses the converging single agent optimal policy for $s \geq \bar{s}$, *i.e.*, each agent behaves as a non-cooperating single agent and when multiple equivalent decisions are possible, chooses the one that minimizes the next agent separation. Hence, only information on whether the agent B is positioned somewhere north or south of agent A needs to be stored in the memory of agent A and vice versa, for $s \geq \bar{s}$. It is only when agents come within separation $\bar{s} - 1$, that edge cost communication is required to take place for optimal cooperative behavior.

In this chapter, we solve the two-agent problem by relying on the solution of the single-agent problem for $s \geq \bar{s}$. Similarly, we believe it is possible to solve n-agents case, but we solve a relatively small $\mathcal{LP}_{in}$-type LP with as solution the optimal n-agent value function where the maximum separation between two neighboring agents equals $\bar{s}$. To $\mathcal{LP}_{in}$ are added a set of constraints so that for each way the agent cluster can split, only the stable modes are excited in the corresponding LTI-system. These LTI-systems are determined from the optimal agent behavior of cluster sizes $n - 1$ and smaller.

8.7 CONCLUSION

In this chapter we study a two-agent navigation problem on a partially unknown graph. The agents observe and share the cost of edges in a local observation zone. Considering an infinitely wide graph avoids unhandy boundary conditions and allows us to devise a computationally efficient method to compute the optimal value function. In particular, we solve a Linear Program for small agent separations and simulate a linear time invariant system for large separations, connecting both parts of the state space with the proper set of additional constraints in the Linear Program. Analysis of the optimal policy gives us insight into the mechanics of cooperation and rigorously quantifies the intuition that for efficient cooperation, the agents should limit their spatial separation. In fact, we compute the probability distribution of the agent separation exactly, for different levels of environment hostility. Furthermore, a quasi-decentralized policy suffices for optimal navigation at relatively large separations and the locations where communication is useful are determined effectively.

Analysis of the optimal policy for different observation and reachable zones in the two-agent case is part of future work, whereby we intend to determine their influence on the cooperation mechanics. We intend to extend the algorithm to tackle the navigation problems with larger agent clusters in a computationally efficient way. A hierarchical method, whereby the solution for n agents employs the problem solutions for all smaller clusters, is a natural extension of the presented algorithm. Future work includes lifting the spatial synchrony assumption, whereby we envision employing a method similar to the one presented here, but adding a horizontal separation to the spatially synchronous case.

ACKNOWLEDGEMENTS

This research was supported by AFOSR/MURI grant F49620-01-1-0361, Air Force/DARPA MURI award 009628-001-03-132, and ONR award N00014-03-1-0171.

REFERENCES

Arai T, Pagello E and Parker L 2002 Editorial: Advances in multi-robot systems. *IEEE Transactions on Robotics and Automation* **18**(5), 655–661.

Beard R and McLain T 2003 Multiple UAV cooperative search under collision avoidance and limited range communication constraints. In *42nd IEEE Conference on Decision and Control*, pp. 25–30.

Bertsekas D 2001 *Dynamic Programming and Optimal Control* 2nd edn. Athena Scientific, Belmont, MA.

Bertsimas D and Tsitsiklis J 1997 *Introduction to Linear Optimization*. Athena Scientific, Belmont, MA.

Burgard W, Fox D, Moors M, Simmons R and Thrun S 2000 Collaborative multi-robot exploration. In *Proceedings of the IEEE International Conference on Robotics and Automation (ICRA)*, San Francisco, CA.

Cassinis R 2000 Multiple single-sensor robots rather then multi-sensor platforms: A reasonable alternative? In *Proc. Crete 2000 International Conference on Explosives and Drug Detection Techniques* (ed. Vourvopoulos G).

Cassinis R, Bianco G, Cavagnini A and Ransenigo P 1999 Strategies for navigation of robot swarms to be used in landmines detection. In *Proc. Eurobot '99*, pp. 211–218.

Choset H 1996 Sensor based motion planning: The hierarchical generalized Voronoi graph, PhD thesis, California Institute of Technology, Pasadera, CA.

Cortes J, Martinez S and Bullo F 2004 Spatially-distributed coverage optimization and control with limited-range interactions. *Submitted to the ESAIM: Control, Optimization, and Calculus of Variations.*

DasGupta B, Hespanha J and Sontag E 2004a Aggregation-based approaches to honey-pot searching with local sensory information. In *Proceedings of the 2004 American Control Conference*, pp. 1202–1207.

DasGupta B, Hespanha J and Sontag E 2004b Computational complexity of honey-pot searching with local sensory information. In *Proceedings of the 2004 American Control Conference*, pp. 2134–2138.

Davey B and Priestley H 1990 *Introduction to Lattices and Order*. Cambridge University Press, Cambridge.

De Mot J and Feron E 2003 Spatial distribution of two-agent clusters for efficient navigation. In *42nd IEEE Conference on Decision and Control*, pp.1029–1034.

De Mot J and Feron E 2005 Spatial distribution statistics for two-agent optimal navigation with cone-shaped local observation. In *Proc American Control Conference*, pp. 1877–1882.

De Mot J, Kulkarni V, Gentry S and Feron E 2002 Spatial distribution results for efficient multi-agent navigation. In *41st IEEE Conference on Decision and Control*, pp. 3776–3781.

Frazzoli E and Bullo F 2004 Decentralized algorithms for vehicle routing in a stochastic time-varying environment. In *43rd IEEE Conference on Decision and Control*, pp. 3357–3363.

Fukada Y 2000 Speed to fly with management of the risk of landing out. *Technical Soaring* **XXV**(3), 88–94.

Haley K and Stone L 1980 *Search Theory and Applications,* vol. 8 of *NATO Conference Series*. Plenum Press, New York.

Hespanha J, Ateskan Y and Kizilocak H 2000 Deception in non-cooperative games with partial information. In *Proc. of the 2nd DARPA-JFACC Symp. on Advances in Enterprise Control.*

Hespanha J, Kim H and Sastry S 1999 Multiple-agent probabilistic pursuit-evasion games. In *38th IEEE Conference on Decision and Control*, pp. 2432–2437.

Hillier F and Lieberman G 1974 *Introduction to Operations Research*. Holden-Day, Inc., San Francisco.

Ho J and Loute E 1980 A comparative study of two methods for staircase linear programs. *ACM Transactions on Mathematical Software* **6**, 17–30.

Ho J and Sundarraj R 1997 Distributed nested decomposition of staircase linear programs. *ACM Transactions on Mathematical Software* **23**(2), 148–173.

Horn R and Johnson C 1985 *Matrix Analysis*. Cambridge University Press, Cambridge.

Jadbabaie A, Lin J and Morse A 2003 Coordination of groups of mobile autonomous agents using nearest neighbor rules. *IEEE Transactions on Automatic Control* **48**(6), 988–1001.

Latombe JC 1991 *Robot Motion Planning*. Kluwer Academic Publishers, Dordrecht.

Liusternik L and Sobolev V 1961 *Elements of Functional Analysis*, Ungar, NY.

Luenberger D 1969 *Optimization by Vector Space Methods*, Wiley, NY.

McCready P 1954 An optimal airspeed selector. *Soaring*, 8.

Ogren P, Fiorelli E and Leonard N 2002 Formations with a mission: Stable coordination of vehicle group maneuvers. In *Proc. Symposium on Mathematical Theory of Networks and Systems*.

Olfati-Saber R 2006 Flocking for multi-agent dynamic systems: Algorithms and theory. *IEEE Transactions on Automatic Control* **51**(3).

Parrilo P 2003 Semidefinite programming relaxations for semialgebraic problems. *Mathematical Programming Ser. B* **96**(2), 293–320.

Prajna S, Papachristodoulou A, Seiler P and Parrilo PA 2004 *SOSTOOLS: Sum of Squares Optimization Toolbox for MATLAB*. URL: http://www.cds.caltech.edu/sostools/

Reichmann H 1978 *Cross-Country Soaring*. Soaring Society of America Inc., 7th edn.

Ribichini G and Frazzoli E 2003 Energy-efficient coordination of multiple-aircraft systems. In *42nd IEEE Conference on Decision and Control*, pp. 1035–1040.

Simmons R, Apfelbaum D, Burgard W, Fox D, Moors M, Thrun S and Younes H 2000 Coordination for multi-robot exploration and mapping. *AAAI/IAAI*, pp. 852–858.

9

Multiagent cooperation through egocentric modeling

Vincent Pei-wen Seah and Jeff S. Shamma

9.1 INTRODUCTION

Many engineering systems can be modeled as a large-scale collection of interacting subsystems – each having access to local information, each making local decisions, and each seeking to optimize local objectives that may well be in conflict with other subsystems. Such models fall under the scope of research on 'multiagent systems'.

There is a vast literature on this subject, cf., the overview of (Weiss 2000), and multiagent models have been employed in a wide variety of application domains including combat systems (Ilachinski 2004), autonomous vehicles (Durfee 1999), robotic exploration (Rekleitis *et al.* 2001), distributed computing (Chow and Kwok 2002), electronic commerce (Greenwald and Kephard 2000), wireless networks (Chang *et al.* 2003), traffic control (Dresner and Stone 2004), and social networks (Sabater and Sierra 2002).

Designing autonomous or semi-autonomous systems is a driving motivation for the multiagent framework. Accordingly, much of the research is focused on *learning* or *adaptation*, e.g., (Wolpert *et al.* 1999). Multiagent learning presents significant challenges (Shoham *et al.* 2007) that distinguish it from conventional single agent (Sutton and Barto 1998; Bertsekas and Tsitsiklis 1996) learning. In single agent learning, there is a stationary environment, and an agent learns to operate within this environment through increased experience. Now consider a multiagent setting, as in Figure 9.1, and in particular, consider the 'environment' from the perspective of *agent N*. It is composed of *other* agents, and since other agents are undergoing a learning process, the environment cannot be modeled as stationary. In other words, by the time *agent N* may have learned the environment from its own perspective, the environment has changed.

There is a parallel line of research that considers similar concerns, namely, the area of 'evolutionary games' or 'learning in games' in economic game theory literature. This line of work also has a substantial body of literature, including several recent monographs

Part of the text is based on "Multiagent Cooperation through Egocentric Modeling", by V.P.-W. Seah and J.S. Shamma, which appeared in the *9th International Control, Automation, Robotics and Vision Conference*. Reproduced with permission. © 2006 IEEE.

Cooperative Control of Distributed Multi-Agent Systems edited by Jeff S. Shamma

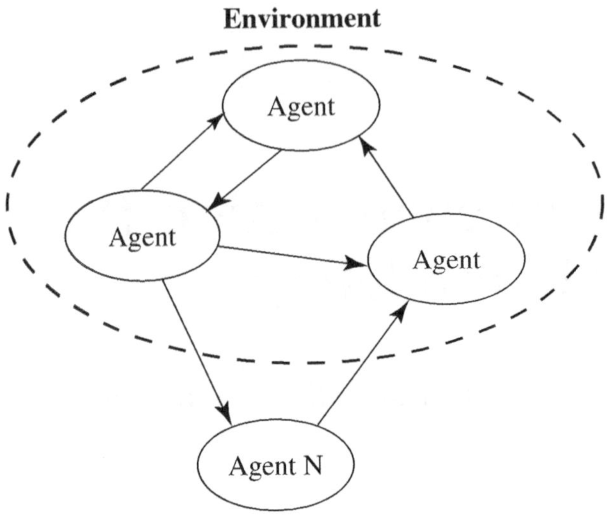

Figure 9.1 Multiagent system schematic illustration.

(Fudenberg and Levine 1998; Hofbauer and Sigmund 1998; Samuelson 1997; Young 1998; Young 2006; Weibull 1995).

There are various shifts of emphasis in this line of work that distinguish it from research in multiagent systems. In particular, attention is shifted away from system design applications in favor of models of learning and the ensuing analysis. Such differences notwithstanding, both lines of research share the same essential features, i.e., multiple entities adjusting their strategies, making local 'self-interested' decisions using only local, possibly overlapping, information.

In this chapter, we also consider a problem of multiple interacting agents. Of particular interest is the following 'decomposition' approach. First, each agent is tasked with a individual performance objective that depends on its own strategy and the strategy of other agents. Through simulation and/or experience, agents revise their strategies in an iterative procedure as follows: (1) agents make a (very) simplified model of the behavior of other agents; (2) agents design 'optimal' strategies that presume such behavior from other agents; and (3) upon deployment of these strategies, agents observe other agents, revise their models, redesign their strategies, and so on.

The implication of convergence is a consistency condition. Namely, each agent's behavior is consistent with how the agent is modeled by others. Furthermore, each agent's local strategy is optimal with respect to how it models other agents.

Such an approach was taken in (Hsu and Shamma 1999; Seah 2002) in the context of manufacturing systems. In that work, simulation studies illustrated the potential of the decomposition approach. Furthermore, simulation studies showed that the iterative procedure exhibits convergence, i.e., agents eventually settle on a set of models and optimized strategies that are consistent. The particular context considered here is a variation of the 'Roboflag drill', introduced in Earl and D'Andrea (2002). In this problem, a team of defenders is to engage an oncoming team of attackers who appear randomly. Figure 9.2 illustrates the setup. The potential of the decomposition approach is to enable

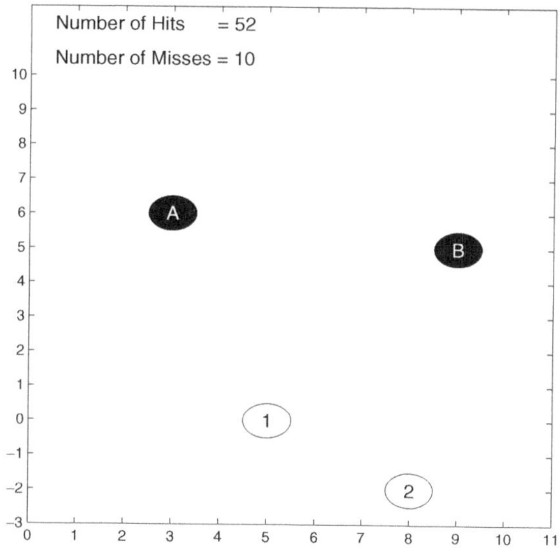

Figure 9.2 Roboflag drill.

the decentralized design of a defense strategy that exhibits effective performance and yet avoids a centralized *a priori* design.

In this chapter, we will present a simulation study of our decomposition approach on the Roboflag drill, compare the performance to a fully centralized and fully decentralized design, and use methods from stochastic approximation (Kushner and Yin 1997; Benaim 1996) to analyze convergence.

The remainder of this chapter is organized as follows. Section 9.2 contains the two attacker/two defender model illustrated in Figure 9.2 and presents both 'fully centralized' and 'fully decentralized' solution to the optimal defense. Section 9.3 presents the decomposition approach of evolutionary coordination. Section 9.4 presents an analytical discussion of convergence. Finally, Section 9.5 contains concluding remarks.

9.2 CENTRALIZED AND DECENTRALIZED OPTIMIZATION

In this section, we specify the model of the Roboflag drill depicted in Figure 9.2 and present two scenarios – a fully centralized optimization and a fully decentralized optimization – as benchmarks for the forthcoming 'evolutionary' scenario.

9.2.1 Markov model

We model a two attacker and two defender scenario as coupled Markov chains.

The state space, S, is all possible positions of the defenders and attackers. Specifically, $s \in S$ is of the form $s = (x_1, x_2, \alpha_A, \alpha_B)$, denoting the positions of Defender 1, Defender 2, Attacker A, and Attacker B, respectively. Each of these takes an integer value in $\{1, 2, \ldots, 10\}$.

To avoid a technical ambiguity, we will impose the convention that if the defenders occupy the same position, then $x1 < x2$. In this regard, it may be convenient to view

$$x_1 \in \{1^-, 2^-, \ldots, 10^-\},$$
$$x_2 \in \{1^+, 2^+, \ldots, 10^+\},$$

where

$$j - \varepsilon = j^- < j < j^+ < j + \varepsilon, \quad j = 1, 2, \ldots 10,$$

for some small $\varepsilon > 0$.

The combined control actions are $u = (u_1, u_2)$, where u_i is the *intended* movement to the ith defender. Each u_i must satisfy $u_i(k) \in U(x_i)$, where

$$U(x_i) = \begin{cases} \{0, 1\}, & x_i = 1; \\ \{-1, 0\}, & x_i = 10; \\ \{-1, 0, 1\}, & \text{otherwise.} \end{cases}$$

In other words, u_i denotes move right, move left, or do not move, with movement restricted at the endpoints.

As stated earlier, the u_i represent the *intended* movement. In this model, the defenders are dynamically coupled in that the intended movement need not be the actual movement. Rather, there is a random 'cohesiveness' that depends on the relative positions of the two defenders.

We can write the discrete-time dynamics of the defenders as

$$x_1(k+1) = x_1(k) + w_1(k) \tag{9.1a}$$

$$x_2(k+1) = x_2(k) + w_2(k). \tag{9.1b}$$

The *actual* moves are determined by the random variables, w_1 and w_2, as opposed to the *intended* moves, u_1 and u_2. The probabilities of w_1 and w_2 depend on (1) the relative positions, x_1 and x_2, and (2) the intended movement, u_1 and u_2, of the two defenders, as follows. If a defender intends, through u_i, to move *towards* the other defender, then the move is realized, through w, with probability p. If a defender intends to move *away* from the other defender, then the move is realized with probability q. Finally, if a defender intends to not move, then the move is realized with probability one.

The values of p and q are chosen to be $1 > p > 1/2$ and $0 < q < 1/2$, although this is not essential for the forthcoming analysis. An interpretation is that an intended move to move *away* from the other defender has a higher probability of being 'vetoed' than an intended move *towards* the other defender. In this way, the dynamics of the two defenders are coupled by their relative positions.

The probabilities of w_i are stated more precisely as follows:

Case 1: $x_1 < x_2$ and u_1 or $u_2 \neq 0$:

$$Pr\,[w_1 = 1 | u_1 = 1] = p \tag{9.2a}$$

$$Pr\,[w_1 = 0|u_1 = 1] = 1 - p \tag{9.2b}$$

$$Pr\,[w_1 = -1|u_1 = -1] = q \tag{9.2c}$$

$$Pr\,[w_1 = 0|u_1 = -1] = 1 - q \tag{9.2d}$$

$$Pr\,[w_2 = 1|u_2 = 1] = q \tag{9.2e}$$

$$Pr\,[w_2 = 0|u_2 = 1] = 1 - q \tag{9.2f}$$

$$Pr\,[w_2 = -1|u_2 = -1] = p \tag{9.2g}$$

$$Pr\,[w_2 = 0|u_2 = -1] = 1 - p \tag{9.2h}$$

Case 2: $x_2 < x_1$ and u_1 or $u_2 \neq 0$:

$$Pr\,[w_1 = 1|u_1 = 1] = q \tag{9.3a}$$

$$Pr\,[w_1 = 0|u_1 = 1] = 1 - q \tag{9.3b}$$

$$Pr\,[w_1 = -1|u_1 = -1] = p \tag{9.3c}$$

$$Pr\,[w_1 = 0|u_1 = -1] = 1 - p \tag{9.3d}$$

$$Pr\,[w_2 = 1|u_2 = 1] = p \tag{9.3e}$$

$$Pr\,[w_2 = 0|u_2 = 1] = 1 - p \tag{9.3f}$$

$$Pr\,[w_2 = -1|u_2 = -1] = q \tag{9.3g}$$

$$Pr\,[w_2 = 0|u_2 = -1] = 1 - q \tag{9.3h}$$

Case 3: u_1 or $u_2 = 0$:

$$Pr\,[w_1 = 0|u_1 = 0] = 1 \tag{9.4a}$$

$$Pr\,[w_2 = 0|u_2 = 0] = 1 \tag{9.4b}$$

Because of the assumed convention that $x_1 \in j^-$ and $x_2 \in j^+$, there is no ambiguity regarding moving 'towards' or 'away'.

The dynamics of the attackers are

$$\alpha_j(k+1) = \begin{cases} \alpha_j(k), & \text{if } x_i(k) \neq \alpha_j(k), i = 1, 2; \\ w_3(k), & \text{otherwise,} \end{cases} \tag{9.5}$$

for $j \in \{A, B\}$. The attacker stays at a location until it is intercepted by either defender. When intercepted, the attacker's reappearance is a random variable, w_3, based on a uniform probability distribution over $\{1, 2, \ldots, 10\}$.

9.2.2 Fully centralized optimization

The objective is to intercept as many attackers as possible. This is reflected in a centralized discounted infinite horizon cost

$$\min E\left\{\sum_{k=1}^{\infty} \rho^k g_{\mathrm{c}}(s(k), u(k))\right\}$$

with $\rho \in (0, 1)$. The minimization is over all stationary policies, μ, which are a function of the state, i.e.,

$$u(k) = \mu(s(k)).$$

The stage cost is

$$g_{\mathrm{c}}(s, u) = \min\left\{E\left\{\left\|x(k+1) - \begin{pmatrix}\alpha_A\\ \alpha_B\end{pmatrix}\right\| \;\Big|\; x(k), u(k)\right\},\; E\left\{\left\|x(k+1) - \begin{pmatrix}\alpha_B\\ \alpha_B\end{pmatrix}\right\| \;\Big|\; x(k), u(k)\right\}\right\},$$

which reflects the expected normed distance of both defenders to both attackers. The minimization reflects an indifference to which defender engages which attacker.

$E\{x(k+1)|x(k), u(k)\}$ refers to the expectation of the location of the defenders at time $(k+1)$ given if $u(k)$ is chosen today.

This is a standard finite state Markov decision problem whose solution can be computed using dynamic programming (Bertsekas 1995). Let $J_{\mathrm{c}}^*(s)$ denote the optimal cost of state s, and let J_{c}^* denote the vector of optimal costs, i.e.,

$$J_{\mathrm{c}}^* = \begin{pmatrix} J_{\mathrm{c}}^*(s^1) \\ J_{\mathrm{c}}^*(s^2) \\ \vdots \end{pmatrix}$$

The Bellman equation can be written as

$$J_{\mathrm{c}}^*(s) = \min_{u \in U} g_{\mathrm{c}}(s, u) + \rho T_{\mathrm{c}}(s, u)^{\mathrm{T}} J_{\mathrm{c}}^*$$

where $T_{\mathrm{c}}(s, u)$ is a vector of state and control dependent state-transition probabilities. More precisely,

$$T_{\mathrm{c}}(s, u) = \begin{pmatrix} p_{ss^1}(u) \\ p_{ss^2}(u) \\ \vdots \end{pmatrix}, \tag{9.6}$$

where $p_{ss^k}(u)$ denotes the u-dependent transition probability from state s to state s^k.

9.2.3 Fully decentralized optimization

In this set-up, each defender independently solves an optimization. Furthermore, a defender is unaware of the location of the co-defender, even though the transition probabilities *still depend* on the relative location of the two defenders. In order to formulate an individual optimization problem, each defender makes a *simplified subjective model* of the other defender's behavior.

In the decentralized optimization, each defender maintains a separate state-space, S_1, and S_2, defined by

$$s_1 = (x_1, \alpha_A, \alpha_B) \in S_1, \quad s_2 = (x_2, \alpha_A, \alpha_B) \in S_2,$$

respectively for Defender 1 and Defender 2. As before, $u_i(k) \in U(x_i(k))$.

The simplified model each defender makes of the other defender is that the other defender is currently to its left or right with some probability and independently of other events. It is understood that this does not reflect the actual behavior. Nonetheless, this representation enables each agent to design an optimal controller independently, given its own subjective model.

Let the parameter θ represent the probability that $x_2(k) > x_1(k)$. The discrete time dynamics still take the form

$$x_1(k+1) = x_1(k) + w_1(k) \tag{9.7a}$$

$$x_2(k+1) = x_2(k) + w_2(k), \tag{9.7b}$$

but now the probabilities for w_i are

$$Pr\,[w_1 = 1|u_1 = 1] = \theta p + (1-\theta)q, \tag{9.8a}$$

$$Pr\,[w_1 = 0|u_1 = 1] = 1 - Pr\,[w_1 = 1|u_1 = 1], \tag{9.8b}$$

$$Pr\,[w_1 = 0|u_1 = 0] = 1, \tag{9.8c}$$

$$Pr\,[w_1 = -1|u_1 = -1] = \theta q + (1-\theta)p, \tag{9.8d}$$

$$Pr\,[w_1 = 0|u_1 = -1] = 1 - Pr\,[w_1 = -1|u_1 = -1]. \tag{9.8e}$$

Likewise,

$$Pr\,[w_2 = 1|u_2 = 1] = \theta q + (1-\theta)p, \tag{9.9a}$$

$$Pr\,[w_2 = 0|u_2 = 1] = 1 - Pr\,[w_2 = 1|u_2 = 1], \tag{9.9b}$$

$$Pr\,[w_2 = 0|u_2 = 0] = 1, \tag{9.9c}$$

$$Pr\,[w_2 = -1|u_2 = -1] = \theta p + (1-\theta)q, \tag{9.9d}$$

$$Pr\,[w_2 = 0|u_2 = -1] = 1 - Pr\,[w_2 = -1|u_2 = -1]. \tag{9.9e}$$

It is important to stress that these are *not* the probabilities for the w_i in the actual evolution of the system. Rather, these are the *presumed* (and incorrect) probabilities, as a function of θ, that each agent uses for the fully decentralized optimization.

The attackers' dynamics, for the sake of decentralized optimization, are the same as the previous set-up. However, neither defender models the other defender intercepting an attacker, i.e.,

$$\alpha_j(k+1) = \begin{cases} \alpha_j(k), & \text{if } x_i(k) \neq \alpha_j(k); \\ w_3(k), & \text{otherwise,} \end{cases} \tag{9.10}$$

for *either* $i = 1$ or $i = 2$.

The stage cost at time k is now defined for each defender as

$$g_{\mathrm{dc}}(s_i, u_i, \theta) = \min_{\alpha_j} E\left\{ \left| x_i(k+1) - \alpha_j \right| \;\middle|\; u_i(k) \right\}. \tag{9.11}$$

The total cost for each defender is the discounted infinite horizon cost, i.e.,

$$\min E \left\{ \sum_{k=1}^{\infty} \rho^k g_{\mathrm{dc}}(s_i(k), u_i(k), \theta) \right\}$$

As before, this optimization can be solved using standard dynamic programming, with the Bellman equation being

$$J^*_{\mathrm{dc},\theta}(s_i) = \min_{u_i \in U(x_i)} g_{\mathrm{dc}}(s_i, u_i, \theta) + \rho T_{\mathrm{dc}}(s_i, u_i, \theta)^{\mathrm{T}} J^*_{\mathrm{dc},\theta}. \tag{9.12}$$

Note that the resulting optimal cost-to-go depends on θ. As before, T_{dc} denotes a vector of transition probabilities,

$$T_{\mathrm{dc}}(s_i, u_i, \theta) = \begin{pmatrix} p_{s_i s^1}(u, \theta) \\ p_{s_i s^2}(u, \theta) \\ \vdots \end{pmatrix} \tag{9.13}$$

where $p_{s_i s^k}(u, \theta)$ denotes the u and θ dependent transition probability from $s_i \in S_i$ to $s^k \in S_i$. Again, it is important to stress that the transition probabilities $T_{\mathrm{dc}}(s_i, u_i, \theta)$ are according to a defender's θ-dependent *internal model* of the overall system (9.7)–(9.9) and attacker dynamics (9.10).

9.3 EVOLUTIONARY COOPERATION

We now look into how defenders with decentralized optimization can achieve cooperation through iterative learning.

The learning will evolve over intervals, $\{I_1, I_2, \ldots\}$, where each interval consists of multiple stages. Over interval I_n, each defender employs an optimal decentralized policy (derived in Section 9.2.3) based on a modeling parameter value $\theta(n)$. Then, a new $\theta(n+1)$ is computed based on (1) $\theta(n)$ and (2) observed date over interval I_n. In particular, define

$$\theta_{\mathrm{obs}}(n) = \frac{\#\text{ times } x_1(k) < x_2(k) \text{ over interval } I_n}{\text{total number of stages in interval } I_n},$$

Table 9.1 Performance comparisons

Setup	Hits	Misses	% Hits	θ^*
Fully Centralized	976	24	97.6%	-
Fully Decentralized	700	90	88.6%	-
Evolutionary Cooperation	862	68	92.7%	0.9113
Evolutionary Cooperation (modified cost)	890	36	96.1%	0.9397

i.e., the percentage of times $x_1(k) < x_2(k)$ over interval I_n. The θ-update equation is now defined as

$$\theta(n+1) = \theta(n) + \frac{1}{n+1}(\theta(n) - \theta_{\text{obs}}(n)).$$

Table 9.1 tabulates the resulting performance of fully centralized, fully decentralized, and evolutionary cooperation. For this table, a 'hit' denotes an attacker being intercepted within 10 steps. After 10 steps, the attacker moves to a randomly selected location and a 'miss' is registered. As anticipated, the fully centralized set-up exhibits the best performance, but the evolutionary cooperation set-up outperforms the fully decentralized setup. The θ values converge to $\theta = 0.9113$ after approximately 50 iterations.

Table 9.1 also shows an additional evolutionary set-up, which uses a modification of the decentralized stage cost defined in (9.11). The idea behind the modified stage cost is as follows. The original stage cost (9.11) reflects that each defender has a greedy policy of intercepting whichever attacker is nearer. The modified stage cost penalizes the defender should it attempt to intercept the attackers in the area where, according to the modeled θ value, a co-defender has a high probability presence. Intuitively, this means that the defender respects the presence of the co-defender and its capability to intercept any attackers in that area *without* modeling the detailed behavior of the co-defender, and thereby maintaining a decentralized optimization. This intuition is inspired from the reference (Dutta and Sen 2003).

The modified stage cost is defined as follows. First, define

$$\gamma_{ij} = E\left\{ \left| x_i(k+1) - \alpha_j \right| \;\middle|\; u_i(k) \right\},$$

where the expectation for $x_i(k+1)$ is according to the (decentralized) probabilities of w_i in (9.8)–(9.9). The stage cost for Defender 1 is

$$g_{\text{mod}}(s_1, u_1, \theta) = \tag{9.14}$$

$$\begin{cases} \min\{(1-\theta)\gamma_{1A}, (1-\theta)\gamma_{1B}\} & x_1 \le \alpha_A, \alpha_B; \\ \min\{\theta\gamma_{1A}, (1-\theta)\gamma_{1B}\} & \alpha_A < x_1 \le \alpha_B; \\ \min\{(1-\theta)\gamma_{1A}, \theta\gamma_{1B}\} & \alpha_B < x_1 \le \alpha_A; \\ \min\{\theta\gamma_{1A}, \theta\gamma_{1B}\} & \alpha_A, \alpha_B < x_1. \end{cases}$$

A similar stage cost is defined for Defender 2. From Defender 1's perspective, the main idea is to give a higher weight to the attacker that has a higher probability of being closer.

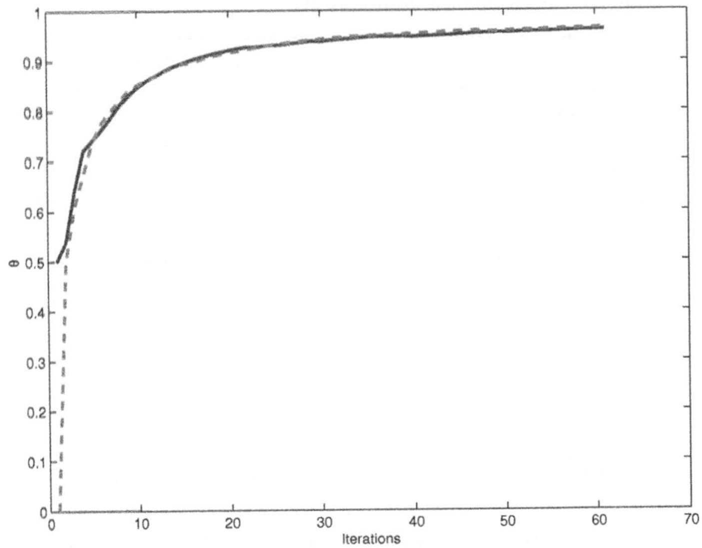

Figure 9.3 Convergence of θ in evolutionary modified cost set-up.

The last row of Table 9.1 shows that the performance with the modified cost approaches that of the optimal centralized set-up. The θ values converge to 0.9397 after approximately 60 iterations.

Figure 9.3 shows the evolution of the $\theta(n)$ for the modified cost set-up. The figure shows two different initializations: $\theta(1) = 0.5$ (solid line) and $\theta(1) = 0.0$ (dashed line). Both appear to converge to the same limiting value of θ.

9.4 ANALYSIS OF CONVERGENCE

The implications of convergence in θ are the following 'consistency' conditions: (1) each defender is employing a policy that is optimal *with respect to* its model of the other defender; and (2) each defender's behavior conforms with the model assumed by the co-defender.

We will analyze a idealized version of the evolutionary cooperation set-up and establish convergence of the θ iterations. The discussion only will be for the stage cost of (9.11), but the analysis for the modified cost (9.14) is similar.

9.4.1 Idealized iterations and main result

The first of two idealizations is that the policy implemented during an interval is a 'smoothed' optimal policy as follows. Let $J_{\mathrm{dc},\theta(n)}$ denote the optimal decentralized cost vector from (9.12) for iteration interval $I(n)$. Accordingly, the optimal policy (for each defender) over iteration interval $I(n)$ at state s_i is

$$\mu_i(s_i) = \arg \min_{u_i \in U(x_i)} g_{\mathrm{dc}}(s_i, u_i, \theta(n)) + \rho T_{\mathrm{dc}}(s_i, u_i, \theta(n))^{\mathrm{T}} J^*_{\mathrm{dc},\theta(n)}.$$

We will replace this policy with a randomized 'softmax' version. For an admissible control action, $a \in U(x_i)$, define

$$v(a; s_i) = g_{\text{dc}}(s_i, a, \theta(n)) + \rho T_{\text{dc}}(s_i, a, \theta(n))^{\text{T}} J^*_{\text{dc},\theta(n)}.$$

For example, if $U(x_i) = \{-1, 0\}$, then

$$\begin{aligned} v &= \begin{pmatrix} v(-1; s_i) \\ v(0; s_i) \end{pmatrix} \\ &= \begin{pmatrix} g(s_i, -1, \theta(n)) + \rho T(s_i, -1, \theta(n))^{\text{T}} J^*_{\text{dc},\theta(n)} \\ g(s_i, 0, \theta(n)) + \rho T(s_i, 0, \theta(n))^{\text{T}} J^*_{\text{dc},\theta(n)} \end{pmatrix}. \end{aligned}$$

The 'smoothed' optimal policy is a randomized version of the optimal policy, where

$$Pr\left[u_i(k) = a \;\middle|\; s_i(k)\right] = \frac{e^{-v(a;s_i(k))/\tau}}{\sum_{\tilde{a}} e^{-v(\tilde{a};s_i(k))/\tau}}. \tag{9.15}$$

Such 'softmax' smoothing is commonly introduced in learning algorithms (e.g., Sutton and Barto 1998; Fudenberg and Levine 1998) to encourage exploration. The (temperature) parameter $\tau > 0$ regulates the degree of exploration. As $\tau \to 0$, the softmax chooses the maximizing (in our case, minimizing) action with probability increasingly close to one.

An important consequence of the smoothed optima l policy is the following theorem.

Proposition 9.4.1 *The attacker dynamics (9.5) and defender dynamics (9.1) under w probabilities (9.2)–(9.4a) and smoothed policies (9.15) form an aperiodic irreducible Markov chain.*

Proof. The smoothed policy (9.15) places a positive probability on any admissible control action. Consequentially, it can be shown that there is a positive probability to transition, over multiple stages, from any state of the overall system (9.1) to any other state. Furthermore, there is a positive probability over a single stage to stay in the same state. □

The second idealization involves $\theta_{\text{obs}}(n)$.

Suppose that each defender constructs an optimal policy according to (9.11)–(9.12) with model parameter θ and employs a smoothed optimal policy as in (9.15). Via standard methods in Markov chains (e.g., Bertsekas and Tsitsiklis 2002, Chapter 6), Proposition 9.4.1 implies that there exists a unique stationary probability distribution, which we denote π_θ. Define

$$\Pi(\theta) = Pr\left[x_2 > x_1\right],$$

where the probability is with respect to the stationary distribution π_θ. We will assume that

$$\theta_{\text{obs}}(n) = \Pi(\theta(n)) + \delta(n) \tag{9.16}$$

where the $\delta(n)$ form a uniformly bounded random sequence with

$$E\left\{\delta(n)|\theta(n)\right\} = 0.$$

In the simulations of the previous section, $\theta_{obs}(n)$ reflected the percentage of times $x_2(k) > x_1(k)$ over stages $k \in I(n)$. The idealization (9.16) assumes that $\theta_{obs}(n)$ is a noisy measurement of the exact steady state probability $\Pi(\theta(n))$.

For clarity of exposition, we now state the complete idealized iterative algorithm: The algorithm is illustrated in Figure 9.4.

1. *Initialization:*
   (a) $k = 0$.
   (b) $\theta(0) = \theta_o$
2. *Iteration n:*
   (a) Based on $\theta(k)$, each defender designs an optimal policy based on the decentralized dynamics, probabilities, and cost function (9.7)–(9.12).
   (b) Over interval $I(k)$, each defender employs a smoothed optimal policy (9.15), for the dynamics defined by equations (9.1), (9.10), and (9.2)–(9.4a).
   (c) At the end of interval $I(n)$, each defender measures $\theta_{obs}(n)$ according to (9.16).
3. *Update:*
   (a) $\theta(n+1) = \theta(n) + \frac{1}{n+1}(\theta_{obs}(n) - \theta(n))$.
   (b) $n = n + 1$.

We are now in a position to state the main result:

Theorem 9.4.2 *In the framework of Algorithm 9.4.1,*

$$\lim_{n\to\infty} \theta(n) = \theta^*,$$

for some θ^*, almost surely.

The remainder of this section is devoted to the proof of Theorem 9.4.2.

9.4.2 Proof of Theorem 9.4.2

The key element of the proof is to show that the function

$$\theta \mapsto \Pi(\theta)$$

is continuous. In the framework of Algorithm 9.4.1, the θ iterations

$$\theta(n+1) = \theta(n) + \frac{1}{n+1}(\Pi(\theta(n)) - \theta(n) + \delta(n)) \tag{9.17}$$

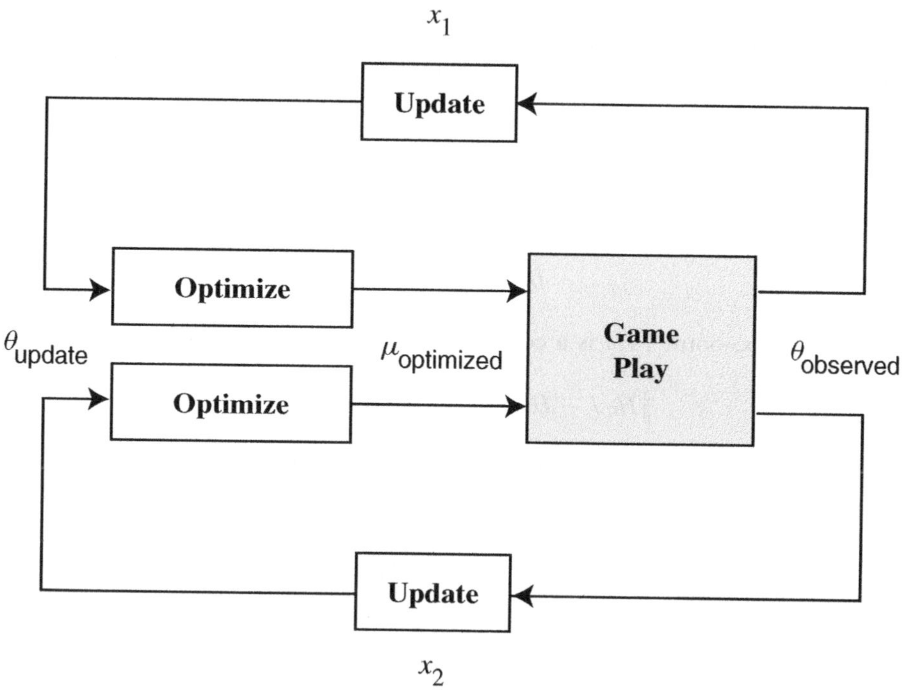

Figure 9.4 Algorithm 9.4.1 flow diagram.

satisfy the stochastic approximation assumptions (e.g., Kushner and Yin 1997; Benaim 1996) so that the limit set of the iterations (9.17) is determined by the continuous differential equation

$$\dot{\theta} = -\theta + \Pi(\theta). \tag{9.18}$$

In our case, the flow induced by (9.18) evolves over [0, 1]. As long as $\Pi(\cdot)$ is continuous, the result of (Benaim 1996, Theorem 1.2) implies that the limit set of (9.17) is almost surely an equilibrium point

$$\theta^* = \Pi(\theta^*).$$

We will prove that $\Pi(\theta)$ is continuous through a series of claims.

Claim 9.4.1 *For any $s_i \in S_i$ and $u_i \in U(x_i)$, the decentralized stage cost (9.11), $g_{\text{dc}}(s_i, u_i, \theta)$ and decentralized transitions probabilities (9.13) $T_{\text{dc}}(s_i, u_i, \theta)$ are both continuous functions of θ.*

Proof. This is clear from the dependence on θ in the decentralized model probabilities (9.8)–(9.9). □

Claim 9.4.2 *For any $s_i \in S_i$, the optimal decentralized cost $J^*_{\text{dc}}(s_i)$ defined by (9.12) is a continuous function of θ.*

Proof. Let J be a vector of the same dimension as the size of the state space in the decentralized optimization, and let H_θ denote the value iteration operator (Bertsekas and Tsitsiklis 1996) which maps J into $H_\theta J$ defined by

$$(H_\theta J)(s_i) = \min_{u_i \in U(x_i)} g_{\mathrm{dc}}(s_i, u_i, \theta) + \rho T_{\mathrm{dc}}(s_i, u_i, \theta)^{\mathrm{T}} J.$$

Then the Bellman equation (9.12) can be written as

$$J^*_{\mathrm{dc},\theta} = H_\theta J^*_{\mathrm{dc},\theta}.$$

Under the vector ∞-norm[1], H_θ is a contraction. In particular, for any two vectors J and J',

$$\left\| H_\theta J - H_\theta J' \right\|_\infty \leq \rho \left\| J - J' \right\|_\infty.$$

Now consider two values θ and θ' and their associated optimal decentralized costs, $J^*_{\mathrm{dc},\theta}$ and $J^*_{\mathrm{dc},\theta'}$. These both satisfy their respective Bellman equations,

$$H_\theta J^*_{\mathrm{dc},\theta} = H_\theta J^*_{\mathrm{dc},\theta}$$

and

$$H_{\theta'} J^*_{\mathrm{dc},\theta'} = H_{\theta'} J^*_{\mathrm{dc},\theta'}.$$

Define

$$J^{1-\mathrm{step}}_{\theta'} = H_{\theta'} J^*_{\mathrm{dc}}.$$

Note that $J^{1-\mathrm{step}}_{\theta'}$ results from applying a one step of value iteration using θ' on the Bellman equation solution corresponding to θ. By standard contraction arguments,

$$J^*_{\mathrm{dc},\theta'} = \lim_{n \to \infty} H^n_{\theta'} J^*_{\mathrm{dc},\theta}.$$

Furthermore,

$$\left\| J^*_{\mathrm{dc},\theta'} - J^*_{\mathrm{dc},\theta} \right\| \leq \frac{1}{1-\rho} \left\| J^{1-\mathrm{step}}_{\theta'} - J^*_{\mathrm{dc},\theta} \right\|. \tag{9.19}$$

We see that the desired continuity would follow from bounding the right-hand side of the above inequality (9.19).

Towards this end, compare

$$J^{1-\mathrm{step}}_{\theta'}(s_i) = \min_{u_i \in U(x_i)} g_{\mathrm{dc}}(s_i, u_i, \theta') + \rho T_{\mathrm{dc}}(s_i, u_i, \theta')^{\mathrm{T}} J^*_{\mathrm{dc},\theta}$$

and

$$J^*_{\mathrm{dc},\theta} = \min_{u_i \in U(x_i)} g_{\mathrm{dc}}(s_i, u_i, \theta) + \rho T_{\mathrm{dc}}(s_i, u_i, \theta)^{\mathrm{T}} J^*_{\mathrm{dc},\theta}.$$

[1] For $x \in \mathbf{R}^n$, $\|x\|_\infty = \max_i |x_i|$.

By the continuity conclusion of Claim 9.4.1, we see that

$$\left\| J_{\theta'}^{1-\text{step}} - J_{\text{dc},\theta}^{*} \right\| \to 0$$

as $|\theta' - \theta| \to 0$, which, via inequality (9.19), implies the desired continuity. □

Claim 9.4.3 *Let $T_c(s;\theta)$ denote the overall (centralized) state transition probabilities (as in (9.6)) induced by smoothed decentralized θ-dependent optimal policies as in (9.15). For any $s \in S$, $T(s;\theta)$ is a continuous function of θ.*

Proof. This claim follows from the continuity conclusion of Claim 9.4.2 combined with the continuity of smoothed policy (9.15). □

Claim 9.4.4 *The function $\Pi(\theta)$ is continuous.*

In the context of Claim 9.4.3, let π_θ denote the stationary probability vector for $T_c(s;\theta)$. It is easy to see that

$$\Pi(\theta) = \sum_{s \in S: x_1 < x_2} \pi_\theta(s).$$

Meyer (1980) shows that the stationary probability vector is a continuous function of the transition matrix. From Claim 9.4.3, we have that $\Pi(\theta)$ is continuous. □

9.5 CONCLUSION

We have developed a framework for cooperation via iterative egocentric modeling. We present both a simulation study and an analytical proof of convergence for an idealized scenario.

There is a strong relationship between the method discussed here and the area of learning in games (e.g., Fudenberg and Levine 1998; Young 1998; Young 2006) as well as (Shamma and Arslan 2005; Arslan and Shamma 2004). In these methods, agents also make a model of opposing agent strategies using empirical data. Unlike the method discussed here, each iteration requires agents to employ an optimal *response* to the empirically characterized strategies of other agents. In our case, we never presumed a policy for the other agent in order to derive an optimal response. Rather, each agent employs a best response to a highly simplified behaviorial model of the other agent, which is more in line with a bounded rationality model of adaptation. Accordingly, the implication of convergence in learning in games is a Nash equilibrium, whereas convergence in our case implies a modeling and optimization consistency condition.

In this work, the evolving parameter, θ, was a scalar. An obvious next step in this work is to consider situations involving multiple parameters, e.g., to allow more sophisticated models of agent interactions. Although such an approach has been demonstrated in (Hsu and Shamma 1999) and (Seah 2002), proof of convergence remains elusive. The obstacle is that one must assess the asymptotic behavior of a multivariate differential equation, e.g., to apply methods from stochastic approximation. In the scalar case, the analysis is straightforward without explicit knowledge of $\Pi(\theta)$ other than continuity and bounded range. In the multivariate case, definite statements regarding asymptotic behavior would be elusive in the absence of additional structural knowledge.

ACKNOWLEDGEMENTS

Research supported by NSF grant #ECS-0501394, AFOSR/MURI grant #F49620-01-1-0361, and ARO grant #W911NF–04–1–0316.

REFERENCES

Arslan G and Shamma JS 2004 Distributed convergence to Nash equilibria with local utility measurements. *43rd IEEE Conference on Decision and Control*, pp. 1538–1543.

Benaim M 1996 A dynamical system approach to stochastic approximations. *SIAM Journal of Control and Optimization* **34**(2), 437–472.

Bertsekas D 1995 *Dynamic Programming and Optimal Control*. Athena Scientific, Belmont, MA.

Bertsekas D and Tsitsiklis J 1996 *Neuro-dynamic Programming*. Athena Scientific, Belmont, MA.

Bertsekas D and Tsitsiklis J 2002 *Introduction to Probability*. Athena Scientific, Belmont, MA.

Chang YH, Ho T and Kaelbling L 2003 Mobilized ad-hoc networks: A reinforcement learning approach. MIT AI Laboratory Memo, AIM-2003-025.

Chow KP and Kwok YK 2002 On load balancing for distributed multiagent computing. *IEEE Transactions on Parallel and Distributed Systems* **13**(8), 787–801.

Dresner K and Stone P 2004 A reservation-based multiagent system for intersection control. *The Fifth IFAC Symposium on Intelligent Autonomous Vehicles*.

Durfee E 1999 Distributed continual planning for unmanned ground vehicle teams. *AI Magazine* **20**(4), 55–61.

Dutta P and Sen S 2003 Forming stable partnerships. *Cognitive Systems Research* **4**, 211–221.

Earl M and D'Andrea R 2002 A study in cooperative control: The RoboFlag drill. In *Proceedings of the American Control Conference*, Anchorage, Alaska.

Fudenberg D and Levine D 1998 *The Theory of Learning in Games*. MIT Press, Cambridge, MA.

Greenwald A and Kephard J 2000 Probabilistic pricebots. In *Proceedings of the Fifth International Conference on Autonomous Agents*, pp. 560–567.

Hofbauer J and Sigmund K 1998 *Evolutionary Games and Population Dynamics*. Cambridge University Press, Cambridge, UK.

Hsu CH and Shamma JS 1999 A decomposition approach to scheduling of failure- prone transfer lines. In *System Theory: Modeling, Analysis, and Control* (ed. Djaferis T and Schick I), Kluwer Academic Publishers, Dordrecht.

Ilachinski A 2004 *Artificial War: Multiagent-Based Simulation of Combat*. World Scientific Publishing, Singapore.

Kushner H and Yin G 1997 *Stochastic Approximation Algorithms and Applications*. Springer-Verlag, New York.

Meyer C 1980 The condition of a finite Markov chain and perturbation bounds for limiting probabilities. *SIAM Journal on Algebraic Discrete Mathematics* **1**, 273–283.

Rekleitis I, Dudek G and Milios E 2001 Multi-robot collaboration for robust exploration. *Annals of Mathematics and Artificial Intelligence* **31**, 7–40.

Sabater J and Sierra C 2002 Reputation and social network analysis in multi-agent systems. In *Proceedings of the First International Joint Conference on Autonomous Agents and Multiagent Systems*, pp. 475–482.

Samuelson L 1997 *Evolutionary Games and Equilibrium Selection*. MIT Press, Cambridge, MA.

Seah VPW 2002 Decomposition based control of failure prone production lines, Master's thesis, UCLA Department of Mechanical and Aerospace Engineering.

Shamma JS and Arslan G 2005 Dynamic fictitious play, dynamic gradient play, and distributed convergence to Nash equilibria. *IEEE Transactions on Automatic Control* **50**(3), 312–327.

Shoham Y, Powers R and Grenager T 2007 If multi-agent learning is the answer, what is the question? forthcoming special issue *Artificial Intelligence*.

Sutton R and Barto A 1998 *Reinforcement Learning*. MIT Press, Cambridge, MA.

Weibull J 1995 *Evolutionary Game Theory*. MIT Press, Cambridge, MA.

Weiss G 2000 *Multiagent Systems: A Modern Approach to Distributed Artificial Intelligence*. MIT Press, Cambridge MA.

Wolpert D, Wheeler K and Tumer K 1999 General principles of learning-based multi-agent systems. *Third International Conference on Autonomous Agents*, pp. 77–83.

Young HP 1998 *Individual Strategy and Social Structure*. Princeton University Press, Princeton, NJ.

Young HP 2006 *Strategic Learning and its Limits*. Oxford University Press, Oxford.

Part III

Adversarial Interactions

10

Multi-vehicle cooperative control using mixed integer linear programming

Matthew G. Earl and Raffaello D'Andrea

10.1 INTRODUCTION

For many problems, a team of vehicles can accomplish an objective more efficiently and more effectively than a single vehicle can. Some examples include target intercept (Beard *et al.* 2002), search (Beard and McLain 2003), terrain mapping (Marco *et al.* 2003), object manipulation (Sugar and Kumar 2002), surveillance, and space-based interferometry. For these problems, it is desirable to design a multi-vehicle cooperative control strategy.

There is a large literature on cooperative control. Work from team theory (Ho and Chu 1972; Marschak and Radner 1972) considers the case where team members have different information and the objective function is quadratic. Cooperative estimation for reconnaissance problems is considered in (Ousingsawat and Campbell 2004). In (Bellingham *et al.* 2002; Earl and D'Andrea 2002a; Richards *et al.* 2002a) mixed integer linear programming is used for multi-vehicle target assignment and intercept problems. Hierarchical methods are used for cooperative rendezvous in (McLain *et al.* 2001) and for target assignment and intercept in (Beard *et al.* 2002; Earl and D'Andrea 2007). A review from the machine learning perspective is presented in (Stone and Veloso 2000). There are several compilations of cooperative control articles in (Arkin and G. A. Bekey 1997; Balch and L. E. Parker 2000; Murphey and Pardalos 2002).

In this chapter, we propose a hybrid systems approach for modeling and cooperative control of multi-vehicle systems. We use the class of hybrid systems called mixed logical dynamical systems (Bemporad and Morari 1999), which are governed by difference equations and logical rules and are subject to linear inequality constraints. The main motivation for using mixed logical dynamical systems is their ability to model a wide variety of multi-vehicle problems and the ease of modifying problem formulations. In our approach, a problem is modeled as a mixed logical dynamical system, which we represent

Cooperative Control of Distributed Multi-Agent Systems edited by Jeff S. Shamma

as a mixed integer linear program (MILP). Then, to generate a cooperative control strategy for the system, the MILP is solved using AMPL (Fourer *et al.* 1993) and CPLEX (ILOG 2000).

Posing a multi-vehicle control problem in a MILP framework involves modeling the vehicle dynamics and constraints, modeling the environment, and expressing the objective. To demonstrate the modeling procedure and our approach, we consider control problems involving Cornell's multi-vehicle system called RoboFlag. For an introduction to RoboFlag, see the papers from the invited session on RoboFlag in the Proceedings of the 2003 American Control Conference (Campbell *et al.* n.d.; D'Andrea and Babish 2003; D'Andrea and Murray 2003).

Our focus is to find optimal solutions using a flexible methodology, which is why we use MILP. However, because MILP is in the NP-hard computation class (Garey and Johnson 1979), the methods may not be fast enough for real-time control of systems with large problem formulations. In this case, the methods can be used to explore optimal cooperative behavior and as benchmarks to evaluate the performance of heuristic or approximate methods. In Section 10.6, we discuss several methods for reducing the computational requirements of our MILP approach so that it can be more readily used in real-time applications.

Our approach for multi-vehicle control, first presented in (Earl and D'Andrea 2002a, b), was developed independently from a similar approach developed by Richards *et al.* (Richards and How 2002). Next, we list some of the noteworthy aspects of our approach. First, the environment which we demonstrate our methods involves an adversarial component. We model the intelligence of the adversaries with state machines. Second, our approach allows multiple, possibly nonuniform, time discretizations. Discretizing continuous variables in time is necessary for MILP formulations. Using many time steps results in large MILPs that require a considerable amount of computation time to solve. Support for nonuniform discretizations in time allows the use of intelligent time step selection algorithms for the generation of more efficient MILP problem formulations (Earl and D'Andrea 2005). Finally, because we include the vehicle dynamics in the problem formulation, the resulting trajectories are feasible, which is advantageous because they can be applied directly to the multi-vehicle system. In order to express the vehicle dynamics efficiently in our MILP formulation, we restrict the control input to each vehicle in a way that allows near-optimal performance, as presented in (Kalmár-Nagy *et al.* 2004).

The chapter is organized as follows: First, we consider vehicle problems that have relatively simple formulations. Then we add features in each section until we arrive at the RoboFlag multi-vehicle problems. In Section 10.2, we introduce the dynamics of the vehicles used to motivate our approach, and we formulate and solve a single vehicle minimum control effort trajectory generation problem. We build upon this in Section 10.3 adding obstacles that must be avoided. In Section 10.4, we show how to generate optimal team strategies for RoboFlag problems. In Section 10.5, we perform an average case computational complexity study on our approach. Finally, in Section 10.6, we discuss our methods and ways in which they can be applied. All files for generating the plots found in this chapter are available online.[1]

[1] See http://arxiv.org/abs/cs.RO/0501092.

10.2 VEHICLE DYNAMICS

Multi-vehicle control problems involving the wheeled robots of Cornell's RoboCup Team (D'Andrea *et al.* 2001; Stone *et al.* 2001) are used to motivate the methods presented in this chapter. In this section, we show how to simplify their nonlinear governing equations using a procedure from (Kalmár-Nagy *et al.* 2004). The result is a linear set of governing equations coupled by a nonlinear constraint on the control input, which admits feasible vehicle trajectories and allows near-optimal performance. We then show how to represent the simplified system in a MILP problem formulation.

Each vehicle has a three-motor omni-directional drive, which allows it to move along any direction irrespective of its orientation. The nondimensional governing equations of each vehicle are given by

$$\begin{bmatrix} \ddot{x}(t) \\ \ddot{y}(t) \\ \ddot{\theta}(t) \end{bmatrix} + \begin{bmatrix} \dot{x}(t) \\ \dot{y}(t) \\ \frac{2mL^2}{I}\dot{\theta}(t) \end{bmatrix} = \mathbf{u}(\theta(t), t), \tag{10.1}$$

where $\mathbf{u}(\theta(t), t) = \mathbf{P}(\theta(t))\mathbf{U}(t)$,

$$\mathbf{P}(\theta) = \begin{bmatrix} -\sin(\theta) & -\sin(\frac{\pi}{3} - \theta) & \sin(\frac{\pi}{3} + \theta) \\ \cos(\theta) & -\cos(\frac{\pi}{3} - \theta) & -\cos(\frac{\pi}{3} + \theta) \\ 1 & 1 & 1 \end{bmatrix}, \tag{10.2}$$

and $\mathbf{U}(t) = (U_1(t), U_2(t), U_3(t)) \in \mathcal{U}$. In these equations, $(x(t), y(t))$ are the coordinates of the vehicle on the playing field, $\theta(t)$ is the orientation of the vehicle, $\mathbf{u}(\theta(t), t)$ is the $\theta(t)$-dependent control input, m is the mass of the vehicle, I is the vehicle's moment of inertia, L is the distance from the drive to the center of mass, and $U_i(t)$ is the voltage applied to motor i. The set of admissible voltages $\mathcal{U}$ is given by the unit cube, and the set of admissible control inputs is given by $P(\theta)\mathcal{U}$.

These governing equations are coupled and nonlinear. To simplify them, we replace the set $P(\theta)\mathcal{U}$ with the maximal θ-independent set found by taking the intersection of all possible sets of admissible controls. The result is a θ-independent control input, denoted $(u_x(t), u_y(t), u_\theta(t))$, that is subject to the inequality constraints $u_x(t)^2 + u_y(t)^2 \leq (3 - |u_\theta(t)|)^2/4$ and $|u_\theta(t)| \leq 3$.

Using the restricted set as the set of allowable control inputs, the governing equations decouple and are given by

$$\begin{bmatrix} \ddot{x}(t) \\ \ddot{y}(t) \\ \ddot{\theta}(t) \end{bmatrix} + \begin{bmatrix} \dot{x}(t) \\ \dot{y}(t) \\ \frac{2mL^2}{I}\dot{\theta}(t) \end{bmatrix} = \begin{bmatrix} u_x(t) \\ u_y(t) \\ u_\theta(t) \end{bmatrix}. \tag{10.3}$$

The constraints on the control input couple the degrees of freedom.

To decouple the θ dynamics we further restrict the admissible control set to a cylinder defined by the following two inequalities: $|u_\theta(t)| \leq 1$ and

$$u_x(t)^2 + u_y(t)^2 \leq 1. \tag{10.4}$$

Now the equations of motion for the translational dynamics of the vehicle are given by

$$\begin{aligned}\ddot{x}(t) + \dot{x}(t) &= u_x(t),\\ \ddot{y}(t) + \dot{y}(t) &= u_y(t),\end{aligned} \tag{10.5}$$

subject to Equation (10.4). In state space form, Equation (10.5) is $\dot{\mathbf{x}}(t) = \mathbf{A}_c\mathbf{x}(t) + \mathbf{B}_c\mathbf{u}(t)$, where $\mathbf{x} = (x, y, \dot{x}, \dot{y})$ is the state and $\mathbf{u} = (u_x, u_y)$ is the control input.

By restricting the admissible control inputs we have simplified the governing equations in a way that allows near optimal performance as shown in (Kalmár-Nagy *et al.* 2004). This procedure allows real-time calculation of many near-optimal trajectories and has been successfully used by Cornell's RoboCup team (D'Andrea *et al.* 2001; Kalmár-Nagy *et al.* 2004; Stone *et al.* 2001).

To represent the governing equations in a MILP framework, we discretize the control input in time and require it be constant between time steps. The result is a set of linear discrete time governing equations.

Let N_u be the number of discretization steps for the control input $\mathbf{u}(t)$, let $t_u[k]$ be the time at step k, and let $T_u[k] > 0$ be the time between steps k and $k+1$, for $k \in \{0, \ldots, N_u - 1\}$. The discrete time governing equations are given by

$$\mathbf{x}_u[k+1] = \mathbf{A}[k]\mathbf{x}_u[k] + \mathbf{B}[k]\mathbf{u}[k], \tag{10.6}$$

where $\mathbf{x}_u[k] = \mathbf{x}(t_u[k])$, $\mathbf{u}[k] = \mathbf{u}(t_u[k])$,

$$\mathbf{A}[k] = \begin{bmatrix} 1 & 0 & 1 - e^{-T_u[k]} & 0 \\ 0 & 1 & 0 & 1 - e^{-T_u[k]} \\ 0 & 0 & e^{-T_u[k]} & 0 \\ 0 & 0 & 0 & e^{-T_u[k]} \end{bmatrix},$$

$$\mathbf{B}[k] = \begin{bmatrix} T_u[k] - 1 + e^{-T_u[k]} & 0 \\ 0 & T_u[k] - 1 + e^{-T_u[k]} \\ 1 - e^{-T_u[k]} & 0 \\ 0 & 1 - e^{-T_u[k]} \end{bmatrix},$$

$\mathbf{x}_u[k] = (x_u[k], y_u[k], \dot{x}_u[k], \dot{y}_u[k])$, and $\mathbf{u}[k] = (u_x[k], u_y[k])$. The coefficients $\mathbf{A}[k]$ and $\mathbf{B}[k]$ are functions of k because we have allowed for nonuniform time discretizations. Because there will be several different time discretizations used in this chapter, we use subscripts to differentiate them. In this section, we use the subscript u to denote variables associated with the discretization in the control input $\mathbf{u}(t)$.

The discrete time governing equations can be solved explicitly to obtain

$$\begin{aligned} x_u[k] &= x_u[0] + \left(1 - e^{-t_u[k]}\right)\dot{x}_u[0] \\ &\quad + \sum_{i=0}^{k-2}\left(\left(T_u[i] - 1 + e^{-T_u[i]}\right)u_x[i]\right. \\ &\quad + \left.\left(1 - e^{t_u[i+1]-t_u[k]}\right)\left(1 - e^{-T_u[i]}\right)u_x[i]\right) \end{aligned}$$

$$
\begin{aligned}
& + \left(T_u[k-1] - 1 + e^{-T_u[k-1]}\right) u_x[k-1], \\
\dot{x}_u[k] &= e^{-t_u[k]}\dot{x}_u[0] \\
& + \sum_{i=0}^{k-2} \left(e^{t_u[i+1]-t_u[k]} \left(1 - e^{-T_u[i]}\right) u_x[i]\right) \\
& + \left(1 - e^{-T_u[k-1]}\right) u_x[k-1],
\end{aligned}
$$

and similarly for $y_u[k]$ and $\dot{y}_u[k]$.

In later sections of this chapter it will be necessary to represent the position of the vehicle at times between control discretization steps, in terms of the control input. Because the set of governing equations is linear, given the discrete state $\mathbf{x}_u[k]$ and the control input $\mathbf{u}[k]$, we can calculate the vehicle's state at any time t using the following equations:

$$
\begin{aligned}
x(t) &= x_u[k] + (1 - e^{t_u[k]-t})\dot{x}_u[k] \\
& + (t - t_u[k] - 1 + e^{t_u[k]-t})u_x[k], \\
\dot{x}(t) &= (e^{t_u[k]-t})\dot{x}_u[k] + (1 - e^{t_u[k]-t})u_x[k],
\end{aligned} \tag{10.7}
$$

where k satisfies $t_u[k] \leq t \leq t_u[k+1]$. If the time discretization of the control input is uniform, $T_u[k_u] = T_u$ for all k_u, then $k_u = \lfloor t/T_u \rfloor$. The other components of the vehicle's state, $y(t)$ and $\dot{y}(t)$, can be calculated in a similar way.

The control input constraint given by Equation (10.4) cannot be expressed exactly in a MILP framework because it is nonlinear. However, it can be incorporated approximately with a set of linear inequalities that define a polygon. The polygon inscribes the region defined by the nonlinear constraint. We take the conservative inscribing polygon to guarantee that the set of allowable controls, defined by the region, is feasible. We define the polygon by the set of M_u linear inequality constraints

$$
\begin{aligned}
& u_x[k] \sin \frac{2\pi m}{M_u} + u_y[k] \cos \frac{2\pi m}{M_u} \leq \cos \frac{\pi}{M_u} \\
& \forall m \in \{1, \ldots, M_u\},
\end{aligned} \tag{10.8}
$$

for each step $k \in \{1, \ldots, N_u\}$.

To illustrate the approach, we consider the following minimum control effort trajectory generation problem. Given a vehicle governed by Equations (10.6) and (10.8), find the sequence of control inputs $\{\mathbf{u}[k]\}_{k=0}^{N_u-1}$ that transfers the vehicle from starting state $\mathbf{x}(0) = \mathbf{x}_s$ to finishing state $\mathbf{x}(t_f) = \mathbf{x}_f$ and minimizes the cost function

$$
J = \sum_{k=0}^{N_u-1} \left(|u_x[k]| + |u_y[k]|\right). \tag{10.9}
$$

To convert the absolute values in the cost function to linear form, we introduce auxiliary continuous variables $z_x[k]$ and $z_y[k]$ and the inequality constraints

$$
\begin{aligned}
& -z_x[k] \leq u_x[k] \leq z_x[k] \\
& -z_y[k] \leq u_y[k] \leq z_y[k].
\end{aligned} \tag{10.10}
$$

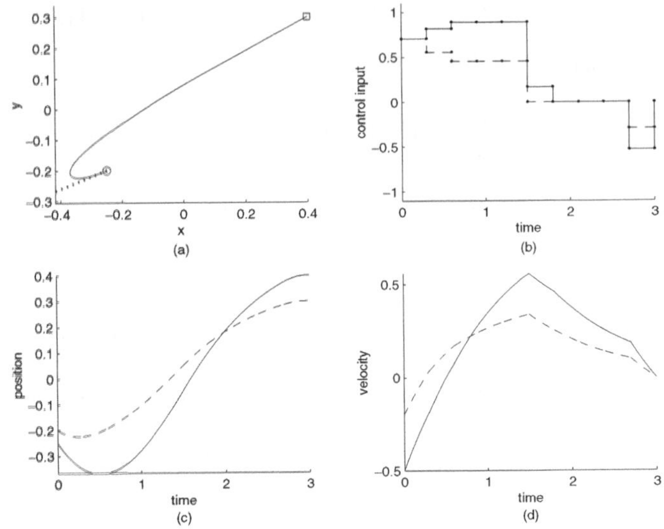

Figure 10.1 Plots of the minimum control effort example. Figure (a) shows the vehicle trajectory in the (x, y) plane. The circle and dotted line denote the initial position and velocity, respectively. The square denotes the final position. Figures (b)–(d) show the time histories of the control inputs, the positions, and the velocities, respectively. The solid lines in Figures (b)–(d) represent x components and the dotted lines represent y components. The values for the parameters are $N_u = 10$, $M_u = 20$, $T_u[k] = 0.3$ for all k, $(x_0, y_0, \dot{x}_0, \dot{y}_0) = (-0.25, -0.2, -0.5, -0.2)$, and $(x_f, y_f, \dot{x}_f, \dot{y}_f) = (0.4, 0.3, 0.0, 0.0)$.

Minimizing $z_x[k]$ subject to the inequalities $u_x[k] \leq z_x[k]$ and $u_x[k] \geq -z_x[k]$ is equivalent to minimizing $|u_x[k]|$ (similarly for $|u_y[k]|$) (Bertsimas and Tsitsiklis 1997). Using the auxiliary variables, the cost function can be written as a linear function,

$$J = \sum_{k=0}^{N_u-1} \left(z_x[k] + z_y[k] \right). \tag{10.11}$$

The resulting optimization problem (minimize (10.11) subject to (10.6), (10.8), (10.10), and the boundary conditions) is in MILP form. Because binary variables do not appear in the problem formulation, it is a linear program and is easily solved to obtain the optimal sequence of control inputs. The solution for an example instance is shown in Figure 10.1.

10.3 OBSTACLE AVOIDANCE

In vehicle control, it is necessary to avoid other vehicles, stationary and moving obstacles, and restricted regions. In this section, we show how to formulate and solve these avoidance problems using MILP. We start by showing a MILP method to guarantee circular obstacle avoidance at N_o discrete times. The version of this method developed in (Richards *et al.* 2002b), and a similar version developed independently in (Earl and D'Andrea 2002a, b), uniformly distributes obstacle avoidance times. Here we present a version of the method that allows nonuniform distributions of obstacle avoidance times.

The subscript o is used to denote variables associated with the time discretization for obstacle avoidance. For step k, taken to be an element of the set $\{1, \ldots, N_o\}$, let $t_o[k]$ be the time at which obstacle avoidance is enforced. Let R_{obst} denote the radius of the obstacle. Let $(x_{obst}[k], y_{obst}[k])$ denote the coordinates of its center at time $t_o[k]$. We approximate the obstacle with a polygon, denoted $\mathcal{O}[k]$, defined by a set of M_o inequalities. The polygon is given by

$$\begin{aligned} \mathcal{O}[k] := \{ &(\overline{x}, \overline{y}) : \\ &(\overline{x} - x_{obst}[k]) \sin \frac{2\pi m}{M_o} \\ &+ (\overline{y} - y_{obst}[k]) \cos \frac{2\pi m}{M_o} \leq R_{obst} \\ &\forall m \in \{1, \ldots, M_o\} \}. \end{aligned} \tag{10.12}$$

To guarantee obstacle avoidance at time $t_o[k]$, the coordinates of the vehicle must be outside the region $\mathcal{O}[k]$. This avoidance condition can be written as $(x_o[k], y_o[k]) \notin \mathcal{O}[k]$, where $(x_o[k], y_o[k])$ are the coordinates of the vehicle at time $t_o[k]$. Here $x_o[k] = x(t_o[k])$ and $y_o[k] = y(t_o[k])$ are expressed in terms of the control inputs using Equation (10.7).

Because at least one constraint defining the region $\mathcal{O}[k]$ must be violated in order to avoid the obstacle, the avoidance condition is equivalent to the following condition:

$$\begin{aligned} &\text{there exists an } m \text{ such that} \\ &(x_o[k] - x_{obst}[k]) \sin \frac{2\pi m}{M_o} \\ &+ (y_o[k] - y_{obst}[k]) \cos \frac{2\pi m}{M_o} > R_{obst}. \end{aligned} \tag{10.13}$$

To express this avoidance constraint in a MILP problem formulation, it must be converted to an equivalent set of linear inequality constraints. We do so by introducing auxiliary binary variable $b_m[k] \in \{0, 1\}$ and the following M_o inequality constraints,

$$\begin{aligned} &(x_o[k] - x_{obst}[k]) \sin \frac{2\pi m}{M_o} \\ &+ (y_o[k] - y_{obst}[k]) \cos \frac{2\pi m}{M_o} > R_{obst} - H b_m[k] \\ &\forall m \in \{1, \ldots, M_o\}, \end{aligned} \tag{10.14}$$

where H is a large positive number taken to be larger than the maximum dimension of the vehicle's operating environment plus the radius of the obstacle. If $b_m[k] = 1$, the right-hand side of the inequality is a large, negative number that is always less than the left-hand side. In this case, the inequality is inactive because it is trivially satisfied. If $b_m[k] = 0$, the inequality is said to be active because it reduces to an inequality from the existence condition above. For obstacle avoidance, at least one of the constraints in

Equation (10.14) must be active. To enforce this, we introduce the following inequality constraint into the problem formulation,

$$\sum_{m=1}^{M_o} b_m[k] \leq M_o - 1. \tag{10.15}$$

Therefore, to enforce obstacle avoidance at time $t_o[k]$, the set of binary variables $\{b_m[k]\}_{m=1}^{M_o}$ and the constraints given by Equations (10.14) and (10.15) are added to the MILP problem formulation.

Consider the example problem from Section 10.2 with obstacles. In this problem, we want to transfer the vehicle from start state $\mathbf{x}_s$ to finish state $\mathbf{x}_f$ in time t_f using minimal control effort while avoiding obstacles. To enforce obstacle avoidance at each time in the set $\{t_o[k]\}_{k=1}^{N_o}$, we augment the MILP formulation in Section 10.2 with the set of binary variables $\{b_m[k]\}_{m=1}^{M_o}$, constraints (10.14), and constraint (10.15) for all k in the set $\{1, \ldots, N_o\}$.

Distributing the avoidance times uniformly (uniform gridding), as in (Earl and D'Andrea 2002a; Richards *et al.* 2002b), results in a trajectory that avoids obstacles at each discrete time in the set, but the trajectory can collide with obstacles between avoidance times. This is shown for an example instance in Figure 10.2(a). In this example, the trajectory intersects the obstacle between the sixth and seventh avoidance time steps. A simple method to eliminate this behavior is to represent the obstacle with a polygon that is larger than the obstacle. Then distribute obstacle avoidance times uniformly such that the sampling time is small enough to guarantee avoidance. In general, this is not a desirable approach because it results in large MILPs that require significant computational effort to solve.

A better approach is to select the avoidance times intelligently. In (Earl and D'Andrea 2005), we have developed an iterative MILP algorithm that does this. We summarize this

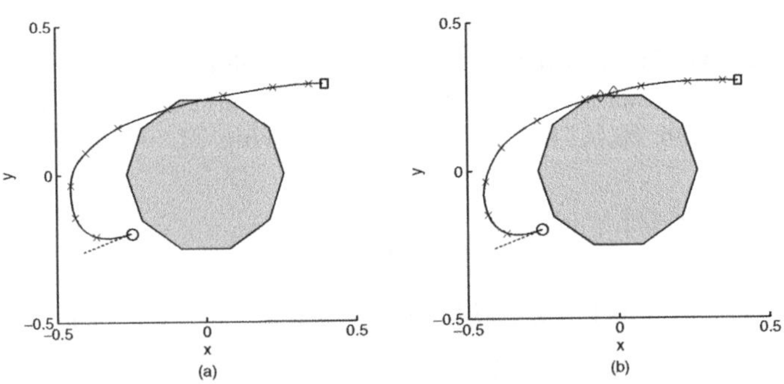

Figure 10.2 Plots of the minimum control effort obstacle avoidance example. The shaded region is the obstacle to be avoided and the ×'s along the trajectory denote the avoidance points $(x_o[k], y_o[k])$. Figure (a) is the original solution. Figure (b) is the solution after two steps of the iterative obstacle avoidance algorithm. The ◇'s are the avoidance points added to the MILP formulation by the iterative algorithm. The values for the parameters are $N_u = 10$, $M_u = 20$, $T_u[k] = 0.3$ for all k, $M_o = 10$, $N_o = 10$, $t_o[k] = kT$, $(x_s, y_s, \dot{x}_s, \dot{y}_s) = (-0.25, -0.2, -0.5, -0.2)$, and $(x_f, y_f, \dot{x}_f, \dot{y}_f) = (0.4, 0.3, 0.0, 0.0)$.

algorithm here. First, pick an initial set of times $\{t_o[k]\}_{k=1}^{N_o}$ at which obstacle avoidance will be enforced. Then, formulate and solve the MILP as described above representing the obstacles with polygons slightly larger than the obstacles. Next, check the resulting trajectory for collisions with any of the obstacles (not the polygons which represent them in the MILP). If there are no collisions, terminate the algorithm. If there is a collision, compute the time intervals for which collisions occur denoting the time interval for collision i by $(t_a^{(i)}, t_b^{(i)})$. For each interval i, pick a time within the interval, such as $(t_a^{(i)} + t_b^{(i)})/2$. At each of these times add obstacle avoidance constraints to the MILP formulation. Then, solve the augmented MILP repeating the procedure above (first checking if the resulting trajectory intersects any obstacles, etc.) until a collision free trajectory is found. Figure 10.2(b) shows the effectiveness of this algorithm after two iterations.

10.4 ROBOFLAG PROBLEMS

To motivate our multi-vehicle methods, we apply them to simplified versions of the RoboFlag game (Campbell *et al.* n.d.; D'Andrea and Babish 2003; D'Andrea and Murray 2003), which we call RoboFlag Drills because they serve as practice for the real game. The drills involve two teams of robots, the defenders and the attackers, on a playing field with a circular region of radius R_{dz} at its center called the Defense Zone, as shown in Figure 10.3. The attackers' objective is to fill the Defense Zone with as many attackers as possible. The defenders' objective is to deny as many attackers from entering the Defense Zone as possible without entering the zone themselves. We consider Defensive Drill problems in which each attacker has a fixed intelligence governed by a state machine.

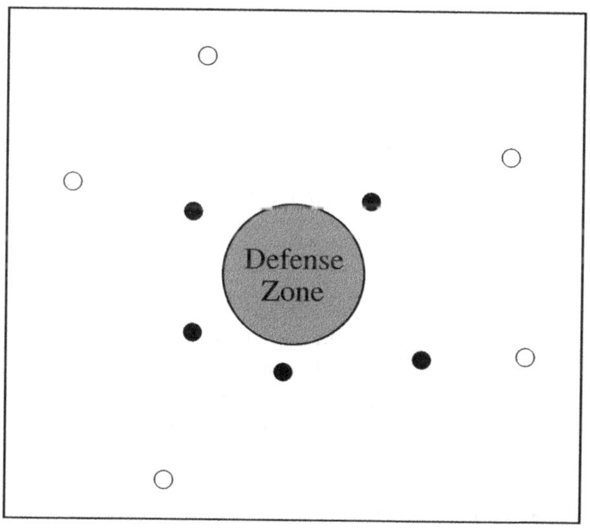

Figure 10.3 The RoboFlag Drill used to motivate the methods presented in this chapter. The drill takes place on a playing field with a Defense Zone at its center. The objective is to design a cooperative control strategy for the team of defending vehicles (black) that minimizes the number of attacking vehicles (white) that enter the Defense Zone.

The goal is to design a team control strategy for the defenders that maximizes the number of attackers denied from the Defense Zone. In this section, we use MILP methods to generate such strategies. We consider two versions of the Defensive Drill, each with a different set of laws governing attacker intelligence.

To start, we consider one-on-one Defensive Drill problems. This is the simplest case and involves one defender and one attacker. Although this case is not particularly interesting, we start with it for notational clarity. Next, we generalize to the case involving N_D defenders and N_A attackers, which is a straightforward extension.

10.4.1 Defensive Drill 1: one-on-one case

The defender is governed by the discrete time dynamical system from Section 10.2

$$\begin{aligned}
&\mathbf{x}_u[k+1] = \mathbf{A}[k]\mathbf{x}_u[k] + \mathbf{B}[k]\mathbf{u}[k] \\
&\mathbf{x}_u[0] = \mathbf{x}_s \\
&u_x[k]\sin\frac{2\pi m}{M_u} + u_y[k]\cos\frac{2\pi m}{M_u} \le \cos\frac{\pi}{M_u} \\
&\forall m \in \{1, \ldots, M_u\} \\
&\forall k \in \{1, \ldots, N_u\}.
\end{aligned} \tag{10.16}$$

The attacker has two discrete modes: attack and inactive. When in attack mode, it moves toward the Defense Zone at constant velocity along a straight line path. The attacker, initially in attack mode at the beginning of play, transitions to inactive mode if the defender intercepts it or if it enters the Defense Zone. Once inactive, the attacker does not move and remains inactive for the remainder of play. These dynamics are captured by the following discrete time equations and state machine

$$\begin{aligned}
p[k+1] &= p[k] + v_p T_a[k] a[k] \\
q[k+1] &= q[k] + v_q T_a[k] a[k]
\end{aligned} \tag{10.17}$$

$$a[k+1] = \begin{cases} 1 & \text{if } (a[k]=1) \\ & \text{and (not in Defense Zone)} \\ & \text{and (not intercepted)} \\ 0 & \text{if } (a[k]=0) \\ & \text{or (in Defense Zone)} \\ & \text{or (intercepted)} \end{cases} \tag{10.18}$$

$$\forall k \in \{1, \ldots, N_a\}$$

with initial conditions

$$p[0] = p_s,\ q[0] = q_s, \text{ and } a[0] = 1, \tag{10.19}$$

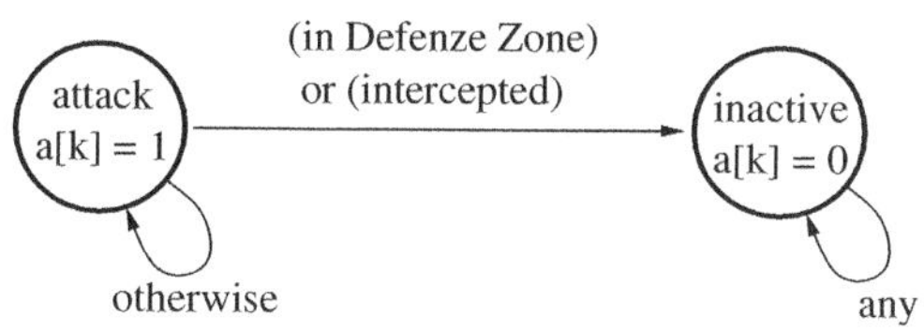

Figure 10.4 The two state (attack and inactive) attacker state machine. The attacker starts in the attack state. It transitions to the inactive state, and remains in this state, if it enters the Defense Zone or if it is intercepted by the defender.

where N_a is the number of samples, $k \in \{1, \ldots, N_a\}$, $T_a[k] > 0$ is the time between samples k and $k+1$, $(p[k], q[k])$ is the attacker's position at time $t_a[k] = \sum_{i=0}^{k-1} T_a[i]$, (v_p, v_q) is its constant velocity vector, and $a[k] \in \{0, 1\}$ is a discrete state indicating the attacker's mode. The attacker is in attack mode when $a[k] = 1$ and inactive mode when $a[k] = 0$. The attacker state machine is shown in Figure 10.4. Here we use the subscript a to denote the time discretization for the attacker's dynamics.

10.4.1.1 Defense zone

Because the defender wants to keep the attacker from entering the Defense Zone, a binary variable indicating whether or not the attacker is inside the Defense Zone is introduced. This variable is used to define the attacker state machine precisely. We denote the binary variable with $\gamma[k] \in \{0, 1\}$. When the attacker is in the Defense Zone at step k, $\gamma[k] = 1$. When the attacker is outside the Defense Zone at step k, $\gamma[k] = 0$.

Similar to the approach used to define obstacles, the Defense Zone is approximated using a polygon $\mathcal{G}$ defined by a set of M_{dz} inequalities

$$\begin{aligned} \mathcal{G} := \{\, & (\bar{x}, \bar{y}) : \\ & \bar{x} \sin \frac{2\pi m}{M_{dz}} + \bar{y} \cos \frac{2\pi m}{M_{dz}} \le R_{dz} \\ & \forall m \in \{1, \ldots, M_{dz}\} \,\}. \end{aligned} \tag{10.20}$$

The association between $\gamma[k]$ and $\mathcal{G}$ is

$$(\gamma[k] = 1) \Leftrightarrow (p[k], q[k]) \in \mathcal{G}. \tag{10.21}$$

If the defender keeps the binary variable $\gamma[k]$ equal to 0 for all $k \in \{1, \ldots, N_a\}$, it has successfully denied the attacker from the region $\mathcal{G}$ and thus from the Defense Zone. However, in order to use the binary variable $\gamma[k]$ in the problem formulation, we must enforce the logical constraint given by Equation (10.21). To enforce this constraint in MILP, we convert it into an equivalent set of inequality constraints.

We introduce the binary variable $g_m[k] \in \{0, 1\}$ to indicate whether or not the mth constraint of $\mathcal{G}$ is satisfied by the attacker with position $(p[k], q[k])$. This association is made by introducing the logical constraint

$$(g_m[k] = 1) \Leftrightarrow \left(p[k] \sin \frac{2\pi m}{M_{dz}} + q[k] \cos \frac{2\pi m}{M_{dz}} \le R_{dz} \right). \tag{10.22}$$

As shown in Appendix 10.7, it is equivalent to the following set of inequalities

$$\begin{aligned} & p[k] \sin \frac{2\pi m}{M_{dz}} + q[k] \cos \frac{2\pi m}{M_{dz}} \\ & \leq R_{dz} + H(1 - g_m[k]) \\ & p[k] \sin \frac{2\pi m}{M_{dz}} + q[k] \cos \frac{2\pi m}{M_{dz}} \\ & \geq R_{dz} + \varepsilon - (H + \varepsilon) g_m[k], \end{aligned} \tag{10.23}$$

where ε is a small positive number and H is a large positive number such that the left-hand sides of the inequalities are bounded from above by H and from below by $-H$.

Using binary variable $g_m[k]$, we can write Equation (10.21) as

$$\begin{aligned} & (\gamma[k] = 1) \Leftrightarrow \\ & (g_m[k] = 1 \quad \forall m \in \{1, \ldots, M_{dz}\}), \end{aligned} \tag{10.24}$$

which is equivalent to the inequality constraints

$$\begin{aligned} & g_m[k] - \gamma[k] \geq 0 \quad \forall m \in \{1, \ldots, M_{dz}\} \\ & \sum_{i=1}^{M_{dz}} (1 - g_i[k]) + \gamma[k] \geq 1, \end{aligned} \tag{10.25}$$

as shown in Appendix 10.7.

The logical constraint given by Equation (10.21) is equivalent to the inequality constraints given by Equations (10.23) and (10.25). Therefore, we can include the indicator variable $\gamma[k]$ in the MILP formulation by also including the binary variables $\{g_m[k]\}_{m=1}^{M_a}$ and constraints (10.23) and (10.25).

10.4.1.2 Intercept region

To define what it means for a defender to intercept an attacker, we introduce an intercept region $\mathcal{I}[k]$ rigidly attached to the defending robot. The intercept region is a polygon defined by a set of M_I inequalities. If an attacker is in this polygon, it is considered intercepted. For the defender with coordinates $(x_a[k], y_a[k])$, the intercept region is given by

$$\begin{aligned} \mathcal{I}[k] := \{ & (\overline{x}, \overline{y}) : \\ & (\overline{x} - x_a[k]) \sin \frac{2\pi m}{M_I} + (\overline{y} - y_a[k]) \cos \frac{2\pi m}{M_I} \leq R_I \\ & \forall m \in \{1, \ldots, M_I\} \}, \end{aligned} \tag{10.26}$$

where $x_a[k] = x(t_a[k])$ and $y_a[k] = y(t_a[k])$ are calculated using Equation (10.7), and R_I is the inscribed radius of the polygon.

The binary variable $\delta[k] \in \{0, 1\}$ is introduced to indicate whether or not the attacker is inside the defender's intercept region. The association is made with the following logical constraint

$$\delta[k] = 1 \Leftrightarrow (p[k], q[k]) \in \mathcal{I}[k]. \tag{10.27}$$

Using techniques similar to those used for $\gamma[k]$, we convert this constraint into an equivalent set of inequality constraints as follows: For each of the inequalities defining the intercept region, we associate a binary variable $d_m[k] \in \{0, 1\}$ by introducing the logical constraint

$$\begin{aligned} &(d_m[k] = 1) \Leftrightarrow \\ &\left((p[k] - x_a[k]) \sin \frac{2\pi m}{M_I}\right. \\ &\left.+ \ (q[k] - y_a[k]) \cos \frac{2\pi m}{M_I} \leq R_I\right), \end{aligned} \tag{10.28}$$

where $(p[k], q[k])$ are the coordinates of the attacking robot. If $d_m[k] = 1$, the coordinates of the attacking robot satisfy the mth inequality defining the intercept region. If $d_m[k] = 0$, the mth inequality is not satisfied. Similar to Equation (10.22), we can express this logic constraint as the set of inequalities

$$\begin{aligned} &(p[k] - x_a[k]) \sin \frac{2\pi m}{M_I} + (q[k] - y_a[k]) \cos \frac{2\pi m}{M_I} \\ &\leq R_I + H(1 - d_m[k]) \\ &(p[k] - x_a[k]) \sin \frac{2\pi m}{M_I} + (q[k] - y_a[k]) \cos \frac{2\pi m}{M_I} \\ &\geq R_I + \varepsilon - (H + \varepsilon) d_m[k]. \end{aligned} \tag{10.29}$$

Using the binary variables $d_m[k]$ we can write Equation (10.27) as

$$\delta[k] = 1 \Leftrightarrow (d_m[k] = 1 \ \forall m \in \{1, \ldots, M_I\}). \tag{10.30}$$

Considering both directions of this equivalence, as done for $\gamma[k]$ above, we find that the statement is equivalent to the following set of inequality constraints

$$\begin{aligned} &d_m[k] - \delta[k] \geq 0 \quad \forall m \in \{1, \ldots, M_I\} \\ &\sum_{i=1}^{M_I} (1 - d_i[k]) + \delta[k] \geq 1. \end{aligned} \tag{10.31}$$

We can use $\delta[k]$ in our problem formulation if we include the binary variables $\{d_m[k]\}_{m=1}^{M_a}$ and the inequalities constraints given by Equations (10.29) and (10.31).

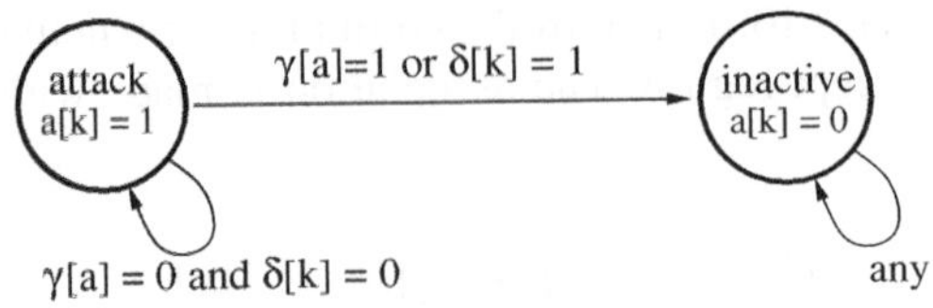

Figure 10.5 The two state attacker state machine, as in Figure 10.4, using binary variables $\delta[k]$ and $\gamma[k]$.

10.4.1.3 State machine and objective function

With the indicator variables $\gamma[k]$ and $\delta[k]$ defined, we revisit the attacker state machine and define it more precisely with the following state equation

$$a[k+1] = \begin{cases} 1 & \text{if } a[k] = 1 \text{ and } \gamma[k] = 0 \\ & \text{and } \delta[k] = 0 \\ 0 & \text{if } a[k] = 0 \text{ or } \gamma[k] = 1 \\ & \text{or } \delta[k] = 1, \end{cases} \tag{10.32}$$

which is shown in Figure 10.5. The state equations can be written as the logical expression

$$\begin{aligned} &(a[k+1] = 1) \Leftrightarrow \\ &(a[k] = 1 \text{ and } \gamma[k] = 0 \text{ and } \delta[k] = 0), \end{aligned} \tag{10.33}$$

which is equivalent to the set of inequality constraints

$$\begin{aligned} &a[k+1] + \delta[k] \leq 1 \\ &a[k+1] - a[k] \leq 0 \\ &a[k+1] + \gamma[k] \leq 1 \\ &a[k] - \delta[k] - \gamma[k] - a[k+1] \leq 0, \end{aligned} \tag{10.34}$$

as shown in Appendix 10.7.

We have defined the dynamics of the defenders and the attackers, and we have converted these dynamics to MILP form. The final step in formulating Defensive Drill 1 is to introduce an objective function that captures the defender's desire to deny the attacker from entering the Defense Zone. This objective is achieved by minimizing the binary variable $\gamma[k]$ over the duration of the drill, which is captured by minimizing the function

$$J = \sum_{k=1}^{N_a} \gamma[k]. \tag{10.35}$$

We set the duration of the drill such that

$$t_a[N_a] \geq \sqrt{\frac{p_s^2 + q_s^2}{v_p^2 + v_q^2}}. \tag{10.36}$$

This allows the attacker enough time to reach the Defense Zone if it is not intercepted.

In addition to intercepting the attacker, we would like to conserve fuel as a secondary objective. One way to achieve this is to add a small penalty on the control effort to the objective function as follows:

$$J = \sum_{k=1}^{N_a} \gamma[k] + \varepsilon \sum_{k=0}^{N_u-1} \left(|u_x[k]| + |u_y[k]|\right), \tag{10.37}$$

where ε is taken to be a small positive number because we want the minimization of the binary variable $\gamma[k]$ to dominate. We use the procedure outlined in Section 10.2 to write Equation (10.37) in MILP form.

10.4.1.4 Summary and example

We have formulated Defensive Drill 1 as the following optimization problem:

minimize (10.37)

such that

(10.16) (defender dynamics)

(10.17), (10.19), (10.34) for all k_a (attacker dynamics)

(10.23), (10.25) for all m_g and k_a (γ indicator variable)

(10.29), (10.31) for all m_I and k_a (δ indicator variable)

(10.14), (10.15) for all m_o and k_o (avoidance constraints)

where $k_a \in \{1, \ldots N_a\}$, $k_o \in \{1, \ldots, N_o\}$, $m_g \in \{1, \ldots, M_{dz}\}$, $m_I \in \{1, \ldots, M_I\}$, and $m_o \in \{1, \ldots, M_o\}$.

The solution to an instance of this problem is shown in Figure 10.6. For this instance, the defender denies the attacker from the Defense Zone (shaded polygon) by intercepting it. Each asterisk along the attacker's trajectory denote the position of the attacker at sample time $t_a[k]$. The white polygons along the defender's trajectory denote the intercept region $\mathcal{I}[k]$ at time $t_a[k]$.

10.4.2 Defensive Drill 2: one-on-one case

The dynamics of the defender are the same as the defender dynamics in Section 10.4.1. The dynamics of the attacker are the same as the attacker dynamics in Defensive Drill 1, but with an additional state called retreat. If a defender is too close to the attacker, the attacker enters the retreat state and reverses direction. While retreating, if the defender is far enough away, the attacker returns to attack mode.

These dynamics are captured by the following discrete time equations and state machine:

$$p[k+1] = p[k] + v_p T_a[k] a_1[k] - \; v_p T_a[k] a_2[k]$$

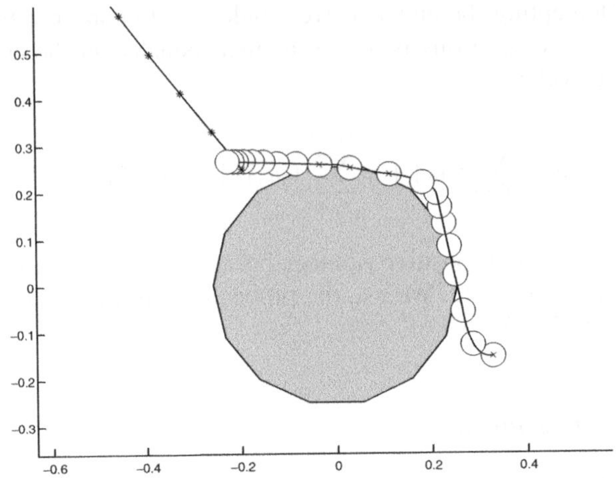

Figure 10.6 The defender denies the attacker from entering the Defense Zone. The $*$'s along the attacker's trajectory denote the attacker's position for each time $t_a[k]$. The polygons along the defender's trajectory denote the intercept region $\mathcal{I}[k]$ for each time $t_a[k]$. The $\times$'s denote obstacle avoidance points.

$$q[k+1] = q[k] + v_q T a[k] a_1[k] - v_q T_a[k] a_2[k] \tag{10.38}$$

$$a_1[k+1] = \begin{cases} 1 \text{ if } [(a_1[k] = 1 \text{ and } a_2[k] = 0) \\ \quad \text{or } (a_1[k] = 0 \text{ and } a_2[k] = 1)] \\ \quad \text{and (not scored)} \\ \quad \text{and (not intercepted)} \\ \quad \text{and (not in warning region)} \\ 0 \text{ otherwise} \end{cases} \tag{10.39}$$

$$a_2[k+1] = \begin{cases} 1 \text{ if } [(a_1[k] = 0 \text{ and } a_2[k] = 1) \text{ or} \\ \quad (a_1[k] = 1 \text{ and } a_2[k] = 0)] \\ \quad \text{and (not scored)} \\ \quad \text{and (not intercepted)} \\ \quad \text{and (in warning region)} \\ 0 \text{ otherwise} \end{cases} \tag{10.40}$$

for all $k \in \{1, \ldots, N_a\}$ and subject to the constraint

$$a_1[k] + a_2[k] \leq 1 \tag{10.41}$$

and the initial conditions

$$p[0] = p_s,\ q[0] = q_s,\ a_1[0] = 1,\ a_2[0] = 0. \tag{10.42}$$

The state machine is shown in Figure 10.7.

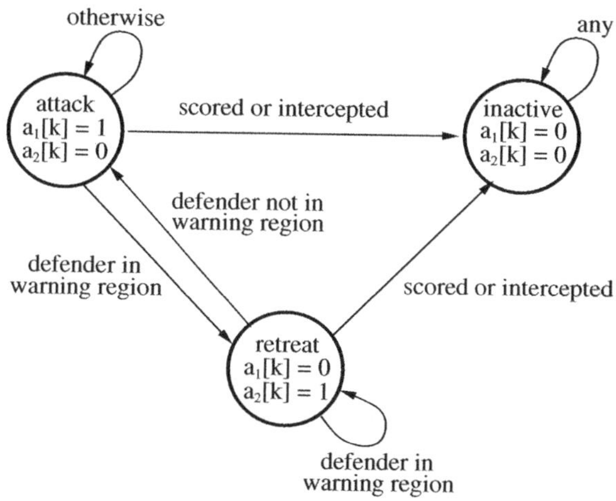

Figure 10.7 The three state (attack, inactive, and retreat) attacker state machine. The attacker starts in the attack state.

The attacker needs two binary state variables because it has three discrete modes of operation: attack, retreat, and inactive. These modes are represented by $(a_1[k], a_2[k]) = (1, 0)$, $(a_1[k], a_2[k]) = (0, 1)$, and $(a_1[k], a_2[k]) = (0, 0)$, respectively. Because the state $(a_1[k], a_2[k]) = (1, 1)$ is not needed, we impose the inequality constraint given by Equation (10.41).

To determine if the defender is too close to the attacker, a warning region is introduced. The warning region $\mathcal{W}[k]$ is a polygon defined by a set of inequalities similar to the intercept region $\mathcal{I}[k]$. We introduce binary auxiliary variables $w_m[k]$ and $\omega[k]$, and we introduce inequality constraints similar to Equations (10.29) and (10.31). The result is the following association for indicator variable $\omega[k]$: If $\omega[k] = 1$, the defender is in the attacker's warning region at step k, otherwise, $\omega[k] = 0$. The attacker state machine can be written as the set of inequality constraints

$$
\begin{aligned}
& a_1[k+1] - a_1[k] + a_2[k] + \gamma[k] + \delta[k] + \omega[k] \geq 0 \\
& a_1[k+1] + a_1[k] - a_2[k] + \gamma[k] + \delta[k] + \omega[k] \geq 0 \\
& a_2[k+1] + a_1[k] - a_2[k] + \gamma[k] + \delta[k] - \omega[k] \geq -1 \\
& a_2[k+1] - a_1[k] + a_2[k] + \gamma[k] + \delta[k] - \omega[k] \geq -1 \\
& a_1[k+1] + a_1[k] + a_2[k] \leq 2 \\
& a_1[k+1] - a_1[k] - a_2[k] \leq 0 \\
& a_1[k+1] + \gamma[k] \leq 1 \\
& a_1[k+1] + \delta[k] \leq 1 \\
& a_1[k+1] + \omega[k] \leq 1 \\
& a_2[k+1] - a_1[k] - a_2[k] \leq 0 \\
& a_2[k+1] + a_1[k] + a_2[k] \leq 2
\end{aligned}
$$

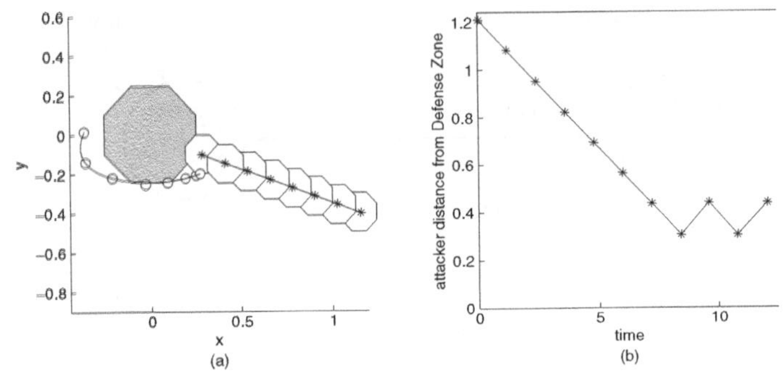

Figure 10.8 The solution to an instance of Defensive Drill 2. Figure (a) shows the playing field. Each polygon along the attacker's trajectory is a warning region. Figure (b) shows the attacker's distance from the center of the Defense Zone versus time. The parameters are $M_o = 8$, $M_u = 4$, $M_I = 8$, $M_w = 8$, $M_{dz} = 8$, $N_u = 4$, $N_o = 4$, $N_a = 10$, $\mathbf{x}_0 = (-.36, .02, -.12, -.22)$, and $(p_0, q_0, v_a) = (1.15, -.4, -.11)$.

$$
\begin{aligned}
&a_2[k+1] + \gamma[k] \leq 1 \\
&a_2[k+1] + \delta[k] \leq 1 \\
&a_2[k+1] - \omega[k] \leq 0.
\end{aligned}
\tag{10.43}
$$

Similar to Defense Drill 1, Defense Drill 2 can be posed as a MILP. The solution of an example instance is shown in Figure 10.8.

10.4.3 N_D-on-N_A case

To generalize the problem formulation to N_D defenders and N_A attackers, we need to add more variables. Defender i, where $i \in \{1, \ldots, N_D\}$, has state $\mathbf{x}_{ui}[k]$, control input $\mathbf{u}_i[k]$, and slack variables $z_{xi}[k]$ and $z_{yi}[k]$. Attacker j, where $j \in \{1, \ldots, N_A\}$, has state $(p_j[k], q_j[k], a_j[k])$ and constant velocity vector (v_{pj}, v_{qj}). The binary variable $\gamma_j[k]$ equals 1 if and only if attacker j is in polygon $\mathcal{G}$ at step k. The binary variable $\delta_{ij}[k]$ equals 1 if and only if attacker j is in defender i's intercept region $\mathcal{I}_i[k]$ at step k. The binary variables $g_{mj}[k]$, $d_{mij}[k]$, and $b_{mi}[k]$ follow a similar trend. For Defensive Drill 2, we also need $\omega_{ij}[k] = 1$ if and only if defender i is in attacker j's warning region $\mathcal{W}_j[k]$ at step k. And similarly for the binary variable $w_{mij}[k]$.

With the variables for the general case defined, inequality constraints are added to the formulation in a similar way as those derived for the one-on-one case. For example, the constraints identifying $\gamma_j[k] = 1$ with $(p_j[k], q_j[k]) \in \mathcal{G}$ are

$$
\begin{aligned}
&g_{mj}[k] - \gamma_j[k] \geq 0 \quad \forall m \in \{1, \ldots, M_{dz}\} \\
&\sum_{l=1}^{M_{dz}} (1 - g_{lj}[k]) + \gamma_j[k] \geq 1,
\end{aligned}
$$

for all $k \in \{1, \ldots, N_a\}$ and for all $j \in \{1, \ldots, N_A\}$.

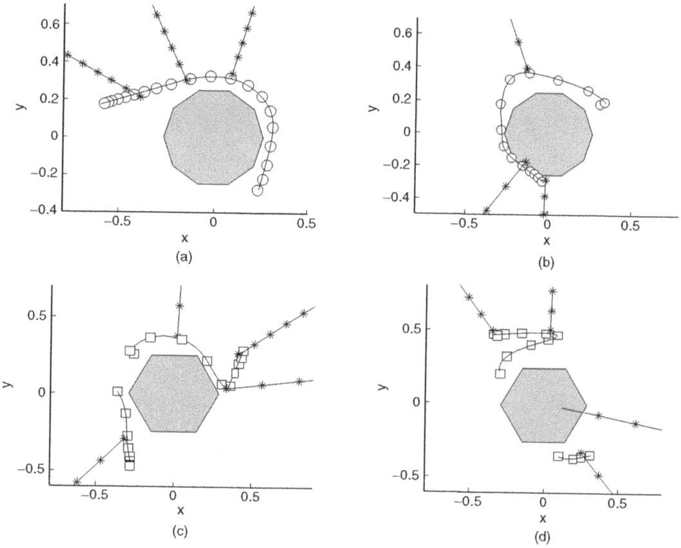

Figure 10.9 The solution to four instances of Defensive Drill 1. For Figures (a) and (b), $N_A = 3$ and $N_D = 1$. For Figures (c) and (d), $N_A = 4$ and $N_D = 2$. In Figures (a) and (c), the defenders deny all attackers from the Defense Zone. In Figures (b) and (d), an attacker enters the Defense Zone.

And finally, the objective function is given by

$$J = \sum_{j=1}^{N_A} \sum_{k=1}^{N_a} \gamma_j[k] + \varepsilon \sum_{i=1}^{N_D} \sum_{k=0}^{N_u-1} \left(|u_{xi}[k]| + |u_{yi}[k]|\right). \tag{10.44}$$

Results for example instances of Defensive Drill 1 are shown in Figure 10.9. Results for Defensive Drill 2 are shown in Figure 10.10.

10.5 AVERAGE CASE COMPLEXITY

In this section, we explore the average case complexity of Defensive Drill 1 by solving randomly generated instances. Each instance is generated by randomly picking parameters from a uniform distribution over the intervals defined below. Each MILP is solved using AMPL (Fourer *et al.* 1993) and CPLEX (ILOG 2000) on a PC with Intel PIII 550MHz processor, 1024KB cache, 3.8GB RAM, and Red Hat Linux. For all instances solved, processor speed was the limiting factor, not memory.

To generate the initial condition (p_s, q_s) and the constant velocity vector (v_p, v_q) for each attacker we pick r_a, θ_a, and v_a randomly from a uniform distribution over the intervals $[r_a^{\min}, r_a^{\max}]$, $[0, 2\pi)$, and $[v_a^{\min}, v_a^{\max}]$, respectively. The parameters are then given by

$$p_s = r_a \cos\theta_a, \quad q_s = r_a \sin\theta_a$$

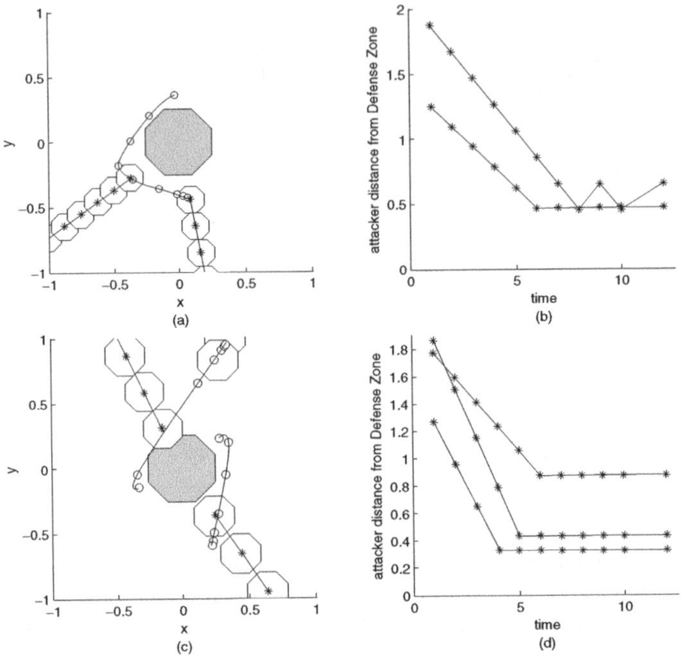

Figure 10.10 The solution to two instances of Defensive Drill 2. Figures (a) and (c) show the playing field. Figures (b) and (d) show each attacker's distance from the center of the Defense Zone versus time.

$$v_p = v_a p_s / \sqrt{p_s^2 + q_s^2}, \ \ v_q = v_a q_s / \sqrt{p_s^2 + q_s^2}. \tag{10.45}$$

To generate the initial condition $(x_s, y_s, \dot{x}_s, \dot{y}_s)$ for each defender, we pick r_d, θ_d, v_d, and θ_v randomly from a uniform distribution over the intervals $[r_d^{\min}, r_d^{\max}]$, $[0, 2\pi)$, $[v_d^{\min}, v_d^{\max}]$, and $[0, 2\pi)$, respectively. The defender's initial condition is then given by

$$\begin{aligned} x_s &= r_d \cos\theta_d, \ \ y_s = r_d \sin\theta_d \\ \dot{x}_s &= v_d \cos\theta_v, \ \ \dot{y}_s = v_d \sin\theta_v. \end{aligned} \tag{10.46}$$

We take the playing field to be circular with radius $R_f = 15$, and we set the radius of the Defense Zone to be $R_{dz} = 2$. The intervals used to generate the instances are $r_d \in [\sqrt{2}R_{dz}, 2\sqrt{2}R_{dz}]$, $v_d \in [0.5, 1.0]$, and $r_a \in [R_f/2, R_f]$. We take the attacker velocity to be $v_a = 1.0$.

In Figure 10.11, we plot the fraction of instances solved versus computation time for the case where the cost function includes the number of attackers that enter the Defense Zone plus a penalty on the control effort. This cost function is given by Equation (10.44) with $\varepsilon = 0.1$. In Figure 10.12, we plot the fraction of instances solved versus computation time for the case where the cost function includes only the number of attackers that enter the Defense Zone. This cost function is given by Equation (10.44) with $\varepsilon = 0$.

As these figures show, the case where the cost function only includes the number of attackers that enter the Defense Zone is less computationally intensive than the case

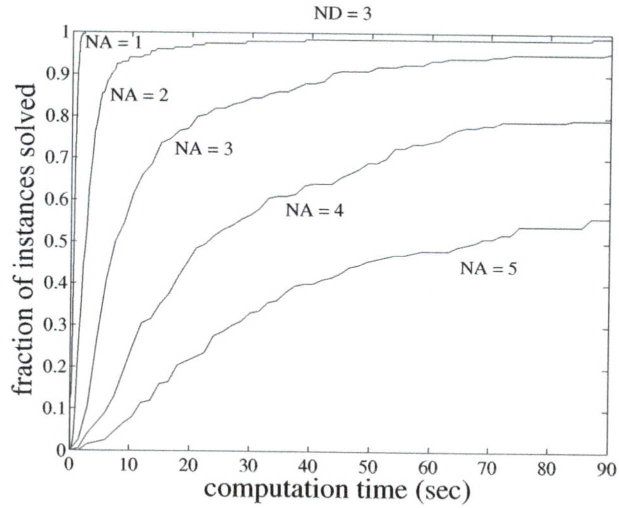

Figure 10.11 Fraction of instances solved versus computation time. The cost function is the number of attackers that enter the Defense Zone plus a penalty on control effort (Equation (10.44) with $\varepsilon > 0$). We consider instances with defender set of size three ($N_D = 3$) and attacker sets of size $N_A = 1, 2, 3, 4, 5$. To generate each curve, 200 random instances were solved. The parameters are $M_u = 4$, $M_I = 4$, $M_{dz} = 4$, $N_u = 15$, and $N_a = 15$.

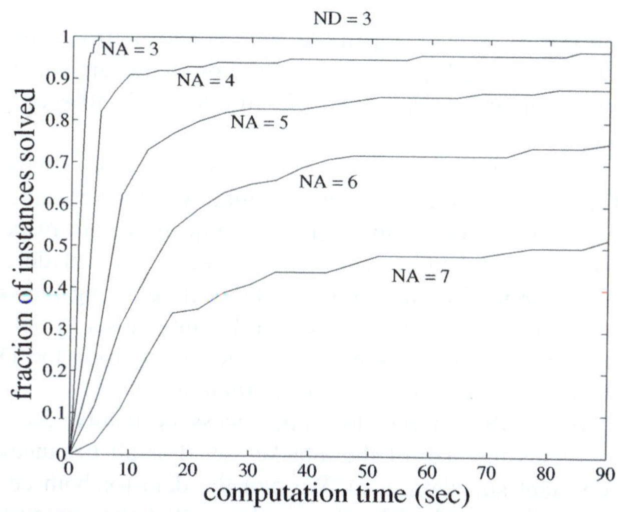

Figure 10.12 Fraction of instances solved versus computation time. The cost function is the number of attackers that enter the Defense Zone (Equation (10.44) with $\varepsilon = 0$). We consider instances with defender set of size three ($N_D = 3$) and attacker sets of size $N_A = 3, 4, 5, 6, 7$. To generate each curve, 200 random instances were solved. The parameters are $M_u = 4$, $M_I = 4$, $M_{dz} = 4$, $N_u = 15$, and $N_a = 15$.

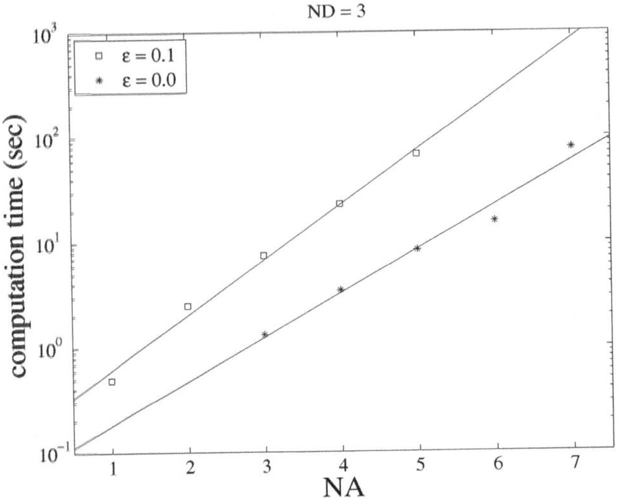

Figure 10.13 Computation time necessary to solve 50% of the randomly generated instances versus the number of attackers. For all instances, the defender set is of size three ($N_D = 3$). Each square denotes a data point for the case where the cost function is the number of attackers that enter the Defense Zone plus a penalty on each defender's control effort ($\varepsilon = 0.1$ case). Each asterisk denotes a data point for the case where the cost function is the number of attackers that enter the Defense Zone ($\varepsilon = 0$ case). The solid lines denote fitted exponential functions to these data.

where the cost function also includes a penalty on the control effort. For example, for the case where $\varepsilon = 0$, $N_D = 3$, and $N_A = 4$, 50% of the problems are solved in 3.5 seconds. However, for the case where $\varepsilon = 0.1$, $N_D = 3$, and $N_A = 4$, 50% of the problems are solved in 78 seconds.

When $\varepsilon = 0$ in the cost function, the solver stops once a set of trajectories for the defenders is found that results in the minimum number of attackers entering the Defense Zone. This trajectory is often not the most efficient trajectory with respect to the control effort of the defenders. For the case where $\varepsilon = 0.1$, once a set of defender trajectories is found that results in the minimum number of attackers entering the Defense Zone the solver continues to search. The solver searches until it finds a set of defender trajectories that not only results in the minimum number of attackers entering the Defense Zone but also uses the defender control effort in the most efficient way.

In Figure 10.13, we plot the computation time necessary to solve 50% of the randomly generated instances versus the size of the attacker set. For all instances considered, the defender set is of constant size ($N_D = 3$). We plot the data for both cost functions considered above ($\varepsilon = 0$ and $\varepsilon = 0.1$). This figure shows that the computation time grows exponentially with the number of vehicles in the attacker set.

10.6 DISCUSSION

In this chapter, we showed how to use MILPs to model and generate strategies for multi-vehicle control problems. The MILP formulation is very expressive and allows all aspects of the problem to be taken into account. This includes the dynamic equations governing the vehicles, the dynamic equations governing the environment, the state machine governing adversary intelligence, logical constraints, and inequality constraints. The solution to the resulting MILP is the optimal team strategy for the problem. As shown by our average case complexity study, the MILP method becomes computationally burdensome for large problems and thus, for these cases, is not suitable for real-time applications.

There are several steps that can be taken to make the MILP approach applicable for real-time multi-vehicle control applications. One approach is to solve the MILPs faster. This can be done by using a more powerful computer or by distributing the computation over a set of processors. For the multi-vehicle control problems, it may be advantageous to distribute the computation over the set of vehicles. Distributed methods for solving MILPs have been considered in (Ralphs 2003).

Another approach is to move all, or some components of, the computational burden offline. This can be done by formulating the problem as a multi-parametric MILP. A multi-parametric MILP is a MILP formulated using a set of parameters each allowed to take on values over an interval. This problem is much more computationally intensive than the MILP problems considered in this chapter. However, because the computation is performed offline, this is not an issue unless the computation takes an unreasonable amount of time. With the solution to the multi-parametric MILP, a table can be formed and used to look up optimal team strategies for the multi-vehicle system in real time. Multi-parametric MILP control problems can be solved using the Multi-Parametric Toolbox (Kvasnica *et al*. 2003).

Finally, we discuss efficient MILP formulations as a way of decreasing computational requirements. The computational effort required to solve a MILP depends most on the number of binary variables used in its MILP problem formulation. Thus, reducing the number of binary variables is advantageous. In this chapter, our motivating problem required three different time discretizations. The discretizations included one for the control input to the vehicles, one for obstacle avoidance, and one for attacker intercept. With each discretization step, a set of binary variables must be added to the MILP formulation. In most of the instances solved in this chapter, we discretized time uniformly. However, we posed the MILP in a general way such that nonuniform time discretizations could be used. This allows for intelligent time distribution selection, which can significantly reduce the required computational effort to solve a problem. In (Earl and D'Andrea 2005), we developed several iterative MILP techniques for intelligent discretization step selection.

10.7 APPENDIX: CONVERTING LOGIC INTO INEQUALITIES

In this appendix, we illustrate how to convert a logic expression into an equivalent set of inequalities using the procedure from (Tyler and Morari 1999). First the logic expression is converted into a conjunction of disjunctions, called conjunctive normal form (CNF), by applying tautologies. For example, let the variables $p_i \in \{0, 1\}$ be binary variables. The expression

$$\begin{aligned} &((p_1 = 1) \text{ or } (p_2 = 1)) \\ &\text{and } ((p_3 = 0) \text{ or } (p_2 = 1)) \end{aligned}$$

$$\text{and } ((p_4 = 1) \text{ or } (p_5 = 1) \text{ or } (p_6 = 0)) \tag{10.47}$$

is in CNF.

Once we have converted the logic expression into CNF we can easily write each disjunction as an inequality constraint that must be satisfied in order for the disjunction to be true. For example, the second disjunction of Equation (10.47), $((p_3 = 0) \text{ or } (p_2 = 1))$, is true if and only if $(1 - p_3) + p_2 \geq 1$. Therefore, Equation 10.47 is true if and only if the following inequalities hold

$$\begin{aligned} &p_1 + p_2 \geq 1 \\ &(1 - p_3) + p_2 \geq 1 \\ &p_4 + p_5 + (1 - p_6) \geq 1. \end{aligned} \tag{10.48}$$

10.7.1 Equation (10.24)

First consider the $(\Rightarrow)$ direction of Equation (10.24). Replacing implication with disjunction we have

$$\begin{aligned} &(\gamma[k_a] = 1) \text{ or} \\ &(g_m[k_a] = 1 \quad \forall m \in \{1, \ldots, M_{dz}\}) \end{aligned} \tag{10.49}$$

and distributing the OR we have

$$(\gamma[k_a] = 0 \text{ or } g_m[k_a] = 1) \quad \forall m \in \{1, \ldots, M_{dz}\} \tag{10.50}$$

which is equivalent to

$$\begin{aligned} &(\gamma[k_a] = 0 \text{ or } g_1[k_a] = 1) \\ &\text{and } (\gamma[k_a] = 0 \text{ or } g_2[k_a] = 1) \\ &\vdots \\ &\text{and } \left(\gamma[k_a] = 0 \text{ or } g_{M_{dz}}[k_a] = 1\right). \end{aligned} \tag{10.51}$$

This expression is in CNF, therefore it is equivalent to the following set of inequality constraints

$$(1 - \gamma[k_a]) + g_m[k_a] \geq 1 \quad \forall m \in \{1, \ldots, M_{dz}\}. \tag{10.52}$$

Now consider the $(\Leftarrow)$ direction of Equation (10.24). Replacing implication with disjunction

$$\begin{aligned} &(\gamma[k_a] = 1) \text{ or} \\ &\neg(g_m[k_a] = 1 \quad \forall m \in \{1, \ldots, M_{dz}\}) \end{aligned} \tag{10.53}$$

and distributing the negation we have

$$(\gamma[k_a] = 1) \text{ or } (\exists m \text{ such that } g_m[k_a] = 0). \tag{10.54}$$

which is equivalent to

$$\begin{aligned} &(\gamma[k_a] = 1) \\ \text{or } &(g_1[k_a] = 0) \\ \text{or } &(g_2[k_a] = 0) \\ &\vdots \\ \text{or } &\left(g_{M_{dz}}[k_a] = 0\right). \end{aligned} \tag{10.55}$$

This expression, which is a disjunction, is in CNF and therefore is equivalent to the following inequality

$$\sum_{i=1}^{M_{dz}} (1 - g_i[k_a]) + \gamma[k_a] \geq 1. \tag{10.56}$$

10.7.2 Equation (10.33)

First consider the ($\Rightarrow$) direction of Equation (10.33). Replacing implication with the equivalent disjunction we have

$$\begin{aligned} &(a[k_a + 1] = 0) \text{ or} \\ &(a[k_a] = 1 \text{ and } \gamma[k_a] = 0 \text{ and } \delta[k_a] = 0) \end{aligned} \tag{10.57}$$

and distributing OR over AND we have

$$\begin{aligned} &(a[k_a + 1] = 0 \text{ or } \delta[k_a] = 0) \\ \text{and } &(a[k_a + 1] = 0 \text{ or } a[k_a] = 1) \\ \text{and } &(a[k_a + 1] = 0 \text{ or } \gamma[k_a] = 0). \end{aligned} \tag{10.58}$$

Now that the expression is in CNF we can easily write it as a set of inequality constraints.

$$\begin{aligned} &(1 - a[k_a + 1]) + (1 - \delta[k_a]) \geq 1 \\ &(1 - a[k_a + 1)) + a[k_a] \geq 1 \\ &(1 - a[k_a + 1)) + (1 - \gamma[k_a]) \geq 1. \end{aligned} \tag{10.59}$$

Now consider the other direction ($\Leftarrow$) of Equation (10.33). Replacing the implication with disjunction we have

$$(a[k_a + 1] = 1) \text{ or } \neg\,(\delta[k_a] = 0 \text{ and } a[k_a] = 1 \text{ and } \gamma[k_a] = 0) \tag{10.60}$$

and distributing the negation we have

$$\delta[k_a] = 1 \text{ or } a[k_a] = 0 \text{ or } \gamma[k_a] = 1 \text{ or } a[k_a + 1] = 1 \tag{10.61}$$

this disjunction is equivalent to the flowing inequality

$$\delta[k_a] + (1 - a[k_a]) + \gamma[k_a] + a[k_a + 1] \geq 1. \tag{10.62}$$

In summary, we have taken the governing equations for the attacker's binary state (10.32) and derived an equivalent set of inequalities: (10.59) and (10.62) which can be simplified to the inequalities given by Equation (10.34).

ACKNOWLEDGEMENTS

We thank J. Ousingsawat for helpful comments and encouragement. This work was supported by the Air Force Office of Scientific Research under Grant F49620-01-1-0361.

REFERENCES

Arkin RC and Bekey GA (eds) 1997 Special issue on robot colonies. *Autonomous Robots.*

Balch T and L. E. Parker E 2000 Special issue on heterogeneous multirobot systems. *Autonomous Robots.*

Beard RW and McLain TW 2003 Multiple UAV cooperative search under collision avoidance and limited range communication constraints. In *Proc. IEEE Conf. Decision and Control*, pp. 25–30, Maui, Hawaii.

Beard RW, McLain TW, Goodrich MA and Anderson E 2002 Coordinated target assignment and intercept for unmanned air vehicles. *IEEE Trans. Robot. Automat.* **18**, 991–922.

Bellingham J, Tillerson M, Alighanbari M and How J 2002 Cooperative path planning for multiple UAVs in dynamic and uncertain environments. In *Proc. IEEE Conf. Decision and Control*, pp. 2816–2822, Las Vegas, Nevada.

Bemporad A and Morari M 1999 Control of systems integrating logic, dynamics, and constraints. *Automatica* **35**, 407–428.

Bertsimas D and Tsitsiklis JN 1997 *Introduction to Linear Optimization*. Athena Scientific, Belmont, MA.

Campbell M, D'Andrea R, Schneider D, Chaudhry A, and Waydo S 2003 RoboFlag games using systems based hierarchical control. In *Proceedings of the American Control Conference*, pp. 661–666.

D'Andrea R and Babish M 2003 The RoboFlag testbed. In *Proceedings of the American Control Conference*, pp. 656–660.

D'Andrea R and Murray RM 2003 The RoboFlag competition. In *Proceedings of the American Control Conference*, pp. 650–655.

D'Andrea R, Kalmár-Nagy T, Ganguly P and Babish M 2001 *The Cornell RoboCup Team*. Springer, New York.

Earl MG and D'Andrea R 2002a Modeling and control of a multi-agent system using mixed integer linear programming. In *Proc. IEEE Conf. Decision and Control*, pp. 107–111, Las Vegas, Neveda.

Earl MG and D'Andrea R 2002b A study in cooperative control: The RoboFlag drill. In *Proceedings of the American Control Conference*, pp. 1811–1812, Anchorage, Alaska.

Earl MG and D'Andrea R 2005 Iterative MILP methods for vehicle control problems. *IEEE Transactions on Robotics* **21**(6), 1158–1167.

Earl MG and D'Andrea R 2007 A decomposition approach to multi-vehicle cooperative control. *Robotics and Autonomous Systems*. **55**(4), 276–291.

Fourer R, Gay DM and Kernighan BW 1993 *AMPL – A Modeling Language for Mathematical Programming*. Boyd & Fraser, Danvers, MA.

Garey MR and Johnson DS 1979 *Computers and Intractability: A Guide to the Theory of NP-Completeness*. W. H. Freeman and Company, New York.

Ho Y and Chu K 1972 Team decision theory and information structures in optimal control problems – part 1. *IEEE Trans. Automatic Control* **AC-17**, 15–22.

ILOG 2000 *ILOG AMPL CPLEX System Version 7.0 User's Guide*.

Kalmár-Nagy T, D'Andrea R and Ganguly P 2004 Near-optimal dynamic trajectory generation and control of an omnidirectional vehicle. *Robotics and Autonomous Systems* **46**, 47–64.

Kvasnica M, Grieder P, Baoti'c M and Morari M 2003 *Multi Parametric Toolbox (MPT)* vol. 2993. Springer Verlag, Pennsylvania, PA.

Marco MD, Garulli A, Giannitrapani A and Vicino A 2003 Simultaneous localization and map building for a team of cooperating robots: A set membership approach. *IEEE Transactions on Robotics and Automation* **19**(2), 238–249.

Marschak J and Radner R 1972 *Econonomic Theory of Teams*. Yale University Press New Haven, CT.

McLain TW, Chandler PR, Rasmussen S and Pachter M 2001 Cooperative control of UAV rendezvous. In *Proc. American Control Conference*, pp. 2309–2314, Arlington, VA.

Murphey R and Pardalos PM (eds) 2002 *Cooperative Control and Optimization*. Kluwer Academic, Boston.

Ousingsawat J and Campbell ME 2004 Establishing optimal trajectories for multi-vehicle reconnaissance. *AIAA Guidance, Navigation and Control Conference*. AIAA Paper #2004–5224.

Ralphs TL 2003 Parallel branch and cut for capacitated vehicle routing. *Parallel Computing* **29**(5), 607–629.

Richards A and How JP 2002 Aircraft trajectory planning with collision avoidance using mixed integer linear programming. In *Proc. American Control Conf.*, pp. 1936–1941.

Richards A, Bellingham J, Tillerson M and How J 2002a Co-ordination and control of multiple UAVs. *AIAA Conf. Guidance Navigation and Control*. AIAA Paper 2002–4588.

Richards A, Schouwenaars T, How J and Feron E 2002b Spacecraft trajectory planning with avoidance constraints using mixed-integer linear programming. *Journal of Guidance, Control, and Dynamics* **25**, 755–764.

Stone P and Veloso M 2000 Multiagent systems: A survey from a machine learning perspective. *Autonomous Robots* **8**, 345–383.

Stone P, Asada M, Balch T, D'Andrea R, Fujita M, Hengst B, Kraetzschmar G, Lima P, Lau N, Lund H, Polani D, Scerri P, Tadokoro S, Weigel T and Wyeth G 2001 Robocup-2000: The fourth robotic soccer world championships. *AI MAGAZINE* **22**(1), 11–38.

Sugar TG and Kumar V 2002 Control of cooperating mobile manipulators. *IEEE Transactions on Robotics and Automation* **18**(1), 94–103.

Tyler ML and Morari M 1999 Propositional logic in control and monitoring problems. *Automatica* **35**, 565–582.

11

LP-based multi-vehicle path planning with adversaries

Georgios C. Chasparis and Jeff S. Shamma

11.1 INTRODUCTION

One problem in autonomous multi-vehicle systems is the real-time derivation of vehicle paths. Often this problem can be formulated as a large-scale optimization. However, environmental conditions are not necessarily stationary and, in many cases, the multi-vehicle system interacts with an uncertain environment. For example, a characteristic dynamic environment could involve a group of adversarial vehicles that try to cause harm. In such environments, the real-time inclusion of these uncertainties to the optimization problem is necessary.

More specifically, such an optimization problem should provide: (a) a team-optimal solution; (b) on-line inclusion of all possible uncertainties of the environment; and (c) computational efficiency. However, the uncertainty of the environment and the complexity of a multi-vehicle control system create several problems in the computation of the optimal trajectories of the vehicles. Therefore, most of the recent work on multi-vehicle path planning focuses on seeking a method that will provide all the above three features.

One of the optimization methods that has been used for multi-vehicle path planning is based on the notion of *coordination variables* and *coordination function* (McLain and Beard 2003). According to this approach, the information that must be shared to facilitate cooperation is collected to a single vector quantity called the *coordination variable*, while the *coordination function* quantifies how changing the coordination variable impacts the individual's myopic objectives. Trajectories are determined so that threat avoidance is ensured and timing constraints are satisfied. However, the locations of the considered threats are deterministically known.

Several papers on mission planning of UAVs construct *Voronoi-based polygonal paths* from the currently known locations of the threats. Taken in different combinations,

Part of the text is based on 'Linear-Programming-Based Multi-Vehicle Path Planning with Adversaries', by G.C. Chasparis and J.S. Shamma, which appeared in the *2005 American Control Conference*. Reproduced with permission. © 2005 IEEE.

Cooperative Control of Distributed Multi-Agent Systems edited by Jeff S. Shamma

the edges of the Voronoi diagram provide a set of paths from the starting point to the target. Among those, we may choose either the lowest-cost flyable path (Chandler and Pachter 1998), or the path that minimizes exposure to radar (Chandler 2001). A probabilistic approach can also be applied, where a probability of a threat or target is assumed to be known (Dogan 2003). In this case, a chosen path minimizes the probability of getting disabled by a threat. Using this method, global strategies may be computationally inefficient, while the path generated by the strategy might get in a limit cycle (Dogan 2003). In a similar setting, the traveling repair-person problem can be formulated as a minimization of the expected waiting time to service targets, where it is assumed that these targets follow a Poisson distribution (Frazzoli and Bullo 2004).

Since several classes of multi-vehicle systems can be modeled as hybrid systems, one of the suggested approaches is based on *model predictive control* (Bemporad and Morari 1999), where an optimization problem is solved based on a prediction of the future evolution of the system. It is preferable to use this method when the prediction of the future evolution of the system cannot be accurate due to several reasons, such as unknown future disturbances. In this case, it is desirable to be able to rerun the optimization and uplink the new solution, thus achieving the benefit of feedback action.

For some classes of multi-vehicle systems such as Unmanned-Aerial-Vehicles (UAVs), this optimization is a *mixed integer linear programming* (MILP) problem. The optimization may combine task assignment subjected to UAV capability constraints, and path planning subjected to dynamics, avoidance and timing constraints (Richards *et al.* 2002; Richards and How 2002; Schouwenaars *et al.* 2001). By collision avoidance constraints, we ensure that vehicles do not collide with each other or with obstacles, which can be fixed or moving according to a predefined motion (Richards and How 2002; Schouwenaars *et al.* 2001).

The utility of MILP has been tested in the RoboFlag competition (Earl and D'Andrea 2002a, b). In this competition, the objective is to compute a set of control inputs that minimizes the number of adversaries that enter a protected zone. The adversaries are considered drones that move on a straight line with constant velocity, thus their motion and future positions are deterministically known. Moreover, the procedure is limited computationally, since the problem does not scale well with increased number of agents, while probabilities of the threats cannot easily be included in the optimization.

Dynamic programming can take into account the probabilities of the threats (Flint *et al.* 2002). Considering a distributed team of UAVs, a stochastic decision process formulation for cooperative control among this team can lead to a solution based on dynamic programming. Dynamic programming can achieve a probably global optimal solution, at least in theory. In practice, this is rarely the case, since this may be computationally infeasible (Flint *et al.* 2002). A limited look-ahead policy can reduce the complexity of the optimization, but produces a sub-optimal solution of the original problem. Also, under the limited look-ahead assumption and in case of more than one vehicle, the probabilities about the next positions of the other vehicles' planning tree need to be computed. However, their calculation is impractical (Flint *et al.* 2002).

In this chapter, we seek a linear formulation of the problem. The potential advantage is that a linear program is computationally appealing. The proposed approach is based on a linear model for resource allocation in an adversarial environment. Both friendly and enemy vehicles are modeled as different resource types in an arena of sectors, and the path planning problem is viewed as a resource allocation problem. However, the resulting linear dynamic model is subject to binary constraints, while the enemy's future locations are unknown. Model simplifications are introduced to allow the use of linear programming in conjunction with a receding horizon implementation for multi-vehicle path planning. Stochastic models based on the current positions and velocities of opposing vehicles are used to describe their possible future trajectories which can be included in the optimization problem in a real-time fashion. The utility of the LP-based algorithm is tested in the RoboFlag drill, where both teams of vehicles have equal path planning capabilities unlike prior work using the proposed algorithm.

11.2 PROBLEM FORMULATION

11.2.1 State-space model

We consider a model that describes the movement and engagement of friendly and enemy resources in an arena of sectors according to (Daniel-Berhe *et al.* 2001). The battlefield is divided into a collection of sectors, $\mathcal{S}$, where the collection of resources, $\mathcal{R}$, evolves in discrete time.

A vehicle is represented as a resource type, $r_j \in \mathcal{R}$. We define its quantity level at sector $s_i \in \mathcal{S}$ to be $x_{s_i,r_j} \in \mathcal{B} \triangleq \{0, 1\}$, so that resource level '1' corresponds to the sector where the vehicle lies in, otherwise it is '0'. Under these assumptions, the state of each resource type $r_j \in \mathcal{R}$ is

$$\mathbf{x}_{r_j} = \begin{pmatrix} x_{s_1,r_j} & x_{s_2,r_j} & \cdots & x_{s_{n_S},r_j} \end{pmatrix}^T \in \mathcal{B}^{n_s},$$

where n_s is the total number of sectors in $\mathcal{S}$. Thus, the state of a collection $\mathcal{R}$ of n_r vehicles will be

$$\mathbf{x} = \begin{pmatrix} \mathbf{x}_{r_1}^T & \mathbf{x}_{r_2}^T & \cdots & \mathbf{x}_{r_{n_r}}^T \end{pmatrix}^T \in \mathcal{B}^{n_x},$$

where $n_x = n_s n_r$ is the total number of states.

Resource level changes represent vehicles movement. They can either remain in the same sector or move to a neighboring sector. Movements within a sector are not modeled. Therefore, the control action includes the transitions of each resource type $r_j \in \mathcal{R}$ from sector $s_i \in \mathcal{S}$ to a neighboring sector $s_k \in \mathcal{N}(s_i, r_j)$, where $\mathcal{N}(s_i, r_j)$ is the set of neighboring sectors of sector s_i that can be reached by resource type r_j within one time-stage.

Define $u_{s_i \leftarrow s_k, r_j} \in \mathcal{B}$ as the level of resource type r_j that is being transferred from sector s_k to sector s_i, where $s_k \in \mathcal{N}(s_i, r_j)$. Then the system evolves according to the following state-space equations:

$$x_{s_i,r_j}^+ = x_{s_i,r_j} + \sum_{s_k \in \mathcal{N}(s_i,r_j)} u_{s_i \leftarrow s_k,r_j} - \sum_{s_k \in \mathcal{N}(s_i,r_j)} u_{s_k \leftarrow s_i,r_j} \tag{11.1}$$

for each $s_i \in \mathcal{S}$ and $r_j \in \mathcal{R}$, where superscript '+' denotes the next time-stage. In order for this set of equations to describe a continuous flow of resources, the constraints:

$$0 \leq \sum_{s_k \in \mathcal{N}(s_i,r_j)} u_{s_k \leftarrow s_i,r_j} \leq x_{s_i,r_j}, \quad \forall s_i \in \mathcal{S}, \quad \forall r_j \in \mathcal{R}, \tag{11.2}$$

must be satisfied, which implies that the level of resource type r_j that exits from sector s_i cannot be larger than the level of the same resource type at sector s_i.

Define the control vector $\mathbf{u}_{s_i,r_j} \in \mathcal{B}^{(n_s-1)}$ as the resource levels of type $r_j \in \mathcal{R}$ that enter $s_i \in \mathcal{S}$, i.e.,

$$\mathbf{u}_{s_i,r_j} = \left(u_{s_i \leftarrow s_1,r_j} \ldots u_{s_i \leftarrow s_{i-1},r_j} \; u_{s_i \leftarrow s_{i+1},r_j} \ldots u_{s_i \leftarrow s_{n_s},r_j} \right)^T,$$

where the control command $u_{s_i \leftarrow s_i,r_j}$ has been omitted since it coincides with the new state $x^+_{s_i,r_j}$. Also, define the control vector $\mathbf{u}_{r_j} \in \mathcal{B}^{n_s(n_s-1)}$ of all transitions of $r_j \in \mathcal{R}$ to be

$$\mathbf{u}_{r_j} = \left(\mathbf{u}^T_{s_1,r_j} \; \mathbf{u}^T_{s_2,r_j} \; \ldots \; \mathbf{u}^T_{s_{n_s},r_j} \right)^T.$$

Then, the control vector of the collection of resource types or vehicles, $\mathcal{R}$, can be defined as:

$$\mathbf{u} = \left(\mathbf{u}^T_{r_1} \; \mathbf{u}^T_{r_2} \; \ldots \; \mathbf{u}^T_{r_{n_r}} \right)^T \in \mathcal{B}^{n_u},$$

where $n_u = n_s n_r (n_s - 1)$ is the total number of controls.

It is not difficult to show that Equations (11.1) and (11.2), which describe the system's evolution, can take on the following form:

$$\begin{cases} \mathbf{x}^+ = \mathbf{x} + \mathbf{B}_{\text{in}}\mathbf{u} - \mathbf{B}_{\text{out}}\mathbf{u} = \mathbf{x} + \mathbf{B}\mathbf{u} \\ \mathbf{0} \leq \mathbf{B}_{\text{out}}\mathbf{u} \leq \mathbf{x} \\ \mathbf{x} \in \mathcal{B}^{n_x}, \quad \mathbf{u} \in \mathcal{B}^{n_u} \end{cases} \tag{11.3}$$

where $\mathbf{B} = \mathbf{B}_{\text{in}} - \mathbf{B}_{\text{out}}$, for some matrices $\mathbf{B}_{\text{in}}$ and $\mathbf{B}_{\text{out}}$ suitably defined. Notice that the entries of both $\mathbf{x}$ and $\mathbf{u}$ take on only binary values.

For example, in case of two sectors and one vehicle, $\mathcal{S} = \{s_1, s_2\}$, $\mathcal{R} = \{r_1\}$, $\mathcal{N}(s_1, r_1) = \{s_2\}$, $\mathcal{N}(s_2, r_1) = \{s_1\}$, and the state-space equations are:

$$\underbrace{\begin{pmatrix} x_{s_1,r_1} \\ x_{s_2,r_1} \end{pmatrix}^+}_{\mathbf{x}^+} = \underbrace{\begin{pmatrix} x_{s_1,r_1} \\ x_{s_2,r_1} \end{pmatrix}}_{\mathbf{x}} + \underbrace{\begin{bmatrix} 1 & -1 \\ -1 & 1 \end{bmatrix}}_{\mathbf{B}} \cdot \underbrace{\begin{pmatrix} u_{s_1 \leftarrow s_2,r_1} \\ u_{s_2 \leftarrow s_1,r_1} \end{pmatrix}}_{\mathbf{u}}.$$

We can also define:

$$\mathbf{B}_{\text{in}} = \begin{bmatrix} 1 & 0 \\ 0 & 1 \end{bmatrix}, \quad \mathbf{B}_{\text{out}} = \begin{bmatrix} 0 & 1 \\ 1 & 0 \end{bmatrix}.$$

11.2.2 Single resource models

An alternative model is to view all vehicles as a single resource type. In this case, the state of the system can be defined as

$$\mathbf{x}_1 = \mathbf{x}_{r_1} + \mathbf{x}_{r_2} + \ldots + \mathbf{x}_{r_{n_r}} \in \mathcal{B}^{n_{x,1}},$$

where $n_{x,1} = n_s$. Similarly, we define

$$\mathbf{u}_1 = \mathbf{u}_{r_1} + \mathbf{u}_{r_2} + \ldots + \mathbf{u}_{r_{n_r}} \in \mathcal{B}^{n_{u,1}},$$

where $n_{u,1} = n_s(n_s - 1)$. Thus, for suitable matrices $\mathbf{B}_1$ and $\mathbf{B}_{\text{out},1}$, we can write

$$\begin{cases} \mathbf{x}_1^+ = \mathbf{x}_1 + \mathbf{B}_1\mathbf{u}_1 \\ \mathbf{0} \leq \mathbf{B}_{\text{out},1}\mathbf{u}_1 \leq \mathbf{x}_1 \\ \mathbf{x}_1 \in \mathcal{B}^{n_{x,1}}, \quad \mathbf{u}_1 \in \mathcal{B}^{n_{u,1}}. \end{cases} \tag{11.4}$$

This state-space representation has fewer states and controls than the one of (11.3).

11.2.3 Adversarial environment

We consider the case that an adversarial team of vehicles also evolves within the same arena of sectors. The two opposing teams are subject to attrition, where attrition could be interpreted as the result of collisions of opponent vehicles. In this case, a possible objective of each team is to cause the largest possible attrition to their enemies.

We model enemy vehicles to follow similar state-space equations as those of friendly vehicles. We also assume that the decisions of both teams are made at the same time instants and that the current state of the opposing resources is known.

Let superscript 'f' denote the friendly team, and superscript 'e' denote the enemy team. Then the evolution of both friendly and enemy vehicles can be described by

$$\begin{cases} (\mathbf{x}^i)^+ = \mathbf{x}^i + \mathbf{B}^i\mathbf{u}^i - \mathbf{d}^i(\mathbf{x}^i, \mathbf{x}^{-i}) \\ \mathbf{0} \leq \mathbf{B}^i_{\text{out}}\mathbf{u}^i \leq \mathbf{x}^i \\ \mathbf{x}^i \in \mathcal{B}^{n^i_x}, \quad \mathbf{u}^i \in \mathcal{B}^{n^i_u} \end{cases}, \quad i \in \{f, e\}, \tag{11.5}$$

where $\mathbf{d}^i$ is the attrition function that depends on the current state of each team and $-i$ is the opposite of i.

11.2.4 Model simplifications

The state-space equations of (11.5) cannot be used to formulate a linear optimization program for future friendly planning. The main obstacles are:

- The attrition function, which is generally nonlinear.
- The unknown control vector of the enemy resources.

In (Daniel-Berhe *et al.* 2001), a similar state-space model, but without the binary constraints, is approximated with a linear model based on two model simplifications that will allow the use of linear programming.

First, we remove the attrition function from the state-space equations of (11.5), i.e.,

$$\begin{cases} (\mathbf{x}^i)^+ = \mathbf{x}^i + \mathbf{B}^i\mathbf{u}^i \\ \mathbf{0} \leq \mathbf{B}^i_{\text{out}}\mathbf{u}^i \leq \mathbf{x}^i \\ \mathbf{x}^i \in \mathcal{B}^{n^i_x}, \quad \mathbf{u}^i \in \mathcal{B}^{n^i_u} \end{cases}, \quad i \in \{f, e\}. \tag{11.6}$$

In other words, we assume that both teams evolve as if no attrition will occur.

Second, since the controls of the enemy team are not known to the friendly team, we assume that the enemy team implements an *assumed* feedback policy $\mathbf{G}^e \in \overline{\mathcal{B}}^{n_u^e \times n_x^e}$, such that

$$\mathbf{u}^e = \mathbf{G}^e \mathbf{x}^e, \tag{11.7}$$

where $\overline{\mathcal{B}} \triangleq [0, 1]$.

Due to these two model simplifications, we can expect that the resulting model of (11.6) and (11.7) will be significantly different from the actual evolution described by (11.5). We overcome this problem by applying a receding horizon strategy (Bemporad and Morari 1999).

11.2.5 Enemy modeling

The enemy's feedback matrix, $\mathbf{G}^e$, contains the *assumed* information about the future states of the enemy resources. It is generally unknown to the friendly resources but introduced for the sake of prediction in optimization. This feedback matrix can be used for modeling several behaviors, such as

- anticipated paths of enemy resources;
- diffusion of enemy resources;
- probability maps of enemy resources.

In particular, for any $s_i \in \mathcal{S}$, $r_j \in \mathcal{R}^e$ and $s_k \in \mathcal{N}^e(s_i, r_j)$, we assume that

$$u^e_{s_k \leftarrow s_i, r_j} = g^e_{s_k \leftarrow s_i, r_j} x^e_{s_i, r_j}, \tag{11.8}$$

where $g^e_{s_k \leftarrow s_i, r_j}$ is the assumed feedback of the enemy resource type r_j.

Setting $g^e_{s_k \leftarrow s_i, r_j} \in \{0, 1\}$, we define an anticipated next destination of the opposing resource type r_j. If we split the resource level $x^e_{s_i, r_j}$ to two or more destination sectors, i.e., $g^e_{s_k \leftarrow s_i, r_j} < 1$, then we create a diffusion of enemy resources. Finally, $g^e_{s_k \leftarrow s_i, r_j}$ can be interpreted as the probability that the opposing resource type r_j will move from sector s_i to sector s_k.

In either case, the following properties must be satisfied

$$\begin{cases} g^e_{s_k \leftarrow s_i, r_j} \in [0, 1] \\ \sum_{s_k \in \mathcal{N}(s_i, r_j)} \{g^e_{s_k \leftarrow s_i, r_j}\} \leq 1 \end{cases} \tag{11.9}$$

which guarantee that the control constraints of (11.2) hold.

In case of two sectors and one opposing vehicle, we have:

$$\underbrace{\begin{pmatrix} u^e_{s_1 \leftarrow s_2, r_1} \\ u^e_{s_2 \leftarrow s_1, r_1} \end{pmatrix}}_{\mathbf{u}^e} = \underbrace{\begin{bmatrix} 0 & g^e_{s_1 \leftarrow s_2, r_1} \\ g^e_{s_2 \leftarrow s_1, r_1} & 0 \end{bmatrix}}_{\mathbf{G}^e} \cdot \underbrace{\begin{pmatrix} x_{s_1, r_1} \\ x_{s_2, r_1} \end{pmatrix}}_{\mathbf{x}^e}.$$

For example, if $g^e_{s_1 \leftarrow s_2, r_1} = 0.3$, then 30% of x_{s_2,r_1} will move from sector s_2 to sector s_1, while the rest of it will remain in sector s_2.

The great advantage of the introduction of the enemy's feedback matrix is that we can model the enemy's intentions based on their current location or even their velocity.

11.3 OPTIMIZATION SET-UP

11.3.1 Objective function

The simplified system of friendly and enemy vehicles is described by a system of linear equations and constraints, (11.6) and (11.7). We now introduce a linear objective function that will allow the use of linear programming in deriving optimal paths. Optimal paths will be described by a sequence of states.

For each team $i \in \{f, e\}$, define the vector of optimized states for a finite optimization horizon, T_p, as

$$\mathbf{X}^i = \left(\mathbf{x}^i [1]^T \ \mathbf{x}^i [2]^T \ \dots \ \mathbf{x}^i \left[T_p\right]^T \right)^T,$$

where $\mathbf{x}^i [t] \in \mathcal{B}^{n^i_x}$ is the state vector at the tth future time-stage. We can also define the vector $\mathbf{X}^i_1$, where all vehicles of the collection $\mathcal{R}^i$ are considered as a single resource type, i.e.,

$$\mathbf{X}^i_1 = \left(\mathbf{x}^i_1 [1]^T \ \mathbf{x}^i_1 [2]^T \ \dots \ \mathbf{x}^i_1 \left[T_p\right]^T \right)^T.$$

Possible objectives in an adversarial environment are:

- Minimization of intercepted friendly vehicles (*evasion*).
- Maximization of intercepted enemy vehicles (*pursuit*).
- Tracking of a reference state vector (*surveillance*).

These objectives can be represented by a linear objective function of the form:

$$\min_{\mathbf{X}^f_1} \left[\alpha^f \cdot \mathbf{X}^e_1 + \beta^f \cdot \mathbf{X}^f_{\text{ref}}\right]^T \cdot \mathbf{X}^f_1. \tag{11.10}$$

The inner product of the vector of optimized states, $\mathbf{X}^f_1$, with the corresponding enemy vector, $\mathbf{X}^e_1$, increases with the number of interceptions between friendly and enemy vehicles. Therefore, in case $\alpha^f < 0$, interceptions of enemy vehicles are encouraged, while $\alpha^f > 0$ causes friendly vehicles to avoid enemy vehicles. Moreover, $\beta^f < 0$ encourages the friendly states to become aligned to the reference ones, $\mathbf{X}^f_{\text{ref}}$. We can always take

$$\left|\alpha^f\right| + \left|\beta^f\right| = 1.$$

11.3.2 Constraints

The objective function of (11.10) is subject to the dynamics of (11.6) and (11.7), throughout the optimization horizon T_p. In particular, the following equations must be satisfied for each $t \in \mathcal{T} \triangleq \{0, 1, \ldots, T_p - 1\}$:

$$\mathbf{x}_1^f[t+1] = \mathbf{x}_1^f[t] + \mathbf{B}_1^f \cdot \mathbf{u}_1^f[t].$$

Define

$$\mathbf{U}_1^f = \left(\mathbf{u}_1^f[0]^T \ \mathbf{u}_1^f[1]^T \ \ldots \ \mathbf{u}_1^f[T_p - 1]^T \right)^T.$$

There exist matrices $\mathbf{T}_{\mathrm{xx}_0}^f$ and $\mathbf{T}_{\mathrm{xu}}^f$ such that the above dynamics are written equivalently as

$$\mathbf{X}_1^f = \mathbf{T}_{\mathrm{xx}_0}^f \cdot \mathbf{x}_1^f[0] + \mathbf{T}_{\mathrm{xu}}^f \cdot \mathbf{U}_1^f. \tag{11.11}$$

The control vector $\mathbf{u}_1^f[t]$ for each $t \in \mathcal{T}$ must also satisfy:

$$\mathbf{B}_{\mathrm{out},1}^f \cdot \mathbf{u}_1^f[t] \leq \mathbf{x}_1^f[t].$$

It is straightforward to construct matrices $\mathbf{T}_{\mathrm{xu,c}}^f$ and $\mathbf{T}_{\mathrm{xx}_0,\mathrm{c}}^f$, such that the above constraints take on the following form:

$$\mathbf{T}_{\mathrm{xu,c}}^f \cdot \mathbf{U}_1^f \leq \mathbf{T}_{\mathrm{xx}_0,\mathrm{c}}^f \cdot \mathbf{x}^f[0]. \tag{11.12}$$

Furthermore, obstacle avoidance easily can be represented by orthogonality constraints, such as

$$(\mathbf{X}_{\mathrm{obs}}^f)^T \cdot \mathbf{X}_1^f = 0, \tag{11.13}$$

where $\mathbf{X}_{\mathrm{obs}}^f \in \mathcal{B}^{T_p n_s}$ is a vector whose entries are equal to '1' if they correspond to the obstacles' locations, and '0' otherwise.

11.3.3 Mixed-integer linear optimization

The objective function of (11.10) in conjunction with the dynamic constraints of (11.11), the control constraints of (11.12) and the obstacle avoidance constraints of (11.13), formulate a *mixed integer linear programming* optimization for friendly planning. This optimization problem can be written as:

$$\begin{aligned}
\text{minimize} \quad & \left[\alpha^f \cdot \mathbf{X}_1^e + \beta^f \cdot \mathbf{X}_{\mathrm{ref}}^f\right]^T \cdot \mathbf{X}_1^f \\
\text{subject to} \quad & \mathbf{T}_{\mathrm{xu,c}}^f \cdot \mathbf{U}_1^f \leq \mathbf{T}_{\mathrm{xx}_0,\mathrm{c}} \cdot \mathbf{x}_1^f[0] \\
& \mathbf{X}_1^f - \mathbf{T}_{\mathrm{xu}}^f \cdot \mathbf{U}_1^f = \mathbf{T}_{\mathrm{xx}_0}^f \cdot \mathbf{x}_1^f[0] \\
& (\mathbf{X}_{\mathrm{obs}}^f)^T \cdot \mathbf{X}_1^f = 0 \\
\text{variables} \quad & \mathbf{X}_1^f \in \mathcal{B}^{T_p n_{x,1}^f}, \quad \mathbf{U}_1^f \in \mathcal{B}^{T_p n_{u,1}^f}.
\end{aligned} \tag{11.14}$$

The vector of states of enemy resources, $\mathbf{X}_1^e$, is not known. As previously discussed, we can assume that enemy resources evolve according to an assumed feedback law, $\mathbf{G}^e$, such that for each $t \in \mathcal{T}$,

$$\mathbf{x}^e[t+1] = \left(\mathbf{I} + \mathbf{B}^e\mathbf{G}^e\right)^{t+1} \cdot \mathbf{x}^e[0]. \tag{11.15}$$

Hence, there exists matrix $\mathbf{T}^e_{\mathrm{xx_0,G}}$ such that the state-space equations of (11.15) are written as

$$\mathbf{X}_1^e = \mathbf{T}^e_{\mathrm{xx_0,G}} \cdot \mathbf{x}_1^e[0]. \tag{11.16}$$

11.4 LP-BASED PATH PLANNING

The optimization of (11.14) is a mixed-integer linear programming problem, where both states and controls take values on $\mathcal{B} = \{0, 1\}$. Although there are several methods that can be used for computing the optimal solution, such as cutting plane methods or branch and bound (Bertsimas and Tsitsiklis 1997), we prefer to solve a linear programming problem instead. The main reason for this preference is the computational complexity of an integer problem.

To this end, we transform the mixed integer program of (11.14) into a *linear-programming-based optimization planning*. This planning includes the following steps:

1. We introduce the *linear programming relaxation* of the *mixed integer programming problem* of (11.14).
2. Given the non-integer optimal solution of the linear programming relaxation, we compute a suboptimal solution of the mixed integer programming problem.
3. We apply this solution according to a *receding horizon implementation*.

11.4.1 Linear programming relaxation

The linear programming relaxation of (11.14) assumes that the vector of optimized states, $\mathbf{X}_1^f$, and controls, $\mathbf{U}_1^f$, can take any value between the vectors $\mathbf{0}$ and $\mathbf{1}$. If an optimal solution to the relaxation is feasible to the mixed integer programming problem, then it is also an optimal solution to the latter (Bertsimas and Tsitsiklis 1997).

In general, the solution of the linear programming relaxation is a non-integer vector, which means that this solution does not belong to the feasible set of the mixed integer programming problem. Therefore, we construct a suboptimal solution to the relaxation that is feasible to the mixed integer program.

11.4.2 Suboptimal solution

Define $\left(\mathbf{u}_1^*\right)^f[t] \in \overline{\mathcal{B}}^{n^f_{u,1}}$, where $\overline{\mathcal{B}} = [0, 1]$, as the optimal control vector to the relaxation for each $t \in \mathcal{T}$, which will generally be a non-integer vector between $\mathbf{0}$ and $\mathbf{1}$. A non-integer control vector results in the splitting of friendly resources to several one-step

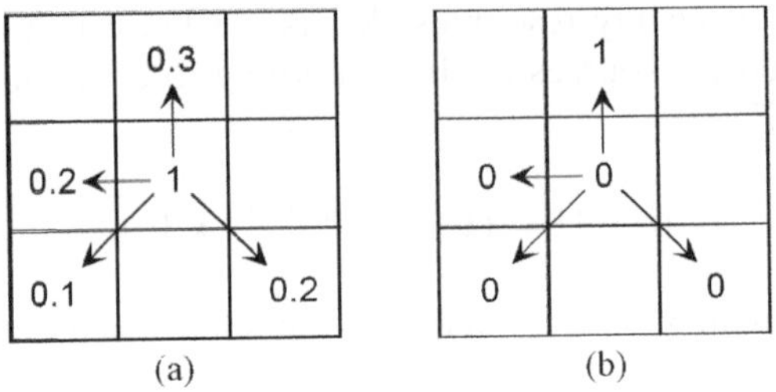

Figure 11.1 (a) A possible optimal solution to the linear programming relaxation. (b) The corresponding integer suboptimal solution to the mixed integer programming problem.

reachable neighboring sectors. An example of a possible optimal solution of the linear programming relaxation is shown in Figure 11.1(a).

Since a vehicle is represented by a binary variable, such a splitting of resources is not feasible. For this reason, among the control quantities that exit from the initial sector or remain in it, we pick up the maximum of them. In Figure 11.1(a) this quantity corresponds to the control '0.3'. We assign value '1' to this control level, while the rest of them are assigned the value '0'. The resulting controls of the suboptimal solution are shown in Figure 11.1(b).

In this way, we define an integer control vector that belongs to the feasible set of the mixed integer program, while, in parallel, the sum of the resource levels remains the same as that of the previous time-stage. We call this solution $\tilde{\mathbf{u}}_1^f[t]$, $t \in \mathcal{T}$.

11.4.3 Receding horizon implementation

Due to the differences between the model used for prediction and optimization, (11.6) and (11.7), and the 'plant' to be controlled (11.5), we should expect significant discrepancies between the responses of these two models. To compensate for this approximation, the optimization result will be implemented according to a receding horizon manner (Bemporad and Morari 1999).

In other words, the following algorithm is implemented:

1. Measure the current state vectors of both teams, $\mathbf{x}^f$ and $\mathbf{x}^e$.
2. Solve the linear programming relaxation of the mixed integer program of (11.14). Let $(\mathbf{u}_1^*)^f[0]$ be the first optimal control of the relaxation.
3. Construct a suboptimal solution, $\tilde{\mathbf{u}}_1^f[0]$, of the initial mixed integer program.
4. Apply only $\tilde{\mathbf{u}}_1^f[0]$ and repeat.

Since the enemy resources, which are the disturbances of the system, are state dependent and the state is bounded, the receding horizon strategy is always stabilizing.

11.5 IMPLEMENTATION

In this chapter, we consider a simplified version of the 'RoboFlag competition' (D'Andrea and Murray 2003), similar to the 'RoboFlag drill' (Earl and D'Andrea 2002a), where two teams of robots are engaged in an arena of sectors with a region at its center called the defense zone. The defenders' goal is the interception of the attackers, in which case the attackers become inactive, while the attackers' goal is the infiltration of the defenders' protected zone, Figure 11.2. Unlike prior work, both teams have equal path planning capabilities using the optimization algorithm proposed here.

11.5.1 Defense path planning

A defense path planning can be designed according to the proposed LP-based path planning. The superscript 'd' denotes defenders and will replace superscript 'f', while 'a' denotes attackers and will replace superscript 'e'.

Since the defenders' objective is the interception of the attackers, the weight α^d must be negative. In addition, a possible reference $\mathbf{X}^d_{\text{ref}}$ for the defenders could be the sectors close to the defense zone, so that the defenders stay always close to it. In the following simulations, we consider the entries of $\mathbf{X}^d_{\text{ref}}$ to be '1' for any sector that belongs to a small zone about the defense zone and '0' otherwise. Moreover, if defenders are not allowed to enter their defense zone, we can set the entries of $\mathbf{X}^d_{\text{obs}}$ to be '1' for sectors in the defense zone and '0' otherwise.

11.5.1.1 A position-based feedback matrix

Defense path planning is complete when the stochastic feedback matrix of the attackers, $\mathbf{G}^a$, is determined. The attackers' first priority is the infiltration of the defense zone.

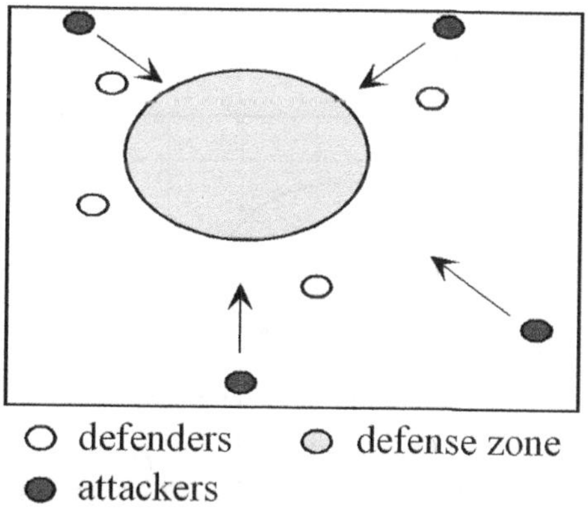

Figure 11.2 The RoboFlag drill.

Thus, a possible feedback matrix can be one that assigns higher probability to an attacker moving closer to the defense zone.

In this case, we can define a function $g^{a,p} : \mathcal{N}^a(\mathcal{S}, \mathcal{R}^a) \times \mathcal{R}^a \times \mathcal{S} \rightarrow [0, 1]$, such that the probability that the attacker $r_j \in \mathcal{R}^a$ will move from sector $s_i \in \mathcal{S}$ to sector $s_k \in \mathcal{N}^a(s_i, r_j)$ is:

$$g^{a,p}_{s_k \leftarrow s_i, r_j} = \frac{\left(\gamma^a_{s_i} + \sum_{s_n} \{\gamma^a_{s_n}\}\right) - \gamma^a_{s_k}}{(n-1) \cdot \left(\gamma^a_{s_i} + \sum_{s_n} \{\gamma^a_{s_n}\}\right)} \tag{11.17}$$

where $\gamma^a_{s_i}$ is the minimum distance from sector s_i to the defense zone, n is the number of one-step reachable destinations (including the current location s_i) and $s_n \in \mathcal{N}^a\left(s_i, r_j\right)$.

According to the system dynamics, (11.1), and the definition of the feedback matrix $\mathbf{G}^a$, (11.8) and (11.9), the probability $g^{a,p}_{s_i \leftarrow s_i, r_j}$ is not included in the feedback matrix. The reason for this choice is that the control of staying in the same sector, $u^a_{s_i \leftarrow s_i, r_j}$, is implicitly included in the system dynamics, since it is always equal to the resource level of the next time-stage, i.e.,

$$u^a_{s_i \leftarrow s_i, r_j} = \left(x^a_{s_i, r_j}\right)^+ .$$

The function $g^{a,p}$ satisfies the properties of (11.9), which implies that a feedback matrix $\mathbf{G}^{a,p}$ can be defined according to (11.8) and (11.17). Moreover, if the position of the defense zone does not change with time, then this feedback matrix $\mathbf{G}^{a,p}$ is time invariant for any sector $s_i \in \mathcal{S}$. Thus, $\mathbf{G}^{a,p}$ can include the transition probabilities for all sectors $s_i \in \mathcal{S}$, based on their relative position with respect to the defense zone according to Equation (11.17). The greatest advantage of such a feedback matrix is that it can be computed off-line, which reduces significantly the complexity of the algorithm.

11.5.1.2 Example

For example, consider the case where an attacker r_j lies in sector s_0, Figure 11.3. We compute the minimum distance from each possible one-step reachable destination,

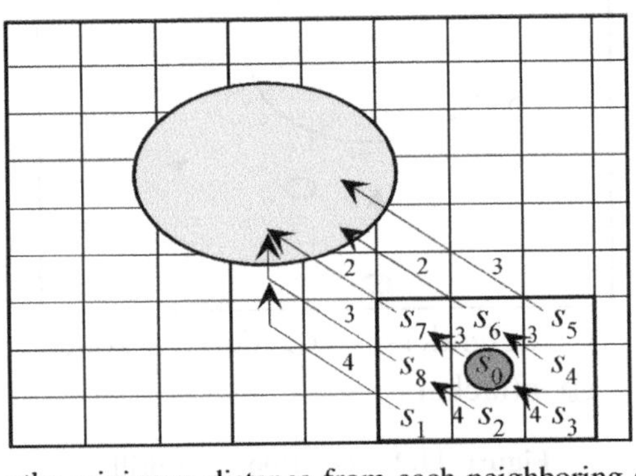

Figure 11.3 Computing the minimum distance from each neighboring sector, $\{s_1, \ldots, s_8\}$, of an attacker that lies in s_0 to the defense zone.

$\{s_0, s_1, \ldots, s_8\}$, of an attacker to the defense zone. In this case, we have

$$\gamma^a_{s_0} = 3, \quad \sum_{s_n} \{\gamma^a_{s_n}\} = \gamma^a_{s_1} + \ldots + \gamma^a_{s_8} = 25.$$

Hence, according to (11.17), the probability that the attacker will move from sector s_0 to sector s_1 is:

$$g^{a,p}_{s_1 \leftarrow s_0, r_j} = \frac{(3+25)-4}{(9-1)\cdot(3+25)} = \frac{24}{224}.$$

It is straightforward to show that

$$\sum_{s_n} g^{a,p}_{s_n \leftarrow s_0, r_j} = \frac{199}{224} < 1$$

and

$$1 - \sum_{s_n} g^{a,p}_{s_n \leftarrow s_0, r_j} = \frac{25}{224} = g^{a,p}_{s_0 \leftarrow s_0, r_j},$$

where $g^{a,p}_{s_0 \leftarrow s_0, r_j}$ is the probability of staying in sector s_0.

11.5.1.3 A velocity-based feedback matrix

Similar functions can be created to include different opponent's objectives. In particular, an alternative stochastic feedback matrix, $\mathbf{G}^{a,v}$, can be defined by computing the previous and current directions of motion of the enemy resources.

Let $\mathbf{v}_{s_i, r_j}$ be the current direction of motion of resource type $r_j \in \mathcal{R}^a$ that lies in sector $s_i \in \mathcal{S}$ and $\mathbf{v}^+_{s_i, r_j}$ be the next direction of motion of the same resource type that moves from sector s_i to $s_k \in \mathcal{N}^a(s_i, r_j) \cup \{s_i\}$. Then, we can define a probability measure that assigns a probability to the inner product of the current and next direction of motion, i.e.,

$$g^{a,v} : \left\{ \langle \mathbf{v}_{s_i, r_j}, \mathbf{v}^+_{s_i, r_j} \rangle \;\middle|\; s_i \in \mathcal{S}, \quad r_j \in \mathcal{R}^a \right\} \rightarrow [0, 1]. \tag{11.18}$$

Assuming that the distance between two neighboring sectors is equal to the unit distance, the vectors $\mathbf{v}_{s_i, r_j}$, $\mathbf{v}^+_{s_i, r_j}$ can be defined according to Figure 11.4. For example, $g^{a,v}$ can be defined as follows:

$$g^{a,v}_{s_k \leftarrow s_i, r_j} = \begin{cases} \kappa \;, & \text{if } \langle \mathbf{v}_{s_i, r_j}, \mathbf{v}^+_{s_i, r_j} \rangle < 0 \\ \lambda \;, & \text{if } \langle \mathbf{v}_{s_i, r_j}, \mathbf{v}^+_{s_i, r_j} \rangle = 0 \\ \mu \;, & \text{if } \langle \mathbf{v}_{s_i, r_j}, \mathbf{v}^+_{s_i, r_j} \rangle > 0 \end{cases},$$

$$s_i \in \mathcal{S}, \quad r_j \in \mathcal{R}^a, \quad s_k \in \mathcal{N}^a(s_i, r_j) \cup \{s_i\}, \tag{11.19}$$

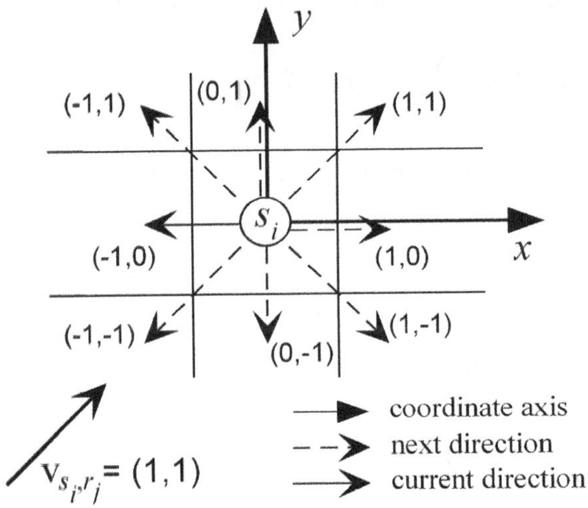

Figure 11.4 Possible next directions of motion, $\mathbf{v}^+_{s_i,r_j}$, of a vehicle lying in sector s_i and moving across $\mathbf{v}_{s_i,r_j}$, which is always assigned the value (1, 1) by appropriately changing the coordinate axes.

where

$$3\kappa + 3\lambda + 3\mu = 1, \quad \text{and} \quad \kappa, \lambda, \mu \geq 0 \tag{11.20}$$

since, according to Figure 11.4, there are three possible next directions of motion for each sign of the inner product $\langle \mathbf{v}_{s_i,r_j}, \mathbf{v}^+_{s_i,r_j} \rangle$.

Assuming that vehicles have non-zero inertia, it is most likely that a vehicle's next velocity corresponds to a positive inner product, i.e., $\kappa \leq \lambda \leq \mu$. The function $g^{a,v}$ satisfies the properties of (11.9), which implies that a feedback matrix $\mathbf{G}^{a,v}$ can be defined according to (11.8) and (11.19). This velocity-based feedback matrix $\mathbf{G}^{a,v}$ can be defined only on-line, which implies that its inclusion in the path planning scheme increases the complexity of the algorithm.

Further, we can combine the velocity-based feedback matrix, $\mathbf{G}^{a,v}$, with the position-based feedback matrix, $\mathbf{G}^{a,p}$, by defining a new feedback matrix as follows:

$$\mathbf{G}^a = \nu \mathbf{G}^{a,v} + \xi \mathbf{G}^{a,p}, \quad \text{where} \quad \nu + \xi = 1 \quad \text{and} \quad \nu, \xi \geq 0. \tag{11.21}$$

11.5.2 Attack path planning

Similar path planning can be designed for the attackers according to the proposed LP-based path planning. Now the superscript 'a' will replace 'f', while 'd' will replace 'e'.

The attackers' objective is to enter the defenders' protected zone. Therefore, the reference state vector, $\mathbf{X}^a_{\text{ref}}$, is defined such that its entries are equal to '1' if the corresponding sectors belong to the defense zone, and '0' otherwise. At the same time, attackers must avoid defenders, which is encouraged by setting $\alpha^a > 0$.

11.5.2.1 A position-based feedback matrix

A stochastic feedback matrix of the defenders, $\mathbf{G}^{d,p}$, is also necessary. The defenders' first priority is the interception of the attackers. Assuming that defenders create a probability distribution of the next attackers' locations given by (11.17), a set of the attackers' most probable future locations can be created, say $\mathcal{L}^a[t]$, for each future time $t \in \mathcal{T}$.

In this case, we can define a function $g^{d,p} : \mathcal{N}^d(\mathcal{S}, \mathcal{R}^d) \times \mathcal{R}^d \times \mathcal{S} \times \mathcal{T} \to [0, 1]$, such that the probability that a defender $r_j \in \mathcal{R}^d$ will move from sector $s_i \in \mathcal{S}$ to sector $s_k \in \mathcal{N}^d(s_i, r_j)$ is:

$$g^{d,p}_{s_k \leftarrow s_i, r_j}[t] = \frac{\left(\gamma^d_{s_i}[t] + \sum_{s_n}\left\{\gamma^d_{s_n}[t]\right\}\right) - \gamma^d_{s_k}[t]}{(n-1)\cdot\left(\gamma^d_{s_i}[t] + \sum_{s_n}\left\{\gamma^d_{s_n}[t]\right\}\right)} \tag{11.22}$$

where $\gamma^d_{s_i}[t]$ is the minimum distance from sector s_i to the set of sectors $\mathcal{L}^a[t]$, n is the number of one-step reachable destinations (including the current location s_i) and $s_n \in \mathcal{N}^d\left(s_i, r_j\right)$.

The function $g^{d,p}$ satisfies the properties of (11.9), which implies that a feedback matrix $\mathbf{G}^{d,p}$ can be defined according to (11.8) and (11.22). However, this feedback matrix depends on the current position of the attackers, which are time-dependent. As a result, it cannot be computed off-line in contrast to the position-based feedback matrix for defense.

11.5.2.2 Example

For example, consider the case where a defender r_j lies in sector s_0, Figure 11.5. We compute the minimum distance from each possible one-step reachable destination of the defender, $\{s_0, s_1, \ldots, s_8\}$, to the most probable future locations of an attacker for the next time-stage, $\mathcal{L}^a[1]$. These locations can be computed using Equation (11.17). For example, $\mathcal{L}^a[1]$ can be defined as the two neighboring sectors of the attacker which are closer to

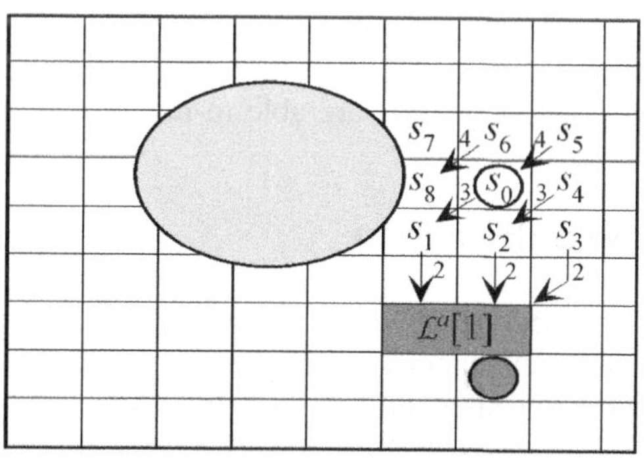

Figure 11.5 Computing the minimum distance from each neighboring sector, $\{s_1, \ldots, s_8\}$, of a defender that lies in s_0 to the set of the most probable future locations of an attacker after one time-stage, $\mathcal{L}^a[1]$.

the defense zone, Figure 11.5. In this case, we have:

$$\gamma_{s0}^{d} = 3, \quad \sum_{s_n} \{\gamma_{s_n}^{d}\} = \gamma_{s_1}^{d} + \ldots + \gamma_{s_8}^{d} = 27.$$

Hence, according to (11.22), the probability that the defender will move from sector s_0 to sector s_1 after one time-stage is:

$$g_{s_1 \leftarrow s_0, r_j}^{d,p}[1] = \frac{(3+27)-2}{(9-1)\cdot(3+27)} = \frac{28}{240}.$$

It is straightforward to show that

$$\sum_{s_n} g_{s_n \leftarrow s_0, r_j}^{d,p}[1] = \frac{213}{240} < 1$$

and

$$1 - \sum_{s_n} g_{s_n \leftarrow s_0, r_j}^{d,p}[1] = \frac{27}{240} = g_{s_0 \leftarrow s_0, r_j}^{d,p}[1],$$

where $g_{s_0 \leftarrow s_0, r_j}^{d,p}[1]$ is the probability of staying in sector s_0. Similarly, we can define transition probabilities for each future time $t \in \mathcal{T}$.

11.5.2.3 A velocity-based feedback matrix

Moreover, a velocity-based feedback matrix, $\mathbf{G}^{d,v}$, can be defined similarly to Equations (11.18) and (11.19), while a combined position and velocity-based feedback matrix, $\mathbf{G}^{d}$, can be defined similarly to Equation (11.21).

Summarizing, in both path planning algorithms for defense and attack, a feedback matrix, $\mathbf{G}$, can incorporate information about future locations of adversarial vehicles for a large optimization horizon, based on current available data, such as the position and velocity of adversarial vehicles. The use of such feedback matrices constitutes an advantageous element of this method in contrast to prior work, which is a path planning scheme versus possibly intelligent opponents that are able to move in an unpredictable way.

11.5.3 Simulations and discussion

The efficiency of the LP-based path planning is tested in a RoboFlag drill created in Matlab which involves three attackers and three defenders in an arena of 300 sectors. According to this scenario, an attacker becomes inactive when it is intercepted by a defender.

At the beginning, the LP-based path planning for defense is implemented with $T_p = 6$, $\alpha^d = -1$ and $\beta^d = 0$, which implies that the defenders' priority is only the interception of the attackers. The attackers follow pre-specified paths towards the defense zone that

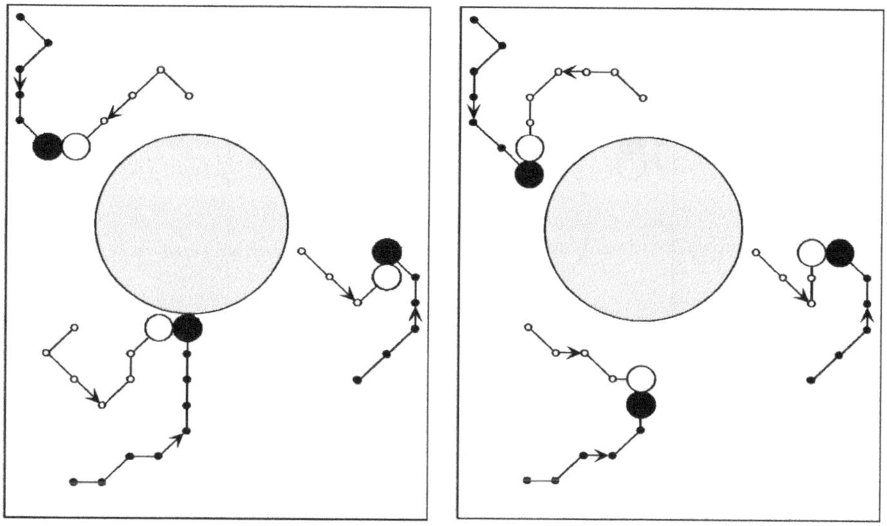

(a) (b)

Figure 11.6 Defenders optimize their paths according to the LP-based path planning against unknown but pre-specified attackers' paths. In (a) a position-based feedback matrix is used to describe the attackers' future locations, while in (b) a combined position- and velocity-based feedback matrix is used.

are unknown to the defenders. For this reason, the defenders use two different stochastic feedback matrices to describe possible future locations of the attackers. First, they use a position-based stochastic feedback matrix for the attackers' future locations according to (11.17), and, second, they use a combined position- and velocity-based feedback matrix according to (11.21).

Figure 11.6(a) shows that the defenders are able to predict the attackers' future locations by using the proposed LP-based path planning and a position-based feedback matrix for the attackers' future locations. At the same time coordination is achieved, since each defender is assigned to a different attacker. The algorithm runs in 3 sec per iteration using Matlab/CPLEX.

In Figure 11.6(b) a combined position- and velocity-based feedback matrix has been used according to (11.21) and with $\nu = \xi = 0.5$. Again the attackers use pre-specified paths towards the defense zone that are unknown to the defenders. In comparison with Figure 11.6(a), we noticed that in this case some defenders move faster towards the attackers due to the fact that their prediction about the attackers' next positions is more accurate.

The efficiency of the LP-based path planning was also tested in a more realistic situation, when both defenders and attackers optimize their paths with $T_p = 6$, Figure 11.7. Both teams make use of a position-based feedback matrix to predict their opponent's future locations. Also, the defenders attach weight $\alpha^d = -0.95$ to getting closer to the attackers and weight $\beta^d = -0.05$ to staying closer to the defense zone. On the other hand, the attackers use $\alpha^a = 0.99$ and $\beta^a = -0.01$, which means that they attach more weight to avoiding the defenders than getting closer to the defense zone. According to Figure 11.7, attackers are now able to infiltrate the defense zone. On the other hand,

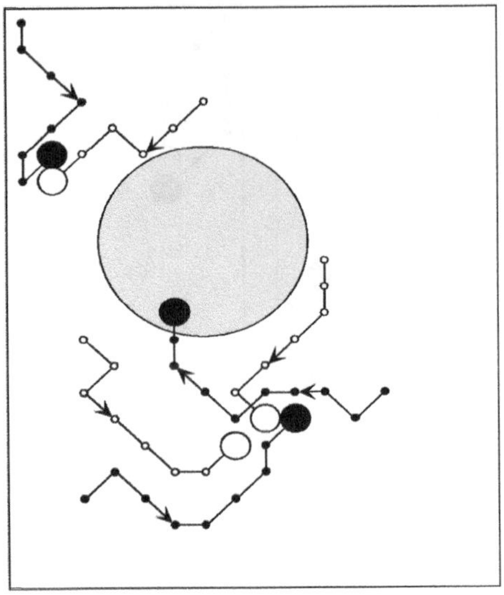

Figure 11.7 Both defenders and attackers optimize their paths according to the LP-based path planning.

defenders make the attackers follow longer paths towards the defense zone, which means that it is more difficult for the attackers to find a clear path towards the defense zone. Hence, the proposed LP-based algorithm can be used for effective defense and attack planning.

11.6 CONCLUSION

In this chapter, the problem of multi-vehicle coordination in an adversarial environment was formulated as a linear programming optimization. Both friendly and enemy vehicles were modeled as different resource types in an arena of sectors, and the path planning problem was viewed as a resource allocation problem. A simplified model allowed the use of linear programming, while enemy forces were assumed to follow stochastic feedback laws based on their current positions and velocities. The solution was implemented according to a receding horizon strategy due to model uncertainties. The utility of the LP-based algorithm was tested in the RoboFlag drill, where both teams of vehicles have equal path planning capabilities using the proposed algorithm. Results showed that the LP-based algorithm can be used for effective multi-vehicle path planning in an adversarial environment.

ACKNOWLEDGEMENTS

Research supported by NSF grant #ECS-0501394, AFOSR/MURI grant #F49620-01-1-0361, and ARO grant #W911NF–04–1–0316.

REFERENCES

Bemporad A and Morari M 1999 Control of systems integrating logic, dynamics and constraints. *Automatica* **35**, 407–428.

Bertsimas D and Tsitsiklis JN 1997 *Introduction to Linear Optimization*, Athena Scientific, Belmont, MA.

Chandler PR 2001 UAV cooperative control. In *Proc. American Control Conference*, Arlington, VA, pp. 50–55.

Chandler PR and Pachter M 1998 Research issues in autonomous control of tactical UAVs. In *Proc. American Control Conference*, Philadelphia, PA, pp. 394–398.

Chasparis GC 2004 Linear-Programming-Based Multi-Vehicle Path Planning with Adversaries, Masters thesis, University of California, Los Angeles, CA.

Chasparis GC and Shamma JS 2005 Linear-programming-based multi-vehicle path planning with adversaries. In *Proc. American Control Conference*, Portland, OR, pp. 296–301.

D'Andrea R and Murray RM 2003 The RoboFlag competition. In *Proc. American Control Conference*, Denver, CO, pp. 650–655.

Daniel-Berhe S, Ait-Rami N, Shamma JS and Speyer J 2001 Optimization based battle management. In *Proc. American Control Conference*, Arlington, VA, pp. 4711–4715.

Dogan A 2003 Probabilistic path planning for UAVs. In *Proc. 2nd AIAA Unmanned Unlimited Systems, Technologies, and Operations*, San Diego, CA.

Earl MG and D'Andrea R 2002a A study in cooperative control: The RoboFlag Drill. In *Proc. American Control Conference*, Anchorage, AK, pp. 1811–1812.

Earl MG and D'Andrea R 2002b Modeling and control of a multi-agent system using mixed-integer linear programming. In *Proc. 41st IEEE Conference on Decision and Control*, Las Vegas, NE, pp. 107–111.

Flint M, Polycarpou M and Fernandez E 2002 Cooperative path-planning. In *Proc. IFAC 15th Triennial World Congress*, Barcelona, Spain.

Frazzoli E and Bullo F 2004 Decentralized algorithms for vehicle routing in a stochastic time-varying environment. In *Proc. 43rd IEEE Conference on Decision and Control*, Paradise Island, Bahamas, pp. 3357–3363.

McLain TW and Beard RW 2003 Cooperative path planning for timing-critical missions. In *Proc. American Control Conference*, Denver, CO, pp. 296–301.

Richards A and How JP 2002 Aircraft trajectory planning with collision avoidance using mixed integer linear programming. In *Proc. American Control Conference*, Anchorage, AK, pp. 1936–1941.

Richards A, Bellingham J, Tillerson M and How J 2002 Coordination and control of multiple UAVs. In *Proc. AIAA Guidance, Navigation and Control Conference*, Monterey, CA. AIAA Paper 2002–4588.

Richards A, Schouwenaars T, How JP and Feron E 2002 Spacecraft trajectory planning with avoidance constraints using mixed-integer linear programming. *AIAA Journal of Guidance, Control and Dynamics* **25**, 755–764.

Schouwenaars T, De Moor B, Feron E and How J 2001 Mixed integer programming for multi-vehicle path planning. In *Proc. European Control Conference*, Porto, Portugal, pp. 2603–2608.

12

Characterization of LQG differential games with different information patterns

Ashitosh Swarup and Jason L. Speyer

12.1 INTRODUCTION

The linear quadratic pursuit-evasion game solved in Bryson and Ho (1979) can be generalized to include uncertainties in the linear dynamical system as well as in the players' measurements (Problem 1). Each player has no knowledge about his opponent's measurement or control history. The conditions on the above problem can be relaxed so that the players make a common, noisy measurement of the state, but do not share control information (Problem 2). A third version of the problem is the one in which one player makes perfect measurements of the state while another makes imperfect, noisy measurements (Problem 3). The continuous-time version of Problem 3 without process noise had previously been thought to have an infinite-dimensional control kernel operating on the measurement history (Rodes and Luenberger 1969; Behn and Ho 1968). To avoid this difficulty, the problem formulation was restricted further by assuming that the perfect measurement player has access to either his opponent's measurement or state estimate (Behn and Ho 1968). Willman (1969) presented a formal numerical solution to Problem 1. No proof of convergence or optimality was presented, and the algorithm was found to converge for only a very small range of parameters based on a discrete-time scalar, two-stage problem. No statement has been made about Problem 2. The problem considered here is the discrete-time version of Problem 3. Process noise is included in the dynamical system, and no assumption is made about what each player knows about his opponent. The problem is first formulated as a discrete-time linear-exponential-Gaussian (LEG) problem with an associated parameter θ multiplying the quadratic argument of the exponential. This class of problems has been shown to be equivalent to a deterministic game (Whittle 1981) where not only the perfect measurement player but also the process noise, measurement noise, and initial conditions can be adversarial to the player with noisy

Cooperative Control of Distributed Multi-Agent Systems edited by Jeff S. Shamma

measurements when the parameter is negative ($\theta < 0$). All these adversaries are assumed to have perfect knowledge of the state, but play against a player with only partial state information. Therefore, our formulation fits into this special class of problems.

To obtain the solution to the discrete-time LQG game problem, the associated parameter θ involved in the LEG solution is allowed to go to zero. It is assumed that the number of noisy measurements (m) is greater than the number of controls of the adversary (q_p) such that the adversary's control coefficient (Γ_p) does not lie in the null space of the measurement matrix (H). The essential features in the development of the discrete-time LQG game problem are: (1) the construction of the strategies of the LEG game adversaries involving an estimator for the noisy measurement adversary; (2) the construction of the LEG game equilibrium cost; and (3) the limit of the LEG game solution to the LQG game solution. The main feature of the LQG game solution is the structure of the filter, henceforth referred to as the LQG Game Filter wherever necessary. The noisy measurement player does not attempt to construct his adversary's strategy in his estimator, since the latter is an intelligent adversary and not a random variable. Therefore, no *a priori* assumption is made by the noisy measurement player about his adversary's control process. Consequently, the *a priori* error variance associated with that direction in the propagation stage is indeterminate. However, the *a posteriori* error variance is finite, allowing all the system states to be estimated. The filter also has specific correlation properties that differ from those of the Kalman filter. Furthermore, the certainty equivalence principle still holds in that the gains for the strategies are the same as those obtained from the deterministic discrete-time LQ game solution.

The chapter is organized in the following way. In Section 12.2 the discrete-time LQG game problem is formulated where each player has different information patterns. We redefine the term "information set" for this class of games to include the corresponding player's control history. In Section 12.3 the solution of the LQG game as the limit of the LEG game is given. The solution process for the LEG problem involves the decomposition of the exponential cost into an element involving a function of the past states, measurements, and controls (where the state estimator is constructed) and a function of future states and controls (where the strategies are constructed). Then, the equilibrium cost function is constructed by averaging over all random variables. Note that no attempt is made to average over the full information adversary's strategy. The LEG to LQG limit of the cost as the parameter $\theta \to 0$ is considered in Section 12.4. In Section 12.5 we present some interesting statistical properties of the filter used by the player with imperfect measurements. In Section 12.6, we look at the equilibrium value of the cost in some detail. In Section 12.7, we compare this solution with other possible solutions. The notion of the deterministic game saddle point is extended to games with uncertainty in Section 12.8; we postulate that the deterministic saddle becomes a saddle interval for the class of finite-dimensional solutions. In Section 12.9 we present the conclusion.

12.2 FORMULATION OF THE DISCRETE-TIME LQG GAME

The discrete-time LQG game with different information patterns is formulated as a zero-sum differential game between two adversaries u_p (called the pursuer) and u_e (called the evader). The discrete-time system dynamics is

$$x_{i+1} = A_i x_i + \Gamma_{e_i} u_{e_i} + \Gamma_{p_i} u_{p_i} + w_i, \quad x(0) = x_0 \tag{12.1}$$

where the dimensions of the state, evader control, pursuer control, and process uncertainty are $x_i \in \Re^n$, $u_{e_i} \in \Re^{q_e}$, $u_{p_i} \in \Re^{q_p}$, $w_i \in \Re^n$, respectively, and $\Gamma_{e_i}, \Gamma_{p_i}$ are rank q_e, q_p, respectively. Assume that w_i is a zero mean white noise sequence with variance W_i and x_0 is Gaussian with mean $\overline{x}_0$ and variance $\overline{P}_0$. It is assumed that the pursuer has access to perfect measurements of the state and that the evader makes noisy partial measurement z_i of the state as $z_i = H_i x_i + v_i$ where $v_i \in \Re^m$ is the zero-mean white measurement noise sequence with variance V_i. Assume that v_i, w_i, and x_0 are independent. Let $X_i = \{x_0, x_1, \ldots, x_i\}$ and $Z_i = \{z_0, z_1, \ldots, z_i\}$ be the respective observation histories of the pursuer and the evader. Also define $U_{p_i} \triangleq \{u_{p_0}, u_{p_1}, \ldots, u_{p_i}\}$, $U_e \triangleq \{u_{e_0}, u_{e_1}, \ldots, u_{e_i}\}$. Then, we define the *information set* of each player as:

$$Y_{p_i} \triangleq \{X_i, U_{p_{i-1}}\}, \qquad Y_{e_i} \triangleq \{Z_i, U_{e_{i-1}}\}.$$

The information sets are not shared. Based on this definition, it is easy to see how Problem 2 falls into the class of stochastic game problems with unshared information. To elaborate, we define an information set of a player as all the information the player knows. The discrete-time LQG game problem with different information patterns is to find the strategy $u_{e_i}(Y_{e_i})$ which minimizes and the strategy $u_{p_i}(Y_{p_i})$ which maximizes the expected cost criterion $J(u_p, u_e) = E[S]$, where $E[\cdot]$ denotes the expectation operator, and S is a quadratic function of the form:

$$S = \frac{1}{2}\left[\|x_N\|^2_{Q_N} + \sum_{i=0}^{N-1} \left(\|x_i\|^2_{Q_i} + \|u_{p_i}\|^2_{R_{p_i}} + \|u_{p_i}\|^2_{R_{e_i}}\right)\right]. \tag{12.2}$$

subject to the linear dynamic equation (12.1). In the cost function the symmetric weightings have the properties $Q_N \geq 0$, $Q_i \geq 0$, $R_{e_i} > 0$, and $R_{p_i} < 0$. The equilibrium controls sought as the solution to this game problem are functions of their respective information sets. Using the superscript " * " to denote the equilibrium strategies, we see that:

$$u^*_{p_i} \triangleq u^*_{p_i}(Y_{p_i}) \qquad u^*_{e_i} \triangleq u^*_{e_i}(Y_{e_i}).$$

Ideally, we would like the equilibrium strategies to be finite-dimensional, i.e. based on only their current observation, or, at worst, a summary of the observation sequence, rather than having to play a controller which is a function of the entire observation history.

12.3 SOLUTION OF THE LQG GAME AS THE LIMIT TO THE LEG GAME

The LEG Game problem is a modification of the LEG control problem. The LEG control problem with imperfect information was first solved by Whittle (1981), which involved a one-step delayed information pattern. The LEG Game problem considered in this chapter is an extension of the results of Fan, Speyer and Jaensch (Fan *et al.* 1994), who solved the LEG control problem with current imperfect information.

12.3.1 Problem formulation of the LEG Game

The LEG Game problem can be stated as $\Psi^* = \min_{U_{e_{N-1}}} \max_{U_{p_{N-1}}} \Psi$, where $\Psi = E[e^{-\theta S}]$, with S being defined as in (12.2). θ is a parameter, which is chosen to be negative, for reasons which will be explained in a future section.

Consider explicitly the expectation operator

$$E[e^{-\theta S}] = E_{x_0,\mathbb{W},\mathbb{V}}[e^{-\theta S}],$$

where x_0,

$$\mathbb{W} = \{w_0, w_1, \ldots, w_{N-1}\}, \quad \text{and} \quad \mathbb{V} = \{v_0, v_1, \ldots, v_N\}$$

are the underlying random variables.

We can also write the expectation explicitly in terms of integrals of density functions w.r.t. the random variables as

$$\begin{aligned} E[e^{-\theta S}] = \alpha \int e^{-\theta S} e^{-1/2\|x_0-\overline{x}_0\|^2_{P_0^{-1}}} \\ \times \prod_{i=0}^{N-1} \left[e^{-1/2\|w_i\|^2_{W_i^{-1}}} \times e^{-1/2\|v_i\|^2_{V_i^{-1}}} \right] \\ \times e^{-1/2\|v_N\|^2_{V_N^{-1}}} dx_0 d\mathbb{W} d\mathbb{V}, \end{aligned} \tag{12.3}$$

where

$$\alpha = \frac{1}{|P_0|^{1/2}} \frac{1}{|V_N|^{1/2}} \prod_{i=0}^{N-1} \frac{1}{|W_i|^{1/2}|V_i|^{1/2}}$$

The key to solving the LEG problem is the fact that the integration w.r.t. the random variable is equivalent to extremizing the argument of the exponent w.r.t. that random variable. Note also that the integration results in the division by the square root of the determinant of the term weighting the norm of that random variable. This fact will be explained further in a subsequent section.

From Equation (12.3), we can define:

$$\begin{aligned} S^c = S + \frac{1}{2} \Bigg[\|x_0 - \overline{x}_0\|^2_{(\theta P_0)^{-1}} + \|v_N\|^2_{(\theta V_N)^{-1}} \\ + \sum_{i=0}^{N-1} \left(\|w_i\|^2_{(\theta W_i)^{-1}} + \|v_i\|^2_{(\theta V_i)^{-1}} \right) \Bigg]. \end{aligned} \tag{12.4}$$

Then, clearly,

$$\min_{U_e} \max_{U_p} E[e^{-\theta S}] = \beta e^{-\theta S^{c^*}}, \tag{12.5}$$

where

$$S^{c^*} = \min_{U_e} \max_{U_p} \operatorname*{ext}_{x_0} \operatorname*{ext}_{\mathbb{W}} \operatorname*{ext}_{\mathbb{V}} S^c. \tag{12.6}$$

β is a constant term which results from the integration, and is evaluated separately from S^{c^*}.

From (12.6) we see that the stochastic problem has been converted into a deterministic game problem with additional players. While solving the game problem, it is convenient to make a change of random variables, with x_0, $\mathbb{W}$ and $\mathbb{V}$ being replaced by X and Z. Thus, finally, the game problem of interest is:

$$S^{c^*} = \min_{U_e} \max_{U_p} \operatorname*{ext}_{X} \operatorname*{ext}_{Z} S^c. \tag{12.7}$$

12.3.2 Solution to the LEG Game problem

In this section, we consider the various aspects involved in finding the equilibrium solutions to the LEG Game Problem. In particular, we see that the quadratic function S^c can be split into two parts, with one part yielding the evader's filter, and the second part yielding the equilibrium strategies for u_p and u_e. It will also be seen that the equilibrium strategies are finite-dimensional.

12.3.2.1 Construction of the past and future stresses

Let us split up S^c as in Fan *et al.* (1994): $S^c = S_{1_i} + S_{2_i}$, where

$$\begin{aligned} S_{1_i} = {} & \frac{1}{2}\|x_0 - \overline{x}_0\|^2_{(\theta \overline{P}_0)^{-1}} + \|z_i - H_i x_i\|^2_{(\theta V_i)^{-1}} \\ & + \frac{1}{2}\sum_{j=0}^{i-1}[\|x_j\|^2_{Q_j} + \|u_{p_j}\|^2_{R_{p_j}} + \|u_{p_j}\|^2_{R_{e_j}} \\ & + \|x_{j+1} - A_j x_j - \Gamma_{e_j} u_{e_j} - \Gamma_{p_j} u_{p_j}\|^2_{(\theta W_j)^{-1}} \\ & + \|z_j - H_j x_j\|^2_{(\theta V_j)^{-1}}], \end{aligned} \tag{12.8}$$

$$\begin{aligned} S_{2_i} = {} & \frac{1}{2}\|x_N\|^2_{Q_N} + \frac{1}{2}\sum_{j=i}^{N-1}[\|x_j\|^2_{Q_j} + \|u_{p_j}\|^2_{R_{p_j}} \\ & + \|u_{p_j}\|^2_{R_{e_j}} \\ & + \|x_{j+1} - A_j x_j - \Gamma_{e_j} u_{e_j} - \Gamma_{p_j} u_{p_j}\|^2_{(\theta W_j)^{-1}}] \\ & + \sum_{j=i+1}^{N} \|z_j - H_j x_j\|^2_{(\theta V_j)^{-1}}. \end{aligned} \tag{12.9}$$

Note that in the above, we have replaced the noise processes w and v by x and z from the state and observation equations, respectively.

Next, we define

$$S_i^p = \underset{X_{i-1}}{\text{ext}} \ \underset{U_{p_{i-1}}}{\max} \ S_{1_i},$$

$$S_i^f = \underset{U_{P_i}^N}{\max} \ \underset{U_{e_i}^N}{\max} \ \underset{X_{i+1}^N}{\text{ext}} \ \underset{Z_{i+1}^N}{\text{ext}} \ S_{2_i},$$

where, for any sequence $\{y_0, y_1, \ldots, y_N\}$,

$$Y_i \triangleq \{y_0, y_1, \ldots, y_i\}, \qquad Y_i^N \triangleq \{y_i, y_{i+1}, \ldots, y_N\}.$$

The terms S^p and S^f are called the Past Stress and the Future Stress respectively. It will be seen that the evader's filter equation results from the past stress, while the equilibrium controllers result from the future stress.

12.3.2.2 Determination of the LEG filter from the past stress

The filter for the LEG Game problem is similar to the filter derived in the LEG control problem (Fan *et al.* 1994), and is based on the following:

Theorem 12.3.1

$$S_i^p = \|x_i - \hat{x}_i\|^2_{(\theta P_i)^{-1}} + \hat{L}_i(Z_i),$$

where the matrix P_i satisfies the equations

$$P_i = (\overline{P}_i^{-1} + H_i^T V_i^{-1} H_i)^{-1}, \quad i = 0, 1, \ldots, N-1,$$

$$\overline{P}_{i+1} = W_i + \Gamma_{p_i}(\theta R_{p_i})^{-1}\Gamma_{p_i}^T + A_i(P_i^{-1} + \theta Q_i)^{-1}A_i^T, \qquad i = 0, 1, \ldots, N-2, \tag{12.10}$$

and the term $\hat{L}_i(Z_i)$ is a function independent of x_i.

The term $\hat{x}_i$ and $\overline{x}_i$ are u_{e_i}'s posteriori *and* a priori *estimates of the state x_i, respectively, and are given by*

$$\hat{x}_i = \overline{x}_i + P_i H_i^T V_i^{-1}(z_i - H_i\overline{x}_i), \tag{12.11}$$

$$\overline{x}_{i+1} = A_i(I + \theta P_i Q_i)^{-1}\hat{x}_i + \Gamma_{e_i} u_{e_i}. \tag{12.12}$$

Proof. The proof follows the proof in (Fan *et al.* 1994), the key difference being the fact that we now have the extra player u_p both in the state equation and in S_1.

Remark 12.3.1 When we extremize S_1 w.r.t. u_p to construct the filter, we see that u_p can take advantage of his perfect state information and inject noise into u_e's filter in the direction of Γ_p. As $\theta \to 0$, u_p can inject infinite noise into u_e's filter. Therefore, u_e adopts a pessimistic viewpoint while designing his filter, which involves constructing the filter based on u_p's worst-case control.

Note that in the limit, it can be shown that the equation to determine the *a priori* estimate of the state becomes indeterminate in a certain direction, due to the singularity of $\overline{P}_i^{-1}$. It is, therefore, necessary to combine both equations into one state estimate equation. From (12.11) and (12.12) and after a bit of algebra, we get:

$$\begin{aligned}\hat{x}_i = {} & P_i \overline{P}_i^{-1} \left[(I + \theta P_{i-1} Q_{i-1})^{-1} A_{i-1} \hat{x}_{i-1} + \Gamma_{e_{i-1}} u_{e_{i-1}}\right] \\ & + P_i H_i^T V_i^{-1} z_i. \end{aligned} \tag{12.13}$$

12.3.2.3 Determination of the LEG Game strategies from the future stress

Once again, following the procedure in (Fan *et al.* 1994), we have:

Theorem 12.3.2 *The future stress S^f is quadratic in the state variable, i.e. $S_i^f = \|x_i\|^2_{\Pi_i}$, where the matrix Π_i satisfies the equation:*

$$\begin{aligned}\Pi_i = {} & Q_i + A_i^T \Pi_{i+1} (I + \Gamma_{e_i} R_{e_i}^{-1} \Gamma_{e_i}^T \Pi_{i+1} + \Gamma_{p_i} R_{p_i}^{-1} \Gamma_{p_i}^T \Pi_{i+1} \\ & + \theta W_i \Pi_{i+1})^{-1} A_i,\end{aligned}$$

with $\Pi_N = Q_N$. The equilibrium controls $u^*_{p_i}$ and $u^*_{e_i}$ are linear controllers, and are functions of the state and the state estimate, respectively.

$$u^*_{p_i} = k_{p_i} x_i, \qquad u^*_{e_i} = k_{e_i} x_i,$$

where

$$\begin{aligned}k_{p_i} = {} & - R_{p_i}^{-1} \Gamma_{p_i}^T \Pi_{i+1} (I + \Gamma_{e_i} R_{e_i}^{-1} \Gamma_{e_i}^T \Pi_{i+1} \\ & + \Gamma_{p_i} R_{p_i}^{-1} \Gamma_{p_i}^T \Pi_{i+1} + \theta W_i \Pi_{i+1})^{-1} A_i, \\ k_{e_i} = {} & - R_{e_i}^{-1} \Gamma_{e_i}^T \Pi_{i+1} (I + \Gamma_{e_i} R_{e_i}^{-1} \Gamma_{e_i}^T \Pi_{i+1} \\ & + \Gamma_{p_i} R_{p_i}^{-1} \Gamma_{p_i}^T \Pi_{i+1} + \theta W_i \Pi_{i+1})^{-1} A_i.\end{aligned}$$

Proof. The result follows readily by extending the result obtained in (Fan *et al.* 1994) to accommodate the player u_p.

12.3.2.4 The worst-case state

We have noted that the state x_i, being a random variable, is also a player in the game. We thus can compute:

$$x_i^* = \arg\min_{x_i}[S_i^p + S_i^f]$$

It can be shown, (Fan *et al.* 1994), that:

$$x_i^* = (I + \theta P_i \Pi_i)\hat{x}_i \tag{12.14}$$

The equilibrium strategies derived in Theorem 12.3.2 are functions of x_i, and after extremizing $[S_i^p + S_i^f]$ w.r.t. x_i, we can express u_{e_i}'s strategy as a function of $\hat{x}_i$, where x_i is replaced by x_i^* using (12.14). Also, we can show that $z_i^* = H_i x_i^*$.

12.3.3 Filter properties for small values of θ

In order to find the limiting filter, it is convenient to first express the filter equation for small values of θ, and then take the limit as $\theta \to 0$. From Equation (12.13), it is clear that for small values of θ,

$$\begin{aligned}\hat{x}_i = P_i \overline{P}_i^{-1}\left[A_{i-1}\hat{x}_{i-1} + \Gamma_{e_{i-1}} u_{e_{i-1}}\right] + P_i H_i^T V_i^{-1} z_i \\ + \mathcal{O}(\theta).\end{aligned}$$

Lemma 12.3.3 *For small values of* θ,

$$\overline{P}_i^{-1}\Gamma_{p_{i-1}} = \mathcal{O}(\theta).$$

Proof. The proof follows readily by using the matrix inversion lemma to get an expression for $\overline{P}_i^{-1}$ using the equation:

$$\begin{aligned}\overline{P}_i = W_{i-1} + \Gamma_{p_{i-1}}(\theta R_{p_{i-1}})^{-1}\Gamma_{p_{i-1}}^T \\ + A_{i-1}(P_{i-1}^{-1} + \theta Q_{i-1})^{-1} A_{i-1}^T.\end{aligned}$$

12.3.3.1 The filter error equation for small values of θ

Lemma 12.3.4 *Let* $e_i \triangleq x_i - \hat{x}_i$. *Then,*

$$e_i = P_i \overline{P}_i^{-1}\left[A_{i-1}e_{i-1} + w_{i-1}\right] - P_i H_i^T V_i^{-1} v_i + \mathcal{O}(\theta).$$

Proof. We know that: $\hat{x}_i = P_i \overline{P}_i^{-1}\left[A_{i-1}\hat{x}_{i-1} + \Gamma_{e_{i-1}} u_{e_{i-1}}\right] + P_i H_i^T V_i^{-1} z_i + \mathcal{O}(\theta)$. To this equation, add the zero term $P_i \overline{P}_i^{-1}\Gamma_{p_{i-1}} u_{p_{i-1}} + w_i - P_i \overline{P}_i^{-1}\Gamma_{p_{i-1}} u_{p_{i-1}} - w_i$, where, by lemma 12.3.3, $P_i\overline{P}_i^{-1}\Gamma_{p_{i-1}}u_{p_{i-1}} = \mathcal{O}(\theta)$. Also, write $z_i = H_i x_i + v_i$. After a bit of algebra, we get:

$$\begin{aligned}\hat{x}_i &= x_i + P_i\overline{P}_i^{-1}\left[-A_{i-1}e_{i-1} - w_{i-1}\right] + P_i H_i^T V_i^{-1} v_i \\ &\quad + \mathcal{O}(\theta), \\ \Rightarrow e_i &= P_i\overline{P}_i^{-1}\left[A_{i-1}e_{i-1} + w_{i-1}\right] - P_i H_i^T V_i^{-1} v_i \\ &\quad + \mathcal{O}(\theta).\end{aligned}$$

12.3.3.2 The Mean and Variance of e_i

In this section, we derive expressions for the first two moments of e_i. We derive these for small values of θ, after which it is straightforward to take the limit as $\theta \to 0$. From the expression for e_i, we have:

$$\begin{aligned}E[e_i e_i^T] &= P_i\overline{P}_i^{-1}\left(A_{i-1}E[e_{i-1}e_{i-1}^T]A_{i-1}^T + W_{i-1}\right)\overline{P}_i^{-1}P_i \\ &\quad + P_i H_i^T V_i^{-1} H_i P_i + \mathcal{O}(\theta). \end{aligned} \tag{12.15}$$

Theorem 12.3.5

$$E[e_i e_i^T] = P_i + \mathcal{O}(\theta), \qquad i.e., \quad \lim_{\theta\to 0} E[e_i e_i^T] = P_i.$$

Proof. In order to prove this theorem, we will make use of Equation (12.15), and find its limiting value. We use an induction argument. The statement of the theorem is certainly true for $i = 0$. Let, if possible, the theorem be true for some $i - 1$, where $i > 1$. Then, Equation (12.15) can be written as:

$$\begin{aligned}E[e_i e_i^T] &= P_i\overline{P}_i^{-1}\left(A_{i-1}P_{i-1}A_{i-1}^T + W_{i-1}\right)\overline{P}_i^{-1}P_i \\ &\quad + P_i H_i^T V_i^{-1} H_i P_i + \mathcal{O}(\theta).\end{aligned}$$

In the above equation, we can add the zero term

$$\begin{aligned}&P_i\overline{P}_i^{-1}\Gamma_{p_{i-1}}\left(\theta R_{p_{i-1}}\right)^{-1}\Gamma_{p_{i-1}}^T\overline{P}_i^{-1}P_i - P_i\overline{P}_i^{-1}\Gamma_{p_{i-1}} \\ &\qquad \times\left(\theta R_{p_{i-1}}\right)^{-1}\Gamma_{p_{i-1}}^T\overline{P}_i^{-1}P_i,\end{aligned}$$

where $P_i\overline{P}_i^{-1}\Gamma_{p_{i-1}}\left(\theta R_{p_{i-1}}\right)^{-1}\overline{P}_i^{-1}P_i\Gamma_{p_{i-1}} = \mathcal{O}(\theta)$ by Lemma 12.3.3. The details of the proof are left to the reader. For small values of θ, we get:

$$\begin{aligned}E[e_i e_i^T] &= P_i + P_i\overline{P}_i^{-1}[-A_{i-1}\left(P_{i-1}^{-1} + \theta Q_{i-1}\right)^{-1}A_{i-1}^T \\ &\quad + A_{i-1}P_{i-1}A_{i-1}^T]\overline{P}_i^{-1}P_i + \mathcal{O}(\theta) \\ &= P_i + \mathcal{O}(\theta).\end{aligned}$$

Clearly, in the limit as $\theta \to 0$, we have $E[e_i e_i^T] = P_i$. Since we have shown that the statement of the theorem holds at time i given that it is true at time $i-1$, for $i > 1$, it follows from the principle of mathematical induction that the theorem is true for all i.

Theorem 12.3.6

$$E[e_i] \triangleq \bar{e}_i = \mathcal{O}(\theta), \qquad i.e. \quad \lim_{\theta \to 0} E[e_i] = 0.$$

Proof. This follows inductively from the recursive error equation, since $E\,[e_0] = 0$, and since all the noise processes are zero mean.

Remark 12.3.2 Note that in the limit, the error equation and the error statistics of the filter remain unchanged for all values of u_p, even if he plays a nonlinear controller at any time. This useful property is lost if the Kalman filter is used instead.

Remark 12.3.3 Even though $\overline{P}^{-1}$ is singular in the limit, the filter still estimates the entire state.

12.3.4 Construction of the LEG equilibrium cost function

From (12.5),

$$\min_{U_e} \max_{U_p} E[e^{-\theta S}] = \beta e^{-\theta S^{c^*}}. \tag{12.16}$$

The quantities β and S^{c^*} are given in the following:

Theorem 12.3.7 *The equilibrium cost function for the LEG Game with partial information and small θ has the form:*

$$J^*(u_p^*, u_e^*) = \frac{1}{|I + \theta \Gamma_y \mathcal{P}_y|^{1/2}} e^{-\frac{1}{2} ||\mu_y||^2_{\theta\left[I + \theta \Gamma_p \mathcal{P}_y^{-1}\right]^{-1} \Gamma_y} + \mathcal{O}(\theta^2)},$$

where the matrix $\mathcal{P}_y$ is finite, and has diagonal terms equal to $\{P_i + \mathcal{O}(\theta)\}$, $\{W_i\}$, and $\overline{P}_0$, and the matrix Γ_y has diagonal terms $\Gamma_i + \mathcal{O}(\theta)$ and off-diagonal terms at least $\mathcal{O}(\theta)$, where:

$$\begin{aligned}
\Gamma_i = & -(I + \theta Q_i P_i)^{-1} k_{e_i}^T \Gamma_{e_i}^T \Pi_{i+1} (I + \theta W_i \Pi_{i+1})^{-1} \\
& \times \left[I + \theta W_i \Pi_{i+1} + \Gamma_{e_i} R_{e_i}^{-1} \Gamma_{e_i}^T \Pi_{i+1}\right] \\
& \times \left[I + \Gamma_{e_i} R_{e_i}^{-1} \Gamma_{e_i}^T \Pi_{i+1} + \Gamma_{p_i} R_{p_i}^{-1} \Gamma_{p_i}^T \Pi_{i+1}\right. \\
& \left. + \theta W_i \Pi_{i+1}\right]^{-1} A_i (I + \theta P_i Q_i)^{-1} + \mathcal{O}(\theta^2).
\end{aligned}$$

Also,

$$\mu_y = \left[\tilde{e}_{N-1}^T \; \tilde{e}_{N-2}^T \; \cdots \; \tilde{e}_0^T \; 0 \; 0 \; \cdots \; 0 \; \overline{x}_0^T\right]^T.$$

□

Remark 12.3.4 There is no $\mathcal{O}(\theta)^2$ term in Γ_{N-1}, otherwise the expression is the same.

12.4 LQG GAME AS THE LIMIT OF THE LEG GAME

The solution to the LQG Game problem is sought as the limit of the LEG Game problem as the parameter $\theta \to 0$.

12.4.1 Behavior of filter in the limit

Before computing the limiting value of the cost, we need to examine the behavior of the filter in the limit. It is clear from the filter equations derived in a previous section that the relevant filter equations are in the limit are:

$$\begin{aligned} \hat{x}_i &= P_i \overline{P}_i^{-1} \left[A_{i-1} \hat{x}_{i-1} + \Gamma_{e_{i-1}} u_{e_{i-1}} \right] + P_i H_i^T V_i^{-1} z_i, \\ e_i &= P_i \overline{P}_i^{-1} \left[A_{i-1} e_{i-1} + w_{i-1} \right] - P_i H_i^T V_i^{-1} v_i, \\ E\left[e_i\right] &= 0, \qquad E\left[e_i e_i^T\right] = P_i. \end{aligned}$$

12.4.2 Limiting value of the cost

At this point, we are ready to take the limit of the LEG game solution as $\theta \to 0$. The quantity we consider is:

$$\Upsilon^* = \frac{J^* - 1}{-\theta}.$$

Then, in the limit, we can apply L'Hospital's rule i.e. differentiate the numerator and denominator w.r.t θ.

$$\lim_{\theta \to 0} \Upsilon^* = -\left(\frac{\partial J^*}{\partial \theta} \right)_{\theta=0}.$$

Thus, we need to find $\frac{\partial J^*}{\partial \theta}$, where:

$$J^* = \frac{1}{|I + \theta \Gamma_y \mathcal{P}_y|^{1/2}} e^{-\frac{1}{2} ||\mu_y||^2_{\theta\left[I + \theta \Gamma_p \mathcal{P}_y^{-1}\right]^{-1} \Gamma_y} + \mathcal{O}(\theta^2)},$$

Let $\beta' = e^{-\frac{1}{2} ||\mu_y||^2_{\theta\left[I + \theta \Gamma_p \mathcal{P}_y^{-1}\right]^{-1} \Gamma_y} + \mathcal{O}(\theta^2)}$. Now, we compute $\frac{\partial J^*}{\partial \theta}$, using the fact that for small θ,

$$|I + \theta Y| \approx 1 + \theta^{\mathrm{T}}(Y) + \mathcal{O}(\theta^2),$$

where ${}^{\mathrm{T}}(\cdot)$ denotes the trace operator. Therefore for small θ,

$$\frac{\partial |I + \theta Y|^{-1/2}}{\partial \theta} = \frac{\partial (1 + \theta^{\mathrm{T}}(Y) + \mathcal{O}(\theta^2))^{-1/2}}{\partial \theta}$$

$$= -\frac{1}{2}(^{\mathrm{T}}(Y) + \mathcal{O}(\theta))(1 + \theta^{\mathrm{T}}(Y) + \mathcal{O}(\theta^2))^{-3/2}.$$

Therefore,

$$\frac{\partial J^*}{\partial \theta} = -\frac{1}{2}\frac{||\mu_y||^2_{\left[I+\theta\Gamma_p\mathcal{P}_y^{-1}\right]^{-1}\Gamma_y} + \mathcal{O}(\theta)}{|I + \theta\Gamma_y\mathcal{P}_y|^{1/2}}\beta' - \frac{1}{2}(^{\mathrm{T}}(\Gamma_y\mathcal{P}_y) + \mathcal{O}(\theta))(1 + \theta^{\mathrm{T}}(\Gamma_y\mathcal{P}_y) + \mathcal{O}(\theta^2))^{-3/2}\beta'. \quad (12.17)$$

Setting $\theta = 0$ in in (12.17) gives:

$$\left(\frac{\partial J^*}{\partial \theta}\right)_{\theta=0} = -\frac{1}{2}||\mu_y^0||^2_{\Gamma_y^0} - \frac{1}{2}^{\mathrm{T}}(\Gamma_y^0\mathcal{P}_y^0).$$

Therefore,

$$\lim_{\theta\to 0} \Upsilon^* = \frac{1}{2}||\mu_y^0||^2_{\Gamma_y^0} + \frac{1}{2}^{\mathrm{T}}(\Gamma_y^0\mathcal{P}_y^0).$$

In the limit, it is clear that:

$$\mu_y^0 = \begin{bmatrix} 0 & 0 & \cdots & 0 & \overline{x}_0^T \end{bmatrix}^T$$

Also,

$$\Gamma_y^0 = diag[\Gamma_{N-1}, \Gamma_{N-2}, \ldots, \Gamma_0, \Pi_N, \Pi_{N-1}, \ldots, \Pi_0],$$

and since Γ_y^0 is block-diagonal, it is easy to verify:

$$^{\mathrm{T}}(\Gamma_y^0\mathcal{P}_y^0) =^{\mathrm{T}} (\overline{P}_0\Pi_0) + \sum_{i=0}^{N-1}\left[^{\mathrm{T}}(\Pi_{i+1}W_i) +^{\mathrm{T}} (\Gamma_i^0 P_i)\right].$$

Hence, we get the LQG equilibrium cost in the limit:

$$\lim_{\theta\to 0} \Upsilon^* = \frac{1}{2}\left[\|\overline{x}_0\|^2_{\Pi_0} +^{\mathrm{T}} (\Pi_0\overline{P}_0) + \sum_{i=0}^{N-1}\left[^{\mathrm{T}}(\Pi_{i+1}W_i) +^{\mathrm{T}} (\Gamma_i^0 P_i)\right]\right].$$

12.4.3 Convexity conditions

For this problem, it can be shown that for bounded controls to exist, the following well-known convexity conditions must be satisfied:

$$R_{p_i} + \Gamma_{p_i}^T \Pi_{i+1} \Gamma_{p_i} < 0, \tag{12.18}$$

$$R_{e_i} + \Gamma_{e_i}^T \Pi_{i+1} \Gamma_{e_i} > 0, \tag{12.19}$$

12.4.4 Results

We have derived the equilibrium cost function in the previous section. The limiting values of the control gains are computed in a straightforward manner. As noted earlier, for the case $H_i \Gamma_{p_{i-1}} \neq 0$, the filter equations are well behaved, and therefore, the game solution as a whole is well behaved. We can summarize all these results in the form of the following:

Theorem 12.4.1 *Consider the following quadratic cost function:*

$$J(u_p, u_e) = \frac{1}{2} E \left[\|x_N\|^2_{Q_N} + \sum_{i=0}^{N-1} \left(\|x_i\|^2_{Q_i} + \|u_{p_i}\|^2_{R_{p_i}} + \|u_{p_i}\|^2_{R_{e_i}} \right) \right],$$

subject to the state equation

$$x(i+1) = A_i x_i + \Gamma_{p_i} u_{p_i} + \Gamma_{e_i} u_{e_i} + w_i.$$

The pursuer u_p makes perfect measurements of the state x, while the evader, u_e, makes imperfect measurements z of the state, given by $z_i = H_i x_i + v_i$. The process noise, w_i, and the observation noise, v_i, are both Gaussian white noise processes, with statistics $(0, W_i)$ and $(0, V_i)$ respectively. The initial state x_0 is Gaussian, with statistics $(\overline{x}_0, \overline{P}_0)$. Then, the solution to the game problem:

$$J^*(u_p^*, u_e^*) = \max_{u_p} \min_{u_e} J(u_p, u_e),$$

has the equilibrium controllers of the form

$$u_{p_i}^* = k_{p_i}^0 x_i, \qquad u_{e_i}^* = k_{e_i}^0 \hat{x}_i,$$

where

$$\begin{aligned} k_{e_i}^0 = & -R_{e_i}^{-1} \Gamma_{e_i}^T \Pi_{i+1} (I + \Gamma_{e_i} R_{e_i}^{-1} \Gamma_{e_i}^T \Pi_{i+1} \\ & + \Gamma_{p_i} R_{p_i}^{-1} \Gamma_{p_i}^T \Pi_{i+1})^{-1} A_i, \\ k_{p_i}^0 = & -R_{p_i}^{-1} \Gamma_{p_i}^T \Pi_{i+1} (I + \Gamma_{e_i} R_{e_i}^{-1} \Gamma_{e_i}^T \Pi_{i+1} \\ & + \Gamma_{p_i} R_{p_i}^{-1} \Gamma_{p_i}^T \Pi_{i+1})^{-1} A_i. \end{aligned}$$

Π_i is found by the following backward equation

$$\Pi_i = Q_i + A_i^T \Pi_{i+1}(I + \Gamma_{e_i} R_{e_i}^{-1} \Gamma_{e_i}^T \Pi_{i+1} + \Gamma_{p_i} R_{p_i}^{-1} \Gamma_{p_i}^T \Pi_{i+1})^{-1} A_i.$$

Also, $\hat{x}_i$ is the estimate generated by the filter equation:

$$\hat{x}_i = P_i \overline{P}_i^{-1} \left[A_{i-1}\hat{x}_{i-1} + \Gamma_{e_{i-1}} u_{e_{i-1}}\right] + P_i H_i^T V_i^{-1} z_i,$$

where the matrix P_i satisfies the following recursions:

$$P_i = (\overline{P}_i^{-1} + H_i^T V_i^{-1} H_i)^{-1},$$

$$\overline{P}_{i+1}^{-1} = N_i^{-1} - N_i^{-1} A_i \left[A_i^T N_i^{-1} A_i + P_i^{-1}\right]^{-1} A_i^T N_i^{-1},$$

$$N_i^{-1} = W_i^{-1}\left[I - \Gamma_{p_i}\left(\Gamma_{p_i}^T W_i^{-1} \Gamma_{p_i}\right)^{-1} \Gamma_{p_i}^T W_i^{-1}\right]$$

The equilibrium cost function for this problem is

$$J^*(u_p^*, u_e^*) = \frac{1}{2}\left[\|\overline{x}_0\|_{\Pi_0}^2 + tr(\Pi_0 \overline{P}_0) + \sum_{i=0}^{N-1}\left(tr(\Pi_{i+1} W_i) + tr(\Phi_i^0 P_i)\right)\right]$$

and $\Phi_i^0 = k_{e_i}^T \left[\Gamma_{e_i}^T \Pi_{i+1} \Gamma_{e_i} + R_{e_i}\right] k_{e_i}$.

12.5 CORRELATION PROPERTIES OF THE LQG GAME FILTER IN THE LIMIT

In this section, we compute the correlation properties of the error with the state and the state estimate. These error correlation properties are derived after decomposing the state space into two directions (components). We start off by writing the estimator and error equations for $\theta = 0$:

$$\hat{x}_i = P_i \overline{P}_i^{-1} \left[A_{i-1}\hat{x}_{i-1} + \Gamma_{e_{i-1}} u_{e_{i-1}}\right] + P_i H_i^T V_i^{-1} z_i,$$

$$e_i = P_i \overline{P}_i^{-1} \left[A_{i-1} e_{i-1} + w_{i-1}\right] - P_i H_i^T V_i^{-1} v_i.$$

In order to decompose the above two equations into two orthogonal spaces, we need to examine the matrix $\overline{P}_i^{-1} P_i$.

12.5.1 Characteristics of the matrix $\overline{P}_i^{-1} P_i$

Let us first find the null space of $\overline{P}_i^{-1} P_i$. Consider:

$$\begin{aligned}\overline{P}_i^{-1} P_i x = 0 &\Rightarrow \left(P_i^{-1} - H_i^T V_i^{-1} H_i\right) P_i x = 0 \\ &\Rightarrow \left(I - H_i^T V_i^{-1} H_i P_i\right) x = 0 \\ &\Rightarrow x = H_i^T V_i^{-1} H_i P_i x.\end{aligned}$$

Thus, the null space of $\overline{P}_i^{-1} P_i$ are all those x that are invariant under the transformation $H_i^T V_i^{-1} H_i P_i$:

$$\mathcal{N}\left(\overline{P}_i^{-1} P_i\right) \equiv \left\{x : H_i^T V_i^{-1} H_i P_i x = x\right\}.$$

Lemma 12.5.1 *Let $\overline{P}_i^{-1} P_i \lambda_i = \mu_i \lambda_i$, where $\mu_i \neq 0$. Then, $\lambda_i \perp \Gamma_{p_{i-1}}$.*

Proof. Since $\Gamma_{p_{i-1}}^T \overline{P}_i^{-1} P_i = 0$, it is clear that $\Gamma_{p_{i-1}}^T \overline{P}_i^{-1} P_i \lambda_i = 0$, or, that $\mu_i \Gamma_{p_{i-1}}^T \lambda_i = 0$. Since $\mu_i \neq 0$, we have $\Gamma_{p_{i-1}}^T \lambda_i = 0$, or, $\Gamma_{p_{i-1}} \perp \lambda_i$.

Thus, the range space of $\overline{P}_i^{-1} P_i$ is orthogonal to $\Gamma_{p_{i-1}}$. It can be shown that μ_i is real.

12.5.2 Transformed filter equations

Let us now define:

$$\begin{aligned}T_i &\triangleq \left[\lambda_i^1 \; \lambda_i^2\right]^T, \qquad \hat{\eta}_i = \left[\hat{\eta}_i^{1^T} \; \hat{\eta}_i^{2^T}\right]^T \triangleq T_i \hat{x}_i, \\ \varepsilon_i &= \left[\varepsilon_i^{1^T} \; \varepsilon_i^{2^T}\right]^T \triangleq T_i e_i,\end{aligned}$$

where λ_i^2 is in the direction of $\mathcal{N}\left(\overline{P}_i^{-1} P_i\right)$, while λ_i^1 is in the direction of $\mathcal{R}\left(\overline{P}_i^{-1} P_i\right)$. We select $(\lambda_i^1, \lambda_i^2)$ such that $T_i T_i^T = I$ and $T_i^T T_i = I$, i.e., $T_i^{-1} \triangleq T_i^T$. Further, since $\lambda_i^1 \varepsilon \mathcal{R}(\overline{P}_i^{-1} P_i)$, let $\overline{P}_i^{-1} P_i \lambda_i^1 = \lambda_i^1 F_i$ where F_i is the matrix of nonzero eigenvalues μ_i. Substituting for $\hat{x}_i$ and e_i using these definitions, we get:

$$\begin{bmatrix}\hat{\eta}_i^1 \\ \hat{\eta}_i^2\end{bmatrix} = \begin{bmatrix} F_i^T \lambda_i^{1^T} T_i^{-1}\left(\tilde{A}_{i-1}\hat{\eta}_{i-1} + \Gamma_{e_{i-1}} u_{e_{i-1}}\right) + \lambda_i^{1^T} P_i H_i^T V_i^{-1} z_i \\ \lambda_i^{2^T} P_i H_i^T V_i^{-1} z_i \end{bmatrix},$$

$$\begin{bmatrix}\varepsilon_i^1 \\ \varepsilon_i^2\end{bmatrix} = \begin{bmatrix} F_i^T \lambda_i^{1^T} T_i^{-1}\left(\tilde{A}_{i-1}\varepsilon_{i-1} + w_{i-1}\right) - \lambda_i^{1^T} P_i H_i^T V_i^{-1} v_i \\ -\lambda_i^{2^T} P_i H_i^T V_i^{-1} v_i \end{bmatrix},$$

where $\tilde{A}_{i-1} = T_i A_{i-1} T_{i-1}^{-1}$, and we have used the fact that $\lambda_i^{2^T} P_i \overline{P}_i^{-1} = 0$. Further, we can rewrite:

$$\lambda_i^{2^T} P_i H_i^T V_i^{-1} z_i = \lambda_i^{2^T} P_i H_i^T V_i^{-1} H_i x_i + \lambda_i^{2^T} P_i H_i^T V_i^{-1} v_i$$
$$= \eta_i^{2^T} + \lambda_i^{2^T} P_i H_i^T V_i^{-1} v_i.$$

In the above equation, if we use the fact that λ_i^2 is invariant under $H_i^T V_i^{-1} H_i P_i$, $\lambda_i^{1^T} T_i^{-1} = [I, 0]$, and $\eta_i^2 \triangleq \lambda_i^2 x_i$, we get the transformed equations for the state estimate and the error as:

$$\begin{bmatrix} \hat{\eta}_i^1 \\ \hat{\eta}_i^2 \end{bmatrix} = \begin{bmatrix} F_i^T [I,0]\left(\tilde{A}_{i-1}\hat{\eta}_{i-1} + \Gamma_{e_{i-1}} u_{e_{i-1}}\right) + \lambda_i^{1^T} P_i H_i^T V_i^{-1} z_i \\ \eta_i^{2^T} + \lambda_i^{2^T} P_i H_i^T V_i^{-1} v_i \end{bmatrix},$$

$$\begin{bmatrix} \varepsilon_i^1 \\ \varepsilon_i^2 \end{bmatrix} = \begin{bmatrix} F_i^T [I,0]\left(\tilde{A}_{i-1}\varepsilon_{i-1} + w_{i-1}\right) - \lambda_i^{1^T} P_i H_i^T V_i^{-1} v_i \\ -\lambda_i^{2^T} P_i H_i^T V_i^{-1} v_i \end{bmatrix}.$$

12.5.3 Correlation properties of ε_i^2

It is easy to see that the $E\left[\varepsilon_i^2 x_i^T\right] = 0$. As a matter of fact, $E\left[\varepsilon_i^2 X_i^T\right] = 0$.

Let us next consider the term $E\left[\varepsilon_i^2 \hat{x}_i^T\right]$. Clearly,

$$E\left[\varepsilon_i^2 \hat{x}_i^T\right] = -\lambda_i^{2^T} P_i H_i^T V_i^{-1} H_i P_i = -\lambda_i^{2^T} P_i,$$

since λ_i^2 is invariant under $H_i^T V_i^{-1} H_i P_i$. Also, since $\lambda_i^2 \varepsilon \mathcal{N}\left(\overline{P}_i^{-1} P_i\right)$, it is easy to see that $E\left[\varepsilon_i^2 \hat{x}_i^T\right] \overline{P}_i^{-1} = 0$. Also, since $\lambda_i^{2^T} P_i \overline{P}_i^{-1} = 0$, it follows that $\lambda_i^{2^T} P_i$ is in the direction of $\Gamma_{p_{i-1}}$. Since $\lambda_i^1 \perp \Gamma_{p_{i-1}}$, it is clear that $\lambda_i^{2^T} P_i \lambda_i^1 = 0$. Using the invariant property of $\lambda_i^{2^T}$, we get $\lambda_i^{2^T} P_i H_i^T V_i^{-1} H_i P_i \lambda_i^1 = 0$. This implies that $E\left[\varepsilon_i^2 \hat{\eta}_i^{1^T}\right] = 0$, and that $E\left[\varepsilon_i^2 \varepsilon_i^{1^T}\right] = 0$. We can summarize these results in the form of the following theorem:

Theorem 12.5.2 *For the decomposed discrete-time LQG pursuit-evasion game filter, we have*

$$E\left[\varepsilon_i^2 X_i^T\right] = 0,$$
$$E\left[\varepsilon_i^2 \hat{\eta}_i^{1^T}\right] = 0,$$
$$E\left[\varepsilon_i^2 \varepsilon_i^{1^T}\right] = 0,$$
$$E\left[\varepsilon_i^2 \hat{x}_i^T\right] \overline{P}_i^{-1} = 0.$$

□

12.5.4 Correlation properties of ε_i^1

The only correlation property of ε_i^1 which readily follows from the above analysis is $E\left[\varepsilon_i^1 \varepsilon_i^{2^T}\right] = 0$. All other correlations are explicit functions of the controls. Since the controls are intelligent strategies and not random variables, it is impossible to assign any statistical properties to them. Therefore, it is impossible to directly compute the correlations without explicitly making an assumption on the controls.

12.6 COST FUNCTION PROPERTIES—EFFECT OF A PERTURBATION IN u_p

In this section, we look at the equilibrium cost function when u_p plays a perturbed control at time i while u_e plays his equilibrium strategy. We will derive an expression for the change in the cost function due to this perturbation. Since the equilibrium value of u_e is a linear strategy, the perturbation should be a linear function of the entire state history X_i due to the fact that all the random variables in this problem are Gaussian. Thus, let $u_{p_i}^p = u_{p_i}^* + \gamma_i X_i$, it can be shown that the difference between the perturbed cost and the equilibrium cost at every stage is:

$$J\left(u_p^p, u_e^*\right)_i - J\left(u_p^*, u_e^*\right)_i = -^{\mathrm{T}}\left[\gamma_i^T \Gamma_{p_i}^T \Pi_{i+1} \Gamma_{e_i} k_{e_i} P_{eX_i}\right]$$
$$+ \frac{1}{2}^{\mathrm{T}}\left[\gamma_i^T\left(\Gamma_{p_i}^T \Pi_{i+1} \Gamma_{p_i} + R_{p_i}\right)\gamma_i P_{XX_i}\right],$$

where $P_{XX_i} = E\left[X_i X_i^T\right]$, and $P_{eX_i} = E\left[e_i X_i^T\right]$.

We omit the computational details in order to keep the treatment concise. Thus, at each stage, we can find a γ_i^* that maximizes the difference in the cost:

$$\gamma_i^* = \left(\Gamma_{p_i}^T \Pi_{i+1} \Gamma_{p_i} + R_{p_i}\right)^{-1} \Gamma_{p_i}^T \Pi_{i+1} \Gamma_{e_i} k_{e_i} P_{eX_i} P_{XX_i}^{-1},$$

and the corresponding improvement in the equilibrium cost for u_p is:

$$\Delta J_i = -\frac{1}{2}^{\mathrm{T}}\left[P_{XX_i}^{-1} P_{eX_i}^T k_{e_i}^T \Gamma_{e_i}^T \Pi_{i+1} \Gamma_{p_i}\left(\Gamma_{p_i}^T \Pi_{i+1} \Gamma_{p_i}\right.\right.$$
$$\left.\left.+R_{p_i}\right)^{-1} \Gamma_{p_i}^T \Pi_{i+1} \Gamma_{e_i} k_{e_i} P_{eX_i} P_{XX_i}^{-1}\right]. \qquad (12.20)$$

Observe that the above quantity is always positive (i.e. the cost increases), since $\left(\Gamma_{p_i}^T \Pi_{i+1} \Gamma_{p_i} + R_{p_i}\right)$ is negative definite, by the convexity condition (12.18). Thus, we have shown that the pursuer can actually take advantage of the cross-correlation between the error and state in order to improve his part of the cost. The following points must be noted:

- Even if the pursuer plays a perturbed strategy at every stage in order to improve on his cost, the improvement he actually gets is bounded, i.e. he cannot make the cost function infinitely large.

- The ε_i^2 component of e is uncorrelated with the state, thus, a component of the cross-correlation P_{eX_i} is zero, i.e., the pursuer can avail himself of the cross-correlation only in a reduced dimension, which further reduces the amount by which he can improve on his cost.
- Playing the perturbed controller involves making an assumption about the evader's control history, and a wrong guess would make it impossible to predict the direction in which the cost function will shift. Furthermore, if the pursuer's guess is caught by the evader, it may be possible for the evader to capitalize on this mistake and deceive the pursuer.
- On the other hand, the evader can try to improve on his estimate of the state by guessing the pursuer's control e.g. by using a Kalman filter, but it can be shown that doing so puts him at a risk of noise injection.
- To conclude, we see that an improvement in cost can be obtained by either player, but such a move makes his position in the game much more vulnerable.

An interesting observation that follows from the above results is that while the LEG game problem for $\theta \neq 0$ has a saddle point solution, the LEG-to-LQG limit as $\theta \to 0$ does not. This inconsistency is currently being investigated by the authors.

12.7 PERFORMANCE OF THE KALMAN FILTERING ALGORITHM

Let us now consider the strategy in which the evader uses a Kalman filter, instead of the LQG Game filter, to estimate the state. While constructing the filter, the evader assumes that the pursuer plays his equilibrium strategy, and averages over it. This leads to two problems. First, u_p can use the cross-correlation between filter error and state to improve on his cost. The improvement obtained by u_p is greater than that obtained with the LQG Game filter, since the entire Kalman filter error vector is correlated with the state. Second, there is the possibility that the convexity condition on the problem can be violated. Consider a two-stage problem in which $u_{p_0} = u_{p_0}^* + \zeta$, where ζ is a white noise process with variance P_ζ. In this case, the difference between the perturbed cost and the equilibrium cost can be shown to be equal to:

$$^{\mathrm{T}}\left(\left[\Gamma_{p_0}^T \Pi_1 \Gamma_{p_0} + R_{p_0}\right] + k_{e_1}^T \left[\Gamma_{e_1}^T \Pi_2 \Gamma_{e_1} + R_{e_1}\right] k_{e_1}\right) P_\zeta .$$

If $\left[\Gamma_{p_0}^T \Pi_1 \Gamma_{p_0} + R_{p_0}\right] + k_{e_1}^T \left[\Gamma_{e_1}^T \Pi_2 \Gamma_{e_1} + R_{e_1}\right] k_{e_1} \geq 0$, then u_p can increase the variance on ζ and make the cost as large as he wants. This happens because the convexity condition (12.18) on u_{p_0} is violated. The above condition can be generalized for an N-stage vector problem. These two problems with the Kalman filter render it unusable for this class of problems.

12.8 COMPARISON WITH THE WILLMAN ALGORITHM

In (Willman 1969), a numerical algorithm is proposed that converges to a saddle point solution for the LQG game problem with imperfect, unshared information. There are two drawbacks that severely limit the use of this algorithm for practical applications. First, the algorithm converges for a very small set of cases. For the cases where it does not converge, it is not clear whether saddle point solutions actually exist or not. The LQG Game filter is found to be stable over a very broad range of system parameters, as long as the opponent's control matrix is in the range space of the observation matrix of the player who constructs the filter. The second drawback of the Willman solution is that it is infinite-dimensional. This results in a strategy that operates over the entire observation history at every time step. In contrast, the LQG Game filter is finite-dimensional, operating on a *summary* of the observation history of the player constructing it.

In order to assess the performance loss due to the suboptimal LQG filter, it was played against the Willman strategy in a numerical simulation. It was found while the Willman algorithm did better in lowering or increasing the value of the cost (depending on the player), the shift from the saddle point value was very small (approximately single-digit percentage values). For this class of problems, the LQG game filter algorithm presents a practically implementable suboptimal strategy with minimum performance compromise. From our discussion about the Kalman filter in the previous section (yet another finite-dimensional suboptimal strategy), it is clear that the LQG Game filter performs better than the Kalman filter in the sense that it is immune to deception. In order to quantify the performance of suboptimal strategies, the notion of a saddle interval is introduced in the next section.

12.9 EQUILIBRIUM PROPERTIES OF THE COST FUNCTION: THE SADDLE INTERVAL

This section extends the notion of a saddle point in dynamic game problems to that of a saddle interval for the class of finite-dimensional solutions in the presence of uncertainties. We have shown how u_p can play a perturbed control, u_p^δ, such that $J(u_p^\delta, u_e^*) \geq J(u_p^*, u_e^*)$, provided all of u_p's guesses about u_e's control are correct. In particular, we found the control $u_p^{\delta^*} = u_p^* + \gamma^* X_i$ which gives the best improvement at $J(u_p^{\delta^*}, u_e^*)$. Clearly, there is a bound on the improvement that u_p can get if u_e plays u_e^*; u_p cannot make the cost infinitely large. The LQG Game filter, therefore, offers some degree of protection against deception by the opponent. Similarly, we can find a $u_e^{\delta^*}$ such that $J(u_p^*, u_e^*) \geq J_e(u_p^*, u_p^{\delta^*})$. Thus, we conclude:

$$J(u_p^{\delta^*}, u_e^*) \geq J(u_p^*, u_e^*) \geq J(u_p^*, u_e^{\delta^*}), \tag{12.21}$$

for the class of finite-dimensional solutions to dynamic game problems with uncertainty. We see that the strategy pair (u_p^*, u_e^*) forms an equilibrium point about which there exists a range over which either player can improve on his cost, given that the other player plays his equilibrium strategy. We define the saddle interval

$$\left[J(u_p^{\delta*}, u_e^*), J(u_p^*, u_e^{\delta*})\right].$$

From the discussion above, it is clear that the saddle interval for the LQG Game filter is smaller than that for the Kalman filter; it is possible for the upper bound of the Kalman filter saddle interval to approach infinity. The notion of a saddle interval thus serves as a useful performance measure of finite-dimensional solutions to dynamic game problems with uncertainty. Further, compared to the infinite-dimensional Willman algorithm which converges for a very limited number of cases, finite dimensional strategies, which have more relaxed existence conditions, are far easier to implement in practice. In such cases, the saddle interval can be used to select a solution that minimizes the possibility of performance loss due to deception. In this respect, the strategy incorporating the LQG game filter offers a significant advantage; the fact that the filter error is independent of the opponent's dynamics ensures a finite upper bound on the saddle interval.

12.10 CONCLUSION

A linear-quadratic-Gaussian (LQG) differential game is considered where the players make decisions based on separate information sets that are not shared. In the problem solved here, one adversary has access to only noisy partial information of the state while the other measures the state perfectly. The equilibrium control gains are finite-dimensional and obey the certainty-equivalence principle. The evader constructs an estimator which ignores the pursuer's control inputs. The estimator error and its first two moments are independent of the players' controls. The implementation of this filter guarantees that the opponent cannot play a malicious input and make the cost function arbitrarily large. For the class of games with uncertainty, the concept of a saddle interval was introduced for the class of finite-dimensional strategies; each player can try to guess the opponent's control strategy/history to improve his cost from the equilibrium value, but doing so puts him in a more vulnerable position in the game if the guess is caught or is wrong. The saddle interval can be used as a measure to evaluate the performance of finite-dimensional strategies. We hypothesize that this theory can be extended to the class of problems where both players have imperfect measurements; each player plays a certainty-equivalence controller operating on an estimate of the state generated by a filter that ignores the opponent's control inputs. The class of admissible strategies can be extended to include mixed strategies, and examining whether a saddle point exists or not. The saddle-point mixed strategies for this class of problems, if they exist, must eliminate the cross-correlations that the opponent can take advantage of.

ACKNOWLEDGEMENTS

This work was sponsored by the Air Force Office of Scientific Research under Grant F49620-01-1-0361 and by NASA/Ames Research Center under Grant NAG2-1484.

REFERENCES

Behn RD and Ho Y 1968 On a class of linear stochastic differential games. *IEEE Transactions on Automatic Control* **13**(3), 227–240.

Bryson A and Ho YC 1979 *Applied Optimal Control*. John Wiley & Sons, Chichester.

Fan CH, Speyer J and Jaensch C 1994 Centralized and decentralized solutions of the linear-exponential-Gaussian problem. *IEEE Transactions on Automatic Control* **39**(10), 1986–2003.

Rhodes I and Luenberger D 1969 Differential games with imperfect state information. *IEEE Transactions on Automatic Control* **14**(1), 29–38.

Whittle P 1981 Risk-sensitive linear/quadratic/Gaussian control. *Adv. Applied Prob.* **13**, 764–777.

Willman W 1969 Formal solutions for a class of stochastic pursuit-evasion games. *IEEE Transactions on Automatic Control* **14**(5), 504–509.

Part IV

Uncertain Evolution

13

Modal estimation of jump linear systems: an information theoretic viewpoint

Nuno C. Martins and Munther A. Dahleh

13.1 ESTIMATION OF A CLASS OF HIDDEN MARKOV MODELS

Consider the estimation set-up depicted in Figure 13.1. That is the archetypal scenario where an external observer intends to estimate $\mathbf{q}_k$, the mode of operation of the system $\mathcal{F}$. The estimator generates, or has access to, a probing signal $\mathbf{V}_k$ which is used to stimulate the system so that the noisy observations of the output are informative. In this framework $\mathbf{V}_k$ is i.i.d., zero mean and Gaussian with covariance matrix Σ_V, which is considered one of the design parameters.

The aforementioned scheme corresponds to problem 2, according to the classification in (Rabiner 2003), but it can also be designated, in the context of Hybrid Systems (Hofbaur 2002), as *mode estimation*. In (Rabiner 2003; Bar-Shalom and Li 1993; Dong 2002; Hofbaur 2002) one finds applications that illustrate the flexibility of this framework and suggest why its employment is so vast. Besides failure detection, other usances are the mode estimation of systems in adversarial environments, where such information might be critical in the prediction of attack maneuvers.

The implications of modal estimation span applications in Adaptive Control and fault detection. Depending on the formulation, modal estimation may be realized as part of a hybrid state estimation paradigm (Sworder and Boyd 1999), comprising discrete (modes) and continuous states. Although the investigation of such problems has generated a vast portfolio of algorithms and methods, Sworder *et al.* (2000) suggests the need to devote more attention to the modal estimation alone. By adopting the framework used in the

Cooperative Control of Distributed Multi-Agent Systems edited by Jeff S. Shamma

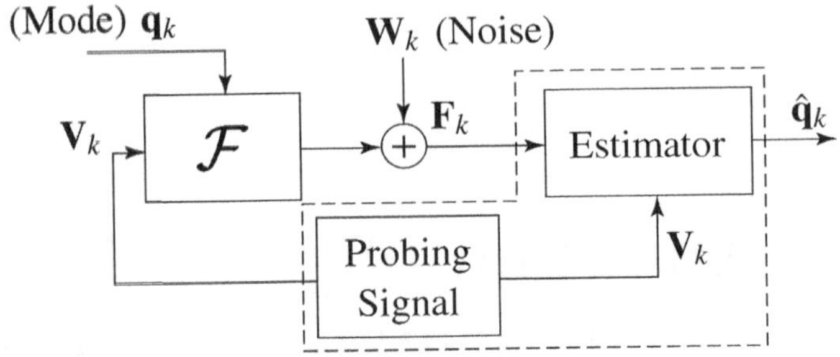

Figure 13.1 Diagram of the estimation set-up.

identification of FIR-LPV systems (Campi 1994), the estimation of the continuous state can be avoided. Consequently, we focus on the modal estimation problem by considering systems that switch among a set of modes which comprise finite impulse response, or moving average, filters. A discrete stochastic process $\mathbf{q}_k$ drives the switching, while the system is excited by a white Gaussian process (probing signal). The optimal modal MAP-estimation problem can be cast as a Bayesian Hypothesis testing. The search space grows with m^k, where m is the number of modes and k is the number of observations.

We recognize that the modal estimation problem is equivalent to a communication set-up in the presence of randomly generated codes (Battail 1995). Under that framework, the input probing signal defines a constrained code and the system is perceived as an encoder of the mode sequence $\mathbf{q}_k$. Motivated by that, we adopt a decoder structure for the mode estimator. The search space is reduced to $m^{k(r^q+\varepsilon)}$ elements, where $r^q \in [0, 1]$ is the entropy rate of $\mathbf{q}_k$ and $\varepsilon > 0$ is a quality parameter that determines the *degree* of optimality of the estimates. Such reduction is achieved by constraining the search to the typical set (Cover and Thomas 1991) of mode sequences. Since such a set is simultaneously a high probability set, no information is lost. This reduction is a unified and asymptotically optimal way to implement the merging and pruning of hypotheses that otherwise would rely on approximations, mode transition detectors, slow variation hypothesis (Gunnarsson 1994) or the use of forgetting factors. We also present an alternative low complexity estimation algorithm that converges in probability, as k tends to infinity, to the decoder proposed. It is based on a binary search that requires, at every step, an optimization using a forward dynamic program of complexity km^2.

Although using a different formulation, we refer to Blom and Bar-Shalom (1988) where the difficulty of computing measures of quality for the available modal estimation algorithms is discussed. In practice, such quality evaluation may have to resort to Monte Carlo simulations. We address that problem by using Shannon's theory to define a measure of distortion $\mathcal{D}_d$ that is suitable for a probabilistic worst-case analysis. Using this distortion, we determine the probability that a sequence of mode estimates $(\hat{q}_1, \ldots, \hat{q}_k)$ (generated by the decoder) is in a ball, of radius β, around the true sequence $(q_1, \ldots, q_k)$. An interesting feature of this framework is that there exists β such that the probability of $\mathcal{D}(q_{1,k}, \hat{q}_{1,k}) < \beta$ converges to 1 as k tends to infinity. We compute such β and show that it is an affine function of r^q, the entropy rate of the switching process. In that sense, as a theoretical result, our work relates to Zames (1994), where it is proven that the uncertainty (Venkatesh 2001) in identifying time-varying systems is directly related to the *speed* of variation. Information theoretically inspired distances have been widely used in system identification and parameter estimation (Arimoto and Kimura 1971; Upadhyaya 1977; Weidemann 1969). The distortion $\mathcal{D}_d$ can also be used as a quality measure on the design of probing signals and can be viewed as a first step to the proper study of the effect of observed inputs on the mode observability of linear hybrid systems. We also stress that our analysis and methods are applicable to other classes of stable switching systems (Liberzon 2003). In these cases, the moving average coefficients should be the truncation of the impulse response of such stable linear systems. Worst-case LPV-FIR system identification was also broached by Belforte and Gay (2000) in the case where the coefficients have a specific functional form.

This chapter is organized as follows: Section 13.1.1 introduces the notation used throughout the text. The problem is stated and its information theoretic equivalence are established in Section 13.2, while Section 13.2.1 provides a guide through the main results. The estimation paradigm is described in Section 13.3, while the performance analysis is carried out in Section 13.4. An efficient estimation algorithm is presented in Section 13.4.1 and numerical examples are provided in Section 13.4.2.

13.1.1 Notation

The following notation is adopted: Large caps letters are used to indicate vectors and matrices. Small caps letters are reserved for real scalars and discrete variables. In addition, p is reserved to represent probability distributions. Discrete-time sequences are indexed by time using integer subscripts, such as x_k. Finite segments of discrete-time sequences are indicated by the range of their time-indexing as in the following example ($k \geq n$):

$$q_{n,k} = (q_n, \dots, q_k) \tag{13.1}$$

Superscripts are reserved for distinguishing different variables and functions according to their meaning. Random variables are represented using boldface letters and follow the conventions above. As an illustration, $\mathbf{x}$ is a valid representation for a scalar random variable, while a sample (or realization) is written as x. Also, a finite segment of a discrete-time sequence of random variables, would be $\mathbf{q}_{1,k}$. A realization of such process would be indicated as $q_{1,k}$. The probability of an *event* is indicated by $\mathcal{P}(event)$. We use the entropy function, of a random variable $\mathbf{Z}$, given by:

$$\mathcal{H}[\mathbf{Z}] = \mathcal{E}[-\ln p^{\mathbf{Z}}(\mathbf{Z})] \tag{13.2}$$

where $p^{\mathbf{Z}}$ is the p.d.f. of $\mathbf{Z}$ and $\mathcal{E}[\cdot]$ is the expected value, taken over $\mathbf{Z}$. Similarly, the conditional entropy is given by $\mathcal{H}[\mathbf{Z}^1|\mathbf{Z}^2] = \mathcal{H}[\mathbf{Z}^1, \mathbf{Z}^2] - \mathcal{H}[\mathbf{Z}^2]$ where, in this case, the expectation is taken with respect to $\mathbf{Z}^1$ and $\mathbf{Z}^2$. The covariance matrix of a random matrix $\mathbf{Z}$, with $Z \in \mathbb{R}^{n^1 \times n^2}$, is given by:

$$\Sigma_Z = \mathcal{E}\left[\left(\vec{\mathbf{Z}} - \mathcal{E}[\vec{\mathbf{Z}}]\right)\left(\vec{\mathbf{Z}} - \mathcal{E}[\vec{\mathbf{Z}}]\right)^T\right], \text{ where } \vec{\mathbf{Z}} = vec(\mathbf{Z}) \tag{13.3}$$

13.2 PROBLEM STATEMENT

Consider the mode alphabet $\mathbb{A} = \{1, \ldots, m\}$, with $m \geq 2$, and the random process $\mathbf{F}_k$, with $F_k \in \mathbb{R}^{n^F}$, described by:

$$\mathbf{F}_k = \mathbf{Y}_k + \mathbf{W}_k, \quad k \geq 1 \tag{13.4}$$

$$\mathbf{Y}_k = \sum_{i=0}^{\alpha} G_i(\mathbf{q}_k)\mathbf{V}_{k-i}, \quad k \geq 1 \tag{13.5}$$

where $k, \alpha \in \mathbb{N}$ and $G_i : \mathbb{A} \to \mathbb{R}^{n^F \times n^V}$ are matrices that specify the switching system. The stochastic processes $\mathbf{V}_k$ (probing signal), $\mathbf{W}_k$ and $\mathbf{q}_k$ are mutually independent and satisfy:

- $\mathbf{V}_k$ and $\mathbf{W}_k$ are Gaussian zero mean i.i.d. processes, with $V_k \in \mathbb{R}^{n^V}$ and $W_k \in \mathbb{R}^{n^F}$. In order to avoid degeneracy problems, we assume $\Sigma_W > 0$. In contrast to $\mathbf{W}_k$, the probing signal $\mathbf{V}_k$ is assumed to be observed by the estimator. Examples where this is a realistic assumption are: when $\mathbf{V}_k$ is generated by the estimator; when it is an exogenous process that can be observed by the estimator or a combination of both.
- $\mathbf{q}_k$ is a discrete, stationary and ergodic Markovian stochastic process with alphabet $\mathbb{A}$. For a given $k \in \mathbb{N}$, we write the p.d.f. of $\mathbf{q}_{1,k}$ as $p^{\mathbf{q}}(q_{1,k})$, regardless of k and $p^{\mathbf{q}}(\mathbf{q}_k|\mathbf{q}_{k-1}) = \frac{p^{\mathbf{q}}(q_k, q_{k-1})}{p^{\mathbf{q}}(q_{k-1})}$. The entropy rate (Cover and Thomas 1991) of $\mathbf{q}_k$, designated by r^q, is computed as:

$$r^q = \mathcal{E}[-log_m p^{\mathbf{q}}(\mathbf{q}_k|\mathbf{q}_{k-1})] \tag{13.6}$$

For a given k, we wish to use the probing signal $\mathbf{V}_{1-\alpha,k}$ and a decision system that, by means of the measurement of $\mathbf{F}_{1,k}$, produces an estimate $\hat{\mathbf{q}}_{1,k}$ of $\mathbf{q}_{1,k}$. The estimation method must have an associated measure of distortion $\mathcal{D}_d(q_{1,k}, \hat{q}_{1,k})$ that allows the specification of a ball around $q_{1,k}$ where the estimates $\hat{q}_{1,k}$ will lie with a given probability. In particular, for a given $\beta > 0$, we are interested in computing $\mathcal{P}(\mathcal{D}_d(\mathbf{q}_{1,k}, \hat{\mathbf{q}}_{1,k}) > \beta)$. Such probability is expected to depend on β and k. By using an information theoretic formulation, this characterization must also reflect the informativity of $\mathbf{V}_k$ and r^q, the entropy rate of the switching process. We would like to stress that, without loss of generality, we develop our analysis using the origin of time as $k = 1$. In real applications, this set-up can be used in a sliding window scheme and k should be viewed as the time horizon of the estimator.

13.2.1 Main results

Among the results presented in this chapter, the following are central to answering the questions posed above: In definition 13.4.2 a measure of distortion $\mathcal{D}_d$ is introduced. It is through this function that the informativity of $\mathbf{V}_k$ can be gauged. If $\mathbf{V}_k$, or only part of it, are generated by the estimator, then that generated portion of Σ_V may be tuned to achieve a desired distortion function. Similar methods for the distance-based design of probing functions were derived by Arimoto and Kimura (1971). In Lemma 13.5.8 we show that, as k increases, $\mathcal{P}(\mathcal{D}_d(\mathbf{q}_{1,k}, \hat{\mathbf{q}}_{1,k}) > \beta)$ decreases. Theorem 13.4.3 proves that if $\beta > r^q \ln(m) + \frac{n^F}{2}$ then $\mathcal{P}(\mathcal{D}_d(\mathbf{q}_{1,k}, \hat{\mathbf{q}}_{1,k}) > \beta)$ can be made arbitrarily small by increasing k.

13.2.2 Posing the problem statement as a coding paradigm

As the generality of the estimation paradigms (Weidemann 1969), the estimation of the mode of (13.4)–(13.5) can be interpreted as a problem of communication through a noisy channel. The message to be transmitted is $q_{1,k}$, the probing signal $\mathbf{V}_{1-\alpha,k}$ specifies the code while the measurement noise $\mathbf{W}_{1,k}$ completes the set-up of such channel. The decoder is bound to use $\mathbf{V}_{1-\alpha,k}$ and the noisy measurements $\mathbf{F}_{1,k}$ to make a decision as to which is the best estimate $\hat{q}_{1,k}$.

13.2.3 Comparative analysis with previous work

In this chapter we introduce a suitable measure of distortion and we provide a unified probabilistic worst case analysis that analytically unveils the importance of the covariance matrix of $\mathbf{V}_k$ as well as the entropy rate of $\mathbf{q}_k$.

There is an extensive list of articles on Signal Processing, Automatic Control and Information Theory that broach the estimation and tracking of HMM. Although these communities are, with a few exceptions, self-referential, the disconnect is artificial. There exists a great deal of intersection in the objectives and motivations. In common they lack a suitable analysis of performance and that is what distinguishes our approach. We briefly discuss the relevant publications according to the following classification:

- **Information theoretic approaches to the estimation and tracking of HMM.** The similarities, between the estimation set-up in Figure 13.1 and the communication scheme in Figure 13.2, suggest that the study of estimation fidelity, in the present context, is

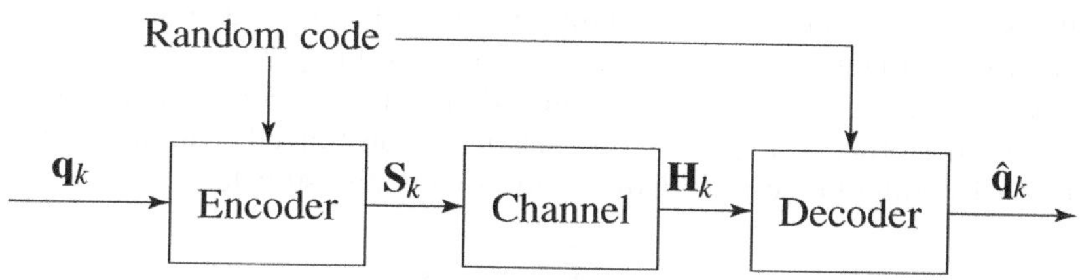

Figure 13.2 Framework for a communication system.

related to rate-distortion theory (Berger 1971). This view has already been explored in the area of pattern recognition, as a way to study the effects of quantization (Dong 2002). The same problem was broached in Ko (1998) for general HMM with discrete measurements. In the aforementioned publication, the rate-distortion tradeoff has to be determined point-wise by numerically solving two non-linear equations. The use of numerical methods in the determination of the rate-distortion tradeoff is a common practice since its introduction by Blahut (1972). The use of entropy-based distortion measures and its relation to the probability of error has also received attention more recently in (Feder and Merhav 1994). When compared to these publications, our work makes a contribution for the constrained coding established by the linear system in Figure 13.1. By suggesting how the input shapes the distortion function, we also provide a unified framework to study the influence of the probing signal. The design of probing signals was studied in a similar framework, for the time-invariant case, by Mosca (1972).

- **Identification of time-varying systems.** Several publications have addressed the tradeoff between tracking and disturbance rejection (Gunnarsson 1990) or, alternatively, the tradeoff between the *rate* of time-variation and uncertainty (Belforte and Gay 2000; Zames 1994). The latter is the deterministic analogous of the rate-distortion tradeoff. Also, in the scope of the design of probing signals, Bittanti *et al.* (2000) note that the state of the art is not satisfactory. A good review of past results can also be found in Campi (1994), where the lack a unified framework becomes apparent as well as the use of assumptions such as *slow variation*.
- **Design of sub-optimal estimators for hybrid linear systems.** Since the first contributions came out (Ackerson and Fu 1970; Chang and Athans 1978) several new algorithms have attempted to tackle the exponential complexity of the hypothesis set. A collection of the main results can be found in Bar-Shalom and Li (1993); Elliot (1995); Sworder and Boyd (1999); and Tugnait (1982). These results rely on approximations that make the study of estimation fidelity very difficult (Blom and Bar-Shalom 1988) and Sworder *et al.* (2000) suggest that more attention should be devoted to the mode estimation problem.

13.3 ENCODING AND DECODING

A specific feature of this communications set-up is that the *encoder* does not know the message to be transmitted. This is a problem if one wants to adopt the approach of coding long words as a way to reduce the probability of error (channel coding theorem). In order to circumvent this difficulty, we follow the procedure in the proof of Shannon's channel coding theorem, i.e., the use of random coding (Battail 1995). Consequently, we consider the white Gaussian process $\mathbf{V}_{1-\alpha,k}$ as establishing a constrained random code specified by Σ_V. In the decoding process we will use the following estimate:

$$\hat{\mathbf{Y}}_k(\hat{q}_k) = \sum_{i=0}^{\alpha} G_i(\hat{q}_k)\mathbf{V}_{k-i} \tag{13.7}$$

The decoding process has a hypothesis testing structure. The likelihood of a given candidate sequence $\hat{q}_{1,k}$ is gauged by means of the estimation error $\mathbf{F}_k - \hat{\mathbf{Y}}_k(\hat{q}_k)$.

13.3.1 Description of the estimator (decoder)

By following the approach that leads to a standard decoder (Cover and Thomas 1991), in this section we construct a mode estimator. In the subsequent analysis we use the following result:

Remark 13.3.1 If $\mathbf{X}$ is a Gaussian, zero mean random variable with covariance matrix $\Sigma_X \in \mathbb{R}^{n^X \times n^X}$, then:

$$\mathcal{H}(\mathbf{X}) = \frac{1}{2} \ln\left((2\pi e)^{n^X} |\Sigma_X|\right) \tag{13.8}$$

The following is the definition of the noise entropy rate r^W. Such quantity is important in the construction of the estimator.

Definition 13.3.1 *Define the noise entropy rate r^W as:*

$$r^W = \mathcal{H}(\mathbf{W}_k) = \frac{\ln\left((2\pi e)^{n^F} |\Sigma_W|\right)}{2} \tag{13.9}$$

The following definitions complete the list of mathematical objects needed to describe the decoder.

Definition 13.3.2 *Consider $\hat{q}_{1,k} \in \mathbb{A}^k$ and realizations $F_{1,k}$ and $\hat{Y}_{1,k}(\hat{q}_{1,k})$. Given parameters $k \in \mathbb{N}$ and $\varepsilon \in (0, 1)$, define the selection indicator $s^{\varepsilon,k} : \mathbb{A}^k \to \{True, False\}$ as:*

$$s^{\varepsilon,k}(\hat{q}_{1,k}) = \begin{cases} True & if \ \frac{-\ln p^{\mathbf{W}_{1,k}}(F_{1,k} - \hat{Y}_{1,k}(\hat{q}_{1,k}))}{kn^F} < \frac{r^W}{n^F} + \varepsilon \\ False & otherwise \end{cases} \tag{13.10}$$

The decoding process is a search in the typical set defined below.

Definition 13.3.3 *Given the parameters $k \in \mathbb{N}$ and $\varepsilon \in (0, 1)$, the set of typical sequences $\mathbb{T}^{\varepsilon,k}$ is defined as:*

$$\mathbb{T}^{\varepsilon,k} = \{\hat{q}_{1,k} \in \mathbb{A}^k : \frac{-\log_m p^{\mathbf{q}}(\hat{q}_{1,k})}{k} < r^q + \varepsilon\} \tag{13.11}$$

It is the cardinality of $\mathbb{T}^{\varepsilon,k}$ that determines the computational complexity of the estimator. According to Cover and Thomas (1991) such quantity is bounded by $m^{k(r^q+\varepsilon)}$. The structure of the decoder is the following:

Definition 13.3.4 ***(Decoder)*** *Given the realizations $F_{1,k}$ and $V_{1,k}$, parameters $k \in \mathbb{N}$ and $\varepsilon \in (0, 1)$, define the decoder as a search in $\mathbb{T}^{\varepsilon,k}$ that generates $\hat{q}_{1,k}$ satisfying:*

$$\hat{q}_{1,k} \in \mathbb{T}^{\varepsilon,k} \text{ and } s^{\varepsilon,k}(\hat{q}_{1,k}) = True \tag{13.12}$$

If, for a given realization, there is no such $\hat{q}_{1,k}$, then the decoder generates an arbitrary $\hat{q}_{1,k} \in \mathbb{T}^{\varepsilon,k}$. The decoding process defines a random variable, which we designate by $\hat{\mathbf{q}}_{1,k}$.

13.4 PERFORMANCE ANALYSIS

The following definition, describing a *conditional* test random variable, will facilitate the performance analysis of the decoder.

Definition 13.4.1 *Given $k \in \mathbb{N}$, $k \geq 1$ and $q_k, \hat{q}_k \in \mathbb{A}$, define the random variable $\mathbf{T}_k$ $(q_k, \hat{q}_k)$ as:*

$$\mathbf{T}_k(q_k, \hat{q}_k) = \sum_{i=0}^{\alpha} (G_i(q_k) - G_i(\hat{q}_k))\,\mathbf{V}_{k-i} + \mathbf{W}_k \tag{13.13}$$

Note that, given $\hat{q}_k \in \mathbb{A}$, the random variable $\mathbf{F}_i - \hat{\mathbf{Y}}(\hat{q}_k)$ conditioned to a fixed q_k is given by $\mathbf{T}_k(q_k, \hat{q}_k)$. If $\hat{q}_k = q_k$, then $\mathbf{T}_k(q_k, \hat{q}_k) = \mathbf{W}_k$. For any given indices k^1, k^2 and $q_{k^1,k^2}, \hat{q}_{k^1,k^2} \in \mathbb{A}^{k^2-k^1+1}$, we adopt the following notation:

$$\mathbf{T}_{k^1,k^2}(q_{k^1,k^2}, \hat{q}_{k^1,k^2}) = \left(\mathbf{T}_{k^1}(q_{k^1}, \hat{q}_{k^1}), \ldots, \mathbf{T}_{k^2}(q_{k^2}, \hat{q}_{k^2})\right) \tag{13.14}$$

The quality evaluation, of the coding/decoding process, is carried out by computing the probability that $\hat{\mathbf{q}}_{1,k}$ is in a ball around $\mathbf{q}_{1,k}$. Such an uncertainty set is specified by means of the distortion $\mathcal{D}_d$ defined below. This function is related to the concept of divergence as in Arimoto and Kimura (1971).

Definition 13.4.2 ***(Measure of distortion)*** *The distortion $\mathcal{D}_d : \mathbb{A}^k \times \mathbb{A}^k \to \mathbb{R}$ is given by:*

$$\mathcal{D}_d(q_{1,k}, \hat{q}_{1,k}) = \frac{\mathcal{H}(\mathbf{T}_{1,k}(q_{1,k}, \hat{q}_{1,k}))}{k} - r^W \tag{13.15}$$

A discussion of the properties of $\mathcal{D}_d$, including a rate-distortion interpretation is given by Martins and Dahleh (2004). The following theorem is the main result of this chapter.

Theorem 13.4.3 *Let $\hat{\mathbf{q}}_{1,k}$ be determined according to the decoding process described in the definition 13.3.4. For any given $\delta > 0$, there exists $k \in \mathbb{N}$ and $\varepsilon \in (0, 1)$ such that:*

$$\mathcal{P}\left(\mathcal{D}_d(\mathbf{q}_{1,k}, \hat{\mathbf{q}}_{1,k}) > \ln(m) r^q + \frac{n^F}{2} + \delta\right) < \delta, \ \textit{if}\ r^q \geq 0 \tag{13.16}$$

$$\mathcal{P}\left(\mathcal{D}_d(\mathbf{q}_{1,k}, \hat{\mathbf{q}}_{1,k}) > \delta\right) < \delta, \ \textit{if}\ r^q = 0 \tag{13.17}$$

Proof. The result for $r^q \geq 0$ follows directly from Lemma 13.5.8 in Section 13.5. □

13.4.1 An efficient decoding algorithm

In this subsection, the main result is Lemma 13.4.5. It shows that, for k sufficiently large, the exhaustive search of definition 13.3.4 can be replaced by a binary search over a non-negative parameter γ. The stopping condition relies on the minimal solution of the following cost functional:

Definition 13.4.4 *Given realizations $q_{1,k}$, $F_{1,k}$, $V_{1,k}$ and $W_{1,k}$, define the following cost function:*

$$\mathcal{J}_\gamma(\overline{q}_{1,k}) = \frac{-\ln\left[p^{\mathbf{W}_{1,k}}\left(F_{1,k} - \hat{Y}_{1,k}(\overline{q}_{1,k})\right)\right]}{k} + \gamma \frac{-\log_m\left(p^{\mathbf{q}}(\overline{q}_{1,k})\right)}{k} \tag{13.18}$$

where $\overline{q}_{1,k} \in \mathbb{A}^k$ is a candidate sequence and γ is a non-negative constant. The first term of the cost function is a sample divergence, while the second is designated as sample entropy rate. Note that the two terms in the cost (13.18) are identical to the ones in (13.10) and (13.11).

Lemma 13.4.5 *Let $\overline{q}^*_{1,k}$ be the optimal solution given by:*

$$\overline{q}^*_{1,k} = \underset{\overline{q}_{1,k} \in \mathbb{A}^k}{argmin} \mathcal{J}_\gamma(\overline{q}_{1,k}) \tag{13.19}$$

and $\hat{\mathbf{q}}_{1,k}$ be the estimate of definition 13.3.4. If $\mathbf{q}_k$ is i.i.d., then the following holds:

$$\lim_{k \to \infty} \mathcal{P}\left(\mathbf{E}_1 \wedge \mathbf{E}_2 \Rightarrow \exists a, b, \forall \gamma \in (a, b)\ \mathbf{E}_3 \wedge \mathbf{E}_4\right) = 1 \tag{13.20}$$

where $\mathbf{E}_1, \mathbf{E}_2, \mathbf{E}_3$ and $\mathbf{E}_4$ are the following events:

$$\mathbf{E}_1 = \mathbf{s}^{\varepsilon,k}(\hat{\mathbf{q}}_{1,k}),\ \mathbf{E}_2 = \left(\hat{\mathbf{q}}_{1,k} \in \mathbb{T}^{\varepsilon,k}\right) \tag{13.21}$$

$$\mathbf{E}_3 = \mathbf{s}^{2\varepsilon,k}(\overline{\mathbf{q}}^*_{1,k}),\ \mathbf{E}_4 = \left(\overline{\mathbf{q}}^*_{1,k} \in \mathbb{T}^{2\varepsilon,k}\right) \tag{13.22}$$

Remark 13.4.1 (Binary search) The two terms in the cost (13.18) analyzed separately, in terms of the events (13.21)–(13.22), lead to the following structure for the binary search:

- A simple optimality argument is sufficient to show that for any $\gamma > 0$, the optimal solution satisfies $E_1 \wedge E_2 \Rightarrow E_3 \vee E_4$.
- If $\gamma = 0$, then the optimal solution is the maximum likelihood estimate over the unrestricted search space $\mathbb{A}^k$. It implies that, for γ sufficiently small, $E_1 \Rightarrow E_3$.
- As γ increases, the first term (divergence) is non-decreasing and the second term (entropy rate) is nonincreasing at the optimal solution. In the limit, the sample entropy rate converges to a constant $c \le r^q$. It shows that for γ sufficiently large, E_4 is true.
- Lemma 13.4.5 shows that, with probability arbitrarily close to 1, there exists an interval (a, b) such that for all $\gamma \in (a, b)$ the optimal solution satisfies $E_1 \wedge E_2 \Rightarrow E_3 \wedge E_4$. The aim of the binary search is to find some γ^* in this interval. Given γ at time k, if $E_3 \wedge \neg E_4$ then, at time $k + 1$, γ must be increased. On the other hand, if $\neg E_3 \wedge E_4$ then γ must be decreased. If $E_1 \wedge E_2$ is satisfied, then, after a finite number of steps, γ will be in (a, b) and $E_3 \wedge E_4$ is satisfied. The optimal sequence is taken as the resulting mode sequence estimate.

- A minimum size for the interval (a, b) must be specified. If the binary search reaches such a limit it must end, declare that the condition $E_3 \wedge E_4$ could not be satisfied and $\overline{q}^*_{1,k}$ be chosen arbitrarily. Since the smaller the minimal size of the interval the better, a safe approach is to simulate the estimator and decrease that quantity until the estimates are within the performance predicted in theory. The minimal size will impact the number of iterations and, as such, the final complexity of the algorithm. We verified empirically that, even in cases where $k > 1000$, $\alpha = 10$ and $n^F = 2$, the search is concluded within a few seconds.

The proof of Lemma 13.4.5 follows by continuity arguments. We also believe that Lemma 13.4.5 extends to general Markovian $\mathbf{q}_k$. So far, we keep it as a conjecture that is empirically verifiable through simulations.

13.4.1.1 *Computational complexity of minimizing $\mathcal{J}_\gamma(\overline{q}_{1,k})$*

In the following analysis we show that the minimization of $\mathcal{J}_\gamma(\overline{q}_{1,k})$ has a computational complexity that grows linearly with k. We start by recalling that since $\mathbf{W}_k$ is i.i.d. and $\mathbf{q}_k$ is Markovian, we can rewrite $\mathcal{J}_\gamma(\overline{q}_{1,k})$ as:

$$\mathcal{J}_\gamma(\overline{q}_{1,k}) = \sum_{i=2}^{k} -\frac{\ln p^{\mathbf{W}_i}(F_i - \hat{Y}_i(\overline{q}_i))}{k} - \gamma\frac{\log_m p^{\mathbf{q}}(\overline{q}_i|\overline{q}_{i-1})}{k} + \\ -\frac{\ln p^{\mathbf{W}_1}(F_1 - \hat{Y}_1(\overline{q}_1))}{k} - \gamma\frac{\log_m p^{\mathbf{q}}(\overline{q}_1)}{k} \tag{13.23}$$

Now notice that (13.23) is suitable for forward dynamic programming (Bertsekas 1995). The optimal $\overline{q}^*_{1,k}$ is the minimal path solution of (13.23). At every step $k^* \in \{1, \ldots, k\}$, the algorithm recursively determines the optimal path $\overline{q}^*_{1,k^*}$ for all end points. The total number of operations is m^2k and, as such, computational complexity grows linearly with k.

13.4.2 Numerical results

In order to illustrate the previously described optimization method, we refer to Figures 13.3 and 13.4 where we depict the results for the following parameters:

- $\mathbb{A} = \{1, 2\}$, $m = 2$
- $G_0(1) = 1$, $G_1(1) = 0.9$, $G_2(1) = 0.81$
- $G_0(2) = 1$, $G_1(2) = 0.3$, $G_2(2) = 0.09$
- $k = 200$

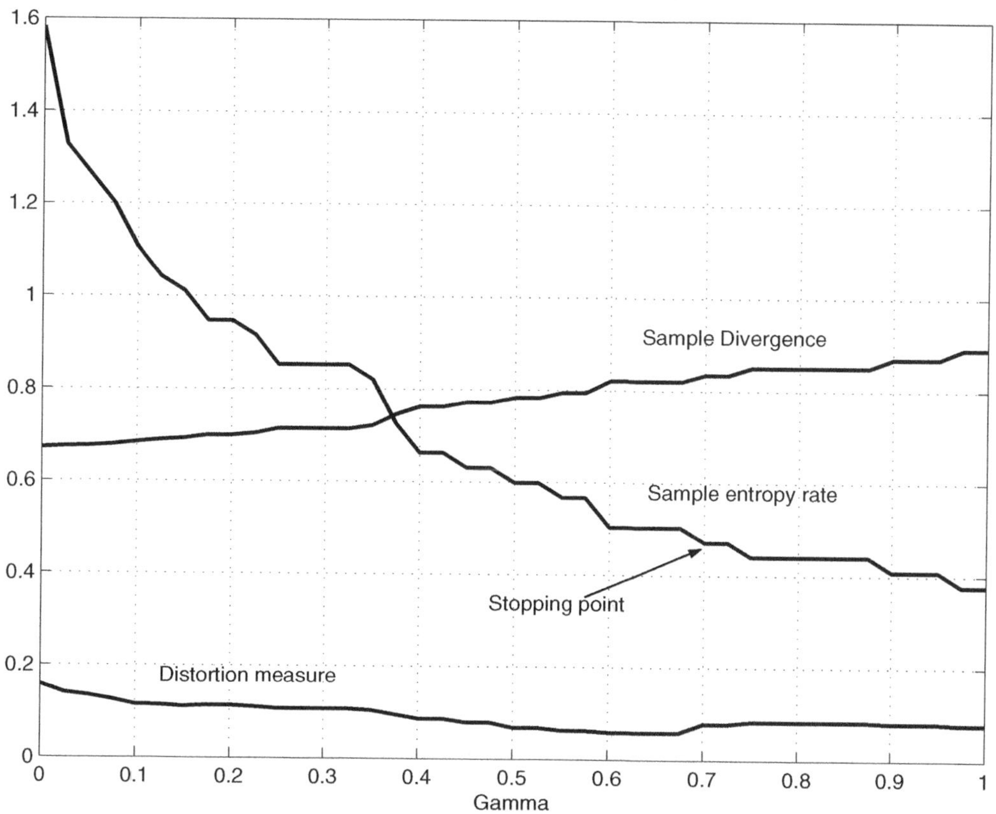

Figure 13.3 Simulation for a switching process described by $p^q(1|1) = p^q(2|2) = 0.9$ and $p^q(1|2) = p^q(2|1) = 0.1$ and entropy rate $r^q = 0.469$.

After inspection, we notice that the distortion measure is much smaller than the upper-bound in 13.16. In the first case, we get $\simeq 0.07$ when the upper-bound dictates $0.469 + 0.5 = 0.969$. In addition, we notice that lower r^q leads to lower distortion measures. After extensive simulations, we found that such conservatism was consistent. Moreover, it should be expected that, as r^q decreases, the distortion measure gets smaller with high probability. Such behavior cannot be captured by Theorem 13.4.3 as the contribution of r^q gets masked by the $1/2$ factor. After analyzing the proof of Theorem 13.4.3, we concluded that the cause was that the upper-bound expressed in Lemma 13.5.6 was too conservative. A more general version of Theorem 13.4.3 will be reported in future publications. The aforementioned extension replaces the $1/2$ factor by a variable β which determines the probability $\mathcal{P}\left(\mathcal{D}_d(\mathbf{q}_{1,k}, \hat{\mathbf{q}}_{1,k}) > \ln(m)r^q + \beta\right)$. If $\beta = 1/2$ then the result in Theorem 13.4.3 is recovered. Lower values of β lead to a non-zero limit probability, as k tends to infinity.

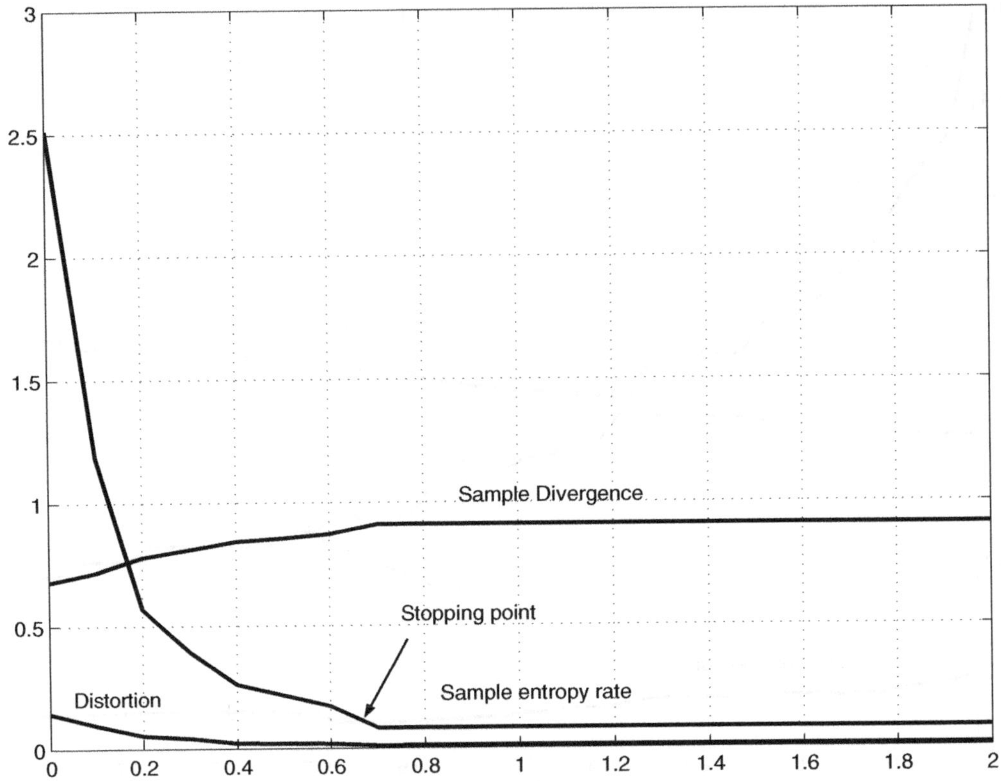

Figure 13.4 Simulation for a switching process described by $p^q(1|1) = p^q(2|2) = 0.99$ and $p^q(1|2) = p^q(2|1) = 0.01$ and entropy rate $r^q = 0.08$.

13.5 AUXILIARY RESULTS LEADING TO THE PROOF OF THEOREM 13.4.3

The main result of this section is Lemma 13.5.8, where explicit expressions are given to bound the probability (13.16). In order to guarantee notational simplicity, we start with the following definitions:

Definition 13.5.1 *We define the following random variable:*

$$\mathbf{z}_k = -\frac{\ln p^{\mathbf{W}_{1,k}}(\mathbf{W}_{1,k})}{kn^F} \tag{13.24}$$

We also adopt $\mathcal{D} = \frac{\mathcal{D}_d}{n^F}$ as a way to make equations smaller.

Lemma 13.5.2 *For any given positive definite $\Sigma_X \in \mathbb{R}^{n^X \times n^X}$, designate by* $\mathbf{X}$ *the zero-mean Gaussian random variable with covariance matrix Σ_X. The following holds:*

$$\sup_{\Sigma_X > 0} \left(\mathcal{E}\left[\left(\frac{\ln p^{\mathbf{X}}(\mathbf{X})}{n^X}\right)^2\right] - \left(\mathcal{E}\left[\frac{\ln p^{\mathbf{X}}(\mathbf{X})}{n^X}\right]\right)^2 \right) < \frac{a}{n^X} \tag{13.25}$$

where $p^{\mathbf{X}}(X) = \frac{e^{-\frac{1}{2}X^T \Sigma_X^{-1} X}}{\left((2\pi)^{n^X} |\Sigma_X|\right)^{1/2}}$ *and* $a = \frac{1}{2}$.

Proof. The result is obtained by evaluating the expected values in (13.25) and applying the change of variables $\tilde{X} = \Sigma_X^{-1/2} X$. □

Lemma 13.5.3 *Given $\varepsilon > 0$ and $k \geq 1$ the following holds:*

$$\mathcal{P}\left(\left| \mathbf{z}_k - \frac{r^W}{n^F} \right| > \varepsilon \right) < \frac{1}{2k\varepsilon^2 n^F} \tag{13.26}$$

Proof. Since $\mathbf{W}_k$ is i.i.d. and $\Sigma_W > 0$, Lemma 13.5.2 and the fact that $W_k \in \mathbb{R}^{kn^F}$ lead to:

$$Var(\mathbf{z}_k) \leq \frac{1}{2kn^F} \tag{13.27}$$

Now notice that $\mathcal{E}[\mathbf{z}_k] = \frac{r^W}{n^F}$, so that the final result is a direct application of the Chebyshev-Bienaymé inequality (Billingsley 1995). □

Definition 13.5.4 *Given $\varepsilon \in (0, 1)$ and $n \in \mathbb{N}$, define Γ_k as the event that the search space of the decoder is empty:*

$$\Gamma_k = \begin{cases} True & \text{if } \{\overline{q}_{1,k} \in \mathbb{T}^{\varepsilon,k} : s^{\varepsilon,k}(\overline{q}_{1,k}) = True\} = \emptyset \\ False & \text{otherwise} \end{cases} \tag{13.28}$$

Lemma 13.5.5 *For any given $\varepsilon \in (0, 1)$, if $k \geq 1$, then the following holds:*

$$\lim_{k \to \infty} P(\Gamma_k) = 0 \tag{13.29}$$

Proof. Notice that, from definition 13.3.4, if a given realization $q_{1,k}$ satisfies $q_{1,k} \in \mathbb{T}^{\varepsilon,k}$ and $s^{\varepsilon,k}(q_{1,k}) = True$, then Γ_k is false because $q_{1,k}$ is itself a valid choice for $\overline{q}_{1,k}$. That leads to the following inequality:

$$\mathcal{P}(\Gamma_k) \leq \mathcal{P}(|\mathbf{z}_k - \frac{r^W}{n^F}| > \varepsilon) + \mathcal{P}(\mathbf{q}_{1,k} \notin \mathbb{T}^{\varepsilon,k}) \tag{13.30}$$

The first term in the RHS of (13.30) can be bounded by means of Lemma 13.5.3. For the second term it suffices to show that $\mathcal{P}(|\frac{-\log_m p^{\mathbf{q}}(\mathbf{q}_{1,k})}{k} - r^q| > \varepsilon) \underset{k\to\infty}{\longrightarrow} 0$. Our analysis starts with the expansion:

$$\log_m p^{\mathbf{q}}(\mathbf{q}_{1,k}) = \sum_{i=2}^{k} \log_m p^{\mathbf{q}}(\mathbf{q}_i|\mathbf{q}_{i-1}) + \log_m p^q(\mathbf{q}_1) \tag{13.31}$$

Since $\mathcal{P}\left(|\frac{\log_m p^q(\mathbf{q}_1)}{k}| > \frac{\varepsilon}{2}\right) \underset{k\to\infty}{\longrightarrow} 0$, we only have to show that:

$$\mathcal{P}\left(|\frac{-\sum_{i=2}^{k} \log_m p^{\mathbf{q}}(\mathbf{q}_i|\mathbf{q}_{i-1})}{k} - r^q| > \frac{\varepsilon}{2}\right) \underset{k\to\infty}{\longrightarrow} 0 \tag{13.32}$$

If $r^q = 0$, then $\mathcal{P}(\log_m p^{\mathbf{q}}(\mathbf{q}_i|\mathbf{q}_{i-1}) = 0) = 1$ and the result is proved. For $r^q > 0$, recall that $\mathbf{q}_i$ is ergodic. Since $\mathbf{q}_k$ has a finite alphabet, there are no unboundedness problems and we conclude that $\log_m p^{\mathbf{q}}(\mathbf{q}_i|\mathbf{q}_{i-1})$ is ergodic, which implies (13.32). □

Lemma 13.5.6 *For any given $q_{1,k}, \hat{q}_{1,k} \in \mathbb{A}^k$ and $k \geq 1$ the following holds:*

$$\max_{X\in\mathbb{R}^{kn^F}} p^{\mathbf{T}_{1,k}(q_{1,k},\hat{q}_{1,k})}(X) = m^{k\frac{n^F}{\ln(m)}(-\mathcal{D}(q_{1,k},\hat{q}_{1,k}) - \frac{r^W}{n^F} + \frac{1}{2})} \tag{13.33}$$

Proof. From the definition of the Gaussian distribution:

$$\max_{X\in\mathbb{R}^{kn^F}} p^{\mathbf{T}_{1,k}(q_{1,k},\hat{q}_{1,k})}(X) = \frac{\left|\Sigma_{\mathbf{T}_{1,k}(q_{1,k},\hat{q}_{1,k})}\right|^{-1/2}}{(2\pi)^{\frac{kn^F}{2}}} \tag{13.34}$$

or equivalently, using (13.8), it can also be written as:

$$\max_{X\in\mathbb{R}^{kn^F}} p^{\mathbf{T}_{1,k}(q_{1,k},\hat{q}_{1,k})}(X) = m^{-\frac{\mathcal{H}(\mathbf{T}_{1,k}(q_{1,k},\hat{q}_{1,k})) + \frac{kn^F}{2}}{\ln(m)}} \tag{13.35}$$

The final result is achieved once we recall that $\mathcal{D}(q_{1,k}, \hat{q}_{1,k}) = \frac{\mathcal{H}(\mathbf{T}_{1,k}(q_{1,k},\hat{q}_{1,k}))}{kn^F} - \frac{r^W}{n^F}$. □

Lemma 13.5.7 *Let $k \geq 1$, $\varepsilon \in (0, 1)$, $q_{1,k} \in \mathcal{A}^k$ and $\hat{q}_{1,k} \in \mathbb{T}^{\varepsilon,k}$. The following is an upper-bound for the conditional probability that $\hat{q}_{1,k}$ satisfies the decoding condition $s^{\varepsilon,k}(\hat{q}_{1,k}) = True$.*

$$P(\mathbf{s}^{\varepsilon,k}(\hat{q}_{1,k}) = True|q_{1,k}) < m^{k\frac{n^F}{\ln(m)}(-\mathcal{D}(q_{1,k},\hat{q}_{1,k}) + \frac{1}{2} + \varepsilon)} \tag{13.36}$$

Proof. The structure of the decoder leads to :

$$P(\mathbf{s}^{\varepsilon,k}(\hat{q}_{1,k}) = True|q_{1,k}) = \int_{\mathbb{S}} p^{\mathbf{T}_{1,k}(q_{1,k},\hat{q}_{1,k})}(X)dX \tag{13.37}$$

where $\mathbb{S}$, the set of realizations leading to (13.10), is given by:

$$\mathbb{S} = \{X \in \mathbb{R}^{kn^F} : \frac{-\ln p^{\mathbf{W}_{1,k}}(X)}{kn^F} < \frac{r^W}{n^F} + \varepsilon\} \tag{13.38}$$

We can use Lemma 13.5.6 to infer that:

$$P(\mathbf{s}^{\varepsilon,k}(\hat{q}_{1,k}) = True|q_{1,k}) \leq \int_{\mathbb{S}} m^{k\frac{n^F}{\ln(m)}\left(-\mathcal{D}(q_{1,k},\hat{q}_{1,k}) - \frac{r^W}{n^F} + \frac{1}{2}\right)} dX \leq$$

$$\leq Vol(\mathbb{S}) m^{n^F \frac{k}{\ln(m)}\left(-\mathcal{D}(q_{1,k},\hat{q}_{1,k}) - \frac{r^W}{n^F} + \frac{1}{2}\right)} \tag{13.39}$$

The proof is completed once we notice that the volume of $\mathbb{S}$ is upper-bounded by $Vol(\mathbb{S}) < m^{k\frac{n^F}{\ln(m)}(\frac{r^W}{n^F}+\varepsilon)}$. □

Lemma 13.5.8 *(Main lemma) Given $\beta \in \mathbb{R}_+$ and $\varepsilon \in (0, 1)$, the following holds:*

$$\mathcal{P}(\mathcal{D}(\mathbf{q}_{1,k}, \hat{\mathbf{q}}_{1,k}) > \beta) < \mathcal{P}(\Gamma_k) + m^{k\frac{n^F}{\ln(m)}\left(\ln(m)\frac{r^q}{n^F} - \beta + \frac{1}{2} + \frac{\ln(m)+n^F}{n^F}\varepsilon\right)} \tag{13.40}$$

where $\mathcal{P}(\Gamma_k)$ is given in Lemma 13.5.5.

Proof. We separate the two events that may generate $\mathcal{D}(\mathbf{q}_{1,k}, \hat{\mathbf{q}}_{1,k}) > \beta$, and write the following bound:

$$\mathcal{P}(\mathcal{D}(\mathbf{q}_{1,k}, \hat{\mathbf{q}}_{1,k}) > \beta) \leq \mathcal{P}(\Gamma_k) + \sum_{q_{1,k}\in\mathbb{A}^k} \sum_{\hat{q}_{1,k}\in\mathbb{D}(q_{1,k})} P(\mathbf{s}^{\varepsilon,k}(\hat{q}_{1,k}) = True|q_{1,k}) p^q(q_{1,k}) \tag{13.41}$$

where $\mathbb{D}(q_{1,k}) = \{\hat{q}_{1,k} \in \mathbb{T}^{\varepsilon,k} : \mathcal{D}(q_{1,k}, \hat{q}_{1,k}) > \beta\}$. Using Lemma 13.5.7 and the inequality above, we get:

$$\mathcal{P}(\mathcal{D}(\mathbf{q}_{1,k}, \hat{\mathbf{q}}_{1,k}) > \beta) \leq \left(\sharp\mathbb{T}^{\varepsilon,k}\right) m^{k\frac{n^F}{\ln(m)}(-\beta+\varepsilon+\frac{1}{2})} + \mathcal{P}(\Gamma_k) \tag{13.42}$$

The fact that $\sharp\mathbb{T}^{\varepsilon,k} \leq m^{k(r^q+\varepsilon)}$ concludes the proof. □

ACKNOWLEDGEMENTS

This work was sponsored by UCLA, MURI project award: 0205-G-CB222. Nuno C. Martins was partially supported by the Portuguese Foundation for Science and Technology and the European Social Fund, PRAXIS BD19630/99.

REFERENCES

Ackerson G A and Fu K S 1970 On state estimation in switching environments. *IEEE Trans. on Automatic Control* **15**(1): 10–17.

Arimoto S and Kimura H 1971 Optimum input test signals for system identification – an information-theoretical approach. *Int. J. Syst. Sci.* **1**(3): 279–90.

Bar-Shalom Y and Li X R 1993 *Estimation and Tracking: Principle, Techniques and Software.* Artech House, Boston, MA.

Battail G 1995 On random-like codes. *Information Theory and Applications II, Lecture notes in Computer Science* **1133**, pp. 76–94.

Belforte G and Gay P 2000 Optimal worst case estimation for LPV-FIR models with bounded errors. In *IEEE Proceedings of the IEEE Conference on Decision and Control* pp. 4573–4577.

Berger T 1971 *Rate Distortion Theory: A Mathematical Basis for Data Compression.* Prentice Hall, Englewood Cliffs, NJ.

Bertsekas D P 1995 *Dynamic Programming and Optimal Control.* Volume 1, Athena Scientific, Belmont, MA.

Billingsley P 1995 *Probability and Measure.* Wiley Inter-Science, Chichester, 3rd Edn.

Bittanti S, Campi MC, and Guo L 2000 Persistence of excitation properties for the identification of time-varying systems. *IEEE Conf. on Decision and Control* pp. 680–684.

Blahut R E 1972 Computation of channel capacity and rate-distortion functions. *IEEE Trans. on Inf. Theory* **18**: 460–473.

Blom H A P and Bar-Shalom Y 1988 The interacting multiple model algorithm for systems with Markovian switching coefficients. *IEEE Trans. A. C.* (8): 780–783.

Campi. M C 1994 Exponentially weighted least squares identification of time-varying systems with white disturbances. *IEEE Tr. Sig. Proc.* **42**: 2906–14.

Chang C B and Athans M 1978 State estimation for discrete systems with switching parameters. *IEEE Trans. on Aerospace and Electronic Systems* **14**(3): 418–425.

Cover T M and Thomas J A 1991 *Elements of Information Theory* Wiley-Interscience, Chichester.

Dong Y 2002 Rate distortion analysis of pose estimation via hidden Markov model. In IEEE, (ed.), *Proc. IEEE ICASP* pp. 2985–2988.

Elliot R 1995 *Hidden Markov Models: Estimation and Control.* Springer-Verlag, New York.

Feder M and Merhav N 1994 Relations between entropy and error probability. *IEEE Trans. on Inf. Theory* **40**: 259–266.

Gunnarsson S 1994 On the quality of recursively identified FIR models. *IEEE Tr. Sig. Proc.* **40**: 679–82.

Gunarsson S and Ljung L 1990 Adaptation and tracking in system identification – a survey. *Automatica* **26**: 7–21.

Ko H and Baran R H 1998 Entropy and information rates for hidden Markov parameters. In *IEEE Proc. IEEE ISIT* p. 374.

Hofbaur M W and Williams B C 2002 Mode estimation of probabilistic hybrid systems. *HSCC 2002, LNCS 2289*, pp. 253–266.

Liberzon D 2003 On stabilization of linear systems with limited information. *IEEE Transactions on Automatic Control* **48**: 304–307.

Martins N C and Dahleh M A 2004, Rate distortion in the modal estimation of switching FIR linear systems, *Proc. IEEE CDC*, pages 3575–3580, Vol. 4.

Mosca E 1972 Probing signal design for linear channel identification. *IEEE Trans. on Information Theory* **18**(4): 481–487.

Rabiner L R 2003 A tutorial on hidden Markov model and selected applications in speech recognition. *Proceedings of the IEEE* **77**: 257–286.

Sworder D D, and Boyd J E 1999 *Estimation Problems in Hybrid Systems*. Cambridge University Press, Cambridge.

Sworder D D, Boyd J E, and Elliott R J 2000 Modal estimation in hybrid systems. *Journal of Math. Analysis. and Ap.* (245): 225–247.

Tugnait J K 1982 Detection and estimation for abruptly changing systems. *IFAC – Automatica* **18**(5): 607–615.

Upadhyaya B R, and Sorenson H W 1977 Bayesian discriminant approach to input signal selection in parameter estimation problems. *Information Sciences*, **1**(12): 61–91.

Venkatesh S R and Dahleh M A 2001 On system identification of complex systems from finite data. *IEEE Trans. A. C.* **46**(2): 235–257.

Weidemann H L 1969 Entropy analysis of parameter estimation. *Inf. and Control* (14): 493–506.

Zames G, Lin L, and Wang L Y 1994 Fast identification n-widths and uncertainty principles for lti and slowly varying systems. *IEEE Trans. A. C.* **39**(9): 1827–1838.

14

Conditionally-linear filtering for mode estimation in jump-linear systems

Daniel Choukroun and Jason L. Speyer

14.1 INTRODUCTION

As opposed to linear optimal filtering theory, featuring the widely used Kalman filter (Kalman 1960), non-linear optimal filtering theory has generated so far few examples of practical algorithms. Although the extended Kalman filtering approach (EKF) (see e.g. Jazwinski (1970)) is very popular because it lends itself to numerous applications and yields practical algorithms, the linearization procedure used in this approach produces a suboptimal filter, which generally lacks any proof of convergence. Nevertheless, motivated by the need for general results, optimal filtering algorithms have been obtained for certain classes of non-linear systems. General optimal non-linear filters were developed for continuous-time systems (Liptser and Shiryayev 1977a; Wong and Hajek 1985) and for discrete-time systems (Elliot *et al.* 1993). In (Elliot *et al.* 1993), a general formula is developed that sequentially yields the conditional probability density function of the unknown components of a Markov sequence. This requires, however, an integration procedure, which might induce computational issues, if the conditional expectation is the sought quantity. On the other hand, (Liptser and Shiryayev 1977a) and (Wong and Hajek 1985) directly developed the conditional expectation estimator. In (Liptser and Shiryayev 1977a), the authors considered scalar known and unknown processes, in which coupled dynamics are perturbed by a Brownian motion and a square integrable martingale process and used square-integrable martingale theory in order to derive the optimal non-linear filtering equations.

A class of systems where practical finite-dimensional optimal non-linear filters can be developed (see e.g. Wonham (1965)) are hybrid jump systems, where discrete state variables characterize the environment in which continuous state variables evolve. Typically, the set of the discrete states, called the mode of the system, switches among a finite number of realizations and the switch time is stochastic (Wonham 1970). For this class of

Cooperative Control of Distributed Multi-Agent Systems edited by Jeff S. Shamma

non-linear systems, the EKF paradigm is clearly inadequate for the task of estimating the mode. Nevertheless, a finite dimensional estimator of the conditional expectation of the mode can be developed (Wonham filter, (Liptser)). This, however, assumes perfect information on the continuous states. Furthermore, in discrete-time, the complete probability distribution of the mode measurement noise is required.

The contribution of this chapter is twofold. A discrete-time filter is developed in order to sequentially approximate the conditional expectation estimate in a very general setting and, as an application, a novel reduced-order mode-estimator for a class of jump-linear systems under partial information on the continuous states is developed.

We consider a general discrete-time non-linear time-varying system that is not necessarily Markov. Furthermore, the dynamics of the unknown state are perturbed by a noise that is neither necessarily Gaussian nor necessarily independent from that state. The suggested suboptimal approach is similar to that of the standard linear filtering approach; namely, the estimate is sought as an orthogonal projection on the space generated by the incoming information. However, as opposed to standard linear filtering, orthogonality is here developed based on *conditional expectations* given the incoming information, rather than unconditional expectations. Hence, the structure of the estimate is *conditionally-linear* (CL) with respect to the incoming information given the past information, and it is better fitted for on-line implementation since the gains of the CL filter are functions of the incoming available information. In the case of an underlying discretization of a continuous-time system with Gaussian additive measurement noises, the CL filter is formally shown to be asymptotically equivalent to the continuous-time optimal non-linear filter as the discretization step becomes very small.

As an application, a novel reduced-order mode-estimator for a class of jump-linear systems under partial information on the continuous states is developed. It will be shown how the state-space equations of the jump-linear system can be recast as a non-Gaussian model that is conditionally-linear given the continuous states. Although applying the general CL filter to this model requires the computation of generally unknown conditional expectations, these quantities can be approximated due to the asymptotic optimality property of the CL filter under the assumption of a Gaussian mode measurement noise. The case of partial information in the continuous states, which cannot be handled by the Wonham filter, is addressed by recasting the model equations via augmentation of the state into a form that is similar to the full information case. A numerical aerospace example illustrates through extensive Monte Carlo simulations the performance of the mode estimator when applied to the detection of gyro failures using noisy but accurate spacecraft attitude measurements.

This chapter is organized as follows. In Section 14.2 the conditionally-linear (CL) filtering approach and the general CL filter are presented. In Section 14.3, the CL filter is tailored to mode-estimation. In Section 14.4, the numerical example of gyro failure detection is presented. Finally, conclusions are given in the last section.

14.2 CONDITIONALLY-LINEAR FILTERING

14.2.1 Short review of the standard linear filtering problem

For the sake of completeness, a short review of the standard linear filtering approach is presented in this section (see e.g. Papoulis 1965 or Luenberger 1969) and its main limitation is emphasized, thus motivating the conditionally-linear filtering approach. Consider the Hilbert space, denoted by $\mathcal{H}$, of the real scalar random variables (r.v.) with zero-mean and finite second-order moment. Given a finite set $\mathcal{V}^k$ of r.v. in $\mathcal{H}$, $\{v_i\}_{i=1}^k$, the linear subspace spanned by $\mathcal{V}^k$, denoted by $\mathcal{L}\{\mathcal{V}^k\}$, is defined as follows

$$\mathcal{L}\{\mathcal{V}^k\} \triangleq \left\{\sum_{i=1}^{k} \lambda_i v_i\right\} \tag{14.1}$$

where $\lambda_i \in \mathbb{R}$. Given two r.v. in $\mathcal{H}$, v_1 and v_2, their inner product in $\mathcal{H}$ is defined as the *unconditional expectation of their product in* $\mathbb{R}$; that is,

$$< v_1, v_2 > \triangleq E\{v_1 v_2\} \tag{14.2}$$

The definitions of norm and orthogonality stem from the inner product. The induced norm of the r.v. v_1 in $\mathcal{H}$ is the square-root of its second-order moment; that is

$$\|v_1\| = E\{v_1^2\}^{1/2}, \tag{14.3}$$

and two r.v. in $\mathcal{H}$, v_1 and v_2, are *orthogonal* if the unconditional expectation of their product is null; that is

$$v_1 \perp v_2 \Leftrightarrow E\{v_1 v_2\} = 0 \tag{14.4}$$

An *orthonormal* set of r.v., $\mathcal{V}^k$, is defined as follows:

$$\forall (v_i, v_j) \in \mathcal{V}^k,\ E\{v_i v_j\} = \delta_{ij} \tag{14.5}$$

where δ_{ij} denotes the Kronecker delta. Based on these premisses, the standard linear optimal filtering problem is stated as follows:

Consider an unknown r.v. $y \in \mathcal{H}$, and an orthonormal sequence of observations, $\mathcal{X}^{k+1} \subset \mathcal{H}$, such that the *unconditional* correlations, $P_{yx_i} \triangleq E\{yx_i\}$, are given. Denote by $\widehat{y}_k$ and by $\widehat{y}_{k+1}$ the sought estimates of y given $\mathcal{X}^k$ and $\mathcal{X}^{k+1}$, respectively. Assuming that $\widehat{y}_k \in \mathcal{L}\{\mathcal{X}^k\}$, and $\widehat{y}_{k+1} \in \mathcal{L}\{\mathcal{X}^{k+1}\}$ as

$$\widehat{y}_{k+1} = \widehat{y}_k + \alpha_{k+1} x_{k+1}, \tag{14.6}$$

find α_{k+1} such that

$$\widetilde{y}_{k+1} \perp \mathcal{L}\{\mathcal{X}^{k+1}\} \tag{14.7}$$

where $\widetilde{y}_{k+1}$ denotes the estimation error, $y - \widehat{y}_{k+1}$ at stage $k+1$.

The solution to this problem consists of the following recursive algorithm for computing $\widehat{y}$:

$$\alpha^\star_{k+1} = P_{yx_{k+1}} \tag{14.8}$$

$$\widehat{y}_{k+1} = \widehat{y}_k + \alpha^\star_{k+1} x_{k+1} \tag{14.9}$$

and, as a by product, of the recursion on $P_k \triangleq E\{\widetilde{y}^2\}$ as

$$P_{k+1} = P_k - \left(P_{yx_{k+1}}\right)^2, \tag{14.10}$$

showing the monotonous decrease of the norm of the estimation error. The estimate $\widehat{y}_k$ is known as the *orthogonal projection* of y onto the subspace $\mathcal{L}\{\mathcal{X}^k\}$. The more general case can be deduced from this simple case using the following standard techniques. If the r.v. x_k have non-zero known means, then new zero-mean observations are defined as

$$\widetilde{x}_k = x_k - E\{x_k\} \tag{14.11}$$

If the x_k are not orthogonal, then they can be orthogonalized using the standard Gram-Schmidt procedure. If they are not unit-norm, then new normalized observations are defined as follows:

$$\widetilde{x}_k = P_{\widetilde{x}}^{-1/2} x_k \tag{14.12}$$

If the r.v. $\mathbf{y}$ and $\mathbf{x}_k$ are m-dimensional and n-dimensional, respectively, then their inner product is defined as the unconditional expectation of their outer product in $\mathbb{R}^{m\times n}$; that is,

$$< \mathbf{y}, \mathbf{x}_k > \triangleq E\{\mathbf{y}\mathbf{x}_k^T\} \tag{14.13}$$

The main drawback in the previously described approach is that the filter gains are computed using unconditional expectations and are thus only functions of the initially available information. For on-line implementation, a better approach should allow the gains to be functions of the growing available information. This is the motivation for the conditionally-linear filtering approach described in the following subsection.

14.2.2 The conditionally-linear filtering problem

14.2.2.1 Conditional orthogonality

Given three r.v in $\mathcal{H}$, v_1, v_2 and v_3, a new inner product is defined as follows:

$$< v_1, v_2 | v_3 > \triangleq E\{v_1 v_2 | v_3\} \tag{14.14}$$

It is straightforward to check that the inner product as defined in Equation (14.14) is a legitimate inner product (see Appendix A 14.6). Using this inner product the concept of *conditional orthogonality* is defined as follows. Two r.v. v_1 and v_2 are conditionally orthogonal given v_3 if the conditional expectation of their product given v_3 is equal to zero. The following notation will be used:

$$v_1 \ c\!\perp v_2 \text{ given } v_3 \Leftrightarrow E\{v_1 v_2 | v_3\} = 0 \tag{14.15}$$

A central point in using this inner product is that any function of the given r.v. v_3 is considered a 'scalar' quantity and can be taken out of the conditional expectation operator.

Particular subspaces of interest are those generated by a r.v., say $\mathbf{x}_{k+1}$, where the 'scalars' are functions of the known past sequence $\mathcal{X}^k$. They will be denoted as follows:

$$\mathcal{C}\{\mathbf{x}_{k+1}\} \stackrel{\triangle}{=} \{\lambda(\mathcal{X}^k)\mathbf{x}_{k+1}\} \tag{14.16}$$

Similarly, the increasing sequence $\mathcal{X}^k$ generates a subspace $\mathcal{C}(\mathcal{X}^k)$ as follows:

$$\mathcal{C}\{\mathcal{X}^{k+1}\} \stackrel{\triangle}{=} \mathcal{C}(\mathcal{X}^k) + \mathcal{C}(\mathbf{x}_{k+1}) \tag{14.17}$$

such that the sequence of subspaces $\mathcal{C}(\mathcal{X}^k)$ satisfies the nested property:

$$\mathcal{C}\{\mathbf{x}_1\} \subseteq \ldots \subseteq \mathcal{C}(\mathcal{X}^k) \subseteq \mathcal{C}(\mathcal{X}^{k+1}) \tag{14.18}$$

14.2.2.2 *Case of a static r.v.* $\mathbf{y} \in \mathbb{R}^{n_y}$ *with zero-mean orthogonal observations* $\mathcal{X}^k \in \mathbb{R}^{n_x \times k}$

Let $\widehat{\mathbf{y}}_k$ and $\widetilde{\mathbf{y}}_k$ denote respectively the estimate and the estimation error at stage k and assume that the *conditional* moments $P_{\widetilde{\mathbf{y}}\mathbf{x}_{k+1}}$ and $P_{\mathbf{x}_{k+1}}$ are known, where

$$P_{\widetilde{\mathbf{y}}\mathbf{x}_{k+1}} = E\{\widetilde{\mathbf{y}}_k \mathbf{x}_{k+1}^T \mid X_k\} \tag{14.19}$$

$$P_{\mathbf{x}_{k+1}} = E\{\mathbf{x}_{k+1}\mathbf{x}_{k+1}^T \mid X_k\} \tag{14.20}$$

Assuming that $\widehat{\mathbf{y}}_k \in \mathcal{C}\{\mathcal{X}^k\}$ and that the estimator's structure is as follows

$$\widehat{\mathbf{y}}_{k+1} = \widehat{\mathbf{y}}_k + \beta_{k+1}(\mathcal{X}^k)\mathbf{x}_{k+1}, \tag{14.21}$$

we consider the problem of finding β_{k+1} such that

$$\widetilde{\mathbf{y}}_{k+1}\ c\bot\ \mathcal{C}\{\mathbf{x}_{k+1}\}\,\text{given}\ \mathcal{X}^k \tag{14.22}$$

The solution to Equation (14.22) is given as follows:

$$\beta_{k+1}^{\star}(\mathcal{X}^k) = P_{\widetilde{\mathbf{y}}\mathbf{x}_{k+1}} P_{\mathbf{x}_{k+1}}^{-1} \tag{14.23}$$

Proof. Proceeding by equivalence from Equation (14.22), one obtains

$$\begin{aligned}
&\widetilde{\mathbf{y}}_{k+1}\ c\bot\ \mathcal{C}\{\mathbf{x}_{k+1}\}\,\text{given}\ \mathcal{X}^k\\
\Leftrightarrow\quad & E\{\widetilde{\mathbf{y}}_{k+1}\mathbf{x}_{k+1}^T \Lambda^T(\mathcal{X}^k) \mid \mathcal{X}^k\} = 0 \qquad \forall \Lambda(\cdot) \in \mathbb{R}^{n_y \times n_x},\\
\Leftrightarrow\quad & E\{[\widetilde{\mathbf{y}}_k - \beta_{k+1}(\mathcal{X}^k)\mathbf{x}_{k+1}]\,\mathbf{x}_{k+1}^T\, \Lambda^T(\mathcal{X}^k) \mid \mathcal{X}^k\} = 0\\
\Leftrightarrow\quad & \left[E\{\widetilde{\mathbf{y}}_k\mathbf{x}_{k+1}^T \mid \mathcal{X}^k\} - \beta_{k+1}(\mathcal{X}^k)\, E\{\mathbf{x}_{k+1}\mathbf{x}_{k+1}^T \mid \mathcal{X}^k\}\right] \Lambda^T(\mathcal{X}^k) = 0\\
\Leftrightarrow\quad & E\{\widetilde{\mathbf{y}}_k\mathbf{x}_{k+1}^T \mid \mathcal{X}^k\} - \beta_{k+1}(\mathcal{X}^k)\, E\{\mathbf{x}_{k+1}\mathbf{x}_{k+1}^T \mid \mathcal{X}^k\} = 0
\end{aligned}$$

$$\Leftrightarrow \quad \beta^{\star}_{k+1}(\mathcal{X}^k) = E\{\widetilde{\mathbf{y}}_k \mathbf{x}^T_{k+1} \,|\, \mathcal{X}^k\}\left[E\{\mathbf{x}_{k+1}\mathbf{x}^T_{k+1} \,|\mathcal{X}^k\}\right]^{-1}$$

$$\Leftrightarrow \quad \beta^{\star}_{k+1}(\mathcal{X}^k) = P_{\widetilde{\mathbf{y}}\mathbf{x}_{k+1}} P^{-1}_{\mathbf{x}_{k+1}} \tag{14.24}$$

where the passage from line four to five in Equation (14.24) is due to the fact that Λ is an arbitrary parameter. A central point in this proof is the use of the conditional expectation, which allowed carrying any expression of the observations $\mathcal{X}^k$ out of the expectation operator. As a result, the estimate $\widehat{\mathbf{y}}_k$, which in expanded form is given as follows,

$$\widehat{\mathbf{y}}_k = \beta_1 \mathbf{x}_1 + \beta_2(\mathbf{x}_1)\,\mathbf{x}_2 + \cdots + \beta_{k+1}(\mathcal{X}^k)\,\mathbf{x}_{k+1} \tag{14.25}$$

was obtained by recursively summing up the successive projections of $\mathbf{y}$ onto the incremental subspaces $\mathcal{C}\{\mathbf{x}_k\}$. The following notation will also be used:

$$\widehat{\mathbf{y}}_k = \Pi\{\mathbf{y}, \mathcal{C}\{\mathcal{X}^k\}\} \tag{14.26}$$

14.2.2.3 Case of a static r.v. $\mathbf{y}$ *with non-orthogonal observations*

Assume that the sequence of observations $\{\widetilde{\mathcal{X}}^k\}$ is zero-mean and orthogonal, and that the observation $\mathbf{x}_{k+1}$ is not orthogonal to $\widetilde{\mathcal{X}}^k$. By analogy with the standard linear filtering theory, we define the residual, $\widetilde{\mathbf{x}}_{k+1}$, as follows:

$$\widetilde{\mathbf{x}}_{k+1} = \mathbf{x}_{k+1} - \widehat{\mathbf{x}}_{k+1} \tag{14.27}$$

where

$$\widehat{\mathbf{x}}_{k+1} = \Pi\{\mathbf{x}_{k+1}, \mathcal{C}\{\widetilde{\mathcal{X}}^k\}\} \tag{14.28}$$

Assuming the following estimator's structure at stage $k+1$,

$$\widehat{\mathbf{y}}_{k+1} = \widehat{\mathbf{y}}_k + \beta_{k+1}(\widetilde{\mathcal{X}}^k)\,\widetilde{\mathbf{x}}_{k+1}, \tag{14.29}$$

the conditionally-linear filtering problem boils down to finding β_{k+1} such that

$$\widetilde{\mathbf{y}}_{k+1}\; c\perp\; \mathcal{C}\{\widetilde{\mathbf{x}}_{k+1}\}\,\text{given}\,\widetilde{\mathcal{X}}^k \tag{14.30}$$

which is similar to the problem of the previous case as stated in Equation (14.22).

14.2.2.4 General case

Consider a joint vector random sequence $\{\mathbf{x}_k, \mathbf{y}_k\}$, $k \in \mathbb{N}$, where $\mathbf{x}_k \in \mathbb{R}^{n_x}$ is observed and $\mathbf{y}_k \in \mathbb{R}^{n_y}$ is unknown. The dynamics of these sequences are governed by the following random difference equations

$$\mathbf{x}_{k+1} = \mathbf{f}_k(\mathcal{X}^k, \mathcal{Y}^k) + \mathbf{w}_k \tag{14.31}$$

$$\mathbf{y}_{k+1} = \mathbf{g}_k(\mathcal{X}^k, \mathcal{Y}^k) + \mathbf{v}_k \tag{14.32}$$

In Equations (14.31) and (14.32), $\mathbf{f}_k$ and $\mathbf{g}_k$ are mappings into $\mathbb{R}^{n_x}$ and $\mathbb{R}^{n_y}$, respectively, that satisfy the conditions of existence and uniqueness of a solution, $\mathcal{X}^k$ and $\mathcal{Y}^k$ denote the past histories of $\mathbf{x}$ and $\mathbf{y}$, respectively, up to stage k as $\mathcal{X}^k = \{\mathbf{x}_0, \mathbf{x}_1, \ldots, \mathbf{x}_k\}$ and $\mathcal{Y}^k = \{\mathbf{y}_0, \mathbf{y}_1, \ldots, \mathbf{y}_k\}$, and the sequence $\{\mathbf{w}_k, \mathbf{v}_k\} \in \mathbb{R}^{n_x} \times \mathbb{R}^{n_y}$ denotes a zero-mean white noise sequence with a known covariance matrix given as

$$\operatorname{cov}\left\{\begin{bmatrix}\mathbf{w}_k\\ \mathbf{v}_k\end{bmatrix}\right\} = \begin{bmatrix} W_k & D_k^T \\ D_k & V_k \end{bmatrix} \tag{14.33}$$

The vector $\mathbf{w}_k$ is assumed to be independent of $\{\mathcal{X}^k, \mathcal{Y}^k\}$, $k \in \mathbb{N}$, and the vector $\mathbf{v}_k$ is assumed to be independent of $\mathcal{X}^k$ and orthogonal to $\mathcal{Y}^k$, $k \in \mathbb{N}$. The initial vector $\mathbf{x}_0$ is known as well as the mean and the covariance matrix of the initial vector $\mathbf{y}_0$ as $E\{\mathbf{y}_0\}$ and $P_{\mathbf{y}_0}$. We consider the conditionally-linear filtering problem that consists in sequentially solving for the projection of $\mathbf{y}_k$ onto $\mathcal{C}\{\widetilde{\mathcal{X}}^k\}$, which is denoted by $\widehat{\mathbf{y}}_{k/k}$. Notice that the system described in Equation (14.31) and (14.32) presents interesting departures from the usual formulation given in a standard estimation problem (see e.g. Jazwinski 1970, p. 174). Here, the dynamics of the estimated state is not necessarily Markov because they depend on the histories $\mathcal{X}^k$ and $\mathcal{Y}^k$. Moreover, the noises are not necessarily Gaussian, the process noise $\mathbf{v}_k$ is not necessarily independent from the current state $\mathbf{y}_k$, and is not necessarily uncorrelated with the observation noise, $\mathbf{w}_k$. Also, the observation model equation of the unknown state; that is, Equation (14.31), is a dynamical equation rather than a static equation.

14.2.2.5 Summary of the algorithm

The following notation will be used throughout the chapter:

$$\widehat{A}_{k/l} \stackrel{\Delta}{=} \Pi\{A_k, \mathcal{C}\{\mathcal{X}^l\}\} \tag{14.34a}$$

$$\widetilde{A}_{k/l} \stackrel{\Delta}{=} A_k - \widehat{A}_{k/l} \tag{14.34b}$$

where A_k denotes any random variable such that the projection in Equation (14.34a) exists. The optimal estimator of $\mathbf{y}_k$ given $\mathcal{X}^k$ is summarized as follows. For the sake of clarity, the related proofs are developed in 14.7.

1. Initialization equation

$$\widehat{\mathbf{y}}_{0/0} = E\{\mathbf{y}_0\} \tag{14.35}$$

2. Smoothing stage equations

$$\widetilde{\mathbf{x}}_{k+1/k} = \mathbf{x}_{k+1} - \widehat{\mathbf{f}}_{k/k} \tag{14.36a}$$

$$P_{\widetilde{\mathbf{f}}_{k/k}} = E\{\widetilde{\mathbf{f}}_{k/k}\widetilde{\mathbf{f}}_{k/k}^T | \mathcal{X}^k\} \tag{14.36b}$$

$$P_{\widetilde{\mathbf{x}}_{k+1/k}} = P_{\widetilde{\mathbf{f}}_{k/k}} + W_k \tag{14.36c}$$

$$P_{\tilde{\mathbf{y}}\tilde{\mathbf{f}}_{k/k}} = E\{\tilde{\mathbf{y}}_{k/k}\tilde{\mathbf{f}}^T_{k/k}|\mathcal{X}^k\} \tag{14.36d}$$

$$\widehat{\mathbf{y}}_{k/k+1} = \widehat{\mathbf{y}}_{k/k} + P_{\tilde{\mathbf{y}}\tilde{\mathbf{f}}_{k/k}} P^{-1}_{\tilde{\mathbf{x}}_{k+1/k}} \tilde{\mathbf{x}}_{k+1/k} \tag{14.36e}$$

3. Time-propagation stage equations

$$\widehat{\mathbf{y}}_{k+1/k+1} = \widehat{\mathbf{g}}_{k/k+1} + D_k P^{-1}_{\tilde{\mathbf{x}}_{k+1/k}} \tilde{\mathbf{x}}_{k+1/k} \tag{14.37}$$

14.2.3 Discussion

14.2.3.1 General comments

The above filter works in a two-steps sequence. There is first a smoothing step (Equation (14.36)). Assuming that the estimates $\widehat{\mathbf{y}}_{k/k}$ and $\widehat{\mathbf{f}}_{k/k}$ have been computed, a new observation $\mathbf{x}_{k+1}$ is acquired and processed at stage $k+1$ yielding a smoothed estimate of $\mathbf{y}_k$ at $k+1$, denoted by $\widehat{\mathbf{y}}_{k/k+1}$. The correction step (Equation (14.36e)) obeys our intuition in the sense that correction of the current estimate is possible thanks to the correlation between the estimated vector $\mathbf{y}_k$ and the predictable part of the observation, $\widehat{\mathbf{f}}_{k/k}$. Using the process equation for $\mathbf{y}_k$, Equation (14.37) propagates the best estimate from stage k to stage $k+1$, yielding thus the filtered estimate $\widehat{\mathbf{y}}_{k+1/k+1}$.

The fact that the observation at stage k, $\mathbf{x}_{k+1}$, is one-step delayed induces the smoothing-than-propagating mechanism instead of the classical propagating-than-filtering mechanism. In spite of the several departures from the standard discrete-time linear state-space model, the general structure of the above algorithm is very similar to that of the standard linear optimal filter. The central difference with the linear filter is that the various covariance matrices in Equations (14.36) and (14.37) are *conditional* covariance matrices given the past $\mathcal{X}^k$, and make, thus, the algorithm highly non-linear with respect to the observations $\mathcal{X}^k$.

The presented algorithm focuses on the state estimate computation and does not show how the projections, $\widehat{\mathbf{f}}_{k/k}$ and $\widehat{\mathbf{g}}_{k/k+1}$, are obtained, nor does it explain how to compute the estimation error covariance matrices. While these quantities are in general very hard to compute, it is well known that the algorithm greatly simplifies when assuming that $\{\mathbf{y}_k\}$ is a linear Gauss-Markov sequence and that $\mathbf{x}_k$ is a linear observation of $\mathbf{y}_k$, without dynamics and with additive Gaussian noise. In the latter case the filtering equations boil down to the classic Kalman filter recursive equations (Kalman 1960) (involving a standard technique to handle the case of correlated noises $\mathbf{w}_k$ and $\mathbf{v}_k$). The appealing simplicity of the filtering equations is only apparent and masks the difficulty attached to the computation of the various conditional expectations.

In the conditionally-linear filtering problem, as stated in Equation (14.22), one only requires orthogonality with respect to the subspace $\mathcal{C}\{\mathbf{x}_{k+1}\}$ rather than to the whole space $\mathcal{C}\{\mathcal{X}^{k+1}\}$. The latter would imply conditions of the following form:

$$E\{\mathbf{x}_{k+1}|\mathcal{X}^k\} = 0 \tag{14.38}$$

which can not in general be satisfied.

14.2.3.2 Approximation of the continuous-time optimal non-linear filter

When applied to a discretized system, and assuming that the measurement noise $\mathbf{w}_k$ is Gaussian, the conditionally-linear filter seems to converge to the continuous-time optimal non-linear filter; that is, to the conditional expectation filter as the discretization time-increment converges to zero. This observation is supported by the following developments and was checked through intensive simulations. For simplicity, without loss of generality, we consider the case of a static unknown non-Gaussian r.v., $\mathbf{y}$, and we assume that its measurement is a scalar process, ξ_t, which is described by the following Itô stochastic differential equation in integral form:

$$\xi_t = \xi_0 + \int_0^t f_s(\mathbf{y}, \xi^s)\, ds + \beta_t \tag{14.39}$$

where β_t is a standard Brownian motion with unit intensity and with independent increments $\Delta\beta_t = \beta_{t+\Delta t} - \beta_t$, where

$$E\{(\Delta\beta_t)^2\} = \Delta t \tag{14.40}$$

Referring to (Wong and Hajek 1985) for the general equations of the optimal filter, we obtain the following measurement update stage for the continuous-time filter:

$$\widehat{\mathbf{y}}_t = \widehat{\mathbf{y}}_0 + \int_0^t \operatorname{cov}\{\mathbf{y}, f_s|\xi^s\}\, dw_s \tag{14.41}$$

where the process w_s is a Brownian motion defined as

$$dw_s = d\xi_s - \widehat{f_s}\, dt \tag{14.42}$$

and such that it has the same statistics as β_t. In particular,

$$E\{(\Delta w_t)^2\} = \Delta t \tag{14.43}$$

By formally discretizing Equation (14.41) between t_k and $t_k + \Delta t$, one obtains the following recursive measurement update stage for the estimate $\widehat{\mathbf{y}}_k$:

$$\widehat{\mathbf{y}}_{k+1} = \widehat{\mathbf{y}}_k + \underbrace{\operatorname{cov}\{\mathbf{y}, f_k|\xi^k\}}_{P_{\mathbf{y}\tilde{\mathbf{f}}}} \underbrace{\frac{1}{(\Delta t)^{-1}}}_{P_{\tilde{x}}^{-1}} \underbrace{\left(\frac{\Delta w_k}{\Delta t}\right)}_{\tilde{x}_{k+1}} + \mathcal{O}\{\Delta t^{3/2}\} \tag{14.44}$$

where, using Equation (14.43), and by abuse of notation, we substituted $\mathcal{O}\{\Delta t^{3/2}\}$ for $\mathcal{O}\{|\Delta w_k|\,\Delta t\}$ in Equation (14.44). Looking at Equation (14.44), one clearly sees the similarity with the update stage equation developed in the conditionally-linear filter, as given in Equation (14.36e). To conclude, under the Gaussian assumption, the discrete-time conditionally-linear (CL) filter provides an approximation of order $\Delta t^{3/2}$ for the optimal *non-linear* filter. In the non-Gaussian case, however, the CL filter still provides a clear interpretation.

14.2.3.3 *Special case: Conditionally-linear systems*

Two known results in optimal filtering are:(1) applying the standard linear filtering approach to a linear system where the unknown states and the noises are jointly Gaussian yields the Minimum-Variance estimator (the Kalman filter);(2) the *same* algorithm remains meaningful even when the Gaussian property does not hold: it represents the minimum-variance algorithm among the linear estimators. Does the conditionally-linear approach bear the same kind of results?

Assume for simplicity that $\mathbf{y}$ is a static parameter, which is observed through a conditionally-linear measurement as follows:

$$\mathbf{x}_{k+1} = H(\mathcal{X}^k)\,\mathbf{y} + \mathbf{w}_k, \tag{14.45}$$

and that $\mathbf{y}$ and $\mathbf{w}_k$ are conditionally jointly Gaussian given $\mathcal{X}^k$. Then, applying the filtering equations to this system yields:

$$\widetilde{\mathbf{x}}_{k+1} = \mathbf{x}_{k+1} - H(\mathcal{X}^k)\,\widehat{\mathbf{y}}_k \tag{14.46}$$

$$P_{\widetilde{\mathbf{x}}_{k+1}} = H(\mathcal{X}^k)\,P_{\widetilde{\mathbf{y}}_k}\,H^T(\mathcal{X}^k) + W_k \tag{14.47}$$

$$P_{\widetilde{\mathbf{y}}\widetilde{\mathbf{x}}_{k+1}} = P_{\widetilde{\mathbf{y}}_k}\,H^T(\mathcal{X}^k) \tag{14.48}$$

$$K_{k+1} = P_{\widetilde{\mathbf{y}}\widetilde{\mathbf{x}}_{k+1}}\,P^{-1}_{\widetilde{\mathbf{x}}_{k+1}} \tag{14.49}$$

$$\widehat{\mathbf{y}}_{k+1} = \widehat{\mathbf{y}}_k + K_{k+1}\,\widetilde{\mathbf{x}}_{k+1} \tag{14.50}$$

Notice that, in developing the expressions for $P_{\widetilde{\mathbf{x}}_{k+1}}$ and $P_{\widetilde{\mathbf{y}}\widetilde{\mathbf{x}}_{k+1}}$ in Equations (14.47) and (14.49), we used the fact that any function of the measurements $\mathcal{X}^k$ is also a function of the residuals $\widetilde{\mathcal{X}}^k$. Thus, given $\widetilde{\mathcal{X}}^k$, the variable $H(\mathcal{X}^k)$ could be extracted from the conditional expectations. With the purpose in mind of developing an update formula for the second-order moment $P_{\widetilde{\mathbf{y}}_k}$, we express the estimation error, $\widetilde{\mathbf{y}}_{k+1}$ as follows:

$$\widetilde{\mathbf{y}}_{k+1} = \left[I - K_{k+1}\,H_k\right]\widetilde{\mathbf{y}}_k - K_{k+1}\mathbf{w}_k \tag{14.51}$$

which, by squaring Equation (14.51), yields

$$\begin{aligned}\widetilde{\mathbf{y}}_{k+1}\widetilde{\mathbf{y}}^T_{k+1} = &\left[I - K_{k+1}\,H_k\right]\widetilde{\mathbf{y}}_k\widetilde{\mathbf{y}}^T_k\left[I - K_{k+1}\,H_k\right]^T + K_{k+1}\mathbf{w}_k\mathbf{w}^T_k K^T_{k+1}\\ &- \left[I - K_{k+1}\,H_k\right]\widetilde{\mathbf{y}}_k\mathbf{w}^T_k K^T_{k+1} - K_{k+1}\mathbf{w}_k\widetilde{\mathbf{y}}^T_k\left[I - K_{k+1}\,H_k\right]^T\end{aligned} \tag{14.52}$$

Applying the operator $E\{\cdot|\widetilde{\mathcal{X}}^k\}$ on both sides of Equation (14.52), and recalling that $\mathbf{w}_k$ is zero-mean and independent from $\widetilde{\mathbf{y}}_k$ yields

$$E\{\widetilde{\mathbf{y}}_{k+1}\widetilde{\mathbf{y}}^T_{k+1}|\widetilde{\mathcal{X}}^k\} = \left[I - K_{k+1}\,H_k\right]E\{\widetilde{\mathbf{y}}_k\widetilde{\mathbf{y}}^T_k|\widetilde{\mathcal{X}}^k\}\left[I - K_{k+1}\,H_k\right]^T + K_{k+1}\,W_k K^T_{k+1} \tag{14.53}$$

where the functions of the measurements could be extracted from the conditional expectations. In the first term on the right-hand side of Equation (14.53), one recognizes the variable $P_{\widetilde{\mathbf{y}}_k}$. Recall now that, at stage $k+1$, the error $\widetilde{\mathbf{y}}_{k+1}$ is conditionally orthogonal to the residual $\widetilde{\mathbf{x}}_{k+1}$ given $\widetilde{\mathcal{X}}^k$ (Equation 14.30). Under the Gaussian assumptions, $\widetilde{\mathbf{y}}_{k+1}$ is conditionally jointly Gaussian with $\widetilde{\mathbf{x}}_{k+1}$ given $\widetilde{\mathcal{X}}^k$ and, thus, $\widetilde{\mathbf{y}}_{k+1}$ is conditionally independent from $\widetilde{\mathbf{x}}_{k+1}$ given $\widetilde{\mathcal{X}}^k$. As a result:

$$\begin{aligned} P_{\widetilde{\mathbf{y}}_{k+1}} &\triangleq E\{\widetilde{\mathbf{y}}_{k+1}\widetilde{\mathbf{y}}_{k+1}^T \mid \widetilde{\mathcal{X}}^{k+1}\} \\ &= E\{\widetilde{\mathbf{y}}_{k+1}\widetilde{\mathbf{y}}_{k+1}^T \mid \widetilde{\mathcal{X}}^{k}, \widetilde{\mathbf{x}}_{k+1}\} \\ &= E\{\widetilde{\mathbf{y}}_{k+1}\widetilde{\mathbf{y}}_{k+1}^T \mid \widetilde{\mathcal{X}}^{k}\} \end{aligned} \tag{14.54}$$

Using Equation (14.54) in Equation (14.53) achieves the recursion:

$$P_{\widetilde{\mathbf{y}}_{k+1}} = \left[I - K_{k+1} H_k\right] P_{\widetilde{\mathbf{y}}_k} \left[I - K_{k+1} H_k\right]^T + K_{k+1} W_k K_{k+1}^T \tag{14.55}$$

The filter consisting of Equations (14.46) to (14.50) and (14.55), called the Conditionally-Gaussian (CG) filter in the literature (Liptser and Shiryayev 1977b), is a conditional expectation estimator. But, if the conditional-Gaussian assumptions do not hold, then Equation (14.54) is not true, and no recursion on $P_{\widetilde{\mathbf{y}}_k}$ can be achieved. In addition, unlike the linear Gaussian case, applying the unconditional expectation operator does not lead to practical expressions because of the dependence of the parameters on the measurements. To conclude, the CG filter does not lend itself to a meaningful algorithm when the Gaussian assumption is dropped. It remains though a good heuristics to apply this algorithm in the case of constant or slowly varying parameters. In the case of abrupt changes, however, divergence is likely to occur. Nevertheless, in the latter case, the conditionally-linear approach together with a specific correction term in the update Equation (14.55) can lend itself to a practical algorithm. This will be illustrated for special jump systems in the next section.

14.3 MODE-ESTIMATION FOR JUMP-LINEAR SYSTEMS

14.3.1 Statement of the problem

A very popular way of modeling dynamical systems in switching environments is by means of hybrid systems, where some state variables are continuous and some are discrete. The discrete variables characterize the environment in which the continuous variables evolve. Typically, the mode, which is the set of the discrete states, switches among a finite number of values and the switch may be deterministic (Liberzon and Morse 1999) or stochastic (Wonham 1970). For the stochastic case, assuming that the continuous state is known, finite-dimensional optimal non-linear mode-estimators can be developed; that is, algorithms which compute the conditional expectation of the mode. In a continuous-time setting these algorithms consist of non-linear stochastic differential equations (see Liptser and Shiryayev 1977a, Chap. 9) and Wonham (1965)). In order to avoid the drawback of numerically integrating them, a recursive optimal non-linear filter can be developed in a discrete-time setting (Liptser). It appears, however, that this algorithm requires a complete knowledge of the probability distribution of the noise in the dynamics equation of the continuous state.

We develop a *distribution-free* recursive mode-estimator for a class of jump-linear systems using the conditionally-linear approach. A novel contribution of this work is our noting that the continuous state process equation can be modelled as linear with respect to the mode. As a result, a linear non-Gaussian state-space model for the mode is developed. The proposed mode-estimator is an application of the general conditionally-linear filter presented in the previous section. In order to achieve recursion for the gain computation, we rely on the assumption, which was discussed in the previous section, that the estimate is a good approximation of the exact conditional expectation in the case of a Gaussian measurement noise.

Consider the discrete-time dynamical system

$$\mathbf{x}_{k+1} = A(y_k)\mathbf{x}_k + B(y_k)\mathbf{u}_k + \mathbf{w}_k \qquad k \in \mathbb{N} \tag{14.56}$$

where $\mathbf{x}_k \in \mathbb{R}^n$ is a known continuous state vector and $\{y_k : k \in \mathbb{N}\}$ is an unknown scalar finite-state homogenous discrete-time Markov chain with state space $S_y = \{\gamma_1, \gamma_2, \ldots, \gamma_\nu\}$ and transition matrix M given as

$$M[i, j] \triangleq Pr(y_{k+1} = \gamma_i \mid y_k = \gamma_j) \tag{14.57}$$

for all $\gamma_i, \gamma_j \in S_y$. The disturbance $\mathbf{w}_k$ is assumed to be a zero-mean white-noise sequence with known covariance matrix W_k, and is assumed to be independent from $\mathbf{x}_k$ and y_k. The control vector $\mathbf{u}_k \in \mathbb{R}^m$ is assumed to be a vector of known inputs that are a function of the available information. It can be shown (Elliot *et al.* 1993., p. 17) that the Markov chain $\{y_k,\ k \in \mathbb{N}\}$ defined on a given probability space $\{\Omega, \mathcal{F}, Pr\}$ can be equivalently replaced by the vector Markov chain $\{\mathbf{y}_k,\ k \in \mathbb{N}\}$ defined over the same probability space, where $\mathbf{y}_k$ is the random vector whose components are the characteristic random variables associated with each of the ν elementary events of $\Omega = \{\omega_i\}_{i=1}^{\nu}$. That is,

$$\begin{aligned} \mathbf{y}_k : \Omega = \{\omega_i\}_{i=1}^{\nu} &\rightarrow S_{\mathbf{y}} = \{\mathbf{e}_1, \mathbf{e}_2, \ldots, \mathbf{e}_\nu\} \\ \omega \mapsto \mathbf{y}_k^T(\omega) &= \left[\delta(\mathbf{y}_k, \mathbf{e}_1)\, \delta(\mathbf{y}_k, \mathbf{e}_2) \ldots \delta(\mathbf{y}_k, \mathbf{e}_\nu)\right] \end{aligned} \tag{14.58}$$

where $\mathbf{e}_i$, for $i = 1, 2, \ldots, \nu$ are the standard unit vectors in $\mathbb{R}^\nu$, and $\delta(\cdot, \cdot)$ represents the Kronecker delta. The dynamics of the Markov chain $\{\mathbf{y}_k \in \mathbb{R}^\nu, k \in \mathbb{N}\}$ is governed by the following linear $\nu \times 1$ vector difference equation

$$\mathbf{y}_{k+1} = M\mathbf{y}_k + \mathbf{v}_k \tag{14.59}$$

where $\{\mathbf{v}_k\}$ is a zero-mean white-noise sequence that is orthogonal to $\mathbf{y}_k$, and independent of $\mathcal{Y}^{k-1}$ and of $\mathcal{X}^k$. Moreover, $\mathbf{v}_k$ and $\mathbf{w}_k$ are assumed to be independent. From the definition of $\mathbf{y}_k$ for $k \in \mathbb{N}$,

$$E\{\mathbf{y}_k\} = [\ldots Pr\{\mathbf{y}_k = \mathbf{e}_i\} \ldots]^T \triangleq \mathbf{p}_k \tag{14.60}$$

where $\mathbf{p}_k \in \mathbb{R}^\nu$ can be recursively computed from the Chapman-Kolmogorov equation of the Markov chain $\mathbf{y}_k$:

$$\mathbf{p}_{k+1} = M\mathbf{p}_k \tag{14.61}$$

starting at $\mathbf{p}_0$. The conditionally-orthogonal projection of $\mathbf{y}_k$ onto $C\{\mathcal{X}^k\}$, denoted by $\widehat{\mathbf{y}}_{k/k}$, is to be determined. It will be assumed in the following that $\widehat{\mathbf{y}}_{k/k}$ is an approximation of the conditional mean of $\mathbf{y}_k$ given $\mathcal{X}^k$. That approximation is of order $\Delta t^{3/2}$, where Δt denotes the discretization time increment which is associated with the difference equation (14.56). Notice that the estimate $\widehat{\mathbf{y}}_{k/k}$ approximates a vector of a posteriori probabilities; that is,

$$\begin{aligned}\widehat{\mathbf{y}}_{k/k}^T &= E\{\mathbf{y}_k^T \,|\mathcal{X}^k\} + \mathcal{O}(\,\Delta t^{3/2}) \\ &= \left[\ldots\, Pr\{\mathbf{y}_k = \mathbf{e}_i \,|\mathcal{X}^k\}\, \ldots\right] + \mathcal{O}(\,\Delta t^{3/2})\end{aligned} \tag{14.62}$$

The following lemma will be used in the sequel of this work. For a proof, see e.g. (Elliot *et al.* 1993., p. 19).

Lemma 14.3.1.1 *Let* $\mathbf{u}$, $\widehat{\mathbf{u}}$, $\widetilde{\mathbf{u}}$, *and* $\mathcal{X}$, *denote, respectively, a random vector defined as in Equation* (14.58), *its conditional expectation given* $\mathcal{X}$, *the associated estimation error, and a collection of conditioning random variables. Then the following identity can be shown:*

$$\operatorname{cov}\{\widetilde{\mathbf{u}}\,|\mathcal{X}\} = \operatorname{diag}\{\widehat{\mathbf{u}}\} - \widehat{\mathbf{u}}\widehat{\mathbf{u}}^T \tag{14.63}$$

14.3.2 State-space model for $\mathbf{y}_k$

14.3.2.1 Reformulation of the state-space equation (14.56)

The governing equations for the dynamics of the discrete-time jump-linear system $\{\mathbf{x}_k, \mathbf{y}_k\}$ described in the previous section can be written as follows:

$$\mathbf{x}_{k+1} = C(\mathbf{x}_k, \mathbf{u}_k)\, \mathbf{y}_k + \mathbf{w}_k \tag{14.64}$$

$$\mathbf{y}_{k+1} = M\mathbf{y}_k + \mathbf{v}_k \tag{14.65}$$

where $C(\mathbf{x}_k, \mathbf{u}_k)$ is an $n \times \nu$ matrix defined as

$$C(\mathbf{x}_k, \mathbf{u}_k) \triangleq \mathcal{A}\left(I_\nu \otimes \begin{bmatrix}\mathbf{x}_k \\ \mathbf{u}_k\end{bmatrix}\right) \tag{14.66}$$

In Equation (14.66), $\otimes$ denotes the Kronecker product, I_ν denotes the $\nu \times \nu$ identity matrix, and $\mathcal{A}$ is the $n \times (n+m)\nu$ matrix expressed as

$$\mathcal{A} \triangleq \left[\ldots \left[A_i\ B_i\right] \ldots\right] \qquad i = 1, 2, \ldots, \nu \tag{14.67}$$

where the matrices $A_i \in \mathbb{R}^n$ and $B_i \in \mathbb{R}^m$ denote, respectively, the matrices $A(\gamma_i)$ and $B(\gamma_i)$, for $i = 1, 2, \ldots, \nu$. The novelty in this state-space model is in Equation (14.64), which is equivalent to Equation (14.56), but is re-written as a linear equation with respect to (w.r.t) the discrete state vector $\mathbf{y}_k$. Moreover, since we assume full knowledge of $\mathbf{x}_k$, Equation (14.64) can be considered a linear observation equation for $\mathbf{y}_k$, where $\mathbf{x}_{k+1}$ is the observation, $C(\mathbf{x}_k, \mathbf{u}_k)$ is the observation matrix, and $\mathbf{w}_k$ is the observation noise. Together with the linear process equation (14.68), this observation equation yields a linear state-space model for $\mathbf{y}_k$.

14.3.2.2 Development of equation (14.64)

The matrices $A(\mathbf{y}_k)$ and $B(\mathbf{y}_k)$ can be rewritten as

$$A(\mathbf{y}_k) = \sum_{i=1}^{\nu} A_i \; \delta(\mathbf{y}_k, \mathbf{e}_i) \tag{14.68}$$

$$B(\mathbf{y}_k) = \sum_{i=1}^{\nu} B_i \; \delta(\mathbf{y}_k, \mathbf{e}_i) \tag{14.69}$$

Using Equations (14.68) and (14.69) in the right-hand-side (RHS) of Equation (14.56), without the noise $\mathbf{w}_k$, yields

$$\begin{aligned}
A(\mathbf{y}_k)\mathbf{x}_k + B(\mathbf{y}_k)\mathbf{u}_k &= \\
&= \left(\sum_{i=1}^{\nu} A_i \delta(\mathbf{y}_k, \mathbf{e}_i)\right)\mathbf{x}_k + \left(\sum_{i=1}^{\nu} B_i \delta(\mathbf{y}_k, \mathbf{e}_i)\right)\mathbf{u}_k \\
&= \sum_{i=1}^{\nu} (A_i \mathbf{x}_k + B_i \mathbf{u}_k)\, \delta(\mathbf{y}_k, \mathbf{e}_i) \\
&= [\ldots [A_i \mathbf{x}_k + B_i \mathbf{u}_k] \ldots] \begin{bmatrix} \vdots \\ \delta(\mathbf{y}_k, \mathbf{e}_i) \\ \vdots \end{bmatrix} \\
&= \left[\cdots [A_i \; B_i] \begin{bmatrix} \mathbf{x}_k \\ \mathbf{u}_k \end{bmatrix} \cdots \right] \begin{bmatrix} \vdots \\ \delta(\mathbf{y}_k, \mathbf{e}_i) \\ \vdots \end{bmatrix} \\
&= [\ldots [A_i \; B_i] \ldots] \begin{bmatrix} \begin{bmatrix} \mathbf{x}_k \\ \mathbf{u}_k \end{bmatrix} & 0 & \cdots \\ 0 & \begin{bmatrix} \mathbf{x}_k \\ \mathbf{u}_k \end{bmatrix} & \cdots \\ \vdots & \vdots & \ddots \end{bmatrix} \begin{bmatrix} \vdots \\ \delta(\mathbf{y}_k, \mathbf{e}_i) \\ \vdots \end{bmatrix} \\
&= \mathcal{A}\left(I_\nu \otimes \begin{bmatrix} \mathbf{x}_k \\ \mathbf{u}_k \end{bmatrix}\right) \mathbf{y}_k
\end{aligned} \tag{14.70}$$

where $\otimes$ denotes the Kronecker product, I_ν is the identity matrix in $\mathbb{R}^\nu$, and the $n \times (n+m)\nu$ matrix $\mathcal{A}$ is defined from Equation (14.70). Finally, defining the $n \times \nu$ matrix $C(\mathbf{x}_k, \mathbf{u}_k)$ as follows:

$$C(\mathbf{x}_k, \mathbf{u}_k) \triangleq \mathcal{A}\left(I_\nu \otimes \begin{bmatrix} \mathbf{x}_k \\ \mathbf{u}_k \end{bmatrix}\right) \tag{14.71}$$

and using Equations (14.70) and (14.71) in Equation (14.56) yields Equation (14.64).

14.3.3 Development of the conditionally-linear filter

14.3.3.1 Algorithm summary

The conditionally-linear mode-estimator is developed by applying the general algorithm presented in the previous section. This algorithm recursively computes $\widehat{\mathbf{y}}_{k/k}$ and $P_{k/k}$, such that

$$\widehat{\mathbf{y}}_{k/k} = E\{\mathbf{y}_k | \mathcal{X}^k\} + \mathcal{O}(\Delta t^{3/2}) \tag{14.72}$$

$$\begin{aligned} P_{k/k} &= \text{cov}\{\widetilde{\mathbf{y}}_{k/k} | \mathcal{X}^k\} \\ &= \text{cov}\{\mathbf{y}_k | \mathcal{X}^k\} + \mathcal{O}(\Delta t^{3/2}) \end{aligned} \tag{14.73}$$

where $\widetilde{\mathbf{y}}$ denotes the estimation error $\mathbf{y} - \widehat{\mathbf{y}}$. The initialization equations are as follows:

$$\widehat{\mathbf{y}}_{0/0} = \mathbf{p}_0 \tag{14.74a}$$

$$P_{0/0} = \text{diag}\{\mathbf{p}_0\} - \mathbf{p}_0 \, \mathbf{p}_0^T \tag{14.74b}$$

14.3.3.2 Smoothing stage

$$\widetilde{\mathbf{x}}_{k+1/k} = \mathbf{x}_{k+1} - C_k \, \widehat{\mathbf{y}}_{k/k} \tag{14.75a}$$

$$S_{k+1} = C_k \, P_{k/k} \, C_k^T + W_k \tag{14.75b}$$

$$K_{k+1} = P_{k/k} \, C_k^T \, S_{k+1}^{-1} \tag{14.75c}$$

$$\delta\widehat{\mathbf{y}}_k = K_{k+1} \, \widetilde{\mathbf{x}}_{k+1/k} \tag{14.75d}$$

$$\widehat{\mathbf{y}}_{k/k+1} = \widehat{\mathbf{y}}_{k/k} + \delta\widehat{\mathbf{y}}_k \tag{14.75e}$$

$$\Delta P_k = \text{diag}\{\delta\widehat{\mathbf{y}}_k\} - \delta\widehat{\mathbf{y}}_k \, \delta\widehat{\mathbf{y}}_k^T - \delta\widehat{\mathbf{y}}_k \, \widehat{\mathbf{y}}_{k/k}^T - \widehat{\mathbf{y}}_{k/k} \, \delta\widehat{\mathbf{y}}_k^T \tag{14.75f}$$

$$\begin{aligned} P_{k/k+1} &= (I_\nu - K_{k+1} C_k)\,(P_{k/k} + \Delta P_k)\,(I_\nu - K_{k+1} C_k)^T \\ &\quad + K_{k+1} \, W_k \, K_{k+1}^T \end{aligned} \tag{14.75g}$$

14.3.3.3 Time-propagation stage

$$\widehat{\mathbf{y}}_{k+1/k+1} = M \, \widehat{\mathbf{y}}_{k/k+1} \tag{14.76a}$$

$$V_{k/k+1} = \text{diag}\{\widehat{\mathbf{y}}_{k+1/k+1}\} - M \, \text{diag}\{\widehat{\mathbf{y}}_{k/k+1}\} M^T \tag{14.76b}$$

$$P_{k+1/k+1} = M \, P_{k/k+1} \, M^T + V_{k/k+1} \tag{14.76c}$$

where $V_{k/k+1}$ denotes $\text{cov}\{\mathbf{v}_k | \mathcal{X}^{k+1}\}$.

14.3.3.4 Algorithm development

14.3.3.5 Smoothing stage

Using Equation (14.72), the vector $\widehat{\mathbf{f}}_{k/k}$ is defined as follows:

$$\begin{aligned} E\{C_k \mathbf{y}_k | \mathcal{X}^k\} &= C_k \widehat{\mathbf{y}}_{k/k} + \mathcal{O}(\Delta t^{3/2}) \\ &= \widehat{\mathbf{f}}_{k/k} + \mathcal{O}(\Delta t^{3/2}) \end{aligned} \tag{14.77}$$

where C_k can be taken out of the expectation since it is function of $\mathbf{x}_k$. At stage $k+1$, the residual $\widetilde{\mathbf{x}}_{k+1/k}$ is computed as

$$\widetilde{\mathbf{x}}_{k+1/k} = \mathbf{x}_{k+1} - \widehat{\mathbf{f}}_{k/k} \tag{14.78}$$

Using Equations (14.64) and (14.72) in Equation (14.78), one can show that the conditional mean of $\widetilde{\mathbf{x}}_{k+1/k}$ given $\mathcal{X}^k$ is in $\mathcal{O}(\Delta t^{3/2})$. An approximation of the conditional covariance matrix of $\widetilde{\mathbf{x}}_{k+1/k}$ given $\mathcal{X}^k$, of order $\Delta t^{3/2}$, can be computed using standard techniques:

$$\begin{aligned} \text{cov}\{\widetilde{\mathbf{x}}_{k+1/k} | \mathcal{X}^k\} &= E\{\widetilde{\mathbf{x}}_{k+1/k} \widetilde{\mathbf{x}}^T_{k+1/k} | \mathcal{X}^k\} + \mathcal{O}(\Delta t^{3/2}) \\ &= C_k P_{k/k} C_k^T + W_k + \mathcal{O}(\Delta t^{3/2}) \end{aligned} \tag{14.79}$$

The cross-terms that are involved in the development of Equation (14.79) cancel out since $\mathbf{w}_k$ is independent of $\widetilde{\mathbf{y}}_{k/k}$ and is zero-mean. The matrix S_{k+1} is defined as the first term on the RHS of Equation (14.79). Let $\widetilde{\mathbf{f}}_{k/k}$ and $P_{\mathbf{y}\widetilde{\mathbf{f}}}(k/k)$ denote, respectively, the estimation error in $\widehat{\mathbf{f}}_{k/k}$, and the conditional cross-covariance matrix of $\mathbf{y}_k$ and $\widetilde{\mathbf{f}}_{k/k}$ given $\mathcal{X}^k$. Their expressions are obtained as follows:

$$\widetilde{\mathbf{f}}_{k/k} = C_k \widetilde{\mathbf{y}}_{k/k} \tag{14.80}$$

$$P_{\mathbf{y}\widetilde{\mathbf{f}}}(k/k) = P_{k/k} C_k^T + \mathcal{O}(\Delta t^{3/2}) \tag{14.81}$$

Using Equations (14.79) and (14.81), we define the gain matrix, K_{k+1}, as

$$K_{k+1} \triangleq P_{\mathbf{y}\widetilde{\mathbf{f}}}(k/k)\, S_{k+1}^{-1} \tag{14.82}$$

which proves Equation (14.75c), and we compute a smoothed estimate of $\mathbf{y}_k$ given $\mathcal{X}^{k+1}$ through the following equation:

$$\widehat{\mathbf{y}}_{k/k+1} = \widehat{\mathbf{y}}_{k/k} + K_{k+1} \widetilde{\mathbf{x}}_{k+1/k} \tag{14.83}$$

which proves Equation (14.75e). The proof of Equation (14.75g) is as follows:

$$P_{k/k+1} = \text{cov}\{\widetilde{\mathbf{y}}_{k/k+1} | \mathcal{X}^{k+1}\} + \mathcal{O}(\Delta t^{3/2})$$

$$= (I_v - K_{k+1} C_k) \operatorname{cov}\{\widetilde{\mathbf{y}}_{k/k} | \mathcal{X}^{k+1}\} (I_v - K_{k+1} C_k)^T$$
$$+ K_{k+1} W_k K_{k+1}^T + \mathcal{O}(\Delta t^{3/2}) \tag{14.84}$$

Furthermore, the conditional covariance matrix on the right-hand-side (RHS) of Equation (14.84) can be computed using the following identity:

$$\operatorname{cov}\{\widetilde{\mathbf{y}}_{k/k} | \cdot\} = E\{\widetilde{\mathbf{y}}_{k/k} \widetilde{\mathbf{y}}_{k/k}^T | \cdot\} - E\{\widetilde{\mathbf{y}}_{k/k} | \cdot\} E\{\widetilde{\mathbf{y}}_{k/k} | \cdot\}^T \tag{14.85}$$

The conditional mean $E\{\widetilde{\mathbf{y}}_{k/k} | \mathcal{X}^{k+1}\}$ is expressed as

$$E\{\widetilde{\mathbf{y}}_{k/k} | \mathcal{X}^{k+1}\} = \widehat{\mathbf{y}}_{k/k+1} - \widehat{\mathbf{y}}_{k/k} + \mathcal{O}(\Delta t^{3/2})$$
$$= \delta\widehat{\mathbf{y}}_k + \mathcal{O}(\Delta t^{3/2}) \tag{14.86}$$

The expression for the conditional second order moment on the RHS of Equation (14.85) is developed as

$$E\{\widetilde{\mathbf{y}}_{k/k} \widetilde{\mathbf{y}}_{k/k}^T | \mathcal{X}^{k+1}\} = E\{\mathbf{y}_k \mathbf{y}_k^T | \mathcal{X}^{k+1}\} - \widehat{\mathbf{y}}_{k/k+1} \widehat{\mathbf{y}}_{k/k}^T - \widehat{\mathbf{y}}_{k/k} \widehat{\mathbf{y}}_{k/k+1}^T + \widehat{\mathbf{y}}_{k/k} \widehat{\mathbf{y}}_{k/k}^T \tag{14.87}$$
$$= \left(\operatorname{diag}\{\widehat{\mathbf{y}}_{k/k}\} - \widehat{\mathbf{y}}_{k/k} \widehat{\mathbf{y}}_{k/k}^T\right) + \operatorname{diag}\{\delta\widehat{\mathbf{y}}_k\} - \delta\widehat{\mathbf{y}}_k \widehat{\mathbf{y}}_{k/k}^T - \widehat{\mathbf{y}}_{k/k} \delta\widehat{\mathbf{y}}_k^T$$
$$= P_{k/k} + \operatorname{diag}\{\delta\widehat{\mathbf{y}}_k\} - \delta\widehat{\mathbf{y}}_k \widehat{\mathbf{y}}_{k/k}^T - \widehat{\mathbf{y}}_{k/k} \delta\widehat{\mathbf{y}}_k^T + \mathcal{O}(\Delta t^{3/2}) \tag{14.88}$$

where the last equality stems from Lemma 14.3.1.1 and Equation (14.73). Using Equations (14.85) to (14.88) in Equation (14.84) yields Equations (14.75f) and (14.75g).

14.3.3.6 Time-propagation stage

Some preliminary results are needed.

Lemma 14.3.3.1 *The random vector $\mathbf{v}_k$ is independent from $\mathcal{X}^k$ and orthogonal to $\{\mathbf{x}_{k+1}\}$. Thus,*

$$E\{\mathbf{v}_k | \mathcal{X}^{k+1}\} = 0 \qquad \forall k \in \mathbb{N} \tag{14.89}$$

Proof. It is known that $\mathbf{v}_k$ is orthogonal to $\mathbf{y}_k$, and is assumed independent from $\mathcal{W}^k \triangleq \{\mathbf{w}_i\}_{i=0}^k$. Furthermore, to ensure the Markov property of the chain $\{\mathbf{y}_k\}$, the vector $\mathbf{v}_k$ is necessarily independent from $\mathcal{X}^k$. As a result, using Equation (14.64), $\mathbf{v}_k$ is orthogonal to $\mathbf{x}_{k+1}$, and, therefore, to $\mathcal{X}^{k+1}$.

Lemma 14.3.3.2 *The random vectors $\mathbf{y}_k$ and $\mathbf{v}_k$ are conditionally orthogonal given $\mathcal{X}^{k+1}$; that is,*

$$E\{\mathbf{y}_k \mathbf{v}_k^T | \mathcal{X}^{k+1}\} = 0 \tag{14.90}$$

Proof.

$$
\begin{aligned}
E\{\mathbf{y}_k \mathbf{v}_k^T | \mathcal{X}^{k+1}\} &= E\{E\{\mathbf{y}_k \mathbf{v}_k^T \mid \mathbf{y}_k, \mathcal{X}^{k+1}\} \mid \mathcal{X}^{k+1}\} \\
&= E\{\mathbf{y}_k \, E\{\mathbf{v}_k^T \mid \mathbf{y}_k, \mathcal{X}^{k+1}\} \mid \mathcal{X}^{k+1}\} \\
&= 0
\end{aligned}
\tag{14.91}
$$

where we used Lemma 14.3.3.1 and the fact that $\mathbf{v}_k$ is orthogonal to $\mathbf{y}_k$ in the passage to the last line.

The propagation equation of the estimate $\widehat{\mathbf{y}}_{k/k}$ as given in Equation (14.76a) is obtained by applying the conditional expectation on both sides of Equation (14.59), by using Lemma 14.3.3.1, and by noting that here $D_k = 0$. The conditional covariance matrix $V_{k/k+1}$ of $\mathbf{v}_k$ given $\mathcal{X}^{k+1}$ is directly obtained by using Lemma 14.3.3.2 in its definition. An expression for $P_{k+1/k+1}$ can be developed as follows:

$$
\begin{aligned}
P_{k+1/k+1} &= E\{\widetilde{\mathbf{y}}_{k+1/k+1} \widetilde{\mathbf{y}}_{k+1/k+1}^T | \mathcal{X}^{k+1}\} + \mathcal{O}(\Delta t^{3/2}) \\
&= M P_{k/k+1} M^T + V_{k/k+1} + \mathcal{O}(\Delta t^{3/2})
\end{aligned}
\tag{14.92}
$$

where it can be shown, using Lemmas 2 and 3, that the cross-terms involving $\widetilde{\mathbf{y}}_{k/k+1}$ and $\mathbf{v}_k$ cancel out.

14.3.4 Discussion

The dynamics equation for the continuous-in-value state $\mathbf{x}_k$ was recast, through a special algebraic manipulation, as a conditionally-linear equation in the discrete state $\mathbf{y}_k$ given $\mathcal{X}^k$ (Equation (14.64)). Adding the linear dynamics equation of $\mathbf{y}_k$ (Equation (14.65)) to Equation (14.64) results in a conventional linear state-space model for $\mathbf{y}_k$. Were $\mathbf{y}_k$, $\mathbf{v}_k$, and $\mathbf{w}_k$ continuous jointly Gaussian processes, we would be in the standard linear conditionally Gaussian case, and the optimal filter would be the standard conditional Gaussian filter (Liptser and Shiryayev 1977b), Chap. 11). The vectors $\mathbf{y}_k$ and $\mathbf{v}_k$ are, however, discrete-valued non-Gaussian random vectors, and $\mathbf{w}_k$ is not necessarily Gaussian, so that the given model differs from the above-mentioned standard case. The linearity of the $\mathbf{y}_k$-model was exploited when applying the conditionally-linear filter. The update stage of the second-order estimation error statistics features a correction term with respect to the standard update stage of the linear filtering approach. In particular, the gain computations in the second filter is highly coupled with the estimate computations.

Simplified computations Notice, using Lemma 14.3.1.1, that there is an alternative and more efficient way of computing the P-matrices in the filter; namely,

$$
P = \operatorname{diag}\{\widehat{\mathbf{y}}\} - \widehat{\mathbf{y}}\widehat{\mathbf{y}}^T \tag{14.93}
$$

In that case, a simplified algorithm is given as follows:

1. Initialization stage:

$$\widehat{\mathbf{y}}_{0/0} = \mathbf{p}_0 \tag{14.94a}$$

$$P_{0/0} = \mathrm{diag}\{\mathbf{p}_0\} - \mathbf{p}_0\,\mathbf{p}_0^T \tag{14.94b}$$

2. Smoothing stage:

$$\widetilde{\mathbf{x}}_{k+1/k} = \mathbf{x}_{k+1} - C_k\,\widehat{\mathbf{y}}_{k/k} \tag{14.95a}$$

$$S_{k+1} = C_k\,P_{k/k}\,C_k^T + W_k \tag{14.95b}$$

$$K_{k+1} = P_{k/k}\,C_k^T\,S_{k+1}^{-1} \tag{14.95c}$$

$$\widehat{\mathbf{y}}_{k/k+1} = \widehat{\mathbf{y}}_{k/k} + K_{k+1}\,\widetilde{\mathbf{x}}_{k+1/k} \tag{14.95d}$$

3. Time-propagation stage:

$$\widehat{\mathbf{y}}_{k+1/k+1} = M\,\widehat{\mathbf{y}}_{k/k+1} \tag{14.96a}$$

$$P_{k+1/k+1} = \mathrm{diag}\{\widehat{\mathbf{y}}_{k+1/k+1}\} - \widehat{\mathbf{y}}_{k+1/k+1}\,\widehat{\mathbf{y}}_{k+1/k+1}^T \tag{14.96b}$$

It is interesting to emphasize that the proposed filter is distribution-free, in the sense that the distribution of $\mathbf{w}_k$ is not explicitly required.

14.3.5 Reduced order filter

The ν-vector $\mathbf{y}_k$ is by definition a unity vector in the sense of the l_1-norm; that is:

$$\|\mathbf{y}_k\|_1 \triangleq \sum_{i=1}^{\nu} y_i(k) = 1 \tag{14.97}$$

By definition of the matrix M, we also know that the columns of M are probability vectors, which components add to one. As a result, the components of the noise vector $\mathbf{v}_k$ in Equation (14.59) add to zero; that is, they satisfy to a linear constraint and are, therefore, perfectly correlated. This leads to a loss of one in the rank of the covariance matrix of $\mathbf{v}_k$. The same issue arises for the estimation errors in the proposed filters. Since the sum of the components of these errors is close to zero, the associated covariance matrix, $P_{k/k}$, is close to be singular. To avoid this, we propose a linear model reduction of the linear state-space equations (14.56) and (14.59) via the linear constraint (14.97). In the following, we only present the development of the reduced model. The application to the conditional-linear filter is straightforward.

14.3.5.1 Reduced-order state-space model

Let $\mathbf{y}_k^r$ denote the $(\nu - 1)$-vector obtained by truncating the last component of $\mathbf{y}_k$; that is,

$$\mathbf{y}_k^r \triangleq \left[Pr\{\mathbf{y}_k = \mathbf{e}_i\}\right]_{i=1}^{\nu-1} \qquad \in \mathbb{R}^{\nu-1} \tag{14.98}$$

The process equation for $\mathbf{y}_k^r$ is given as the following $(\nu - 1)$-vector equation:

$$\mathbf{y}_{k+1}^r = M^r \mathbf{y}_k^r + \mathbf{m}_\nu^r + \mathbf{v}_k^r \tag{14.99}$$

where

$$\mathbf{m}_\nu^r \triangleq \left[m_{i,\nu}\right]_{i=1}^{\nu-1} \qquad \in \mathbb{R}^{\nu-1} \tag{14.100}$$

$$\mathbf{v}_k^r \triangleq \left[v_i(k)\right]_{i=1}^{\nu-1} \qquad \in \mathbb{R}^{\nu-1} \tag{14.101}$$

$$M^r \triangleq M_{\nu-1} - \mathbf{m}_\nu^r \mathbf{1}^T \qquad \in \mathbb{R}^{\nu-1\times\nu-1} \tag{14.102}$$

and where $m_{i,\nu}$ denotes the element i, ν of the matrix M, $v_i(k)$ denotes the ith component of the vector $\mathbf{v}_k$, the vector $\mathbf{1}$ has all its $\nu - 1$ components equal to one, and the $\nu - 1$ dimensional matrix $M_{\nu-1}$ is the principal submatrix of M obtained by extracting the $\nu - 1$ first rows and columns. In Equation (14.99), the vector $\mathbf{m}_\nu^r$ is a vector of known deterministic inputs. The stochastic properties of the reduced noise vector $\mathbf{v}_k^r$ can be directly deduced from those of the full vector $\mathbf{v}_k$. In particular, an expression for the covariance matrix of $\mathbf{v}_k^r$ is obtained by deleting the last row and column of the covariance matrix of $\mathbf{v}_k$.

The observation equation for the truncated vector $\mathbf{y}_k^r$ is given as the following $n \times 1$ vector equation:

$$\mathbf{z}_{k+1} = C_k^r \mathbf{y}_k^r + \mathbf{w}_k \tag{14.103}$$

where

$$\mathbf{z}_{k+1} \triangleq \mathbf{x}_{k+1} - \mathbf{c}_\nu(k) \qquad \in \mathbb{R}^n \tag{14.104}$$

$$C_k^r \triangleq \left[\mathbf{c}_i(k) - \mathbf{c}_\nu(k)\right]_{i=1}^{\nu-1} \qquad \in \mathbb{R}^{n\times\nu-1} \tag{14.105}$$

and $\mathbf{c}_j(k)$, for $j = 1, 2, \ldots, \nu$, denote the following $n \times 1$ vectors

$$\mathbf{c}_j(k) \triangleq A_j \mathbf{x}_k + B_j \mathbf{u}_k \tag{14.106}$$

14.3.5.2 Development of the reduced-order model

Consider the $\nu \times 1$ process equation for $\mathbf{y}_k$, which in expanded form is written as

$$\begin{bmatrix} y_0(k+1) \\ \vdots \\ y_\nu(k+1) \end{bmatrix} = \begin{bmatrix} m_{1,1} & \cdots & m_{1,\nu} \\ \vdots & & \vdots \\ m_{\nu,1} & \cdots & m_{\nu,\nu} \end{bmatrix} \begin{bmatrix} y_0(k) \\ \vdots \\ y_\nu(k) \end{bmatrix} + \begin{bmatrix} v_1(k) \\ \vdots \\ v_\nu(k) \end{bmatrix} \tag{14.107}$$

Then, substituting Equation (14.97) in Equation (14.107) and rearranging yields

$$\begin{bmatrix} y_0(k+1) \\ \vdots \\ y_\nu(k+1) \end{bmatrix} = $$
$$= \begin{bmatrix} m_{1,1} - m_{1,\nu} & \dots & m_{1,\nu-1} - m_{1,\nu} & 0 \\ \vdots & & \vdots & \vdots \\ m_{\nu,1} - m_{\nu,\nu} & \dots & m_{\nu,\nu-1} - m_{\nu,\nu} & 0 \end{bmatrix} \begin{bmatrix} y_0(k) \\ \vdots \\ y_\nu(k) \end{bmatrix}$$
$$+ \begin{bmatrix} m_{1,\nu} \\ \vdots \\ m_{\nu,\nu} \end{bmatrix} + \begin{bmatrix} v_1(k) \\ \vdots \\ v_\nu(k) \end{bmatrix} \tag{14.108}$$

and deleting the last equation from Equation (14.108) yields the sought equation (14.99). The observation equation for $\mathbf{y}_k$ is developed as follows:

$$\mathbf{x}_{k+1} = \left[\mathbf{c}_1(k) \dots \mathbf{c}_{\nu-1}(k)\, \mathbf{c}_\nu(k)\right] \begin{bmatrix} y_0(k) \\ \vdots \\ y_{\nu-1}(k) \\ 1 - \sum_{i=1}^{\nu-1} y_i(k) \end{bmatrix} + \mathbf{w}_k$$
$$= \left[\dots \mathbf{c}_i(k) - \mathbf{c}_\nu(k) \dots\right] \begin{bmatrix} y_0(k) \\ \vdots \\ y_{\nu-1}(k) \end{bmatrix} + \mathbf{c}_\nu(k) + \mathbf{w}_k$$
$$= C_k^r \mathbf{y}_k^r + \mathbf{c}_\nu(k) + \mathbf{w}_k \tag{14.109}$$

which, using Equations (14.104) and (14.105) yields Equation (14.103).

14.3.6 Comparison with Wonham filter

14.3.6.1 Wonham filter

During the writing of this chapter, the authors became aware that, as referred in (Liptser), the optimal non-linear estimator of a finite-state homogenous Markov chain, $\{y_k\}$, can be developed in discrete-time. Central assumptions along the development of that algorithm, called Wonham filter (Liptser), are that the mode is observed via the following model:

$$x_k = y_k + w_k \tag{14.110}$$

and that the distribution of w_k is known via its probability density function (p.d.f.). It is possible to adapt that algorithm to the state-space model as given in Equations (14.64) and (14.65):

$$\mathbf{x}_{k+1} = C(\mathbf{x}_k, \mathbf{u}_k)\, \mathbf{y}_k + \mathbf{w}_k \tag{14.111}$$

$$\mathbf{y}_{k+1} = M\mathbf{y}_k + \mathbf{v}_k \tag{14.112}$$

In the following, the p.d.f. of $\mathbf{w}_k$ will be denoted by $f_{\mathbf{w}_k}(\boldsymbol{\alpha})$, where $\boldsymbol{\alpha}$ is a realization of the r.v. $\mathbf{w}_k$. Recalling that the realizations of $\mathbf{y}_k$ are the standard unit vectors in $\mathbb{R}^\nu$, denoted by $\mathbf{e}_1, \mathbf{e}_2, \ldots, \mathbf{e}_\nu$, the Wonham filter, which computes the conditional expectation of $\mathbf{y}_k$ given $\mathcal{X}^k$, is summarized as follows:

1. Initialization stage:

$$\overline{\mathbf{y}}_{0/0} = \mathbf{p}_0 \tag{14.113a}$$

$$\overline{P}_{0/0} = \mathrm{diag}\{\mathbf{p}_0\} - \mathbf{p}_0\,\mathbf{p}_0^T \tag{14.113b}$$

2. Smoothing stage:

$$\boldsymbol{\alpha}_i = \mathbf{x}_{k+1} - C_k\,\mathbf{e}_i \qquad i = 1, 2, \ldots, \nu \tag{14.114a}$$

$$\mathbf{f}(\mathbf{x}_{k+1}) = \begin{bmatrix} \vdots \\ f_{\mathbf{w}_k}(\boldsymbol{\alpha}_i) \\ \vdots \end{bmatrix} \in \mathbb{R}^\nu \tag{14.114b}$$

$$\overline{\mathbf{y}}_{k/k+1} = \frac{\mathrm{diag}\{\mathbf{f}(\mathbf{x}_{k+1})\}\,\overline{\mathbf{y}}_{k/k}}{\mathbf{f}^T(\mathbf{x}_{k+1})\,\overline{\mathbf{y}}_{k/k}} \tag{14.114c}$$

3. Time-propagation stage:

$$\overline{\mathbf{y}}_{k+1/k+1} = M\,\overline{\mathbf{y}}_{k/k+1} \tag{14.115}$$

The Wonham filter seems computationally simpler than the conditionally-linear (CL) filter and it explicitly preserves the unit l_1-norm among the components of $\overline{\mathbf{y}}_{k/k}$. It, however, requires the full knowledge of the distribution of $\mathbf{w}_k$. The CL-filter, on the other hand, is suboptimal, and only approximates the conditional mean when $\mathbf{w}_k$ is Gaussian. Nevertheless, it still provides a meaningful approximation in the non-Gaussian case, while requiring the first two moments of $\mathbf{w}_k$ only.

14.3.6.2 Numerical example

For simplicity, we consider a static mode $\mathbf{y}$ with 3 possible values,1, 2, 3, and a scalar measurement equation:

$$x_{k+1} = a(y_k)\,x_k + w_k, \qquad x_0 = 0 \tag{14.116}$$

where $a(1) = 0.2$, $a(2) = 0.5$, $a(3) = 0.9$, the noise w_k has a zero mean and a variance $W = \sigma^2\,\Delta t$, with $\sigma = 10^{2.5}$ and $\Delta t = 10^{-7}$. The true mode is assumed to be 2. Four different cases are considered for the distribution of w_k: Gaussian, Uniform, Exponential, and Rayleigh. The various p.d.f are parameterized such as to maintain the consistency on w_k among the four cases; namely, identical means and variances (see Table 14.1).

Table 14.1 Probability Density Functions of w_k such that $E\{w_k\} = 0$ and $E\{w_k^2\} = W$. The step function is denoted by $s(\cdot)$

Distribution	$f_w(\alpha)$	λ
Gaussian	$\frac{1}{2\pi\lambda} e^{-\frac{\alpha^2}{2\lambda^2}}$	$\sqrt{W}$
Uniform	$\frac{1}{2\lambda}\,[s(\alpha+\lambda) - s(\alpha-\lambda)]$	$\sqrt{3W}$
Exponential	$\lambda\, e^{-\lambda\alpha}\, s(\alpha + \frac{1}{\lambda})$	$\frac{1}{\sqrt{W}}$
Rayleigh	$\frac{\alpha}{\lambda^2} e^{-\frac{\alpha^2}{2\lambda^2}}\, s(\alpha + \lambda\sqrt{\frac{\pi}{2}})$	$\sqrt{\frac{W}{2-\pi/2}}$

Both filter estimates, denoted by $\bar{\mathbf{y}}$ and $\hat{\mathbf{y}}$, respectively for the Wonham filter and for the CL-filter, are initialized with a priori equiprobabilities:

$$\bar{\mathbf{y}}_{0/0}^T = \hat{\mathbf{y}}_{0/0}^T = [1/3,\ 1/3,\ 1/3] \tag{14.117}$$

Extensive Monte Carlo (MC) simulations (1000) were performed and the results are summarized in Figure 14.1. Figure 14.1a depicts the MC averages of the estimates from the Wonham filter (black line) and from the CL-filter with $\Delta t = 10^{-7}$ sec (blue line) in the Gaussian case. As expected, the CL-estimates are close to the optimal estimates. As seen from Figures 14.1b to 14.1d, the CL-filter still converges in the non-Gaussian cases, but it may present significant departures from the optimal filter.

14.3.7 Case of noisy observations of $\mathbf{x}_k$

14.3.7.1 Model equations

Considering the case of partial information in $\mathbf{x}_k$, it is assumed that a noisy measurement of $\mathbf{x}_k$, denoted by $\mathbf{z}_k$, is available. The model equations for the system $(\mathbf{x}_k, \mathbf{y}_k)$ are now as follows:

$$\mathbf{x}_{k+1} = C_k(\mathbf{x}_k)\,\mathbf{y}_k + \mathbf{v}_k \tag{14.118}$$

$$\mathbf{y}_{k+1} = M\mathbf{y}_k + \mathbf{w}_k \tag{14.119}$$

$$\mathbf{z}_{k+1} = \mathbf{x}_{k+1} + \mathbf{n}_{k+1} \tag{14.120}$$

where $\mathbf{n}_k$ is a zero-mean white sequence with covariance matrix N_k, and $\mathbf{n}_k$ is independent of $(\mathbf{x}_k, \mathbf{y}_k)$. The assumptions on the noises $\mathbf{v}_k$, $\mathbf{w}_k$ remain unchanged.
Next, the model, as described in Equations (14.118)–(14.120), will be recast to fit the formulation used earlier (see Equations14.64–14.65). The following notation will be used:

$$\overline{A}_k \stackrel{\Delta}{=} \sum_{i=1}^{i=\nu} A_i Pr\{\mathbf{y}_k = \mathbf{e}_i\} \tag{14.121}$$

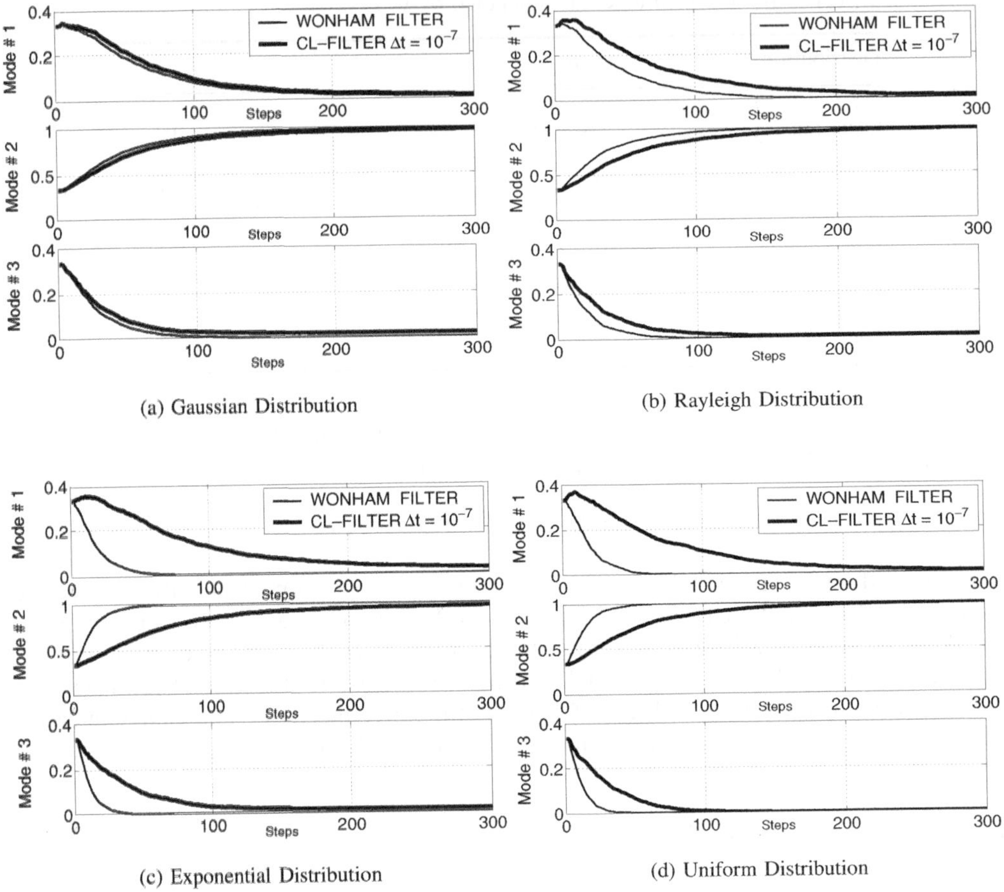

Figure 14.1 Comparison of the MC-averages of the mode estimates using the CL-filter and the Wonham filter (1000 runs).

Lemma 14.3.7.1 *Consider the random sequence* $\mathbf{u}_k$ *defined as follows:*

$$\mathbf{u}_k = \mathbf{n}_{k+1} - A(\mathbf{y}_k)\,\mathbf{n}_k \tag{14.122}$$

where $A(\mathbf{y}_k)$ *denotes the random variable with* ν *realizations,* A_i. *The sequence* $\mathbf{u}_k$ *satisfy the following properties:*

$$E\{\mathbf{u}_k\} = 0 \tag{14.123}$$

$$E\{\mathbf{u}_k\mathbf{u}_k^T\} = N_{k+1} + \overline{A}_k N_k \overline{A}_k^{\,T} \tag{14.124}$$

$$E\{\mathbf{u}_{k+1}\mathbf{u}_k^T\} = -\overline{A}_{k+1} N_{k+1} \tag{14.125}$$

$$E\{\mathbf{u}_k\mathbf{u}_{k+j}^T\} = 0 \qquad j = 2, 3, \ldots \tag{14.126}$$

Proof. Equations (14.123)–(14.126) are proved by direct computation where the independence of $\mathbf{n}_k$ and $\mathbf{y}_k$ has a central function. As it will be shown, the sequence $\mathbf{u}_k$ has the function of a colored noise in the reformulated model equations. Following a common approach, which seeks to reformulate the system equations using only white noise as inputs, the colored sequence $\mathbf{u}_k$ is modeled as a Markov sequence. Here we choose to approximate it as a first-order Markov sequence, which is consistent with the statistics given in Equations (14.123)–(14.125). Higher-order approximation would yield a better consistency, but, on the other hand, would increase the computational burden. In the following, for the sake of notational simplicity, the modeled first-order Markov sequence will be denoted by $\mathbf{u}_k$.

Lemma 14.3.7.2 *Consider the following first-order Markov sequence* $\mathbf{u}_k$*:*

$$\mathbf{u}_{k+1} = F_k \mathbf{u}_k + \boldsymbol{\lambda}_k \tag{14.127}$$

where

$$F_k = -\overline{A}_{k+1} N_{k+1} \left(N_{k+1} + \overline{A}_k N_k \overline{A}_k{}^T \right)^{-1} \tag{14.128}$$

and $\boldsymbol{\lambda}_k$ *is a zero-mean white sequence with covariance matrix* Λ_k*, which is expressed as follows:*

$$\Lambda_k = N_{k+2} + \overline{A}_{k+1} N_{k+1} \overline{A}_{k+1}{}^T - \overline{A}_{k+1} N_{k+1} \left(N_{k+1} + \overline{A}_k N_k \overline{A}_k{}^T \right)^{-1} N_{k+1} \overline{A}_{k+1}{}^T \tag{14.129}$$

Assume furthermore that $\mathbf{u}_0$ *is zero-mean. Then, the sequences* $\{\boldsymbol{\lambda}_k\}$ *and* $\{\mathbf{u}_k\}$ *are orthogonal, and* $\{\mathbf{u}_k\}$ *is consistent with the properties described in Equations* (14.123)–(14.125).

Proof. The proof is straightforward and can be done by induction.

The main result of this section is formulated in the following Proposition.

Proposition 14.3.7.1 *Let the augmented state vector* $\boldsymbol{\theta}_k$ *be defined as:*

$$\boldsymbol{\theta}_k = \begin{bmatrix} \mathbf{y}_k \\ \mathbf{u}_k \end{bmatrix} \tag{14.130}$$

then the model described in Equations (14.118)–(14.120) *is equivalent to the following model:*

$$\mathbf{z}_{k+1} = H_k(\mathbf{z}_k)\,\boldsymbol{\theta}_k + \mathbf{w}_k \tag{14.131}$$

$$\boldsymbol{\theta}_{k+1} = \Phi_k \boldsymbol{\theta}_k + \boldsymbol{\nu}_k \tag{14.132}$$

where the augmented variables $\boldsymbol{\nu}_k$, Φ_k *and* H_k *are defined as follows:*

$$\boldsymbol{\nu}_k = \begin{bmatrix} \mathbf{v}_k \\ \boldsymbol{\lambda}_k \end{bmatrix} \tag{14.133}$$

$$\Phi_k = \begin{bmatrix} M & 0 \\ 0 & F_k \end{bmatrix} \tag{14.134}$$

$$H_k(\mathbf{z}_k) = \begin{bmatrix} C_k(\mathbf{z}_k) & I_n \end{bmatrix} \tag{14.135}$$

In Equation (14.132), *the augmented noise process,* $\boldsymbol{\nu}_k$, *is an orthogonal sequence; it is orthogonal to* $\boldsymbol{\theta}_k$ *and is independent from the sequence* $\boldsymbol{\theta}^{k-1}$. *In Equation* (14.135), $C_k(\mathbf{z}_k)$ *is defined from Equation* (14.66), *and* I_n *denotes the identity matrix in* $\mathbb{R}^n$.

Proof. Using Equation (14.120) in Equation (14.118) yields

$$\begin{aligned} \mathbf{z}_{k+1} - \mathbf{n}_{k+1} &= C_k(\mathbf{z}_k - \mathbf{n}_k)\mathbf{y}_k + \mathbf{w}_k \\ &= C_k(\mathbf{z}_k)\mathbf{y}_k - C_k(\mathbf{n}_k)\mathbf{y}_k + \mathbf{w}_k \\ &= C_k(\mathbf{z}_k)\mathbf{y}_k - A(\mathbf{y}_k)\mathbf{n}_k + \mathbf{w}_k \end{aligned} \tag{14.136}$$

Then, taking $\mathbf{n}_{k+1}$ to the RHS of Equation (14.136) and using Equation (14.122) yields

$$\begin{aligned} \mathbf{z}_{k+1} &= C_k(\mathbf{z}_k)\mathbf{y}_k + \mathbf{n}_{k+1} - A(\mathbf{y}_k)\mathbf{n}_k + \mathbf{w}_k \\ &= C_k(\mathbf{z}_k)\mathbf{y}_k + \mathbf{u}_k + \mathbf{w}_k \\ &= \begin{bmatrix} C_k(\mathbf{z}_k) & I_n \end{bmatrix} \begin{bmatrix} \mathbf{y}_k \\ \mathbf{u}_k \end{bmatrix} + \mathbf{w}_k \\ &= H_k(\mathbf{z}_k)\boldsymbol{\theta}_k + \mathbf{w}_k \end{aligned} \tag{14.137}$$

which proves Equation (14.131). Equation (14.132) is simply obtained by appending Equations (14.65) and (14.127) in order to form the augmented process equation. The statistical properties of the augmented process stem from the statistical properties of the individual noises $\mathbf{v}_k$ and $\boldsymbol{\lambda}_k$.

As a concluding remark, one notices by looking at Equations (14.131), (14.132) that the model was recast in a form that fits Equations (14.64)–(14.65), where $\mathbf{z}_k$ and $\boldsymbol{\theta}_k$ are replacing $\mathbf{x}_k$ and $\mathbf{y}_k$, respectively.

14.3.7.2 Application of the CL-filter

Since the state vector $\boldsymbol{\theta}_k$ is composed of discrete and continuous variables, the recursion in the augmented P-matrix cannot be achieved as in the previous case. In order to alleviate this difficulty, we first notice that the continuous state variables, $\mathbf{u}_k$, are defined as a difference of noise signals: estimating the vector $\mathbf{u}_k$ is not critical in this application, where the main goal is to estimate the mode vector $\mathbf{y}_k$. Furthermore, we assumed that the correlation between $\mathbf{u}_k$ and $\mathbf{y}_k$ is relatively small; that is, the dominant information used in the estimation of $\mathbf{y}_k$ comes from the correlation between the measurement $\mathbf{z}_{k+1}$ and $\mathbf{y}_k$. Thus, in the algorithm development, the smoothing effect of $\mathbf{z}_{k+1}$ on the covariances of the estimation errors involving $\mathbf{u}_k$ will be neglected. This yields the following recursion for the augmented P-matrix. Given $P_{k/k}$, where

$$P_{k/k} = \begin{bmatrix} P_{yy}(k/k) & P_{yu}(k/k) \\ P_{uy}k/k & P_{uu}(k/k) \end{bmatrix} \tag{14.138}$$

then a correction term $\Delta P_{yy}(k/k)$ is computed, similarly to Equation (14.75f); that is,

$$\Delta P_{yy}(k/k) = \operatorname{diag}\{\delta\widehat{\mathbf{y}}_k\} - \delta\widehat{\mathbf{y}}_k\, \delta\widehat{\mathbf{y}}_k^T - \delta\widehat{\mathbf{y}}_k\, \widehat{\mathbf{y}}_{k/k}^T - \widehat{\mathbf{y}}_{k/k}\, \delta\widehat{\mathbf{y}}_k^T \tag{14.139}$$

where

$$\delta\widehat{\mathbf{y}}_k = \widehat{\mathbf{y}}_{k/k+1} - \widehat{\mathbf{y}}_{k/k} \tag{14.140}$$

An intermediary P-matrix is computed as follows:

$$\overline{P}_{k/k+1} = \begin{bmatrix} P_{yy}(k/k) + \Delta P_{yy}(k/k) & P_{yu}(k/k) \\ P_{uy}(k/k) & P_{uu}(k/k) \end{bmatrix} \tag{14.141}$$

and Equation (14.141) is used in the measurement update stage for computing $P_{k/k+1}$. For the sake of completeness, the CL-algorithm for the case of noisy measurement of $\mathbf{x}_k$ is given in the following.

14.3.7.3 Algorithm summary

14.3.7.4 Smoothing stage

$$\widetilde{\mathbf{z}}_{k+1/k} = \mathbf{z}_{k+1} - H_k\, \widehat{\boldsymbol{\theta}}_{k/k} \tag{14.142a}$$

$$S_{k+1} = H_k P_{k/k} H_k^T + W_k \tag{14.142b}$$

$$K_{k+1} = P_{k/k} H_k^T S_{k+1}^{-1} \tag{14.142c}$$

$$\widehat{\boldsymbol{\theta}}_{k/k+1} = \widehat{\boldsymbol{\theta}}_{k/k} + K_{k+1}\, \widetilde{\mathbf{z}}_{k+1/k} \tag{14.142d}$$

$$\delta\widehat{\mathbf{y}}_k = [I_v, 0]\left(\widehat{\boldsymbol{\theta}}_{k/k+1} - \widehat{\boldsymbol{\theta}}_{k/k}\right) \tag{14.142e}$$

$$\Delta P_{yy}(k/k) = \operatorname{diag}\{\delta\widehat{\mathbf{y}}_k\} - \delta\widehat{\mathbf{y}}_k\, \delta\widehat{\mathbf{y}}_k^T - \delta\widehat{\mathbf{y}}_k\, \widehat{\mathbf{y}}_{k/k}^T - \widehat{\mathbf{y}}_{k/k}\, \delta\widehat{\mathbf{y}}_k^T \tag{14.142f}$$

$$\overline{P}_{k/k+1} = \begin{bmatrix} P_{yy}(k/k) + \Delta P_{yy}(k/k) & P_{yu}(k/k) \\ P_{uy}(k/k) & P_{uu}(k/k) \end{bmatrix} \tag{14.142g}$$

$$P_{k/k+1} = \left(I - K_{k+1} H_k\right) \overline{P}_{k/k+1} \left(I - K_{k+1} H_k\right)^T + K_{k+1} W_k K_{k+1}^T \tag{14.142h}$$

14.3.7.5 Time-propagation stage

$$\widehat{\mathbf{y}}_{k+1/k+1} = M\, \widehat{\mathbf{y}}_{k/k+1} \tag{14.143a}$$

$$\widehat{\mathbf{u}}_{k+1/k+1} = F_k\, \widehat{\mathbf{u}}_{k/k+1} \tag{14.143b}$$

$$V_{k/k+1} = \text{diag}\{\widehat{\mathbf{y}}_{k+1/k+1}\} - M\,\text{diag}\{\widehat{\mathbf{y}}_{k/k+1}\}M^T \tag{14.143c}$$

$$\Phi_k = \begin{bmatrix} M & 0 \\ 0 & F_k \end{bmatrix} \tag{14.143d}$$

$$P_{k+1/k+1} = \Phi_k P_{k/k+1}\,\Phi_k^T + \begin{bmatrix} V_{k/k+1} & 0 \\ 0 & \Lambda_k \end{bmatrix} \tag{14.143e}$$

14.4 NUMERICAL EXAMPLE

14.4.1 Gyro failure detection from accurate spacecraft attitude measurements Description

Description. In this example, we consider a rigid-body spacecraft that undergoes free rotation around its center of mass, and that is equipped with a high accuracy attitude sensing package delivering single-frame measurements of the rotation quaternion, $\mathbf{q}_k$. The rotation quaternion is, here, a four-dimensional parameter, with a vector part, denoted by $\mathbf{e}_k$ and a scalar part, denoted by q_k. It is widely used in the aeronautical and astronautical communities for representing the angular orientation of the instantaneous spacecraft-body Cartesian frame, $\mathcal{B}_k$, with respect to a given reference Cartesian coordinate frame, $\mathcal{R}$ (Wertz 1984). The quaternion measurement unit (QMU) typically achieves a very high equivalent angular accuracy of 1 arcsec (e.g. the Goodrich-HD 1003 star-tracker). This high-level accuracy allows us, for the sake of the example, to consider the true quaternion as a known vector. In addition to the QMU, the spacecraft is assumed to be equipped with three low-grade body-mounted Rate Integrating Gyroscopes (RIG). Each gyro, labelled RIG #1, #2, and #3, is measuring the small incremental angular rotation of $\mathcal{B}_k$ around each axis between two successive RIG readouts. For simplicity, it is assumed that the output of each RIG is corrupted with electronic noise, which is modelled as an additive zero-mean white Gaussian noise with a known standard deviation. Furthermore, each gyro is subject to failure at a random time. Failure of RIG #i, $i = 1, 2, 3$, is here modelled by adding a given bias μ_i, $i = 1, 2, 3$, to the output of the failed gyro. The assumed scenario is that one gyro, at most, can fail during the simulation time, and that, if this happened, the gyro remains in a failed status. The system can therefore be at each time t_k in either one of four possible modes, labelled Mode #i, $i = 0, 1, 2, 3$; that is, one healthy mode (Mode #0), and three failed modes (Mode #i, $i = 1, 2, 3$). The failure triggering mechanism is modelled via a four-states jump Markov parameter with known initial and transition probabilities. Thus, the purpose of the filter developed in this example is to identify a failed RIG from the accurate observations of the quaternion. Note that the general concept of compensating low-grade gyro, e.g. of MEMS type, by estimating their drifts from camera-quaternion outputs was used in the development of operational attitude sensing hardware (Inertial Stellar Compass of Draper's Laboratory [Draper])

Let $\boldsymbol{\theta}_k^o$ denote the true incremental rotation of the body frame during the time increment $\Delta t = t_{k+1} - t_k$. It is assumed that

$$\boldsymbol{\theta}_k^o = [1 \;\; 1 \;\; 1]^T \sin(2\pi t/150)\,\Delta t \quad [deg] \tag{14.144}$$

The rigid-body discrete-time kinematics of the spacecraft, described in terms of the quaternion, $\mathbf{q}_k$, are governed by the following difference equation (Wertz 1984, pp. 511, 512)

$$\mathbf{q}_{k+1} = \Phi_k(\boldsymbol{\theta}_k^o)\,\mathbf{q}_k \qquad k = 0, 1, \ldots \tag{14.145}$$

where $\Phi_k(\boldsymbol{\theta}_k^o)$ denotes a four-dimensional orthogonal transition matrix that is defined as follows

$$\Phi_k(\boldsymbol{\theta}_k^o) = \exp\left(\frac{1}{2}\,\Theta_k^o\right) \tag{14.146}$$

where

$$\Theta_k^o = \begin{bmatrix} -\,[\boldsymbol{\theta}_k^o \times] & \boldsymbol{\theta}_k^o \\ -\boldsymbol{\theta}_k^{oT} & 0 \end{bmatrix} \tag{14.147}$$

and the matrix $[\boldsymbol{\theta}_k^o \times]$ denotes the so-called cross-product matrix. Let Δt_{RIG} be the time between two successive RIG readouts, then the output of the triad of RIG, denoted by $\boldsymbol{\theta}_k$, is expressed as follows

$$\boldsymbol{\theta}_k = \boldsymbol{\theta}_k^o + \boldsymbol{\mu}(\mathbf{y}_k) + \mathbf{n}_k \tag{14.148}$$

where $\mathbf{n}_k$ is a zero-mean Gaussian white noise with standard deviation $\sigma_1\sqrt{\Delta t_{RIG}}I_3$, where $\sigma_1 = 100$ arcsec, $\Delta t_{RIG} = 0.1$ sec, and I_3 is the three-dimensional identity matrix. The 3×1 vector $\boldsymbol{\mu}(\mathbf{y}_k)$ is a discrete random variable with the four possible outcomes, $\boldsymbol{\mu}_i$, $i = 0, 1, 2, 3$,

$$\boldsymbol{\mu}_0 = \begin{bmatrix} 0 \\ 0 \\ 0 \end{bmatrix} [\tfrac{deg}{hour}] \quad \boldsymbol{\mu}_1 = \begin{bmatrix} 1 \\ 0 \\ 0 \end{bmatrix} [\tfrac{deg}{hour}] \quad \boldsymbol{\mu}_2 = \begin{bmatrix} 0 \\ 1 \\ 0 \end{bmatrix} [\tfrac{deg}{hour}] \quad \boldsymbol{\mu}_3 = \begin{bmatrix} 0 \\ 0 \\ 1 \end{bmatrix} [\tfrac{deg}{hour}] \tag{14.149}$$

and $\{\mathbf{y}_k, k = 0, 1, \ldots\}$ is a four-state stationary Markov chain which state space is the set of the four columns of the identity matrix in $\mathbb{R}^4$. The transition probability matrix of $\mathbf{y}_k$ over the time increment Δt is computed as follows (Wonham 1970), p. 142:

$$M = \exp(Q\,\Delta t) \tag{14.150}$$

where Q denotes the 4×4 rate probability matrix

$$Q = \begin{bmatrix} -0.015 & 0 & 0 & 0 \\ 0.005 & 0 & 0 & 0 \\ 0.005 & 0 & 0 & 0 \\ 0.005 & 0 & 0 & 0 \end{bmatrix} \tag{14.151}$$

The three columns of zeros in Q indicate that the failure modes (#1, #2, and #3) are absorbing modes; that is, the system cannot recover from any failure and is trapped in these modes. On the other hand, the first column indicates that, at any time, there is equiprobability for the system to switch to any of the failure modes. The jump-linear process equation for the system under consideration is obtained as follows. Using Equation (14.148) into Equation (14.145) yields

$$\begin{aligned}\mathbf{q}_{k+1} &= \Phi_k(\boldsymbol{\theta}_k^o)\mathbf{q}_k \\ &= \Phi_k[\boldsymbol{\theta}_k - \boldsymbol{\mu}(\mathbf{y}_k) - \mathbf{n}_k]\mathbf{q}_k\end{aligned} \tag{14.152}$$

It can be shown (Choukroun *et al.* 2006) that, for $\mathbf{n}_k$ small enough, an approximation to first order in $\mathbf{n}_k$, is expressed as follows

$$\mathbf{q}_{k+1} = \Phi_k[\boldsymbol{\theta}_k - \boldsymbol{\mu}(\mathbf{y}_k)] - \frac{1}{2}\Xi_k\mathbf{n}_k \tag{14.153}$$

where the 3×4 matrix Ξ_k is defined as

$$\Xi_k = \begin{bmatrix}[\mathbf{e}_k\times] + q_k I_3 \\ -\mathbf{e}_k^T\end{bmatrix} \tag{14.154}$$

and $\mathbf{e}_k$ and q_k are the vector and scalar part, respectively, of $\mathbf{q}_k$. Let $\mathbf{x}_k$, $A(\mathbf{y}_k)$, and $\mathbf{w}_k$, be defined as follows:

$$\mathbf{x}_k \triangleq \mathbf{q}_k \tag{14.155}$$

$$A(\mathbf{y}_k) \triangleq \Phi_k[\boldsymbol{\theta}_k - \boldsymbol{\mu}(\mathbf{y}_k)] \tag{14.156}$$

$$\mathbf{w}_k \triangleq -\frac{1}{2}\Xi_k\mathbf{n}_k \tag{14.157}$$

then, using Equations (14.155) to (14.157) into Equation (14.153) yields the following jump-linear process equation, which fits the discrete-time dynamical system described in Equation (14.56) (with $\mathbf{u}_k = \mathbf{0}$):

$$\mathbf{x}_{k+1} = A(\mathbf{y}_k)\mathbf{x}_k + \mathbf{w}_k \tag{14.158}$$

Notice that the independence assumption between $\mathbf{y}_k$ and $\mathbf{n}_k$ implies that $\mathbf{y}_k$ and $\mathbf{w}_k$ are also independent. Furthermore, the dependence of $\mathbf{w}_k$ on $\mathbf{q}_k$ does not impair the validity of the filter presented in this work.[1] The covariance matrix needed for the filter implementation is the conditional covariance matrix of $\mathbf{w}_k$ given $\mathbf{q}_k$; that is,

$$\begin{aligned}W_k &= \frac{1}{4}\sigma_1^2\Xi_k\Xi_k^T \\ &= \frac{1}{4}\sigma_1^2(I_4 - \mathbf{q}_k\mathbf{q}_k^T)\end{aligned} \tag{14.159}$$

[1] This is essentially due to the fact that the filter development involves expectations conditioned on the state history – here $\mathbf{q}_k$.

where the second equality can be verified by direct computation. The dyadic matrix in the parenthesis of the right-hand-side (RHS) of Equation (14.159) is a rank three four-dimensional matrix, and is, thus, singular. In order to avoid numerical instability of the filter due to matrix inversion for the gain computation, we simply add a regularization term to the RHS of Equation (14.159); that is,

$$W_k = \frac{1}{4}\sigma_1^2(I_4 - \mathbf{q}_k\mathbf{q}_k^T) + \alpha I_4 \tag{14.160}$$

where α is chosen as 5.10^{-12} $[\text{rad}]^2$. This value, which was chosen by simulations, is of the order of 10^4 less than the nominal value of W_k, and therefore does not impair the filter performance for all practical purposes. Since $\mathbf{q}_k$ is actually the output of a QMU we simulate this measurement by computing the true quaternion from Equations (14.144) and (14.145), and by adding to it a zero-mean white Gaussian noise with standard deviation $\sigma_2/\sqrt{\Delta t_{QMU}}I_4$, where $\sigma_2 = 1$ arcsec and $\Delta t_{QMU} = 1$ sec. This value for σ_2 is chosen such as to achieve a typical 1 arcsec angular accuracy for a 1 second observation time span. The dependence on $1/\sqrt{\Delta t_{QMU}}$ describes the decrease in the QMU performance when the observation time is decreased. The diagonal structure of the QMU error covariance matrix is assumed here for simplicity and could be replaced by a more realistic one. Each of the 1000 Monte Carlo simulation runs lasted 150 seconds. The time increment Δt in Equation (14.144) was chosen equal to Δt_{RIG}.

In order to investigate the performance of the filter as a function of the QMU sampling time, Δt_{QMU}, and of the RIG sampling time, Δt_{RIG}, the following fixed scenario was chosen: the system starts in the healthy mode, Mode #0, and switches to Mode #2, due to a failure of RIG #2, at the epoch time $t = 75$ sec. Extensive simulations showed that this epoch time was a likely failure time for the given rate probability matrix Q [see Equation (14.151)].

Results analysis and interpretation. We ran 1000 Monte Carlo simulations with four different values of Δt_{QMU}; that is, 0.1 sec, 0.5 sec, 1 sec, and 1.5 sec, starting with an equiprobability distribution of the initial filter mode probabilities. Mode #0 was active until $t = 75$ sec and Mode #2 became active after that time. The results are summarized in Figure 14.2 which depicts the Monte Carlo averages of the a posteriori estimates for the mode probabilities. There are similarities in the filter performance for the four cases. We observe a monotonic convergence of the a posteriori probabilities averages. Before the failure, Mode #0 (solid lines) clearly appears as the most probable, but its steady-state probability is smaller than one. After the failure, Mode #2 (dashed lines) is identified and its steady-state probability equals one. This may be interpreted as follows. The probability steady-states differ from 1 before the failure and equal 1 after the failure because Mode #0 is not an absorbing mode while Mode #2 is an absorbing mode. More precisely, the mechanism in the increase of the probability for Mode #0 is related to the measurement-update equations; in addition, the mechanism in the decrease of this probability is related to the time-propagation equations, where the given Q matrix ensures non-zero probabilities of transiting from Mode #0 to any other mode. The trade-off between these two mechanisms yields a steady-state value that is less than one. On the other hand, that trade-off does not exist after the failure happens because the transition probabilities from Mode #2 to any other mode are assumed to be zero. There are dissimilarities between the four mentioned

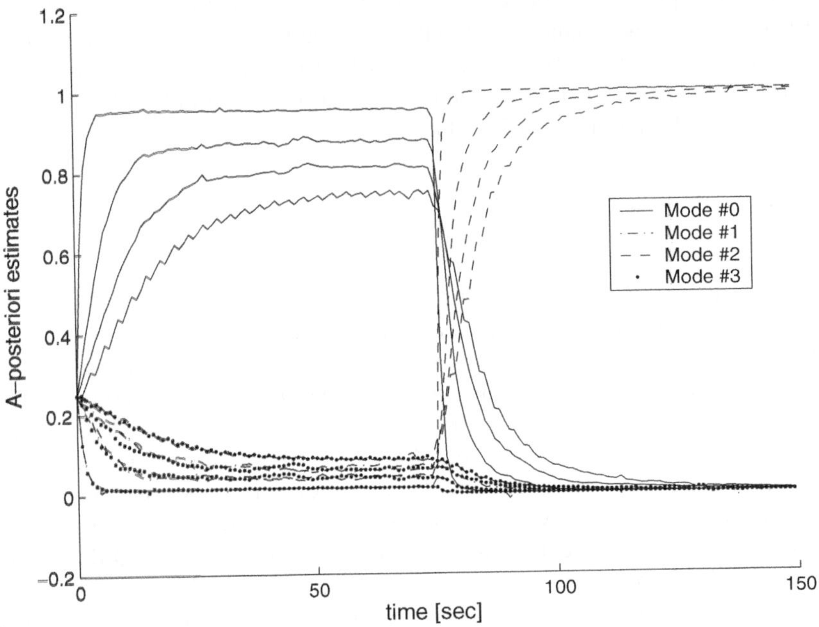

Figure 14.2 Monte Carlo averages over 1000 runs of the a-posteriori estimates of the mode probabilities for various values of the sampling time in the quaternion observation – $\Delta t_{QMU} = [0.1, 0.5, 1, 1.5] sec$.

cases: the smaller the sampling time Δt_{QMU} is, the higher are the convergence rates of the probabilities, and, before the failure, the higher is the steady-state of the probability for Mode #0. Thus, as expected, the transient and steady-state performance of the filter are improved when increasing the sampling rate of the quaternion measurement process. This would imply lower detection delays if the a-posteriori probabilities were to be processed by a mode detector.

The filter performance for various values of the gyro noise standard deviations were investigated for the same scenario, keeping the quaternion measurement sampling time constant. Figure 14.3 summarizes the results for $\sigma_{RIG} = 0.5$ arcsec, 1 arcsec, 2 arcsec, and 4 arcsec, with $\Delta t_{QMU} = 0.1$ sec. As expected, the transient filter performance decrease as σ_{RIG} increases, since the filter gain decreases as σ_{RIG} increases. The fact that the steady-state error in the healthy phase also increases can be explained similarly to the previous simulation.

14.5 CONCLUSION

This chapter introduced an extension of the standard linear filtering approach for estimation of non-linear non-Gaussian systems. The proposed discrete-time algorithm is formally shown to be asymptotically equivalent to the continuous-time optimal non-linear filter, as the discretization step becomes very small, in the case of Gaussian measurement noise.

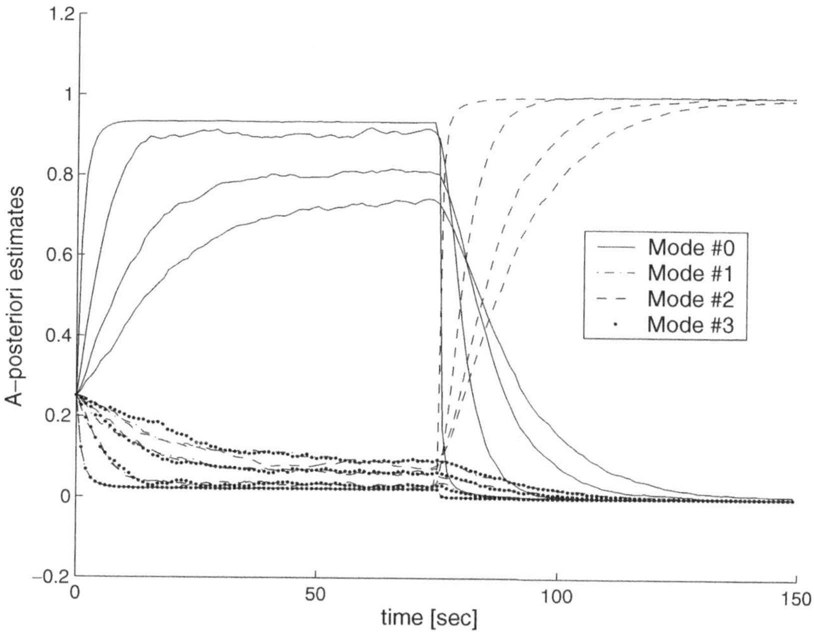

Figure 14.3 Monte Carlo averages over 1000 runs of the a-posteriori estimates of the mode probabilities for various values of the standard deviation in the gyro noise – $\sigma_{RIG} = [0.5, 1, 2, 4]$ arcsec.

For conditionally-Gaussian systems, this filter coincides with the standard conditionally-Gaussian filter. Moreover, its application to conditionally-linear non-Gaussian systems is straightforward. The conditionally-linear approach was applied in order to develop an estimator of the mode for a class of jump-linear systems. One highlight of the estimator is that it only requires the first two moments of the mode-measurement noise, while the optimal non-linear filter (Wonham filter) requires a complete knowledge of the probability distribution of that noise. Furthermore, the case of partial information on the continuous states, which cannot be handled by the Wonham filter, was also considered and an approximate conditionally-linear filter for mode-estimation was developed. As an example, the problem of gyro failure detection from accurate spacecraft attitude measurements was considered and the filter performance were illustrated via extensive Monte Carlo simulations.

14.6 APPENDIX A: INNER PRODUCT OF EQUATION (14.14)

The linearity properties stem from the linearity of the expectation operator:

$$
\begin{aligned}
< v_1 + v_2, u|v_3 > &= E\{(v_1 + v_2)\, u|v_3\} \\
&= E\{v_1 u|v_3\} + E\{v_2 u|v_3\} \\
&=< v_1, u|v_3 > + < v_2, u|v_3 >
\end{aligned}
\tag{14.161}
$$

For any scalar $\lambda \in \mathbb{R}$,

$$\begin{aligned} < \lambda u, v|w > &= E\{\lambda\, u\, v|w\} \\ &= \lambda E\{u\, v|w\} \\ &= \lambda < u\, v|w > \end{aligned} \tag{14.162}$$

Notice that λ in Equation (14.162) may be any function of a given realization of the r.v. w. The symmetry property is shown as follows:

$$\begin{aligned} < u, v|w > &= E\{u\, v|w\} \\ &= E\{u\, v|w\} \\ &=< v\, u|w > \end{aligned} \tag{14.163}$$

Finally, the positiveness property is straightforward since $E\{u^2|v\} \geq 0$. Furthermore,

$$u = 0 \Rightarrow E\{u^2|v\} = 0 \Rightarrow < u, u|v >= 0, \tag{14.164}$$

and conversely,

$$< u, u|v >= 0 \Rightarrow E\{u^2|v\} = 0 \Rightarrow E\{u^2\} = 0 \Rightarrow u = 0 \tag{14.165}$$

14.7 APPENDIX B: DEVELOPMENT OF THE FILTER EQUATIONS (14.36) TO (14.37)

Consider the observation equation (14.31),

$$\mathbf{x}_{k+1} = \mathbf{f}_k + \mathbf{w}_k \tag{14.166}$$

and assume that

$$\forall i \leq k \quad \widehat{\mathbf{x}}_{i/i-1} = \widehat{\mathbf{f}}_{i/i} \tag{14.167}$$

Then the residuals $\widetilde{\mathbf{x}}_{i/i-1}$ can be expressed as follows

$$\begin{aligned} \widetilde{\mathbf{x}}_{i/i-1} &= \mathbf{x}_i - \widehat{\mathbf{x}}_{i/i-1} \\ &= \widetilde{\mathbf{f}}_{i/i} + \mathbf{w}_i \end{aligned} \tag{14.168}$$

Next, we show that the projection of $\mathbf{w}_k$ onto $\mathcal{C}\{\widetilde{\mathcal{X}}^k\}$ is zero. Indeed,

$$\Pi\{\mathbf{w}_k, \mathcal{C}\{\widetilde{\mathcal{X}}^k\}\} = \sum_{i=1}^{k} \gamma_i (\widetilde{\mathcal{X}}^{i-1}\} \widetilde{\mathbf{x}}_{i/i-1} \tag{14.169}$$

where

$$\gamma_i (\widetilde{\mathcal{X}}^{i-1}\} = P_{\mathbf{w}_k \widetilde{\mathbf{x}}_{i/i-1}} P^{-1}_{\widetilde{\mathbf{x}}_i}$$

$$= \left[E\{\mathbf{w}_k \widetilde{\mathbf{f}}^T_{i/i} | \widetilde{\mathcal{X}}^{i-1}\} + E\{\mathbf{w}_k \mathbf{w}^T_{i-1} | \widetilde{\mathcal{X}}^{i-1}\} \right] P^{-1}_{\widetilde{\mathbf{x}}_i}$$
$$= 0 \tag{14.170}$$

The last line in Equation (14.170) stems from the assumptions of zero-mean and independence of the sequence $\mathbf{w}_k$. So, applying the projection operator $\Pi\{\cdot, \mathcal{C}\{X_{k+1}\}\}$ on both sides of Equation (14.166) yields:

$$\widehat{\mathbf{x}}_{k+1/k} = \widehat{\mathbf{f}}_{k/k} \tag{14.171}$$

which proves Equation (14.36a).
Equations (14.36b) to (14.36e) are developed as follows:

$$\begin{aligned} P_{\widetilde{\mathbf{y}}\widetilde{\mathbf{x}}} &= E\{\widetilde{\mathbf{y}}_{k/k} \widetilde{\mathbf{x}}^T_{k+1/k} | \widetilde{\mathcal{X}}^k\} \\ &= E\{\widetilde{\mathbf{y}}_{k/k} \widetilde{\mathbf{f}}^T_{k/k} | \widetilde{\mathcal{X}}^k\} \\ &= P_{\widetilde{\mathbf{y}}\widetilde{\mathbf{f}}_{k/k}} \end{aligned} \tag{14.172}$$

where the second line is again due to the zero-mean and independence properties of $\mathbf{w}_k$. Similarly, since $\widetilde{\mathbf{f}}_{k/k}$ and $\mathbf{w}_k$ are independent, and $\mathbf{w}_k$ is zero-mean:

$$\begin{aligned} P_{\widetilde{\mathbf{x}}_{k+1/k}} &= E\{\widetilde{\mathbf{x}}_{k+1/k} \widetilde{\mathbf{x}}^T_{k+1/k} | \widetilde{\mathcal{X}}^k\} \\ &= E\{\widetilde{\mathbf{f}}_{k/k} \widetilde{\mathbf{f}}^T_{k/k} | \widetilde{\mathcal{X}}^k\} + E\{\mathbf{w}_k \mathbf{w}^T_k | \widetilde{\mathcal{X}}^k\} \\ &= P_{\widetilde{\mathbf{f}}_{k/k}} + W_k \end{aligned} \tag{14.173}$$

which proves the smoothing stage equations. The time-propagation stage is proven as follows. Consider the process equation (14.32),

$$\mathbf{y}_{k+1} = \mathbf{g}_k + \mathbf{v}_k \tag{14.174}$$

Applying the projection operator $\Pi\{\cdot, \mathcal{C}\{\widetilde{\mathcal{X}}^{k+1}\}\}$ on both sides of Equation (14.174) yields

$$\Pi\{\mathbf{y}_{k+1}, \mathcal{C}\{\widetilde{\mathcal{X}}^{k+1}\}\} = \Pi\{\mathbf{g}_k, \mathcal{C}\{\widetilde{\mathcal{X}}^{k+1}\}\} + \Pi\{\mathbf{v}_k, \mathcal{C}\{\widetilde{\mathcal{X}}^{k+1}\}\} \tag{14.175}$$

The second term on the right-hand side of Equation (14.175) can readily be shown to be expressed as

$$\Pi\{\mathbf{v}_k, \mathcal{C}\{\widetilde{\mathcal{X}}^{k+1}\}\} = D_k P^{-1}_{\widetilde{\mathbf{x}}_{k+1/k}} \widetilde{\mathbf{x}}_{k+1/k} \tag{14.176}$$

First, we notice that

$$\begin{aligned} \forall i < k \quad E\{\mathbf{v}_k \widetilde{\mathbf{x}}^T_{i/i-1} | \widetilde{\mathcal{X}}^{i-1}\} &= E\{\mathbf{v}_k \widetilde{\mathbf{f}}^T_{i/i} | \widetilde{\mathcal{X}}^{i-1}\} + E\{\mathbf{v}_k \mathbf{w}^T_i | \widetilde{\mathcal{X}}^{i-1}\} \\ &= 0 \end{aligned} \tag{14.177}$$

where the last line stems from the fact that $\mathbf{v}_k$ is zero-mean and independent from $\mathbf{w}_i$ $\forall i < k$ and from $\widetilde{\mathbf{f}}_{i/i}$, $\forall i \leq k$. When $i = k$, we know by assumption that

$$E\{\mathbf{v}_k \mathbf{w}_i^T\} = D_k \tag{14.178}$$

Using Equations (14.177) and (14.178) proves Equation (14.176). Using Equation (14.176) in Equation (14.175) proves Equation (14.37).

ACKNOWLEDGEMENTS

This work was sponsored by AFOSR under Grant F49620-01-1-0361 and NASA-Ames Research Center under Grant NAG2-1484:4.

REFERENCES

Choukroun D, Bar-Itzhack I Y and Oshman Y 2006 Novel quaternion Kalman filter. *IEEE Transactions in Aerospace and Electronic Systems* **42**(1), 174–190.

Draper *www.draper.com/publications/digest03/paper303.pdf*

Elliot R J 1993 A general recursive discrete-time filter. *Journal of Applied Probability*, Vol. 30, pp. 575–588.

Elliot R J, Aggoun L and Moore J B 1995 *Hidden Markov Models, Estimation, and Control.* Springer-Verlag, New York.

Jazwinski A H 1970 *Stochastic Processes and Filtering Theory*. Academic, New York.

Kalman R E 1960 A new approach to linear filtering and prediction problems. *Trans. ASME, J. Basic Eng.*, Ser. D. **82**, 35–45.

Liberzon D and Morse A S 1999 Basic problems in design and stability of switched systems. *IEEE Control Systems Letters*, Oct., pp. 59–70.

Liptser R S *Lecture Notes in Stochastic Processes*, Lect. 8, Tel-Aviv University, Israel.

Liptser R S and Shiryayev A N 1977a *Statistics of Random Processes I: General Theory*, Chap. 8. Springer-Verlag, New York.

Liptser R S and Shiryayev A N 1977b *Statistics of Random Processes II: Applications*. Springer-Verlag, New York.

Luenberger D 1969 *Optimization by Vector Space Methods*. Wiley, New York.

Papoulis A 1965 *Probability, Random Variables and Stochastic Processes*. Mc-Graw Hill, New York.

Wertz J R (ed.) 1984 *Spacecraft Attitude Determination and Control*. D Reidel, Dordrecht, The Netherlands.

Wong E and Hajek B 1985 *Stochastic Processes in Engineering Systems*. Springer-Verlag, New York.

Wonham W M 1965 Some applications of stochastic differential equations to optimal non-linear filtering. *J. SIAM Control*, Ser. A, **2**, 347–368, 575–588.

Wonham W M 1970 Random differential equations in control theory. In A. Bharucha-Reid (ed.) *Probabilistic Methods in Applied Mathematics*, Vol. 2, pp. 192–194. Academic Press, New York.

15

Cohesion of languages in grammar networks

Y. Lee, T.C. Collier, C.E. Taylor and E.P. Stabler

15.1 INTRODUCTION

The long-term goal of our research is to produce a collection of heterogeneous agents that learn about their environment, communicate with one another about it, affect the environment in ways that take advantage of one another's special abilities and circumstances, and report to human observers. A principal challenge of this research is to define the cognitive and linguistic features that allow agents to develop a high-level language among themselves. For most purposes we would like all agents to agree on a common language so that each can communicate with all others. This is in some ways analogous to wishing convergent control of distributed adaptive agents.

There are several reasons why it is desirable for autonomous agents to acquire their own language (Collier and Taylor 2004; Lee *et al.* 2004). First, an acquired language can evolve over time to adopt its structure to various types of noise and different types of information to be transmitted, thus providing a truly flexible communication channel. Second, agents with heterogeneous sensors can map the same symbol to an environmental stimulus despite having very different internal representations. This feature also allows agents to express observations that were not anticipated by the designer of the system. Third, with a learned high-level language, agents can communicate using very short transmissions as opposed to sending the raw observation data. Finally, this acquired language will provide a logical manipulation facility for cognitive reasoning.

From a linguistic perspective, the primary problems in designing collaborative agents are: symbol grounding, language expression and learnability, and language evolution. The details of our approach has been explained elsewhere (Collier and Taylor 2004; Lee *et al.* 2004; Wee *et al.* 2001). In this chapter we will focus our attention on the third of these language problems – language evolution.

Cooperative Control of Distributed Multi-Agent Systems edited by Jeff S. Shamma

15.2 EVOLUTIONARY DYNAMICS OF LANGUAGES

A *language* is a set of linguistic expressions which are typically gestural complexes, and often symbol sequences, interpreted in a certain way. This set of expressions may be infinite, and so a finite language user must use some finite representation of the infinite set. This finite representation is typically called a *grammar*. The grammar typically comprises a finite set of atoms, the *lexicon*, together with a set of functions (rules) for building complexes from the elements (Keenan and Stabler 2003). The language is then the closure of the lexicon with respect to the rules. Since these notions are so closely related, we will indifferently refer to populations of languages, populations of grammars, or populations of speakers, except when the distinctions matter. And we sometimes refer to a language user as 'speaking with grammar G,' which means of course that the language user recognizes and produces the expressions defined by G. When the space of possible grammars is finite (though the language defined by each grammar may still be infinite), it is common to represent the grammar by a finite vector of *parameter* values. In models where the grammar rules are fixed, these parameters may be simply *lexical items*. Such representations of grammars and the languages they define induce new metrics of language similarity: two languages may be classified as similar when their finite vector representations are similar (even when the languages they define may differ on infinitely many expressions).

The problem of language evolution involves determining how the languages used by a collection of agents will change over time, and whether an initially homogeneous language will diverge into mutually unintelligible dialects, or will stay uniformly understandable. Here we address the dynamics of a population of language learners as a less abstract instantiation of the general case of agents learning from one another.

In the last decade, much progress has been made in the field of robotics to investigate the social and/or cognitive factors contributing to the emergence of coherent communication among autonomous agents (Arita and Taylor 1996; Collier and Taylor 2004; Lee *et al.* 2005b, 2004; Marocco and Nolfi 2006; Marocco *et al.* 2003; Steels 2001, 2006). Nowak and his colleagues have studied the dynamics of languages in various simple settings in a mathematical framework. We shall informally refer to these as *NKN models*. For example, in simple regimes, it has been possible to determine things like the maximum error rate a communication channel can withstand before the agents lose their ability to communicate in a coherent language (Komarova and Nowak 2001; Komarova *et al.* 2001; Nowak *et al.* 2001).

The NKN models are simplified in many respects. For example, in most of these models it is assumed that each individual in a population has full access to the other members. Such a condition is often not guaranteed in wireless networks. Also in most NKN models a population learns languages from a set where each grammar is equally related to every other grammar. In a setting where language is modular and the unit of transmission is a module rather than a whole language, the evolutionary dynamics of language convergence is different from what one would expect from NKN model (Lee 2006; Lee *et al.* 2005a, b), and this will be the focus of this chapter.

First, we discuss how the structure of a population – who talks to whom – can be conceptualized as a kind of network structure that affects the degree of linguistic coherence attained in the limit (Lee *et al.* 2005b). Second, we demonstrate how language structure influences convergence (Lee *et al.* 2005a), and how this can be conceptualized as inducing a network structure on the population, represented by a graph of the similarity relations among grammars. This work is inspired by consensus behavior studies in engineering using graph theory (Olfati-Saber 2005; Olfati-Saber and Murray 2004).

Let x_i denote the proportion of a population of constant size speaking grammar G_i with n possible grammars existing. We assume that each individual uses only one grammar, thus we have $\sum_{j=1}^{n} x_j = 1$.

The fitness of individuals with grammar G_i is $f_i = f_0 + \sum_j a_{ij}x_j$ where f_0 is the base fitness which does not depend on the language, and $\sum_j a_{ij}x_j$ is the language contribution to fitness. Note that this fitness equation is frequency dependent.

The evolutionary dynamics of this population is of the form

$$\dot{x}_i = \sum_j x_j f_j q_{ji} - \phi x_i, \qquad 1 \leq j \leq n \tag{15.1}$$

where $\phi = \sum_i x_i f_i$ is the average fitness, and $Q = [q_{ij}]$ is the learning fidelity matrix. The value ϕ is also used as a measure of linguistic coherence.

This dynamic system can be thought of either as having individuals that produce offspring to replace a randomly chosen individual, or as having individuals that change their grammars by learning a teacher's language.

The learning model is given by the matrix Q, which is a function of the mutual similarities A:

$$q_{ii} = q, \qquad q_{ij} = (1-q)\frac{a_{ij}}{\sum_{j \neq i} a_{ij}} \text{ for all } i \neq j. \tag{15.2}$$

The *learning fidelity* q is the probability that a learner acquires the same grammar as its teacher. Q satisfies the condition $\sum_j q_{ij} = 1$ for all i. The special case of this transition matrix where $a_{ij} = a$ for all $i \neq j$ was analyzed by Nowak, Komarova, and Niyogi (Komarova and Nowak 2001; Komarova *et al.* 2001).

15.3 TOPOLOGIES OF LANGUAGE POPULATIONS

When all individuals have an equal probability of learning from each other, we can regard the population as fully-connected, and the fitness of each language is a function of its frequency in the entire population. Breaking the symmetry that a fully-connected population provides makes finding an analytical solutions much more difficult. Therefore, we used simulations to explore the convergence properties where each individual exists within a topology defining a set of neighboring individuals.

Using a simplified system where the fitness of individual k, f_k, is the base fitness, f_0, plus a linguistic merit proportional to the probability that individual k could successfully communicate with its neighbors is:

$$f_k = f_0 + \frac{1}{2}\sum_{m}^{n_c}(a_{mk} + a_{km}) \tag{15.3}$$

At each time step, an individual is chosen to reproduce, where the probability of being chosen is proportional to its fitness. Reproduction can be thought of as the production

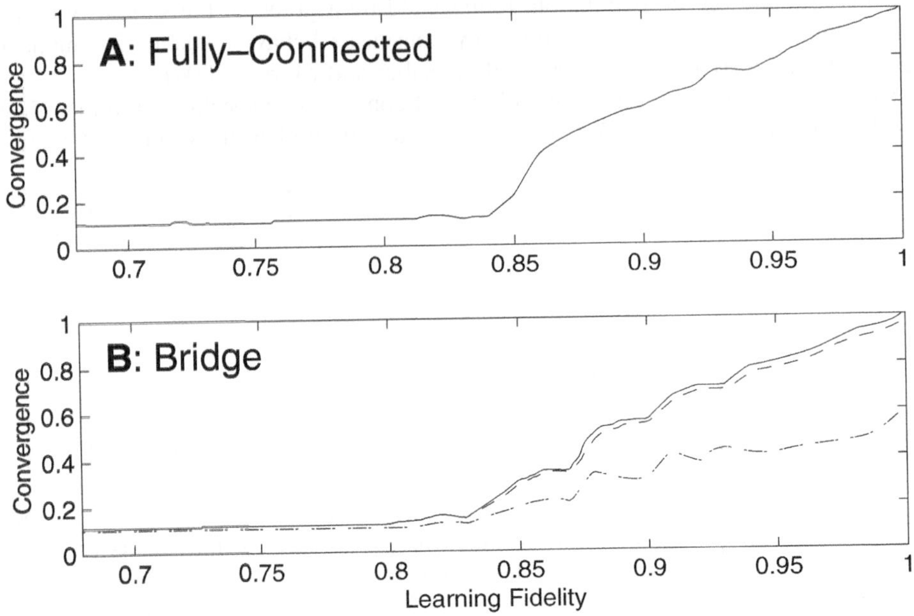

Figure 15.1 Various levels of linguistic coherence ϕ for a fully-connected topology (A) and a bridge topology (B). The solid line(–) for ϕ_0, the dash line(--) for ϕ_1, and the dot-dashed line(·-) for global coherence ϕ_∞.

of an offspring which inherits the parent's grammar and replaces one of the parent's neighbors. The offspring learns the parent's grammar with a certain *learning fidelity*, q. This learning fidelity is properly a function of the specifics of the learning method the child uses and the complexity of the grammar, but in the simplified system the learning fidelity is reducible to a transition probability function between grammar G_i and grammar G_j equal to q for $i = j$, and $(1 - q)/(n - 1)$ for $i \neq j$. For details and the algorithms of this simulation, see (Lee 2006; Lee *et al.* 2005b).

We plot the linguistic coherence, denoted as ϕ on Figure 15.1. The ϕ is measured using the following equation:

$$\phi = \frac{1}{N} \sum_{k=1}^{N} f_k \tag{15.4}$$

where N represents the population size. Various different 'levels' of coherence exist as defined by the set of neighboring individuals of an individual k. *Local coherence*, ϕ_0, only counts for the neighbors of each individual and is proportional to mean fitness (equal if $f_0 = 0$). ϕ_1 is the coherence measured over the set of neighbor's neighbors, and generally, ϕ_i is measured using the set of (neighbor's)^i neighbors. *Global coherence*, ϕ_∞, corresponds to convergence level over the entire population.

In the fully-connected topology, all of these convergence levels appear to reduce to the same value, as shown in Figure 15.1 A. The bridge topology with two subpopulations of equal size and a limited random connections between subpopulations (Figure 15.1 B) demonstrates a very similar learning fidelity threshold as the fully-connected run shown in Figure 15.1 A. However, global convergence is never achieved reliably in excess to the probability that both subpopulations individually converge to the same grammar by chance. Additionally, ϕ_1 is extremely close to ϕ_0 while ϕ_∞ only rises to approximately 0.5, indicating that there is a large degree of separation between many individuals.

The above example, however, clearly shows that topology of population networks is critical in determining the degree of linguistic coherence. Other topologies are discussed in (Lee 2006).

15.4 LANGUAGE STRUCTURE

In most NKN models, grammars are mutationally equidistant from each other with arbitrarily assigned similarities. It seems however, that changes in natural languages are more systematic and small. The language of a child is usually similar to that of the linguistic community. This suggests an approach where the similarity between languages is correlated with their distance from each other in mutational space.

We introduce a regularity in the language space by viewing the locus of language transmission as a series of learned parameters and calculating the similarity between languages as the proportion of parameters that agree.

Consider a fully-connected finite population of N individuals, each of whom possesses a language which is encoded as a sequence of l linguistic 'components' or 'parameters'. Each parameter can take only a limited number d of values. For example, a language L with 10 parameters each taking three values (A, B, C) can be represented by a linear sequence like AABABCBABA.[1] This will correspond to parameters in the Chomskian sense just in case these latter parameters are appropriately relevant to linguistic transmission (Chomsky 1965, 1980).

Representing a language as a vector, a finite sequence of parameter values, we define the language similarity between individual k and j, denoted a_{kj}, as the proportion of parameters on which the two individuals agree. For example, the language similarity between an individual k whose language is represented as AAA and an individual k whose language is represented as ABA is $2/3$ and $a_{kj} = a_{jk}$. Other than the language structure and how language similarity is calculated, reproduction cycle is the same as previous population network model. Actual choice of parameters and algorithm is described elsewhere (Lee 2006; Lee *et al.* 2005a)

Figure 15.2 shows the language convergence as a function of learning fidelity. Panel A is the result of a single parameter which has 64 different options. We set a uniform similarity of $a = 0.5$ for setting A to facilitate comparison with NKN models. Panel B plots the evolution of six parameter languages, each of which has two options available. Both language structures have similar mean similarity $\overline{a}$ values ($\overline{a_A} \simeq \overline{a_D}$). Counting the actual number of languages that exist in the population may disguise the degree of variation when some languages disproportionately dominate. Consequently, we used an analog to the effective number of alleles in a population, which we will refer to as the

[1] This string is not an example of a statement from the language, but rather represents the language itself.

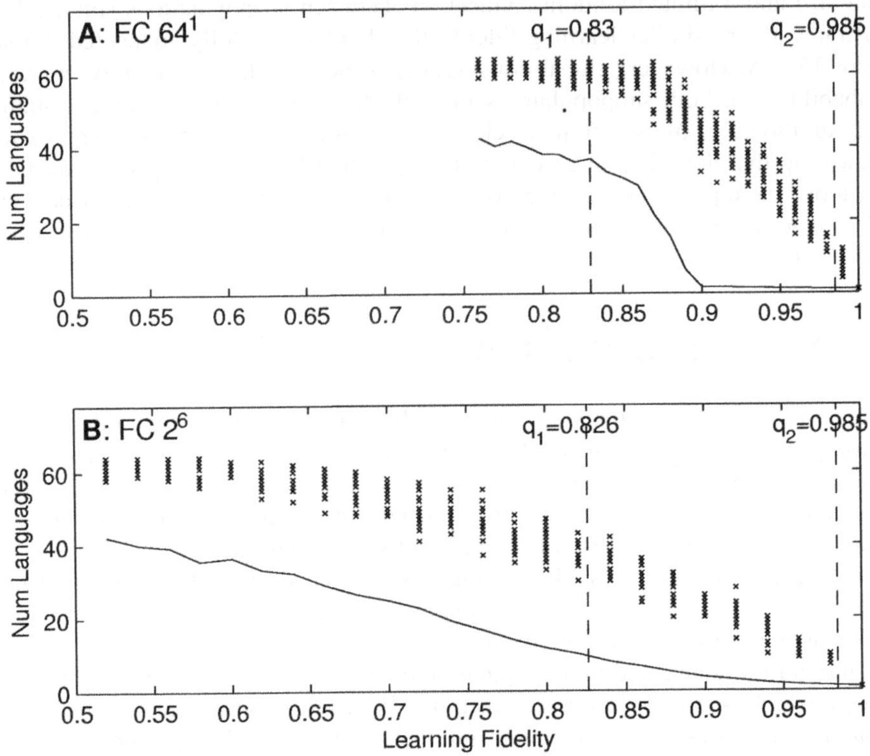

Figure 15.2 The number of languages (×) and the average effective number of languages n_e (—). Panel A is the result of a 1 parameter language with 64 possible options. Panel B is the outcome when language has 6 parameters and 2 options for each parameter. FC represents the fully-connected topology of a population network.

effective number of languages in the population, n_e (Crow and Kimura 1970):

$$n_e = \left(\sum_{i=1}^{N} p_i^2 \right)^{-1} \tag{15.5}$$

where p_i is the frequency of each language.

In the case of single-parameter languages, all the possible languages exist in the population in the region where $q < q_{1A}$. On the other hand, the 6-parameter case B has only half of the all possible languages at $q = q_{1B}$. This shows that substantial levels of linguistic coherence can be achieved with lower learning fidelity if language structure is introduced.

In set A, all possible languages are a single step away in sequence space; in other words, all possible languages are reachable by a single incorrect/incomplete learning event. In set B, however, only a small subset of possible languages are producible as single step variants from the dominant language. These single-step variants of the dominant languages account for the majority of non-dominant languages in the population. Additionally, these variants have a high fitness relative $\bar{a}$, and a higher equilibrium frequency in mutation-selection balance.

15.5 NETWORKS INDUCED BY STRUCTURAL SIMILARITY

Natural languages are organized hierarchically, with language families, subgroups, and even dialects (Cavalli-Sforza *et al.* 1988; Comrie 2001). Some pairs of languages share more features than other pairs. These observations lead researchers to conceptualize language similarity with grammar networks (Lee *et al.* 2005a; Matsen and Nowak 2004; Olfati-Saber 2005; Olfati-Saber and Murray 2004). Inspired by the consensus studies using graph theory in engineering communities (Olfati-Saber 2005; Olfati-Saber and Murray 2004), we explore the behavior of linguistic convergence in grammar networks.

Consider a network of grammars with nodes $U = \{G_1, G_2, \ldots, G_n\}$. A link (i, j) is established between grammars G_i and G_j if a learner learning from a teacher who speaks grammar G_i might end up speaking grammar G_j. We define the adjacency weights of the grammar network as $a_{ij} = (s_{ij} + s_{ji})/2$, where s_{ij} denote the similarity of G_i to G_j. The matrix of mutual similarities $A = [a_{ij}]$ represents the interconnection topology of the grammar network. In our setting, A is defined as:

$$a_{ij} = \begin{cases} 1 & \text{if } i = j \\ a & \text{if } (i, j) \in E \\ 0 & \text{if } (i, j) \notin E \end{cases}$$

Notice again the distinction between two types of similarities, structural similarity (defined over the representations of the grammars) and expressive similarity (defined over the sentences of the languages). Structural similarity is how many grammatical rules or lexical items or parameter settings two grammars share. Expressive similarity relates to the probability that a sentence generated from one grammar is similar to a sentence generated from another grammar. Structural similarity is analogous to genotype similarity, and expressive similarity is analogous to phenotype similarity. In this experiment, all grammars in a network are positioned on a polygon where their positions are indicative of structural similarity. If two grammars are positioned side by side, they share many common rules for generating sentences.

For our model, the grammar network is completely specified by the matrix of mutual similarities A. Note that a mutual similarity a_{ij} is equivalent to the expressive similarity of two language hypotheses. As long as two grammars are connected ($(i, j) \in E$), they have some degree of mutual intelligibility ($a_{ij} > 0$).

Within this setting, a single mutation or language learning step will not generally yield a grammar that is structurally similar to its teacher's grammar. Sometimes the learner can acquire a set of rules that are completely different from its parent's and yet generate sentences that are very close to its input. Thus, the grammar network defines the space which an individual explores while acquiring a grammar.

Four canonical types of graphs with substantially different structural properties are considered in this experiment: the complete graph, a regular ring lattice, a random graph, and a small-world graph, depicted in Figure 15.3. Less than complete graphs make sense

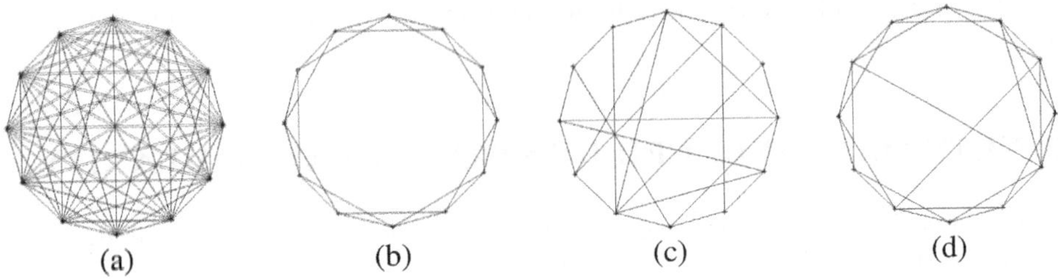

Figure 15.3 Different types of graphs with 12 nodes. (a) A complete graph, (b) a ring lattice with $k = 2$, (c) a random graph with $k = 1$ and $r = 12$, (d) a small-world graph with $k = 2$ and $r = 3$. The examples shown here have different densities. The experiments, however, are designed in such way that all the graphs except the complete graph have the same density.

from a biological or sociological standpoint since learning a language that is similar to what it has already seems much easier than learning a drastically different language. For each graph, the dynamics are different. For full details, see (Lee 2006).

15.5.1 Three equilibrium states

In Figure 15.4 we show the effective number of grammars over a range of q values for a complete graph and regular lattices with fixed n and a. Each point is the result from a single run, but the q interval ($= 10^{-4}$) is small enough that the points appear as a line in places. The grammar diversity n_e is the maximum when a population is in symmetric state, while n_e is the minimum ($\simeq 1$) when a single dominant grammar emerges.

Figure 15.4(a) shows two classes of equilibrium states as well as bi-stability region over a range of learning fidelities $q_a \leq q \leq q_s$ for the complete graph. The learning fidelity threshold for asymmetric solution, q_a, is the highest for the complete graph. It may be possible to verify these results analytically, in future work.

Figure 15.4(b) suggests a third class of solutions which occurs at q values between q_s and q_a for ring lattice networks. This class of stable attractors is characterized by a nearly linear decrease in n_e when $q_s \leq q \leq q_a$, as shown in Figure 15.4(b).

For the regular ring lattice with $k = 10$, the symmetric state does not exist for $q \geq 0$ as shown in Figure 15.4(c). The symmetric solution can still be obtained using negative q values, but the interpretation of such values is not obvious. Figure 15.4(c) also shows that some level of coherence can be achieved even with the learning fidelity of 0 when the graph is sparse ($k << n$).

In the regular ring lattice with $k = 15$, the frequencies of each grammar x_i at approximate equilibrium state for different learning fidelities are distributed as shown in Figure 15.5. We cut the ring at the point opposite to the dominant grammar and spread it along the x axis, so that the dominant grammar is always at the center. If the grammar is positioned close to the dominant, the grammar index is close to the index of the dominant, indicating that they are structurally similar.

Figure 15.5(a) shows the grammar frequency distribution when $q = 0$ as an example of the symmetric state. If the learning fidelity q is greater than q_a, only one grammar dominates as shown in Figure 15.5(c). We call this phase *strong cohesion*. When the learning fidelity is between q_s and q_a, the grammar frequencies form a smooth curve reminiscent

(a)

(b)

(c)

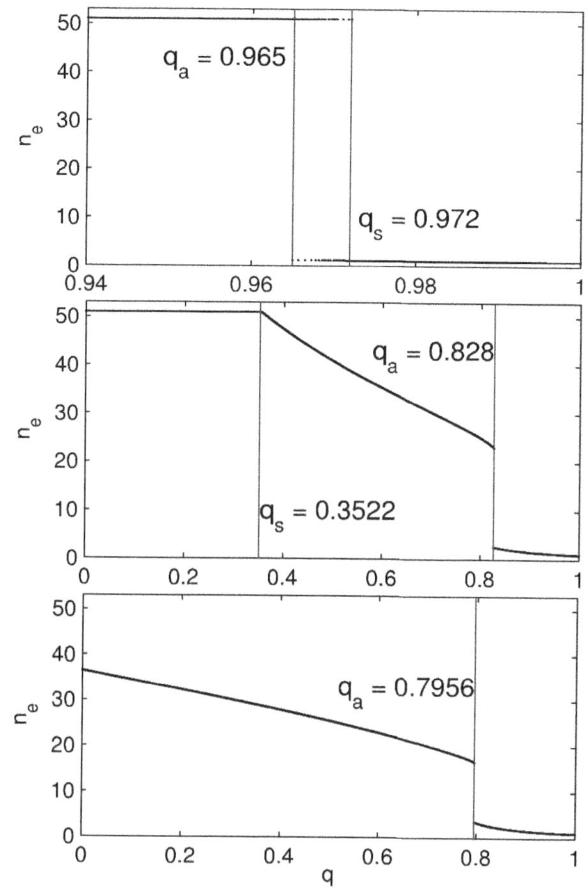

Figure 15.4 The effective number of grammars in response to learning fidelity q in a ring lattice with $n = 51$ and $a = 0.5$: (a) $k = 25$ (complete graph), (b) $k = 15$, (c) $k = 10$.

of a Gaussian as shown in Figure 15.5(b). We call this phase *weak cohesion*. In this phase, learning fidelity is too low for a single grammar to dominate by faithfully reproducing itself, however structure in grammar space allows for a collection of closely structurally similar grammars to rise in frequency. Since learning errors produce similar grammars to the teacher's grammar, the effective learning fidelity for the group of grammars is higher. This is analogous to the formation of a quasi-species in molecular evolution (Eigen and Schuster 1979; Eigen *et al.* 1989; Fontana and Schuster 1987).

The grammar frequency distribution over the range of q for various topologies is shown in Figure 15.6. For each q value the grammar frequencies are sorted so that the adjusted grammar index starts from 1 for the least frequent grammar to the maximum of 51 for the most frequent grammar. For all types of grammar network, a population is in strong cohesion state if learning fidelity exceeds the error threshold as shown in Figure 15.6. A population is in the symmetric state if the learning fidelity is sufficiently low. The observed error threshold is the highest for the complete graph and the lowest for the random graph.

(a)

(b)

(c)

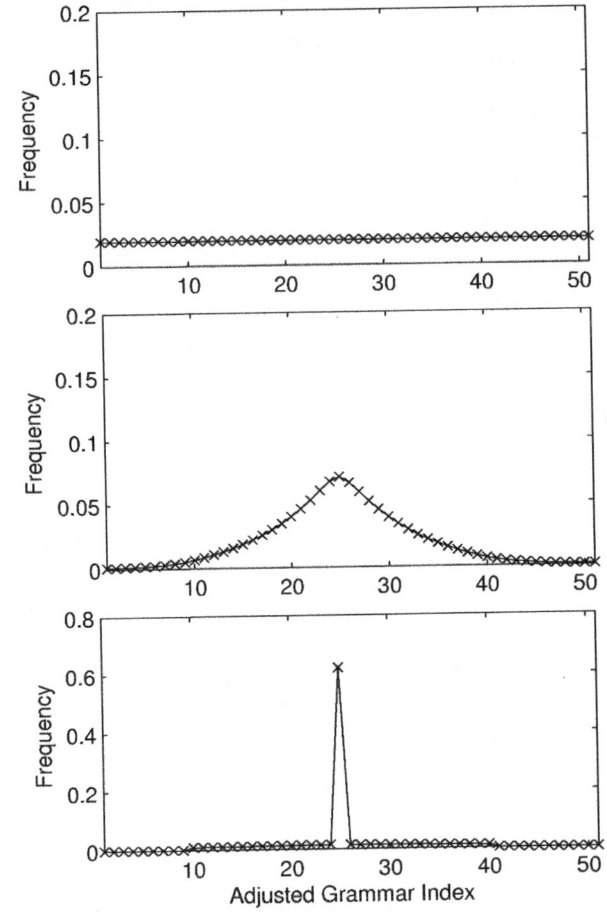

Figure 15.5 Frequency of all grammars, sorted by index number, of a single run at the steady state for a ring lattice with $n = 51$, $k = 15$, and $a = 0.5$: (a): $q = 0$, (b): $q = 0.82$, and (c): $q = .83$.

In the regular ring lattice with $k = 15$, a weak cohesion phase is observed in mid-range learning fidelities. A small-world network shows a narrower range of learning fidelities where weak cohesion can occur. We were unable to detect any obvious weak cohesion for a random graph as shown in Figure 15.6(c).

15.5.2 Density of grammar networks and language convergence

Figure 15.7 shows the effect of graph density on the level of coherence for a regular ring lattice. We plot (a) grammar diversity, and (b) minimum grammar frequency for a ring lattice with $n = 200$ and $a = 0.5$ given a fixed learning fidelity. Notice that the learning fidelity here $q = 0.8$ is smaller than $q_a = 0.829$ for a complete graph. The grammar diversity and dominant frequency change in non-linear fashion in response to network density.

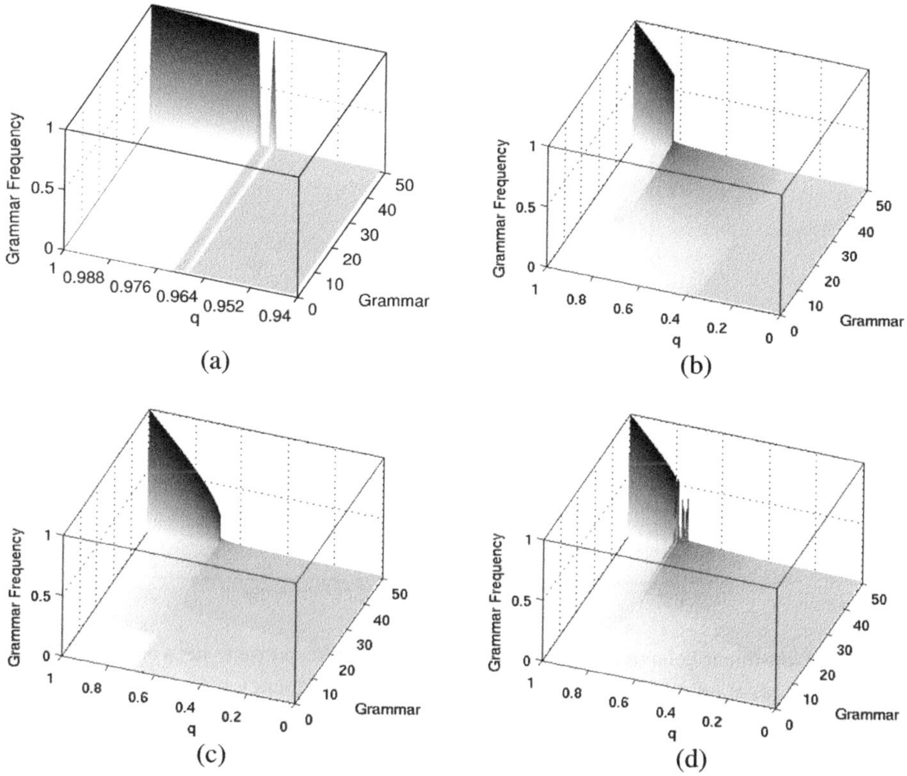

Figure 15.6 Distribution of grammar frequencies at the steady state with $n = 51$ and $a = 0.5$ over the range of $0 \leq q \leq 1$. For each q value, the grammar frequencies are sorted. (a) complete graph, (b) ring lattice with $k = 15$, (c) random network with $k = 1$ and $r = 714$ and (d) small word with $k = 13$ and $r = 102$.

When a grammar network forms a complete graph ($k = 100$), the population is in the symmetric state ($n_e = 200$) as expected. When the density of a ring lattice is sufficiently high ($k \geq 83$), the population is also in the symmetric state, and both x_{max} and x_{min} are equal to $1/n = 5 \times 10^{-3}$ as expected.

As the density of the graph decreases, grammar diversity n_e decreases reflecting an increased level of linguistic coherence, indicating a weak cohesion state. When the network density is sufficiently small ($k \leq 10$), the grammar diversity is close to 1 and the dominant frequency is over 0.5, which is indicative of strong cohesion.

Figure 15.8 shows the effect of graph density on the convergence for a random network. We were unable to find weak cohesion conditions for a random graph. We label d_a as the maximum graph density where only strong cohesion is a stable attractor. d_b is the graph density where strong cohesion becomes a stable attractor but at least one other equilibrium state is also stable.

Given a fixed learning fidelity value of $q = 0.6$, a random graph exhibits a sort of symmetric state if graph density is sufficiently high. In other words, if network density is sufficiently small, strong cohesion can be observed. In grammar network context, if

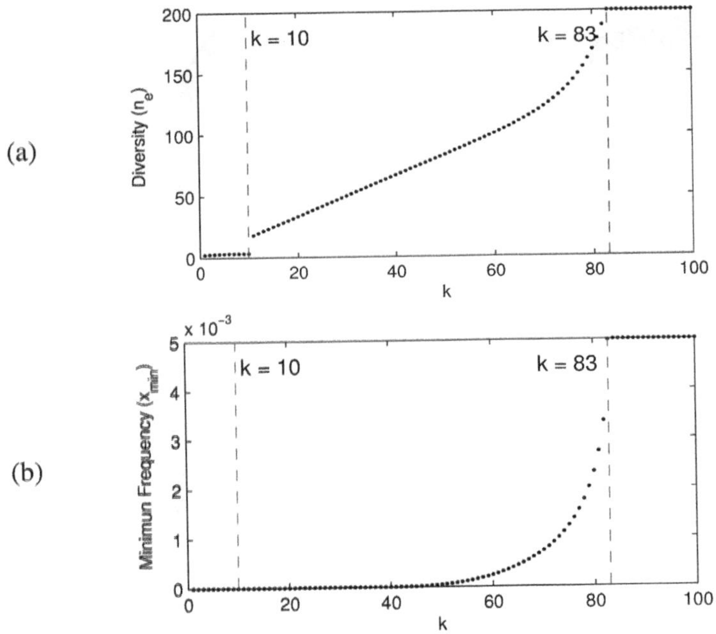

Figure 15.7 Non-linear behavior of language convergence in response to network density. Various measures such as (a) grammar diversity, n_e, and (b) minimum frequency x_{min} are plotted for varying $k = [1, 100]$ given a ring lattice with $n = 200$, $a = 0.5$, and a learning fidelity $q = 0.8$.

a grammar has a limited number of grammars that it can mutate into and its number of transitions is relatively small compared to the number of overall possible transitions, population may reach a common consensus.

Unlike a ring lattice, a random graph does not appear to have a perfect symmetric state; the x_{max} remains small but x_{min} do not remain at $1/n = 5 \times 10^{-3}$.

15.5.3 Rate of language convergence in grammar networks

Figure 15.9 shows the convergence time measured in elapsed time for each grammar network to reach the approximate equilibrium state. The q regions where convergence time is relatively long closely match the transition points between phases seen from monitoring x_{max} and n_e.[2]

Although the actual elapsed time may vary depending on the choice of integration step size or integration method, the relative time clearly shows that it take more time to reach steady-state when the learning fidelity resides near a transition point.

Figure 15.9(b) shows a sharp peak in convergence time for a ring lattice with $k = 15$ around $q = 0.35$, where transition from symmetric to weak cohesion occurs, and around $q = 0.82$, where transition from weak cohesion to strong cohesion occurs. In contrast, a random network shows only one peak, a transition to strong cohesion, as shown in

[2] Compare the convergence time for the complete graph and ring lattice with the study presented in Figure 15.4.

(a)

(b)

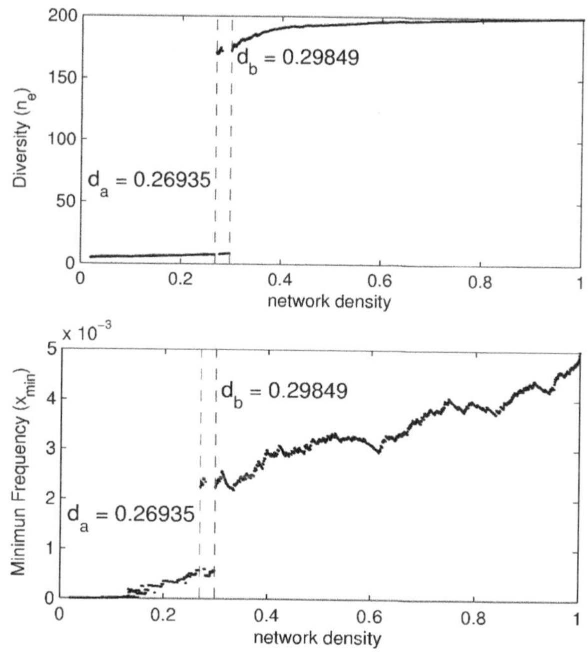

Figure 15.8 Grammar diversity, dominant frequency, minimum frequency in response to network density of a random graph with $k = 1$ and $r = 714$ at the learning fidelity of $q = 0.6$.

Table 15.1 A brief summary of convergence time for various grammar networks

Network type	Network density	Mean relative convergence time $\pm\delta$
A complete graph	1	145.0 ± 200.5
A regular ring lattice	0.6	548.3 ± 1414.1
A random graph	0.6	205.6 ± 269.2
A small-world	0.6	885.0 ± 1154.1

Note: A population reaches the steady state quickest in complete graph and slowest in a regular ring lattice. Convergence time is longer if a learning fidelity falls near transition points.

Figure 15.9(c). For a small world network, the transition from symmetric to weak cohesion and the transition from weak cohesion to strong cohesion overlap over broad region of q values as shown in Figure 15.9(d), yet it is evident that two distinct peaks correspond to the transition point from the symmetric state to weak cohesion and the transition point from weak cohesion to strong cohesion.

Mean convergence time over the range of $q = [0, 1]$ for each grammar network are presented in Table 15.1. Overall, the mean convergence time is the shortest for a complete graph and longest for a small-world network. Among three graphs that have the same density, a random graph reaches at equilibrium fastest.

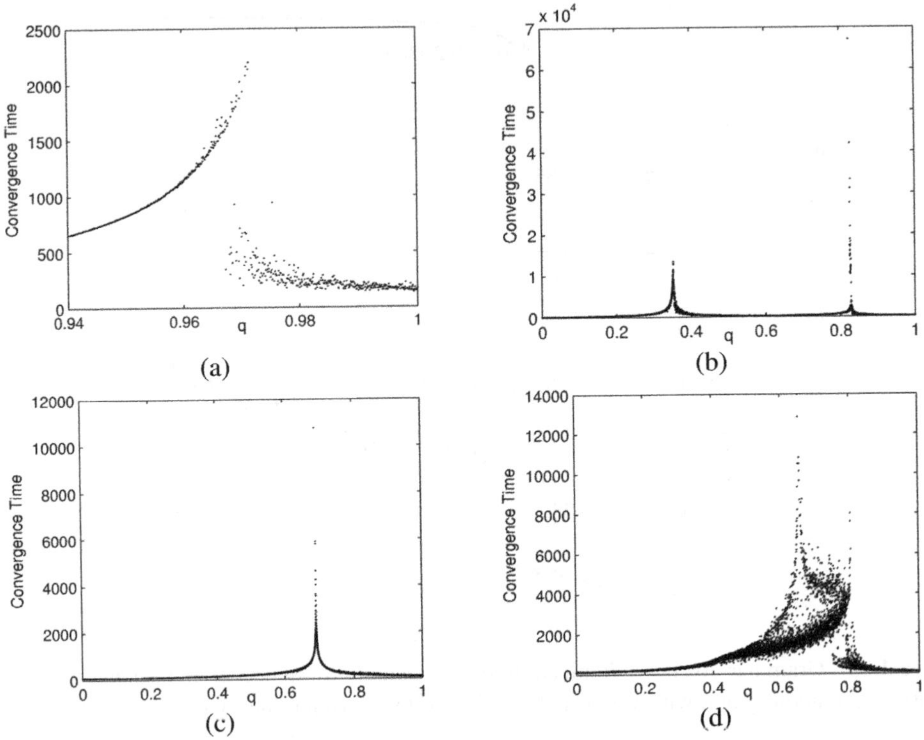

Figure 15.9 Time (or the number of replication cycles) to reach a equilibrium state for each topology: (a) complete graph, (b) regular graph with $k = 15$, (c) random network with $k = 1$ and $r = 714$, and (d) small-world network with $k = 13$ and $r = 102$. Numerical summary of this graph is presented in Table 15.1.

15.6 CONCLUSION

This study shows that the evolutionary dynamics of linguistic convergence is affected not only by learning fidelity, but also other factors such as topology of a population network, language structure, and grammar networks. These factors are just a few aspects that contribute to the understanding of complex language convergence dynamics.

The grammar network perspective provides a means for studying language structure in terms of mutational similarity of grammars. In this framework, we are able to identify three possible equilibrium states: (1) the symmetric state ($q \leq q_s$) where $x_i = 1/n$ for all i; (2) a weak cohesion state where the symmetry in grammar frequencies breaks and the distribution of grammar frequencies forms roughly a Gaussian shape centered around the most frequent grammar; and (3) a strong cohesion state ($q \geq q_a$) where a single predominant grammar emerges.

In the various grammar networks studied here we find distinct patterns of language convergence, with varying composition of three equilibrium states. For the same density, a random network will generally have a much smaller mean path length than a regular graph. Thus the evolutionary dynamics of relatively dense random graphs much more closely resemble the complete graph than a regular graph with the same density.

For a grammar space defined by a mid to low density regular ring lattice, a weak cohesion phase of equilibria can be identified at learning fidelities between q_s and q_a. This region is below the error threshold for a complete graph, where no cohesion or evolution can take place. The existence of a weak cohesion phase is dependent on the structure of the grammar space. Structured spaces allow the formation of a peaked quasi-species of related grammars, a peaked distribution of elements generated by a mutation-selection process (Nowak 2002). While the learning fidelity in this region is too low to allow a single grammar to faithfully reproduce itself well enough to maintain a higher frequency than other grammars, the effective learning fidelity of the quasi-species as a whole is sufficiently high for the collection of grammars within it to dominate.

The existence of the weak cohesion phase suggests that a group of related grammars can emerge with lower learning fidelities than is required to establish a single consensus. Weak cohesion is also characterized by a large amount of standing heritable variation within the population which is particularly intriguing from an evolutionary perspective.

Our results of varying graph density indicate that there are threshold density values for convergence. In general, as the density of the grammar network decreases, the level of linguistic convergence increases for a fixed learning fidelity. We expect that this result holds generally for other structured graphs, although the mean path length or clustering coefficient may be more indicative metrics of structure in grammar space.

The regular ring lattice is just one simple example of many different network structures which may yield qualitatively different dynamics, where convergence to a set of closely related languages is facilitated relative to the fully-connected model. When the grammar space is completely connected, high fidelity learner-driven change, such as the sort exhibited by human languages, can only occur just above a critical error threshold where the bifurcation of strong cohesion and the symmetric state begins. When the grammar space has more structure, strong cohesion can be achieved in lower learning fidelity regions.

Different networks not only exhibit different patterns of language convergence, but also reach an equilibrium state at different rates. The relative convergence time indicates that a population reaches a steady state more slowly when learning fidelity is close to a transition point, regardless of the topology of grammar networks. The convergence time results suggest that there may be another critical error threshold where weak cohesion becomes a stable attractor but the symmetric solution is also stable in technological applications where agents learn from each other. The study is far from predicting the state or course of evolution of natural languages.

In technological applications where agents learn from each other, where it is desirable for the overall system to converge, these results may provide a guide to designing properties of the language or state representation depending on the degree of convergence desired. If it is sufficient that agents of the system just mostly agree, i.e. converge to close variants of a dominant grammar, then a structured state space may provide a way to achieve faster convergence at higher mutation values. However, if absolute convergence is required, the state space must be designed so that minor variants are strongly selected against, producing a sharp fitness peak. This constraint also implies that a critical mutation/learning fidelity threshold exists.

ACKNOWLEDGEMENTS

This work was supported in part by AFOSR/MURI Grant #49620-01-1-0361, and NSF Grant #EF-0410438. Any opinions, findings, and conclusions or recommendations expressed in this publication are those of the authors and do not necessarily reflect the views of the sponsoring agencies.

REFERENCES

Arita T and Taylor CE 1996 A simple model for the evolution of communication In *The Fifth Annual Conference on Evolutionary Programming* (ed. Fogel L, Angeline PJ and Bäck T), MIT Press, Cambridge, MA, pp. 405–410.

Cavalli-Sforza L, Piazza A and Mountain J 1988 Reconstruction of human evolution: Bringing together genetic, archaeological and linguistic data. In *Proceedings of the National Academy of Sciences of the United States of America* **85**, 6002–6006.

Chomsky N 1965 *Aspects of the Theory of Syntax*. MIT Press, Cambridge, MA.

Chomsky N 1980 *Rules and Representations*. Basil Blackwell, London.

Collier TC and Taylor CE 2004 Self-organization in sensor networks. *Journal of Parallel and Distributed Computing* **64**(7), 866–873.

Comrie B 2001 Typology and the history of language. In *Aspects of Typology and Universals* (ed. Bisang W) Akademie Verlag, Berlin, pp. 21–35.

Crow JF and Kimura M 1970 *An Introduction to Population Genetics Theory*. Harper & Row Publishers, New York, Evanston and London.

Eigen M and Schuster P 1979 *The Hypercycle: A Principle of Natural Self-organization*. Springer Verlag, Berlin.

Eigen M, McCaskill J and Schuster P 1989 The molecular quasi-species. *Adv. Chem. Phys.* **75**, 149–263.

Fontana W and Schuster P 1987 A computer model of evolutionary optimization. *Biophysical Chemistry* pp. 123–147.

Keenan EL and Stabler EP 2003 *Bare Grammar*. CSLI Publications, Stanford, California.

Komarova NL and Nowak MA 2001 The evolutionary dynamics of the lexical matrix. *Bulletin of Mathematical Biology* **63**(3), 451–485.

Komarova NL, Niyogi P and Nowak MA 2001 Evolutionary dynamics of grammar acquisition. *Journal of Theoretical Biology* **209**(1), 43–59.

Lee Y 2006 The effect of structure in population and grammar space on language divergence. Ph.D. thesis. University of California, Los Angeles.

Lee Y, Collier TC, Kobele GM, Stabler EP and Taylor CE 2005a Grammar structure and the dynamics of language evolution. In *Advances in Artificial Life: 8th European Conference, ECAL 2005, Canterbury, UK, September 5-9, 2005, Proceedings*, pp. 624–633. Springer-Verlag, Berlin.

Lee Y, Collier TC, Stabler EP and Taylor CE 2005b The role of population structure in language evolution. *Artificial Life and Robotics*.

Lee Y, Riggle J, Collier TC, Stabler EP and Taylor CE 2004 Adaptive communication among collaborative agents: Preliminary results with symbol grounding. *Artificial Life and Robotics* **8**, 127–132.

Marocco D and Nolfi S 2006 Emergence of communication in teams of embodied and situated agents. In *Proceedings of the 6th Evolution of Language Conference*, pp. 198–205.

Marocco D, Cangelosi A and Nolfi S 2003 The emergence of communication in evolutionary robots. *Philosophical Transactions: Mathematical, Physical and Engineering Sciences* **361**(1811), 2397–2421.

Matsen FA and Nowak MA 2004 Win-stay, lose-shift in language learning from peers. *PNAS* **101**(52), 18053–18057.

Nowak MA 2002 What is a quasispecies?. *Trends in Ecology and Evolution* **7**, 118–121.

Nowak MA, Komarova NL and Niyogi P 2001 Evolution of universal grammar. *Science* **291**, 114–118.

Olfati-Saber R 2005 Ultrafast consensus in small-world networks. In *Proc. of American Control Conference,* pp. 2371–2378.

Olfati-Saber R and Murray RM 2004 Consensus problems in networks of agents with switching topology and time-delays. *IEEE Trans. Automatic Control* **49**(9), 1520–1533.

Steels L 2001 Language games for autonomous robots. *IEEE Intelligent Systems* **16**, 16–22.

Steels L 2006 Semiotic dynamics for embodied agents. *IEEE Intelligent Systems* **21**(3), 32–38.

Wee K, Collier T, Kobele G, Stabler E and Taylor C 2001 Natural language interface to an intrusion detection system. In *Proceedings, International Conference on Control, Automation and Systems,* ICCAS.

Part V

Complexity Management

16

Complexity management in the state estimation of multi-agent systems

Domitilla Del Vecchio and Richard M. Murray

16.1 INTRODUCTION

Logic and decision-making are playing increasingly large roles in modern control systems, and virtually all modern control systems are implemented using digital computers. Examples include aerospace systems, transportation systems (air, automotive, and rail), communication networks (wired, wireless, and cellular), and supply networks (electrical power and manufacturing). The evolution of these systems is determined by the interplay of continuous dynamics and logic. The continuous variables can represent quantities such as position, velocity, acceleration, voltage, current, etc., while the discrete variables can represent the state of the decision and communication protocol that is used for coordination and control. Most of these systems are also multi-agent, in which an agent can be, for example, a wireless device, a micro-controller, a robot, a piece of machinery, a piece of hardware or software, or even a human. The need for understanding and analyzing the behavior of these systems is compelling. However, the coupling of continuous dynamics and logics and the multi-agent nature of these systems render the study of these systems interesting and complicated enough that new tools are needed for the sake of analysis and control. In particular, multi-agent systems are usually affected by the combinatorial explosion of the state space that renders most of the existing state estimation algorithms inapplicable.

The problem of estimating the state of a decision and control system has been addressed by several authors for control or as a means for solving monitoring or surveillance problems in distributed environments. In the hybrid systems literature, Bemporad *et al.* (1999) propose the notion of incremental observability for piecewise affine systems and construct

Part of the text and figures are based on D. Del Vechhio, RM Murray, and E. Klavins 2006 *Automatica*, **42**, 271–285. Reproduced with permission.

Cooperative Control of Distributed Multi-Agent Systems edited by Jeff S. Shamma

a deadbeat observer that requires large amounts of computation. Balluchi *et al.* (2002) combine a *location* observer with a Luenberger observer to design hybrid observers that identify the location in a finite number of steps and converge exponentially to the continuous state. However, if the number of locations is large, as in the systems that we consider, such an approach is impracticable. In Balluchi *et al.* (2003), sufficient conditions for a linear hybrid system to be final state determinable are given. In Alessandri and Coletta (2001, 2003), Luenberger-like observers are proposed for hybrid systems where the system location is known. Vidal *et al.* (2002) derive sufficient and necessary conditions for observability of discrete time jump-linear systems, based on a simple rank test on the parameters of the model. In later work (Vidal *et al.* 2003), these notions are generalized to the case of continuous time jump linear systems. For jump Markov linear systems, Costa and do Val (2002) derive a test for observability, and Cassandra *et al.* (1994) propose an approach to optimal control for partially observable Markov decision processes. For continuous time hybrid systems, Santis *et al.* (2003) propose a definition of observability based on the possibility of reconstructing the system state, and testable conditions for observability are provided.

In the discrete event literature, observability has been defined by Ramadge (1986), for example, who derives a test for current state observability. Oishi *et al.* (2003) derive a test for immediate observability in which the state of the system can be unambiguously reconstructed from the output associated with the current state and last and next events. Ozveren and Willsky (1990), Caines *et al.* (1991), and Caines and Wang (1995) propose discrete event observers based on the construction of the current-location observation tree that is impracticable when the number of locations is large, which is our case. Observability is also considered in the context of distributed monitoring and control in industrial automation, where agents are cooperating to perform system-level tasks such as failure detection and identification on the basis of local information (Rudie *et al.* (2003)). Diaz *et al.* (1994) consider observers for formal on-line validation of distributed systems, in which the on-line behavior is checked against a formal model. In the context of sensor networks, state estimation covers a fundamental role when solving surveillance and monitoring tasks in which the state usually has several components, such as the position of an agent, its identity, and its intent (see, for example, Collins *et al.* (2001) or Bui *et al.* (2002)).

The main contribution of this work is to design state estimators for multi-agent systems that overcome severe complexity issues encountered in previous work (Balluchi *et al.* (2002); Caines and Wang (1995); Caines *et al.* (1991)). These complexity issues render prohibitive the estimation problem for systems with a large discrete state space, which is often the case in multi-agent systems. Our point of view is that some of the complexity issues, such as those encountered in Caines *et al.* (1991) or Balluchi *et al.* (2002), can be avoided by finding a good way of representing the sets of interest and by finding a good way of computing maps on them. In this work, this is achieved by using partial order theory. Partial order theory has been historically used in theoretical computer science to prove properties about convergence of algorithms (Cousot and Cousot 1977). It has also been used for studying controllability properties of finite state machines (Caines and Wei 1996) and for tackling the state explosion problem in the verification of concurrent systems in Godefroid (1996). In this work, we exploit partial order theory to estimate the state in systems with a large discrete space resulting from the multi-agent nature of the system. In particular, given a system Σ defined on its space of variables, we extend it to a larger space of variables that has lattice structure to obtain an extended system $\tilde{\Sigma}$. Under certain properties verified by the extension $\tilde{\Sigma}$, an observer for system Σ can be constructed, which updates at each step only two variables. It updates the least and greatest element of the set of all values of variables compatible with the output sequence and with the dynamics of Σ. The structure of the obtained observer resembles the structure of the Luenberger observer (Luenberger 1971) or a Kalman filter (Kalman 1960) as it is obtained by 'copying' the dynamics of the system Σ and by correcting it according to the measured output values. This work is concerned with the estimation of the discrete state in case the continuous state is measured, and with the estimation of the whole system state in case a cascade structure of the estimator is possible. We also show that a system is observable if and only if there is a lattice in which the extended system satisfies the requirements for the construction of the proposed estimator. Thus, our approach to state estimation is general.

The contents of this work are organized as follows. In Section 16.2, the RoboFlag Drill is introduced as our motivating example. In Section 16.3, basic definitions on partial orders and transition systems are reviewed. Section 16.4 formulates the discrete state estimation problem and Section 16.5 proposes a solution. In Section 16.6, the RoboFlag Drill is revisited. In Section 16.7, the generality of the approach is investigated and in Section 16.8 the proposed state estimation approach is extended to estimation of continuous and discrete variables for the case of a cascade form of the estimator.

16.2 MOTIVATING EXAMPLE

As a motivating example, we consider a task that represents a defensive maneuver for a robotic 'capture the flag' game (D'Andrea *et al.* 2003). We do not propose to devise a strategy that addresses the full complexity of the game. Instead, we examine the following very simple *drill* or exercise that we call 'RoboFlag Drill.' Some number of blue robots with positions $(z_i, 0) \in \mathbb{R}^2$ (denoted by open circles) must defend their zone $\{(x, y) \in \mathbb{R}^2 \mid y \leq 0\}$ from an equal number of incoming red robots (denoted by filled circles). The positions of the red robots are $(x_i, y_i) \in \mathbb{R}^2$. An example for 8 robots is illustrated in Figure 16.1. The red robots move straight toward the blue robots' defensive zone. The blue robots are each assigned to a red robot, and they coordinate to intercept the red robots. Let N represent the number of robots in each team. The robots start with an arbitrary (bijective) assignment $\alpha : \{1, \ldots, N\} \rightarrow \{1, \ldots, N\}$, where α_i is the red robot that blue robot i is required to intercept. At each step, each blue robot communicates with its neighbors and decides to either switch assignments with its left or right neighbor or keep its assignment. It is possible to show that the α assignment reaches the equilibrium value $(1, \ldots, N)$ (see Klavins and Murray (2004) or Klavins (2003) for details). We consider the problem of estimating the current assignment α given the motions of the blue robots, which might be of interest to, for example, the red robots in that they may use such information to determine a better strategy of attack. We do not consider the problem of how they would change their strategy in this work.

The RoboFlag Drill system can be specified by the following rules:

$$y_i(k+1) = y_i(k) - \delta \text{ if } y_i(k) \geq \delta \tag{16.1}$$

$$z_i(k+1) = z_i(k) + \delta \text{ if } z_i(k) < x_{\alpha_i(k)} \tag{16.2}$$

$$z_i(k+1) = z_i(k) - \delta \text{ if } z_i(k) > x_{\alpha_i(k)} \tag{16.3}$$

$$(\alpha_i(k+1), \alpha_{i+1}(k+1)) = (\alpha_{i+1}(k), \alpha_i(k)) \text{ if } x_{\alpha_i(k)} \geq z_{i+1}(k) \wedge x_{\alpha_{i+1}(k)} \leq z_{i+1}(k), \tag{16.4}$$

where we assume $z_i \leq z_{i+1}$ and $x_i < z_i < x_{i+1}$ for all k. Also, if none of the 'if' statements above are verified for a given variable, the new value of the variable is equal to the old one. This system is a slight simplification of the original system described in Klavins (2003). In such a work in fact, two close robots might decide to swap their assignments even if they are moving in the same direction, while in the present case, two close robots swap their assignments only if they are moving one toward the other. Also, in Klavins (2003), the decisions are taken sequentially first from the robots on the left and then from the robots on the right, and the decisions are coordinated by a token that moves from left to right. In the present case, the decision protocol is completely decentralized.

Equation (16.4) establishes that two robots trade their assignments if the current assignments cause them to go toward each other. The question we are interested in is the following: given the evolution of the measurable quantities z, x, y, can we build an estimator that tracks on-line the value of the assignment $\alpha(k)$? The value of $\alpha \in \text{perm}(N)$ determines the discrete state, i.e., $S = \text{perm}(N)$. The discrete state α determines also what has been called in previous work the location of the system (see Balluchi *et al.* (2002)). The number of possible locations is $N!$, that is, $|S| = N!$. This for $N \geq 8$ renders prohibitive the application of location observers based on the current location observation tree of Caines *et al.* (1991), used in Balluchi *et al.* (2002), and in Ozveren and Willsky (1990). At each step, the set of possible α values compatible with the current output and with the previously seen outputs can be so large as to render impractical its computation. As an example, we consider the situation depicted in Figure 16.1 (left) where $N = 8$. We see the blue robots 1, 3, 5 going right and the others going left. From equations (16.2)–(16.3) with $x_i < z_i < x_{i+1}$ we deduce that the set of all possible $\alpha \in \text{perm}(N)$ compatible with this observation is such that $\alpha_i \geq i + 1$ for $i \in \{1, 2, 3\}$ and $\alpha_i \leq i$ for $i \in \{2, 4, 6, 7, 8\}$.

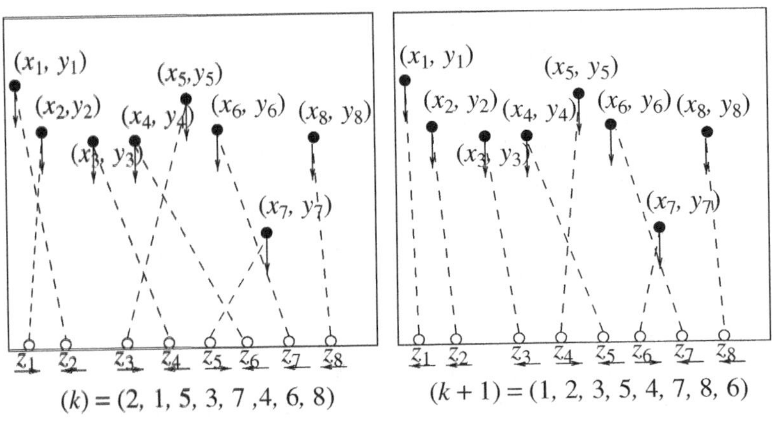

Figure 16.1 Example of the RoboFlag Drill with 8 robots per team. The dashed lines represent the assignment of each blue robot to red robot. The arrows denote the direction of motion of each robot.

$$\underbrace{\left[\begin{pmatrix}2\\1\\4\\1\\6\\1\\1\\1\end{pmatrix},\begin{pmatrix}8\\2\\8\\4\\8\\6\\7\\8\end{pmatrix}\right]}_{O_y(k)} \xrightarrow[\tilde{f}]{} \underbrace{\left[\begin{pmatrix}1\\2\\1\\4\\1\\6\\1\\1\end{pmatrix},\begin{pmatrix}2\\8\\4\\8\\6\\8\\7\\8\end{pmatrix}\right]}_{\tilde{f}(O_y(k))} \cap \underbrace{\left[\begin{pmatrix}1\\1\\1\\5\\1\\7\\1\\1\end{pmatrix},\begin{pmatrix}1\\2\\3\\8\\5\\8\\7\\8\end{pmatrix}\right]}_{O_y(k+1)} = \left[\begin{pmatrix}1\\2\\1\\5\\1\\7\\1\\1\end{pmatrix},\begin{pmatrix}1\\2\\3\\8\\5\\8\\7\\8\end{pmatrix}\right]$$

Figure 16.2 The observation of the z motion at step k gives the set of possible α, $O_y(k)$. At each step, the set is described by the lower and upper bounds of an *interval sublattice* in an appropriately defined lattice. Such a set is then mapped through the system dynamics $\tilde{f}$ to obtain at step $k+1$ the set of α that are compatible also with the observation at step k. Such a set is then intersected with $O_y(k+1)$, which is the set of α compatible with the z motion observed at step $k+1$.

The size of this set is 40320. According to the enumeration methods, this set needs to be mapped forward through the dynamics of the system to see what the values of α are at the next step that correspond to this output. Such a set is then intersected with the set of α values compatible with the new observation. To overcome the complexity issue that comes from the need of listing 40320 elements for performing such operations, we propose to represent a set by a lower L and an upper U elements according to some partial order. Then, we can perform the previously described operations only on L and U, two elements instead of 40320. This idea is developed in the following paragraph.

For this example, we can view $\alpha \in \mathbb{N}^N$. The set of possible assignments compatible with the observation of the z motion deduced from the equations (16.2)–(16.3), denoted $O_y(k)$, can be represented as an interval with the order established component-wise, see Figure 16.2. The function $\tilde{f}$ that maps such a set forward, specified by the equation (16.4) with the assumption that $x_i < z_i < x_{i+1}$, simply swaps two adjacent robot assignments if these cause the two robots to move toward each other. Thus, it maps the set $O_y(k)$ to the set $\tilde{f}(O_y(k))$ shown in Figure 16.2, which can still be represented as an interval. When the new output measurement becomes available (Figure 16.1, right) we obtain the new set $O_y(k+1)$ reported in Figure 16.2. The sets $\tilde{f}(O_y(k))$ and $O_y(k+1)$ can be intersected by simply computing the supremum of their lower bounds and the infinum of their upper bounds. This way, we obtain the system that updates L and U, being L and U the lower and upper bounds of the set of all possible α compatible with the output sequence:

$$\begin{aligned} L(k+1) &= \tilde{f}\left(\sup(L(k), \inf O_y(k))\right) \\ U(k+1) &= \tilde{f}\left(\inf(U(k), \sup O_y(k))\right). \end{aligned} \tag{16.5}$$

The variables $L(k)$ and $U(k)$ represent the lower and upper bound, respectively, of the set of all possible discrete state values compatible with the output sequence and with the system dynamics. The computational burden of this implementation is of the order of N

if N is the number of robots. This computational burden is to be compared to $N!$, which is the computation requirement that we have with the enumeration approach.

In the next section, we introduce some basic notions on partial order theory and deterministic transition system models.

16.3 BASIC CONCEPTS

In this chapter, we review some basic notions that will be used throughout this work. First, we give some background on partial order and lattice theory in Section 16.3.1 (for more details, the reader is referred to Davey and Priesteley (2002)). The theory of partial orders, while standard in computer science, may be less well known to the intended audience of this work. The class of deterministic transition systems is introduced in Section 16.3.2.

16.3.1 Partial order theory

A partial order is a set χ with a partial order relation '$\leq$', and we denote it by the pair $(\chi, \leq)$. For any $x, w \in \chi$, $\sup\{x, w\}$ is the smallest element that is larger than both x and w. In a similar way, $\inf\{x, w\}$ is the largest element that is smaller than both x and w. We define the *join* '$\curlyvee$' and the *meet* '$\curlywedge$' of two elements x and w in χ as $x \curlyvee w := \sup\{x, w\}$ and $x \curlywedge w := \inf\{x, w\}$ Also, if $S \subseteq \chi$, we have $\bigvee S := \sup\ S$, and $\bigwedge S := \inf\ S$.

Let $(\chi, \leq)$ be a partial order. If $x \curlywedge w \in \chi$ and $x \curlyvee w \in \chi$ for any $x, w \in \chi$, then $(\chi, \leq)$ is a *lattice*. In Figure 16.3, we illustrate Hasse diagrams (Davey and Priesteley 2002) showing partially ordered sets. From the diagrams, it is easy to tell when one element is less than another: $x < w$ if and only if there is a sequence of connected line

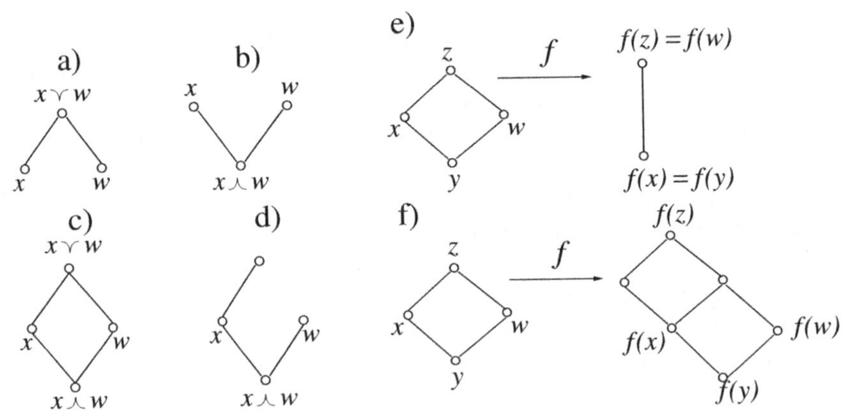

Figure 16.3 Left figure: In a) and b), x and w are not related, but they have a join and a meet, respectively; in c), we show a complete lattice; in d), we show a partially ordered set that is not a lattice, since the elements x and w have a meet, but not a join. Right figure: In e), we show a map that is, order preserving but not order embedding; in f), we show an order embedding that is, not an order isomorphism (it is not onto).

segments moving upward from x to w. The partial order $(\chi, \leq)$ is a *chain* if for all $x, w \in \chi$, either $x \leq w$ or $w \leq x$, that is, any two elements are comparable. If instead any two elements are not comparable, i.e., $x \leq y$ if and only if $x = y$, $(\chi, \leq)$ is said to be an *anti-chain*. If $x < w$ and there is no other element in between x and w, we write $x \ll w$. Let $(\chi, \leq)$ be a lattice and let $S \subseteq \chi$ be a non-empty subset of χ. Then, $(S, \leq)$ is a *sublattice* of χ if $a, b \in S$ implies that $a \curlyvee b \in S$ and $a \curlywedge b \in S$. If any sublattice of χ contains its least and greatest elements, then $(\chi, \leq)$ is called *complete*. Any finite lattice is complete, but infinite lattices may not be complete, and hence the significance of the notion of a complete partial order (Abramsky 1994). Given a complete lattice $(\chi, \leq)$, we will be concerned with a special kind of a sublattice called an *interval sublattice* defined as follows. Any interval sublattice of $(\chi, \leq)$ is given by $[L, U] = \{w \in \chi | L \leq w \leq U\}$ for $L, U \in \chi$. That is, this special sublattice can be represented by two elements only. For example, the interval sublattices of $(\mathbb{R}, \leq)$ are just the familiar closed intervals on the real line.

Let $(\chi, \leq)$ be a lattice with least element $\perp$ (the bottom). Then, $a \in \chi$ is called an *atom* of $(\chi, \leq)$ if $a > \perp$ and there is no element b such that $\perp < b < a$. The set of atoms of $(\chi, \leq)$ is denoted $\mathcal{A}(\chi, \leq)$. The *power lattice* of a set $\mathcal{U}$, denoted $(\mathcal{P}(\mathcal{U}), \subseteq)$, is given by the power set of $\mathcal{U}$, $\mathcal{P}(\mathcal{U})$ (the set of all subsets of $\mathcal{U}$), ordered according to the set inclusion $\subseteq$. The meet and join of the power lattice is given by intersection and union. The bottom element is the empty set, that is, $\perp = \emptyset$, and the top element is $\mathcal{U}$ itself, that is, $\top = \mathcal{U}$. Note that $\mathcal{A}(\mathcal{P}(\mathcal{U}), \subseteq) = \mathcal{U}$. Given a set P, we denote by $|P|$ its cardinality.

Definition 16.3.1 *Let $(P, \leq)$ and $(Q, \leq)$ be partially ordered sets. A map $f : P \to Q$ is*

(i) an order preserving map if $x \leq w \Rightarrow f(x) \leq f(w)$;

(ii) an order embedding if $x \leq w \Longleftrightarrow f(x) \leq f(w)$;

(iii) an order isomorphism if it is order embedding and it maps P onto Q.

These different types of maps are shown in (e) and (f) of Figure 16.3. Every order isomorphism faithfully mirrors the structure of P onto Q.

Definition 16.3.2 *If $(P, \leq)$ and $(Q, \leq)$ are lattices, then a map $f : P \to Q$ is said to be a homomorphism if f is join-preserving and meet-preserving, that is, for all $x, w \in P$ we have that $f(x \curlyvee w) = f(x) \curlyvee f(w)$ and $f(x \curlywedge w) = f(x) \curlywedge f(w)$.*

Proposition 16.3.3 *(See Davey and Priesteley (2002)) If $f : P \to Q$ is a bijective homomorphism, then it is an order isomorphism.*

A partial order induces a notion of distance between elements in the space. Define the distance function on a partial order in the following way.

Definition 16.3.4 *(Distance on a partial order) Let $(P, \leq)$ be a partial order. A distance d on $(P, \leq)$ is a function $d : P \times P \to \mathbb{R}$ such that the following properties are verified:*

(i) $d(x, y) \geq 0$ for any $x, y \in P$ and $d(x, y) = 0$ if and only if $x = y$;

(ii) $d(x, y) = d(y, x)$;

(iii) if $x \leq y \leq z$ then $d(x, y) \leq d(x, z)$;

(iv) $d(x, z) \leq d(x, y) + d(y, z)$ (triangular inequality).

Since we will deal with a partial order on the space of the discrete variables and with a partial order on the space of the continuous variables, it is useful to introduce the Cartesian product of two partial orders as it can be found in Abramsky (1994).

Definition 16.3.5 *(Cartesian product of partial orders) Let $(P_1, \leq)$ and $(P_2, \leq)$ be two partial orders. Their Cartesian product is given by $(P_1 \times P_2, \leq)$, where $P_1 \times P_2 = \{(x, y) \mid x \in P_1 \text{ and } y \in P_2\}$, and $(x, y) \leq (x', y')$ if and only if $x \leq x'$ and $y \leq y'$. For any $(p_1, p_2) \in P_1 \times P_2$ the standard projections $\pi_1 : P_1 \times P_2 \rightarrow P_1$ and $\pi_2 : P_1 \times P_2 \rightarrow P_2$ are such that $\pi_1(p_1, p_2) = p_1$ and $\pi_2(p_1, p_2) = p_2$.*

One can easily verify that the projection operators preserve the order. In this work we will also deal with *approximations* of sets and elements of a partial order. We thus give the following definition.

Definition 16.3.6 *(Upper and lower approximation) Let P_1 and P_2 be two sets with $P_1 \subseteq P_2$ and $(P_2, \leq)$ a partial order. For any $x \in P_2$, we define the lower and upper approximations of x in P_1 as $a_L(x) := \max_{(P_2, \leq)}\{w \in P_1 \mid w \leq x\}$ and $a_U(x) := \min_{(P_2, \leq)}\{w \in P_1 \mid w \geq x\}$. If such lower and upper approximations exist for any $x \in P_2$, then the partial order $(P_2, \leq)$ is said to be closed with respect to P_1.*

One can verify that the lower and upper approximation functions are order preserving. This means that for any $x_1, x_2 \in P_2$ with $x_1 \leq x_2$, then $a_L(x_1) \leq a_L(x_2)$ and $a_U(x_1) \leq a_U(x_2)$. In this section, we have given some basic definitions on partial order and lattice theory. In the next section, we introduce the class of models that we are going to consider in this work. These are transition systems with output.

16.3.2 Deterministic transition systems

The class of systems we are concerned with are deterministic, infinite state systems with output. The following definition introduces such a class.

Definition 16.3.7 *(Deterministic transition systems) A deterministic transition system (DTS) is the tuple $\Sigma = (S, \mathcal{Y}, F, g)$, where S is a set of states with $s \in S$; $\mathcal{Y}$ is a set of outputs with $y \in \mathcal{Y}$; $F : S \rightarrow S$ is the state transition function; $g : S \rightarrow \mathcal{Y}$ is the output function.*

An execution of Σ is any sequence $\sigma = \{s(k)\}_{k \in \mathbb{N}}$ such that $s(0) \in S$ and $s(k+1) = F(s(k))$ for all $k \in \mathbb{N}$. The set of all executions of Σ is denoted $\mathcal{E}(\Sigma)$. An output sequence of Σ is denoted $y = \{y(k)\}_{k \in \mathbb{N}}$, with $y(k) = g(\sigma(k))$, for $\sigma \in \mathcal{E}(\Sigma)$.

Definition 16.3.8 *Given a deterministic transition system $\Sigma = (S, \mathcal{Y}, F, g)$, two executions σ_1, σ_2 in $\mathcal{E}(\Sigma)$ are distinguishable if there exists a k such that $g(\sigma_1(k)) \neq g(\sigma_2(k))$.*

Definition 16.3.9 *(Observability) The deterministic transition system $\Sigma = (S, \mathcal{Y}, F, g)$ is said to be observable if any two different executions $\sigma_1, \sigma_2 \in \mathcal{E}(\Sigma)$ are distinguishable.*

From this definition, we deduce that if a system Σ is observable, any two different initial states will give rise to two executions σ_1 and σ_2 with different output sequences. Thus, the initial states can be distinguished by looking at the output sequence.

16.4 PROBLEM FORMULATION

The deterministic transition systems Σ we defined in the previous section are quite general. In this section, we restrict our attention to systems with a specific structure. In particular, for a system $\Sigma = (S, \mathcal{Y}, F, g)$ we suppose that (i) $S = \mathcal{U} \times \mathcal{Z}$ with $\mathcal{U}$ a finite set and $\mathcal{Z}$ a finite dimensional space; (ii) $F = (f, h)$, where $f : \mathcal{U} \times \mathcal{Z} \to \mathcal{U}$ and $h : \mathcal{U} \times \mathcal{Z} \to \mathcal{Z}$; (iii) $y = g(\alpha, z) := z$, where $\alpha \in \mathcal{U}$, $z \in \mathcal{Z}$, $y \in \mathcal{Y}$, and $\mathcal{Y} = \mathcal{Z}$. The set $\mathcal{U}$ is a set of logic states and $\mathcal{Z}$ is a set of measured states or physical states, as one might find in a robot system. In the case of the example given in Section 16.2, $\mathcal{U} = \text{perm}(N)$ and $\mathcal{Z} = \mathbb{R}^N$, the function f is represented by equations (16.4) and the function h is represented by equations (16.2)–(16.3). In the sequel, we will denote with abuse of notation this class of deterministic transition systems by $\Sigma = (\mathcal{U}, \mathcal{Z}, f, h)$, in which we associate to the tuple $(\mathcal{U}, \mathcal{Z}, f, h)$, the equations:

$$\begin{aligned} \alpha(k+1) &= f(\alpha(k), z(k)) \\ z(k+1) &= h(\alpha(k), z(k)) \\ y(k) &= z(k), \end{aligned} \tag{16.6}$$

in which $\alpha \in \mathcal{U}$ and $z \in \mathcal{Z}$. An execution of the system Σ in equations (16.6) is a sequence $\sigma = \{\alpha(k), z(k)\}_{k\in\mathbb{N}}$. The output sequence is $\{y(k)\}_{k\in\mathbb{N}} = \{z(k)\}_{k\in\mathbb{N}}$. Given an execution σ of the system Σ, we denote the α and z sequences corresponding to such an execution by $\{\sigma(k)(\alpha)\}_{k\in\mathbb{N}}$ and $\{\sigma(k)(z)\}_{k\in\mathbb{N}}$, respectively.

From the measurement of the output sequence, which in our case coincides with the evolution of the continuous variables, we want to construct a discrete state estimator: a system $\hat{\Sigma}$ that takes as input the values of the measurable variables and asymptotically tracks the value of the variable α. We thus define in the following definition a deterministic transition system with input.

Definition 16.4.1 *(Deterministic transition system with input) A deterministic transition system with input is a tuple $(S, \mathcal{I}, \mathcal{Y}, F, g)$ in which S is a set of states; $\mathcal{I}$ is a set of inputs; $\mathcal{Y}$ is a set of outputs; $F : S \times \mathcal{I} \to S$ is a transition function; $g : S \times \mathcal{I} \to \mathcal{Y}$ is an output function.*

An alternative to simply maintaining a list of all possible values for α is next proposed. Specifically, if the set $\mathcal{U}$ can be immersed in a larger set χ whose elements can be related by an order relation $\leq$, we can represent a subset of $(\chi, \leq)$ as an interval sublattice $[L, U]$. Let 'id' denote the identity operator. We formulate the discrete state estimation problem on a lattice as follows.

Problem 16.4.1 *(Discrete state estimator on a lattice)* Given the deterministic transition system $\Sigma = (\mathcal{U}, \mathcal{Z}, f, h)$, find a deterministic transition system with input $\hat{\Sigma} = (\chi \times$

$\chi, \mathcal{Y} \times \mathcal{Y}, \chi \times \chi, (f_1, f_2), \text{id})$, with $f_1 : \chi \times \mathcal{Y} \times \mathcal{Y} \to \chi$, $f_2 : \chi \times \mathcal{Y} \times \mathcal{Y} \to \chi, \mathcal{U} \subseteq \chi$, with $(\chi, \leq)$ a lattice, represented by the equations

$$L(k+1) = f_1(L(k), y(k), y(k+1))$$
$$U(k+1) = f_2(U(k), y(k), y(k+1)),$$

with $L(k) \in \chi$, $U(k) \in \chi$, $L(0) := \bigwedge \chi$, $U(0) := \bigvee \chi$, *such that*

(i) $L(k) \leq \alpha(k) \leq U(k)$ *(correctness);*

(ii) $|[L(k+1), U(k+1)]| \leq |[L(k), U(k)]|$ *(non-increasing error);*

(iii) There exists $k_0 > 0$ *such that for any* $k \geq k_0$ *we have* $[L(k), U(k)] \cap \mathcal{U} = \alpha(k)$ *(convergence).*

16.5 PROBLEM SOLUTION

For finding a solution to Problem 16.4.1, we need to find the functions f_1 and f_2 defined on a lattice $(\chi, \leq)$ such that $\mathcal{U} \subseteq \chi$ for some finite lattice χ. We propose in the following definitions a way of extending a system Σ defined on $\mathcal{U}$ to a system $\tilde{\Sigma}$ defined on χ with $\mathcal{U} \subseteq \chi$. Moreover, as we have seen in the motivating example, we want to represent the set of possible α values compatible with an output measurement as an interval sublattice in $(\chi, \leq)$. We thus define the $\tilde{\Sigma}$ transition classes, with each transition class corresponding to a set of values in χ compatible with an output measurement. We define the partial order $(\chi, \leq)$ and the system $\tilde{\Sigma}$ to be interval compatible if such equivalence classes are interval sublattices and $\tilde{\Sigma}$ preserves their structure.

Definition 16.5.1 *(Extended system) Given the deterministic transition system* $\Sigma = (\mathcal{U}, \mathcal{Z}, f, h)$, *an extension of* Σ *on* χ, *with* $\mathcal{U} \subseteq \chi$ *and* $(\chi, \leq)$ *a finite lattice, is any system* $\tilde{\Sigma} = (\chi, \mathcal{Z}, \tilde{f}, \tilde{h})$ *in which* $\tilde{f} : \chi \times \mathcal{Z} \to \chi$ *is such that* $\tilde{f}|_{\mathcal{U}\times\mathcal{Z}} = f$ *and* $\tilde{h} : \chi \times \mathcal{Z} \to \mathcal{Z}$ *is such that* $\tilde{h}|_{\mathcal{U}\times\mathcal{Z}} = h$.

Definition 16.5.2 *(Transition sets) Let* $\tilde{\Sigma} = (\chi, \mathcal{Z}, \tilde{f}, \tilde{h})$ *be a deterministic transition system. The non empty sets* $T_{(z^1,z^2)}(\tilde{\Sigma}) = \{w \in \chi \mid z^2 = \tilde{h}(w, z^1)\}$, *for* $z^1, z^2 \in \mathcal{Z}$, *are named the* $\tilde{\Sigma}$*-transition sets.*

Each $\tilde{\Sigma}$-transition set contains all of $w \in \chi$ values that allow the transition from z^1 to z^2 through $\tilde{h}$. It will also be useful to define the transition class $\mathcal{T}_i(\tilde{\Sigma})$, which corresponds to multiple transition sets, as transition sets obtained by different pairs (z^1, z^2) can define the same set in χ.

Definition 16.5.3 *(Transition classes)* The set $\mathcal{T}(\tilde{\Sigma}) = \{\mathcal{T}_1(\tilde{\Sigma}), \ldots, \mathcal{T}_M(\tilde{\Sigma})\}$, with $\mathcal{T}_i(\tilde{\Sigma})$ such that

(i) for any $\mathcal{T}_i(\tilde{\Sigma}) \in \mathcal{T}(\tilde{\Sigma})$ *there are* $z^1, z^2 \in \mathcal{Z}$ *such that* $\mathcal{T}_i(\tilde{\Sigma}) = T_{(z^1,z^2)}(\tilde{\Sigma})$;

(ii) for any $T_{(z^1,z^2)}(\tilde{\Sigma})$ *there is* $j \in \{1, \ldots, M\}$ *such that* $T_{(z^1,z^2)}(\tilde{\Sigma}) = \mathcal{T}_j(\tilde{\Sigma})$;

is the set of $\tilde{\Sigma}$*-transition classes.*

Note that $T_{(z^1,z^2)}$ and $T_{(z^3,z^4)}$ might be the same set even if $(z^1, z^2) \neq (z^3, z^4)$: in the RoboFlag Drill example introduced in Section 16.2, if robot j is moving right, the set of possible values of α_j is $[j+1, N]$ independently of the values of $z_j(k)$. Thus, $T_{(z^1,z^2)}$ and $T_{(z^3,z^4)}$ can define the same set that we call $\mathcal{T}_i(\tilde{\Sigma})$ for some i. Also, the transition classes $\mathcal{T}_i(\tilde{\Sigma})$ are not necessarily equivalence classes as they might not be pairwise disjoint. However, for the RoboFlag Drill it is the case that the transition classes are pairwise disjoint, and thus they partition the lattice $(\chi, \leq)$ in equivalence classes.

Definition 16.5.4 *(Output set) Given the extension $\tilde{\Sigma} = (\chi, \mathcal{Z}, \tilde{f}, \tilde{h})$ of the deterministic transition system $\Sigma = (\mathcal{U}, \mathcal{Z}, f, h)$ on the lattice $(\chi, \leq)$, and given an output sequence $\{y(k)\}_{k\in\mathbb{N}}$ of Σ, the set $O_y(k) := \{w \in \chi \mid \tilde{h}(w, y(k)) = y(k+1)\}$ is the output set at step k.*

Note that by definition, for any k, $O_y(k) = T_{(y(k),y(k+1))}(\tilde{\Sigma})$, and thus it is equal to $\mathcal{T}_i(\tilde{\Sigma})$ for some $i \in \{1, \ldots, M\}$. By definition of the extended functions ($\tilde{h}|_{\mathcal{U}\times\mathcal{Z}} = h$), this output set contains also all of the values of α compatible with the same output pair.

Definition 16.5.5 *(Interval compatibility) Given the extension $\tilde{\Sigma} = (\chi, \mathcal{Z}, \tilde{f}, \tilde{h})$ of the system $\Sigma = (\mathcal{U}, \mathcal{Z}, f, h)$ on the lattice $(\chi, \leq)$, the pair $(\tilde{\Sigma}, (\chi, \leq))$ is said to be interval compatible if*

(i) each $\tilde{\Sigma}$-transition class, $\mathcal{T}_i(\tilde{\Sigma}) \in \mathcal{T}(\tilde{\Sigma})$, is an interval sublattice of $(\chi, \leq)$, that is, $\mathcal{T}_i(\tilde{\Sigma}) = [\bigwedge\mathcal{T}_i(\tilde{\Sigma}), \bigvee\mathcal{T}_i(\tilde{\Sigma})]$;

(ii) $\tilde{f} : (\mathcal{T}_i(\tilde{\Sigma}), z) \to [\tilde{f}(\bigwedge\mathcal{T}_i(\tilde{\Sigma}), z), \tilde{f}(\bigvee\mathcal{T}_i(\tilde{\Sigma}), z)]$ is an order isomorphism for any $i \in \{1, \ldots, M\}$ and for any $z \in \mathcal{Z}$.

The following theorem gives the main result, which proposes a solution to Problem 16.4.1.

Theorem 16.5.6 *Assume that the deterministic transition system $\Sigma = (\mathcal{U}, \mathcal{Z}, f, h)$ is observable. If there is a lattice $(\chi, \leq)$, such that the pair $(\tilde{\Sigma}, (\chi, \leq))$ is interval compatible, then the deterministic transition system with input $\hat{\Sigma} = (\chi \times \chi, \mathcal{Z} \times \mathcal{Z}, \chi \times \chi, (f_1, f_2), id)$ with*

$$f_1(L(k), y(k), y(k+1)) = \tilde{f}\left(L(k) \curlyvee \bigwedge O_y(k), y(k)\right)$$
$$f_2(U(k), y(k), y(k+1)) = \tilde{f}\left(U(k) \curlywedge \bigvee O_y(k), y(k)\right)$$

solves Problem 16.4.1.

Proof. In order to prove the statement of the theorem, we need to prove that the system

$$\begin{aligned} L(k+1) &= \tilde{f}(L(k) \curlyvee \bigwedge O_y(k), y(k)) \\ U(k+1) &= \tilde{f}(U(k) \curlywedge \bigvee O_y(k), y(k)) \end{aligned} \tag{16.7}$$

with $L(0) = \bigwedge \chi$, $U(0) = \bigvee \chi$ is such that properties (i)–(iii) of Problem 16.4.1 are satisfied. For simplicity of notation, we omit the dependence of $\tilde{f}$ on its second argument.

Proof of (i): This is proved by induction on k. Base case: for $k = 0$ we have that $L(0) = \bigwedge \chi$ and that $U(0) = \bigvee \chi$, so that $L(0) \leq \alpha(0) \leq U(0)$. Induction step: we assume that $L(k) \leq \alpha(k) \leq U(k)$ and we show that $L(k+1) \leq \alpha(k+1) \leq U(k+1)$. Note that $\alpha(k) \in O_y(k)$. This, along with the assumption of the induction step, implies that

$$L(k) \curlyvee \bigwedge O_y(k) \leq \alpha(k) \leq U(k) \curlywedge \bigvee O_y(k).$$

This last relation also implies that there is x such that $x \geq L(k) \curlyvee \bigwedge O_y(k)$ and $x \leq \bigvee O_y(k)$. This in turn implies that

$$L(k) \curlyvee \bigwedge O_y(k) \leq \bigvee O_y(k).$$

This in turn implies that $L(k) \curlyvee \bigwedge O_y(k) \in O_y(k)$. Because of this, because (by analogous reasonings) $U(k) \curlywedge \bigvee O_y(k) \in O_y(k)$, and because the pair $(\tilde{\Sigma}, (\chi, \leq))$ is interval compatible, we can use the isomorphic property of $\tilde{f}$ (property (ii) of Definition 16.5.5), which leads to

$$\tilde{f}(L(k) \curlyvee \bigwedge O_y(k)) \leq \alpha(k+1) \leq \tilde{f}(U(k) \curlywedge \bigvee O_y(k)).$$

This relationship combined with Equation (16.7) proves (i).

Proof of (ii): This can be shown by proving that for any $w \in [L(k+1), U(k+1)]$ there is $z \in [L(k), U(k)]$ such that $w = \tilde{f}(z)$. By Equation (16.7), $w \in [L(k+1), U(k+1)]$ implies that

$$\tilde{f}(L(k) \curlyvee \bigwedge O_y(k)) \leq w \leq \tilde{f}(U(k) \curlywedge \bigvee O_y(k)). \tag{16.8}$$

In addition, we have that

$$\bigwedge O_y(k) \leq L(k) \curlyvee \bigwedge O_y(k)$$

and

$$U(k) \curlywedge \bigvee O_y(k) \leq \bigvee O_y(k).$$

Because the pair $(\tilde{\Sigma}, (\chi, \leq))$ is interval compatible, by virtue of the isomorphic property of $\tilde{f}$ (property (ii) of Definition 16.5.5), we have that

$$\tilde{f}(\bigwedge O_y(k)) \leq \tilde{f}(L(k) \curlyvee \bigwedge O_y(k))$$

and

$$\tilde{f}(U(k) \curlywedge \bigvee O_y(k)) \leq \tilde{f}(\bigvee O_y(k)).$$

This, along with relation (16.8), implies that

$$w \in [\tilde{f}(\bigwedge O_y(k)), \tilde{f}(\bigvee O_y(k))].$$

From this, using again the order isomorphic property of $\tilde{f}$, we deduce that there is $z \in O_y(k)$ such that $w = \tilde{f}(z)$. This with relation (16.8) implies that

$$L(k) \curlyvee \bigwedge O_y(k) \leq z \leq U(k) \curlywedge \bigvee O_y(k),$$

which in turn implies that $z \in [L(k), U(k)]$.

Proof of (iii): We proceed by contradiction. Thus, assume that for any k_0 there exists a $k \geq k_0$ such that $\{\alpha(k), \beta_k\} \subseteq [L(k), U(k)] \cap \mathcal{U}$ for some $\beta_k \neq \alpha(k)$ and $\beta_k \in \mathcal{U}$. By the proof of part (ii) we also have that β_k is such that $\beta_k = \tilde{f}(\beta_{k-1})$ for some $\beta_{k-1} \in [L(k-1), U(k-1)]$.

We want to show that in fact $\beta_{k-1} \in [L(k-1), U(k-1)] \cap \mathcal{U}$. If this is not the case, we can construct an infinite sequence $\{k_i\}_{i \in \mathbb{N}^+}$ such that $\beta_{k_i} \in [L(k_i), U(k_i)] \cap \mathcal{U}$ with $\beta_{k_i} = \tilde{f}(\beta_{k_i - 1})$ and $\beta_{k_i - 1} \in [L(k_i - 1), U(k_i - 1)] \cap (\chi - \mathcal{U})$. Notice that $|[L(k_1 - 1), U(k_1 - 1)] \cap (\chi - \mathcal{U})| = M < \infty$. Also, we have

$$|[L(k_1), U(k_1)] \cap (\chi - \mathcal{U})| < |[L(k_1 - 1), U(k_1 - 1)] \cap (\chi - \mathcal{U})|.$$

This is due to the fact that $\tilde{f}(\beta_{k_1 - 1}) \notin [L(k_1), U(k_1)] \cap (\chi - \mathcal{U})$, and to the fact that each element in $[L(k_1), U(k_1)] \cap (\chi - \mathcal{U})$ comes from one element in $[L(k_1 - 1), U(k_1 - 1)] \cap (\chi - \mathcal{U})$ (proof of (ii) and because $\mathcal{U}$ is invariant under $\tilde{f}$). Thus we have a strictly decreasing sequence of natural numbers $\{|[L(k_i - 1), U(k_i - 1)] \cap (\chi - \mathcal{U})|\}$ with initial value M. Since M is finite, we reach the contradiction that $|[L(k_i - 1), U(k_i - 1)] \cap (\chi - \mathcal{U})| < 0$ for some i. Therefore, $\beta_{k-1} \in [L(k-1), U(k-1)] \cap \mathcal{U}$.

Thus for any k_0 there is $k \geq k_0$ such that $\{\alpha(k), \beta_k\} \subseteq [L(k), U(k)] \cap \mathcal{U}$, with $\beta_k = f(\beta_{k-1})$ for some $\beta_{k-1} \in [L(k-1), U(k-1)] \cap \mathcal{U}$. Also, from the proof of part (ii) we have that $\beta_{k-1} \in O_y(k-1)$. As a consequence, there exists $\bar{k} > 0$ such that $\{\beta_{k-1}, z(k-1)\}_{k \geq \bar{k}} = \sigma_1$ and $\{\alpha(k-1), z(k-1)\}_{k \geq \bar{k}} = \sigma_2$ are two executions of Σ sharing the same output. This contradicts the observability assumption.

Corollary 16.5.7 *If the extended system $\tilde{\Sigma}$ of an observable system Σ is observable, then the estimator $\hat{\Sigma}$ given in Theorem 16.5.6 solves Problem 16.4.1 with $L(k) = U(k) = \alpha(k)$ for $k \geq k_0$.*

16.6 EXAMPLE: THE ROBOFLAG DRILL

The RoboFlag Drill has been described in Section 16.2. In this section, we revisit the example by showing that it is observable with measurable variables z, and by finding a lattice and a system extension that can be used for constructing the estimator proposed in Theorem 16.5.6. The system specification is given in Equations (16.1, 16.2, 16.3, 16.4). In particular, the rules in Equations (16.2, 16.3) model the function $h : \mathcal{U} \times \mathcal{Z} \to \mathcal{Z}$ that updates the continuous variables, and the rules in Equation (16.4) model the function $f : \mathcal{U} \times \mathcal{Z} \to \mathcal{U}$ that updates the discrete variables. In this example, we have $\mathcal{U} = \text{perm}(N)$ the set of permutations of N elements, and $\mathcal{Z} = \mathbb{R}^N$. Thus, the RoboFlag system is given by $\Sigma = (\text{perm}(N), \mathbb{R}^N, f, h)$, in which the variables $z \in \mathbb{R}^N$ are measured.

RoboFlag Drill Observation Problem. *Given initial values for x and y and the values of z corresponding to an execution of $\Sigma = (perm(N), \mathbb{R}^N, f, h)$, determine the value of α during that execution.*

Before constructing the estimator for the system $\Sigma = (\text{perm}(N), \mathbb{R}^N, f, h)$, we show in the following proposition that such a system is observable.

Proposition 16.6.1 *The system $\Sigma = (perm(N), \mathbb{R}^N, f, h)$ represented by the rules (16.2–16.3–16.4) with measurable variables z is observable.*

Proof. Given any two executions σ_1 and σ_2 of Σ, for proving observability, it is enough to show that if $\{\sigma_1(k)(\alpha)\}_{k\in\mathbb{N}} \neq \{\sigma_2(k)(\alpha)\}_{k\in\mathbb{N}}$, then $\{\sigma_1(k)(z)\}_{k\in\mathbb{N}} \neq \{\sigma_2(k)(z)\}_{k\in\mathbb{N}}$. Since the measurable variables are the z_is, their direction of motion is also measurable. Thus, we consider the vector of directions of motion of the z_i as output. Let $g(\sigma(k))$ denote such a vector at step k for the execution σ. It is enough to show that there is a k such that $g(\sigma_1(k)) \neq g(\sigma_2(k))$. Note that, in any execution of Σ, the α trajectory reaches the equilibrium value $[1, \ldots, N]$, and therefore there is a step $\bar{k}$ at which $f(\sigma_1(\bar{k})) = f(\sigma_2(\bar{k}))$ and $\sigma_1(\bar{k})(\alpha) \neq \sigma_2(\bar{k})(\alpha)$. As a consequence, the system is observable if $g(\sigma_1(\bar{k})) \neq g(\sigma_2(\bar{k}))$. Therefore it is enough to prove that for any $\alpha \neq \beta$, for $\alpha, \beta \in \mathcal{U}$, we have $g(\alpha, z) = g(\beta, v) \implies f(\alpha, z) \neq f(\beta, v)$, where $z, v \in \mathbb{R}^N$. Thus, $g(\alpha) = g(\beta)$ by (16.2–16.3) implies that (1) $z_i < x_{\alpha_i} \iff v_i < x_{\beta_i}$ and (2) $z_i \geq x_{\alpha_i} \iff v_i \geq x_{\beta_i}$. This implies that $x_{\alpha_i} \geq z_{i+1} \wedge x_{\alpha_{i+1}} \leq z_{i+1} \Leftrightarrow x_{\beta_i} \geq v_{i+1} \wedge x_{\beta_{i+1}} \leq v_{i+1}$. By denoting $\alpha' = f(\alpha, z)$ and $\beta' = f(\beta, v)$, we have that $(\alpha'_i, \alpha'_{i+1}) = (\alpha_{i+1}, \alpha_i) \Leftrightarrow (\beta'_i, \beta'_{i+1}) = (\beta_{i+1}, \beta_i)$. Hence if there exists an i such that $\alpha_i \neq \beta_i$, then there exists a j such that $\alpha'_j \neq \beta'_j$, and therefore $f(\alpha, z) \neq f(\beta, v)$.

16.6.1 RoboFlag Drill estimator

We next construct the estimator proposed in Theorem 16.5.6 in order to estimate and track the value of the assignment α in any execution. To accomplish this, we find a lattice $(\chi, \leq)$ in which to immerse the set $\mathcal{U}$ and an extension $\tilde{\Sigma}$ of the system Σ to χ, so that the pair $(\tilde{\Sigma}, (\chi, \leq))$ is interval compatible. We choose as χ the set of vectors in $\mathbb{N}^N$ with coordinates $x_i \in [1, N]$, that is,

$$\chi = \{x \in \mathbb{N}^N : x_i \in [1, N]\}. \tag{16.9}$$

For the elements in χ, we use the vector notation, that is, $x = (x_1, \ldots, x_N)$. The partial order that we choose on such a set is given by

$$\forall x, w \in \chi,\ x \leq w \text{ if } x_i \leq w_i\ \forall i. \tag{16.10}$$

As a consequence, the join and the meet between any two elements in χ are given by

$$\forall\, x, w\ \in \chi,\ v = x \curlyvee w \text{ if } v_i = \max\{x_i, w_i\},$$

$$\forall\, x, w\ \in \chi,\ v = x \curlywedge w \text{ if } v_i = \min\{x_i, w_i\}.$$

With this choice,we have $\bigvee \chi = (N, \ldots, N)$ and $\bigwedge \chi = (1, \ldots, 1)$. The pair $(\chi, \leq)$ with the order defined by (16.10) is clearly a lattice. The set $\mathcal{U}$ is the set of all permutations of N elements and it is a subset of χ. All of the elements in $\mathcal{U}$ form an anti-chain of the lattice, that is, any two elements of $\mathcal{U}$ are not related by the order $(\chi, \leq)$. In the reminder of this section, we denote by w the variables in χ not specifying whether w is in $\mathcal{U}$, and we denote by α the variables in $\mathcal{U}$. The function $h : \text{perm}(N) \times \mathbb{R}^N \to \mathbb{R}^N$ can be naturally extended to χ as

$$\begin{aligned} z_i(k+1) &= z_i(k) + \delta \text{ if } z_i(k) < x_{w_i(k)} \\ z_i(k+1) &= z_i(k) - \delta \text{ if } z_i(k) > x_{w_i(k)} \end{aligned} \tag{16.11}$$

for $w \in \chi$. The rules (16.11) specify $\tilde{h} : \chi \times \mathbb{R}^N \to \mathbb{R}^N$, and one can check that $\tilde{h}|_{\mathcal{U}\times\mathcal{Z}} = h$. In an analogous way $f : \text{perm}(N) \times \mathbb{R}^N \to \text{perm}(N)$ is extended to χ as

$$\begin{aligned} &(w_i(k+1), w_{i+1}(k+1)) = (w_{i+1}(k), w_i(k)) \\ &\text{if } x_{w_i(k)} \geq z_{i+1}(k) \wedge x_{w_{i+1}(k)} \leq z_{i+1}(k), \end{aligned} \tag{16.12}$$

for $w \in \chi$. The rules (16.12) model the function $\tilde{f} : \chi \times \mathbb{R}^N \to \chi$, and one can check that $\tilde{f}|_{\mathcal{U}\times\mathcal{Z}} = f$. Therefore, the system $\tilde{\Sigma} = (\chi, \mathbb{R}^N, \tilde{f}, \tilde{h})$ is the extended system of $\Sigma = (\text{perm}(N), \mathbb{R}^N, f, h)$ (see Definition 16.5.1). The following proposition shows that the pair $(\tilde{\Sigma}, (\chi, \leq))$ is interval compatible.

Proposition 16.6.2 *The pair $(\tilde{\Sigma}, (\chi, \leq))$, where $\Sigma = (\text{perm}(N), \mathbb{R}^N, f, h)$ is represented by the rules (16.2–16.3–16.4), and $(\chi, \leq)$ is given by (16.9)–(16.10), is interval compatible.*

Proof. We need to show properties (i) and (ii) of Definition 16.5.5. To simplify notation, we neglect the dependence of $\tilde{f}$ on its second argument.

Proof of property (i): By (16.11) we have that $T_{(z^1,z^2)}(\tilde{\Sigma})$ is not empty if for any i we have $z_i^2 = z_i^1 + \delta$, $z_i^2 = z_i^1 - \delta$, or $z_i^2 = z_i^1$. Thus

$$T_{(z^1,z^2)}(\tilde{\Sigma}) = \begin{cases} \{w \mid x_{w_i} > z_i^1, \}, & \text{if } z_i^2 = z_i^1 + \delta \\ \{w \mid x_{w_i} < z_i^1, \}, & \text{if } z_i^2 = z_i^1 - \delta \\ \{w \mid x_{w_i} = z_i^1, \}, & \text{if } z_i^2 = z_i^1. \end{cases} \tag{16.13}$$

Because we assumed that $x_i < z_i < x_{i+1}$, we have that $x_{w_i} > z_i$ if and only if $w_i > i$ and $x_{w_i} < z_i$ if and only if $w_i < i$. This, along with relations (16.13) and Definition 16.5.3, imply (i).

Proof of property (ii): To show that $\tilde{f} : T_i(\tilde{\Sigma}) \to [\tilde{f}(\bigwedge T_i(\tilde{\Sigma})), \tilde{f}(\bigvee T_i(\tilde{\Sigma}))]$ is an order isomorphism we show: (a) that it is onto; and (b) that it is order embedding. (a) To show that it is onto, we show directly that $f(T_i(\tilde{\Sigma})) = [\tilde{f}(\bigwedge T_i(\tilde{\Sigma})), \tilde{f}(\bigvee T_i(\tilde{\Sigma}))]$. We omit the dependence on $\tilde{\Sigma}$ to simplify notation. From the proof of (i), we deduce that the sets T_i are of the form $T_i = (T_{i,1}, \ldots, T_{i,N})$, with $T_{i,j} \in \{[1, j], [j+1, N], [j, j]\}$. Denote by $\tilde{f}(T_i)_j$ the jth coordinate set of $\tilde{f}(T_i)$. By Equation (16.12) we derive that $\tilde{f}(T_i)_j \in \{T_{i,j}, T_{i,j-1}, T_{i,j-1}\}$. We consider the case where $\tilde{f}(T_i)_j = T_{i,j-1}$; the other cases can be treated in analogous way. If $\tilde{f}(T_i)_j = T_{i,j-1}$ then $\tilde{f}(T_i)_{j-1} = T_{i,j}$. Denoting $\bigwedge T_i = l$ and $\bigvee T_i = u$, with $l = (l_1, \ldots, l_N)$ and $u = (u_1, \ldots, u_N)$, we have also that $\tilde{f}(l)_j = l_{j-1}$, $\tilde{f}(l)_{j-1} = l_j$, $\tilde{f}(u)_j = u_{j-1}$, $\tilde{f}(u)_{j-1} = u_j$. Thus, $\tilde{f}(T_i)_j = [\tilde{f}(l)_j, \tilde{f}(u)_j]$ for all j. This in turn implies that $\tilde{f}(T_i) = [\tilde{f}(l), \tilde{f}(u)]$, which is what we wanted to show. (b) To show that $\tilde{f} : T_i \to [\tilde{f}(\bigwedge T_i), \tilde{f}(\bigvee T_i)]$ is order embedding, it is enough to note again that

$\tilde{f}(\mathcal{T}_i)$ is obtained by switching $\mathcal{T}_{i,j}$ with $\mathcal{T}_{i,j+1}$, $\mathcal{T}_{i,j-1}$, or leaving it as $\mathcal{T}_{i,j}$. Therefore if $w \leq v$ for $w, v \in \mathcal{T}_i$ then $\tilde{f}(w) \leq \tilde{f}(v)$ since coordinate-wise we will compare the same numbers. By the same reasoning the reverse is also true, that is, if $\tilde{f}(w) \leq \tilde{f}(v)$ then $w \leq v$.

The estimator $\hat{\Sigma}$ given in Theorem 16.5.6 is thus represented by the following rules

$$l_i(k+1) = i+1 \text{ if } z_i(k+1) = z_i(k) + \delta \tag{16.14}$$

$$l_i(k+1) = 1 \text{ if } z_i(k+1) = z_i(k) - \delta \tag{16.15}$$

$$L_{i,y}(k+1) = \max\{L_i(k), l_i(k+1)\} \tag{16.16}$$

$$(L_i(k+1), L_{i+1}(k+1)) = (L_{i+1,y}(k+1), L_{i,y}(k+1))$$
$$\text{if } x_{L_{i,y}(k+1)} \geq z_{i+1}(k) \wedge x_{L_{i+1,y}(k+1)} \leq z_{i+1}(k) \tag{16.17}$$

$$u_i(k+1) = N \text{ if } z_i(k+1) = z_i(k) + \delta \tag{16.18}$$

$$u_i(k+1) = i \text{ if } z_i(k+1) = z_i(k) - \delta \tag{16.19}$$

$$U_{i,y}(k+1) = \min\{U_i(k), u_i(k+1)\} \tag{16.20}$$

$$(U_i(k+1), U_{i+1}(k+1)) = (U_{i+1,y}(k+1), U_{i,y}(k+1))$$
$$\text{if } x_{U_{i,y}(k+1)} \geq z_{i+1}(k) \wedge x_{U_{i+1,y}(k+1)} \leq z_{i+1}(k) \tag{16.21}$$

initialized with $L(0) = \bigwedge \chi$ and $U(0) = \bigvee \chi$. Rules (16.14–16.15) and (16.18–16.19) take the output information z and set the lower and upper bound of $O_y(k)$, respectively. Rules (16.16) and (16.20) compute the lower and upper bound of the intersection $[L(k), U(k)] \cap O_y(k)$, respectively. Finally, rules (16.17) and (16.21) compute the lower and upper bound of the set $\tilde{f}([L(k), U(k)] \cap O_y(k))$, respectively. Also note that the rules in (16.14–16.21) are obtained by 'copying' the rules in (16.12) and correcting them by means of the output information.

16.6.2 Complexity of the RoboFlag Drill estimator

The amount of computation required for updating L and U according to rules (16.14)–(16.21) is proportional to the amount of computation required for updating the variables α in system Σ. In fact, we have $2N$ rules, $2N$ variables, and $2N$ computations of 'max' and 'min' of values in $\mathbb{N}$. Therefore, the computational complexity of the algorithm that generates $L(k)$ and $U(k)$ is proportional to $2N$, which is of the same order as the complexity of the algorithm that generates the α trajectories. As established by property (iii) of Problem 16.4.1, the function of k given by $|[L(k), U(k)] \cap \mathcal{U} - \alpha(k)|$ tends to zero. This function is useful for analysis purposes, but it is not necessary to compute it at any point in the estimation algorithm proposed in equation (16.14–16.21). However, since $L(k)$ does not converge to $U(k)$ once the algorithm has converged, i.e., when $|[L(k), U(k)] \cap \mathcal{U}| = 1$, we cannot find the value of $\alpha(k)$ from the values of $U(k)$ and $L(k)$ directly. Instead of computing directly $[L(k), U(k)] \cap \mathcal{U}$, we carry out a simple algorithm, which in the case of the RoboFlag Drill example takes at most $(N^2 + N)/2$ steps and takes as inputs $L(k)$ and $U(k)$ and gives as output $\alpha(k)$ if the algorithm has converged. This is formally explained in the following paragraph.

(Refinement algorithm) Let $c_i = [L_i, U_i]$. Then the algorithm $(m_1, \ldots, m_N) = Refine(c_1, \ldots, c_N)$, which takes assignment sets $c_1, \ldots, c_N$ and produces assignment sets $m_1, \ldots, m_N$, is such that if $m_i = \{k\}$ then $k \notin m_j$ for any $j \neq i$.

This algorithm takes as input the sets c_i and removes singletons occurring at one coordinate set from all of the other coordinate sets. By construction, it follows that $m_i \subseteq c_i$. It does this iteratively: if in the process of removing one singleton, a new one is created in some other coordinate set, then such a singleton is also removed from all of the other coordinate sets. The refinement algorithm has two useful properties. First, the sets m_i are equal to the α_i when $[L, U] \cap \mathcal{U} = \alpha$. Second, the cardinality of the sets $m_i(k)$ is non-increasing with the time step k. As a final remark, note that the sets m_i are not used in the update laws of the estimation algorithm, they are just computed at each step from the set $[L, U]$ in order to extract α from it when the algorithm has converged.

16.6.3 Simulation results

The RoboFlag Drill system represented in the rules (16.2), (16.3), and (16.4) has been implemented in Matlab together with the estimator reported in the rules (16.14–16.21). Figure 16.4 shows the behavior of the quantities $E(k) = \frac{1}{N}\sum_{i=1}^{N} |\alpha_i(k) - i|$ and $W(k) = \frac{1}{N}\sum_{i=1}^{N} |m_i(k)|$. The function $E(k)$ is a function of α, and it is not increasing along the executions of the system $\Sigma = (\text{perm}(N), \mathbb{R}^N, f, h)$. This quantity is showing the rate of convergence of the α assignment to its equilibrium $(1, \ldots, N)$. We show in Figure 16.4 the logarithm of $E(k)$ and the logarithm of $W(k)$, which is non-increasing and converging to one, that is, the sets $(m_1(k), \ldots, m_N(k))$ converge to $\alpha(k) = (\alpha_1(k), \ldots, \alpha_N(k))$. In the same figure, we notice that when $W(k)$ converges to one, $E(k)$ has not converged to zero yet. This shows that the estimator is faster than the dynamics of the system under study.

16.7 EXISTENCE OF DISCRETE STATE ESTIMATORS ON A LATTICE

In this section, we show that if the system $\Sigma = (\mathcal{U}, \mathcal{Z}, f, h)$ is observable, there always exists a lattice $(\chi, \leq)$ such that the pair $(\tilde{\Sigma}, (\chi, \leq))$ is interval compatible. The size of the set χ is singled out as a cause of complexity, and a worst case size is computed. In particular, the worst case size of the lattice never exceeds the size of the observer tree proposed in Caines *et al.* (1991). For systems in which $\mathcal{U}$ can be immersed in a space equipped with algebraic properties, as is the case for the RoboFlag Drill, a preferred lattice

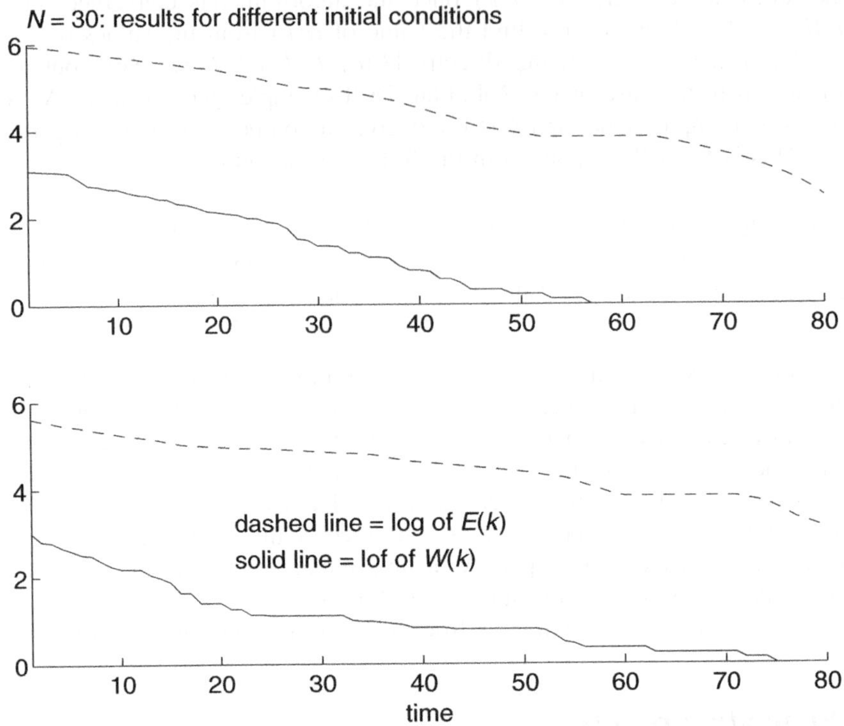

Figure 16.4 Example with $N = 30$: note that the function $W(k)$ is always non-increasing, and its logarithm is converging to zero.

structure $(\chi, \leq)$ exists where joins and meets can be efficiently computed and represented by exploiting the algebra. For these systems, the estimation methodology proposed in this work reduces complexity with respect to enumeration methods. Consider the deterministic transition system $\Sigma = (\mathcal{U}, \mathcal{Z}, f, h)$. In order to show the link between the original system and its extension, it is useful to define the Σ-transition sets and the Σ-transition classes as defined for the extended system $\tilde{\Sigma} = (\mathcal{U}, \mathcal{Z}, \tilde{f}, \tilde{h})$ in Definition 16.5.2 and Definition 16.5.3, respectively.

Definition 16.7.1 *The non-empty sets* $T_{(z^1,z^2)}(\Sigma) = \{\alpha \in \mathcal{U} \mid z^2 = h(\alpha, z^1)\}$, *for* $z^1, z^2 \in \mathcal{Z}$, *are named the* Σ*-transition sets.*

Definition 16.7.2 *The set* $\mathcal{T}(\Sigma) = \{\mathcal{T}_1(\Sigma), \ldots, \mathcal{T}_M(\Sigma)\}$, *with* $\mathcal{T}_i(\Sigma)$ *such that*

(i) *for any* $\mathcal{T}_i(\Sigma) \in \mathcal{T}(\Sigma)$ *there is* $(z^1, z^2) \in \mathcal{Z}$ *such that* $\mathcal{T}_i(\Sigma) = T_{(z^1,z^2)}(\Sigma)$;

(ii) *for any* $z^1, z^2 \in \mathcal{Z}$ *for which* $T_{(z^1,z^2)}(\Sigma)$ *is not empty, there is* $j \in \{1, \ldots, M\}$ *such that* $T_{(z^1,z^2)}(\Sigma) = \mathcal{T}_j$;

is the set of Σ*-transition classes.*

Note that the set $\mathcal{T}(\Sigma)$ is finite as we assumed at the beginning that the set $\mathcal{U}$ is finite. Each Σ-transition set $T_{(z_1,z^2)}(\Sigma)$ contains all of α values in $\mathcal{U}$ that allow the transition from z^1 to z^2 through the function h. Note also that for any $z^1, z^2 \in \mathcal{Z}$ we have $T_{(z^1,z^2)}(\Sigma) \subseteq T_{(z^1,z^2)}(\tilde{\Sigma})$ because $\tilde{h}|_{\mathcal{U}\times\mathcal{Z}} = h$ and $\mathcal{U} \subseteq \chi$. This in turn implies that $\mathcal{T}_i(\Sigma) \subseteq \mathcal{T}_i(\tilde{\Sigma})$.

Lemma 16.7.3 *Consider the deterministic transition system $\Sigma = (\mathcal{U}, \mathcal{Z}, f, h)$. If Σ is observable, then $f : (\mathcal{T}_j(\Sigma), z) \to f(\mathcal{T}_j(\Sigma), z)$ is one to one for any $j \in \{1, .., M\}$ and for any $z \in \mathcal{Z}$.*

Proof. We have to show that if $\alpha_a \neq \alpha_b$ and $\alpha_a, \alpha_b \in \mathcal{T}_j(\Sigma)$ for some j, then $f(\alpha_a, z) \neq f(\alpha_b, z)$. Suppose instead that $f(\alpha_a, z) = f(\alpha_b, z)$, this means that the two executions σ_a, σ_b starting at $\sigma_a(0) = (\alpha_a, z)$ and $\sigma_b(0) = (\alpha_b, z)$ have the same output sequence, but they are different. This means that they are not distinguishable, and therefore the system is not observable. This contradicts the assumption.

This lemma is used in the following theorem, which shows the link between observability and extensibility of the system Σ into a system $\tilde{\Sigma}$ that is interval compatible with a lattice $(\chi, \leq)$.

Theorem 16.7.4 *Consider the deterministic transition system $\Sigma = (\mathcal{U}, \mathcal{Z}, f, h)$. If Σ is observable, then there exists a complete lattice $(\chi, \leq)$ with $\mathcal{U} \subseteq \chi$, such that the extension $\tilde{\Sigma} = (\tilde{f}, \tilde{h}, \chi, \mathcal{Z})$ of Σ on χ is such that $(\tilde{\Sigma}, (\chi, \leq))$ is interval compatible.*

Proof. We show the existence of a lattice $(\chi, \leq)$ and of an extended system $\tilde{\Sigma} = (\chi, \mathcal{Z}, \tilde{f}, \tilde{h})$ with $(\tilde{\Sigma}, (\chi, \leq))$ an interval compatible pair by construction. Define $\chi := \mathcal{P}(\mathcal{U})$, and $(\chi, \leq) := (\mathcal{P}(\mathcal{U}), \subseteq)$. To define $\tilde{h}$, define the sublattices $(\mathcal{T}_i(\tilde{\Sigma}), \leq)$ of $(\chi, \leq)$ for $i \in \{1, \ldots, M\}$, by $(\mathcal{T}_i(\tilde{\Sigma}), \leq) := (\mathcal{P}(\mathcal{T}_i(\Sigma)), \subseteq)$ as shown in Figure 16.5. As a consequence, for any given $z^1, z^2 \in \mathcal{Z}$ such that $z^2 = h(\alpha, z^1)$ for $\alpha \in \mathcal{T}_i(\Sigma)$ for some i, we define $z^2 = \tilde{h}(w, z^1)$ for any $w \in \mathcal{T}_i(\tilde{\Sigma})$. Clearly, $\tilde{h}|_{\mathcal{U}\times\mathcal{Z}} = h$, and $\mathcal{T}_i(\tilde{\Sigma})$ for any i is an

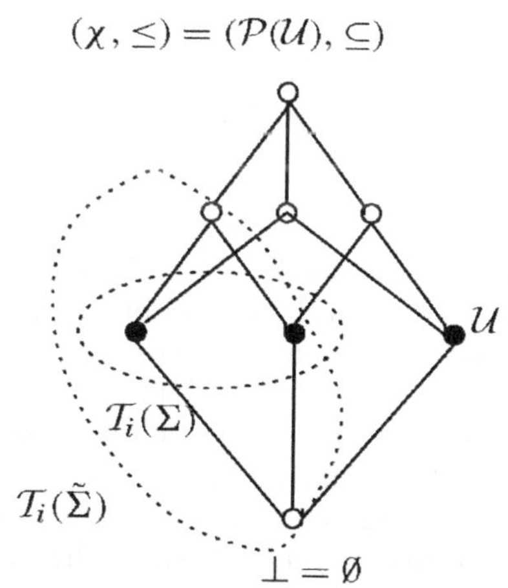

Figure 16.5 Example of $(\mathcal{P}(\mathcal{U}), \subseteq)$ with $\mathcal{U}$ composed by three elements (solid circles).

interval sublattice of the form $\mathcal{T}_i(\tilde{\Sigma}) = [\bot, \bigvee \mathcal{T}_i(\tilde{\Sigma})]$. The function $\tilde{f}$ is defined in the following way. For any $x, w \in \chi$ and $\alpha \in \mathcal{U}$ we define

$$\begin{cases} \tilde{f}(x \curlyvee w) & := \tilde{f}(x) \curlyvee \tilde{f}(w) \\ \tilde{f}(x \curlywedge w) & := \tilde{f}(x) \curlywedge \tilde{f}(w) \\ \tilde{f}(\bot) & := \bot \\ \tilde{f}(\alpha) & := f(\alpha), \end{cases} \tag{16.22}$$

in which we have omitted the dependence on z to simplify notation. We prove first that $\tilde{f} : \mathcal{T}_i(\tilde{\Sigma}) \to [\bot, \tilde{f}(\bigvee \mathcal{T}_i(\tilde{\Sigma}))]$ is onto. We have to show that for any $w \in [\bot, \tilde{f}(\bigvee \mathcal{T}_i(\tilde{\Sigma}))]$, with $w \neq \bot$, there is $x \in [\bot, \bigvee \mathcal{T}_i(\tilde{\Sigma})]$ such that $w = \tilde{f}(x)$. Since $\bigvee \mathcal{T}_i(\tilde{\Sigma}) = \alpha_1 \curlyvee \ldots \curlyvee \alpha_p$ for $\{\alpha_1, \ldots, \alpha_p\} = \mathcal{T}_i(\Sigma)$, we have also that $\tilde{f}(\bigvee \mathcal{T}_i(\tilde{\Sigma})) = f(\alpha_1) \curlyvee \ldots \curlyvee f(\alpha_p)$ by virtue of equations (16.22). Because $w \leq \tilde{f}(\bigvee \mathcal{T}_i(\tilde{\Sigma}))$, we have that $w = f(\alpha_{j_1}) \curlyvee \cdots \curlyvee f(\alpha_{j_m})$ for $j_k \in \{1, \ldots, p\}$ and $m < p$. This in turn implies, by Equation (16.22), that $w = \tilde{f}(\alpha_{j_1} \curlyvee \cdots \curlyvee \alpha_{j_m})$. Since $x := \alpha_{j_1} \curlyvee \cdots \curlyvee \alpha_{j_m} < \bigvee \mathcal{T}_i(\tilde{\Sigma})$, we have proved that $w = \tilde{f}(x)$ for $x \in \mathcal{T}_i(\tilde{\Sigma})$. Second, we notice that $\tilde{f} : \mathcal{T}_i(\tilde{\Sigma}) \to [\bot, \tilde{f}(\bigvee \mathcal{T}_i(\tilde{\Sigma}))]$ is one to one because of Lemma 16.7.3. Thus, we have proved that $\tilde{f} : \mathcal{T}_i(\tilde{\Sigma}) \to [\bot, \tilde{f}(\bigvee \mathcal{T}_i(\tilde{\Sigma}))]$ is a bijection, and by equations (16.22) it is also a homomorphism. We then apply Proposition 16.3.3 to obtain the result.

Theorem 16.7.4 shows that an observable system admits a lattice and a system extension that satisfy interval compatibility by constructing them in a way similar to the way one shows that a stable dynamical system has a Lyapunov function. However, the lattice constructed in the proof of the theorem is impractical for the implementation of the estimator of Theorem 16.5.6 when the size of $\mathcal{U}$ is large because the size of the representation of the elements of χ is large as well. However, one does not need to have $\chi = \mathcal{P}(\mathcal{U})$, but it is enough to have in χ the elements that the estimator needs. With this consideration, the following proposition computes the worst case size of χ.

Proposition 16.7.5 *Consider the system $\Sigma = (\mathcal{U}, \mathcal{Z}, f, h)$, with $f : \mathcal{U} \to \mathcal{U}$. Assume that the sets $\{\mathcal{T}_1(\Sigma), \ldots, \mathcal{T}_m(\Sigma)\}$ are all disjoint. Then there exist an extension $\tilde{\Sigma} = (\chi, \mathcal{Z}, \tilde{f}, \tilde{h})$ with $|\chi| \leq 2|\mathcal{U}|^2$.*

For the proof, see (DelVecchio and Murray 2004).

The size of χ gives an idea of how many values of joins and meets need to be stored. In the case of the RoboFlag example with $N = 4$ robots per team, the size of $\mathcal{P}(\mathcal{U})$ is 16778238, while the worst case size given in Proposition 16.7.5 is 576, and the size of the lattice χ proposed in Section 16.6.1 is $4^4 = 256$. Thus the estimate given by Proposition 16.7.5 significantly reduces the size of χ given by $\mathcal{P}(\mathcal{U})$. Note that the size of the lattice proposed in Section 16.6.1 is smaller than 576 because there are pairs of elements that have the same join, for example, the pairs $(3, 1, 4, 2)$, $(4, 2, 1, 3)$, and $(4, 2, 1, 3)$, $(2, 1, 4, 3)$ have the same join, that is, $(4, 2, 4, 3)$. In the next section, we extend the result of Theorem 16.5.6 to the estimation of continuous and discrete variables in case an estimator in cascade form is possible.

16.8 EXTENSIONS TO THE ESTIMATION OF DISCRETE AND CONTINUOUS VARIABLES

In this section, we consider systems in which the continuous variables are not directly measured as it was in the previous sections. As an example, consider the RoboFlag Drill, in which now the blue robots move according to a second order dynamics, of which only the position is measured. In order to formally define the structure of the systems under study, we define the feedback interconnection of two systems with input.

Definition 16.8.1 *Consider the two systems with input* $\Sigma_1 = (S_1, \mathcal{I}_1, \mathcal{Y}_1, F_1, g_1)$ *and* $\Sigma_2 = (S_2, \mathcal{I}_2, \mathcal{Y}_2, F_2, g_2)$*, in which* $\mathcal{I}_1 = \mathcal{Y}_2$, $\mathcal{I}_2 = \mathcal{Y}_1$*, and* $g_1 : S_1 \to \mathcal{Y}_1$*. The feedback interconnection of* Σ_1 *with* Σ_2*, denoted by* $\Sigma_1 \circ_f \Sigma_2$*, is the deterministic transition system given by* $\Sigma_1 \circ_f \Sigma_2 := (S_1 \times S_2, \mathcal{Y}_2, (F_1', F_2'), g_2')$*, in which for any* $s_1 \in S_1$ *and any* $s_2 \in S_2$*, we have* $F_1'(s_1, s_2) = F_1(s_1, g_2(s_2, g_1(s_1)))$*,* $F_2'(s_1, s_2) = F_2(s_2, g_1(s_1))$*, and* $g_2'(s_1, s_2) = g_2(s_2, g_1(s_1))$*.*

The output of the feedback interconnection $\Sigma_1 \circ_f \Sigma_2$ is the output of Σ_2. Let $\mathcal{U}$ be a finite discrete set, $\mathcal{Z}$ an infinite possibly dense set, and $\mathcal{Y}$ a finite or infinite set. In this section, we consider systems that are the feedback interconnection of a system that updates the discrete variable dynamics and of a system that updates the continuous variables dynamics. These systems have the form $\Sigma = \Sigma_1 \circ_f \Sigma_2$, in which $\Sigma_1 = (\mathcal{U}, \mathcal{Y}, \mathcal{U}, f, \mathrm{id}_1)$ and $\Sigma_2 = (\mathcal{Z}, \mathcal{U}, \mathcal{Y}, h, g)$. We have denoted by id_1 the function that attaches s to a pair $(s, u) \in S \times \mathcal{I}$. Thus, we will have that $\Sigma_1 \circ_f \Sigma_2 = (\mathcal{U} \times \mathcal{Z}, \mathcal{Y}, (f', h'), g')$, in which for any $\alpha, z \in \mathcal{U} \times \mathcal{Z}$ we have that $f'(\alpha, z) = f(\alpha, g(z, \alpha))$, $h'(\alpha, z) = h(z, \alpha)$, and $g'(\alpha, z) = g(z, \alpha)$. We represent these systems by means of the following update equations

$$\alpha(k+1) = f(\alpha(k), y(k)) \tag{16.23}$$

$$z(k+1) = h(z(k), \alpha(k)) \tag{16.24}$$

$$y(k) = g(z(k), \alpha(k)).$$

The Σ-*transition sets* (Definition 16.7.1) take the form

$$T_{(y_1, y_2)}(\Sigma) = \{\alpha \in \mathcal{U} \mid \exists\, z \in \mathcal{Z} \text{ such that } y_1 = g(z, \alpha) \text{ and } y_2 = g(f(\alpha, y_1), h(z, \alpha))\}$$

with $y_1, y_2 \in \mathcal{Y}$. We denote the property that allows to distinguish two different initial values of the discrete state α of the Σ independently of the continuous state z by independent discrete state observability.

Definition 16.8.2 *The system* $\Sigma = \Sigma_1 \circ_f \Sigma_2$ *is said to be independent discrete state observable if it is observable and for any output sequence* $\{y(k)\}_{k \in \mathbb{N}}$*, we have that for any two executions* $\sigma_1, \sigma_2 \in \mathcal{E}(\Sigma)$ *such that* $\{\sigma_1(k)(\alpha)\}_{k \in \mathbb{N}} \neq \{\sigma_2(k)(\alpha)\}_{k \in \mathbb{N}}$*, there is* $k > 0$ *such that* $\sigma_1(k)(\alpha) \in T_{(y(k), y(k+1))}(\Sigma)$ *and* $\sigma_2(k)(\alpha) \notin T_{(y(k), y(k+1))}(\Sigma)$*.*

This property will allow to construct a discrete-continuous state estimator that is a cascade interconnection of a discrete state estimator as we have developed in Section 16.5, and a continuous state estimator. Before introducing such an estimator, we define the cascade interconnection of two systems with input.

Definition 16.8.3 *Consider the two systems with input* $\Sigma_1 = (S_1, \mathcal{I}_1, \mathcal{Y}_1, F_1, g_1)$ *and* $\Sigma_2 = (S_2, \mathcal{I}_2, \mathcal{Y}_2, F_2, g_2)$*, in which* $\mathcal{I}_2 = \mathcal{Y}_1$*. The cascade interconnection, denoted* $\Sigma_1 \circ_c \Sigma_2$*, is the deterministic transition system with input given by* $\Sigma_1 \circ_c \Sigma_2 := (S_1 \times S_2, \mathcal{I}_1, \mathcal{Y}_2, (F_1', F_2'), g_2')$*, such that for any* $s_1 \in S_1$*,* $s_2 \in S_2$ *and* $u_1 \in \mathcal{I}_1$ *we have that* $F_1'(s_1, s_2, u_1) = F_1(s_1, u_1)$*,* $F_2'(s_1, s_2, u_1) = F_2(s_2, g_1(s_1, u_1))$ *and* $g_2'(s_1, s_2, u_1) = g_2(s_2, g_1(s_1, u_1))$*.*

Consider the deterministic transition system $\Sigma = \Sigma_1 \circ_f \Sigma_2$, with output sequence $\{y(k)\}_{k\in\mathbb{N}}$. From the measurement of the output sequence, we construct a cascade state estimator: A system $\hat{\Sigma} = \hat{\Sigma}_1 \circ_c \hat{\Sigma}_2$, in which $\hat{\Sigma}_1$ takes as input the values of the output of Σ and asymptotically tracks the value of the variables α, while $\hat{\Sigma}_2$ takes as input the discrete state estimates and the output of Σ to asymptotically track the value of z. The cascade state estimation problem is formally introduced as follows.

Problem 16.8.1 *(Cascade state estimator) Given the deterministic transition system* $\Sigma = \Sigma_1 \circ_f \Sigma_2$*, find the cascade interconnection* $\hat{\Sigma} = \hat{\Sigma}_1 \circ_c \hat{\Sigma}_2$*, in which* $\hat{\Sigma}_1$ *is as given in Theorem 16.5.6 and* $\hat{\Sigma}_2 = (\mathcal{L} \times \mathcal{L}, \chi \times \chi \times \mathcal{Y} \times \mathcal{Y}, \chi \times \chi \times \mathcal{Z}_E \times \mathcal{Z}_E, (f_3, f_4), (g_1, g_2, g_3, g_4))$ *where* $f_3 : \mathcal{L} \times \chi \times \chi \times \mathcal{Y} \times \mathcal{Y} \to \mathcal{L}$*,* $f_4 : \mathcal{L} \times \chi \times \chi \times \mathcal{Y} \times \mathcal{Y} \to \mathcal{L}$*,* $g_1 : \chi \to \chi$*,* $g_2 : \chi \to \chi$*,* $g_1 = g_2 = id$*,* $g_3 : \mathcal{L} \to \mathcal{Z}_E$*, and* $g_4 : \mathcal{L} \to \mathcal{Z}_E$*,* $\mathcal{U} \subseteq \chi$*,* $(\chi, \leq)$ *a lattice,* $\mathcal{Z} \subseteq \mathcal{Z}_E$ *with* $(\mathcal{Z}_E, \leq)$ *a lattice,* $\chi \times \mathcal{Z}_E \subseteq \mathcal{L}$*,* $(\mathcal{L}, \leq)$ *a lattice, such that for any execution* $\sigma = \{(\alpha(k), z(k))\}_{k\in\mathbb{N}}$ *of* Σ *with output sequence* $\{y(k)\}_{k\in\mathbb{N}}$ *the update laws*

$$
\begin{aligned}
L(k+1) &= f_1(L(k), y(k), y(k+1)) \\
U(k+1) &= f_2(U(k), y(k), y(k+1)) \\
q_L(k+1) &= f_3(q_L(k), L(k), U(k), y(k), y(k+1)) \\
q_U(k+1) &= f_4(q_U(k), L(k), U(k), y(k), y(k+1)),
\end{aligned} \tag{16.25}
$$

in which $L(0) := \bigwedge \chi$*,* $U(0) := \bigvee \chi$*,* $q_L(0) = \bigwedge \mathcal{L}$*,* $q_U(0) = \bigvee \mathcal{L}$*, and* $z_L(k) = g_3(q_L(k))$*, and* $z_U(k) = g_4(q_U(k))$*, have properties (i)–(iii) of Problem 16.4.1 and*

(i') $z_L(k) \leq z(k) \leq z_U(k)$ (correctness);

(ii') $d(z_L(k), z_U(k)) \leq \gamma(|[L(k), U(k)]|)$, with γ a monotonically decreasing function of its argument (non-increasing error);

(iii') there exists $k_0' > 0$ such that $d(z_{L'}(k), z_{U'}(k)) = 0$ for any $k \geq k_0'$, where $L'(k) = \bigwedge([L(k), U(k)] \cap \mathcal{U})$, $U'(k) = \bigvee([L(k), U(k)] \cap \mathcal{U})$, and

$$
\begin{aligned}
q_{L'}(k+1) &= f_3(q_{L'}(k), L'(k), U'(k), y(k), y(k+1)) \\
q_{U'}(k+1) &= f_4(q_{U'}(k), L'(k), U'(k), y(k), y(k+1))
\end{aligned}
$$

$$z_{L'}(k) = \bigwedge g_3\left([q_{L'}(k), q_{U'}(k)] \cap (\mathcal{U} \times \mathcal{Z})\right) \tag{16.26}$$

$$z_{U'}(k) = \bigvee g_4\left([q_{L'}(k), q_{U'}(k)] \cap (\mathcal{U} \times \mathcal{Z})\right) \tag{16.27}$$

with $q_{L'}(0) = q_L(0)$ and $q_{U'}(0) = q_U(0)$, for some distance function 'd' (convergence).

□

The variables L and U have the same meaning as they had in Section 16.5. The variables z_L and z_U instead represent the lower and upper bounds in $(\mathcal{Z}_E, \leq)$ of the set of all possible continuous variable values that are compatible with the output sequence, with the system dynamics established by Σ, and with the set of possible discrete variable values. The variables q_L and q_U are auxiliary variables that are needed to model the coupling of the continuous and discrete state dynamics. They represent the lower and upper bounds in $(\mathcal{L}, \leq)$ of the set of all possible pairs (α, z) compatible with the output sequence, with the system dynamics, and with the set of possible discrete variable values. The distance function 'd' has been left unspecified for the moment, as its form depends on the particular partial order chosen $(\mathcal{Z}_E, \leq)$. In the case in which $\mathcal{Z}_E = \mathcal{Z}$ and $\mathcal{Z} = \mathbb{R}^n$ with the order established component-wise, the distance can be the classical euclidean distance, for example. To determine a solution to Problem 16.8.1, we introduce the notion of extension of a system to a joint continuous-discrete partial order.

Definition 16.8.4 *Consider the system $\Sigma = \Sigma_1 \circ_f \Sigma_2$ with $\Sigma_1 = (\mathcal{U}, \mathcal{Y}, \mathcal{U}, f, id_1)$ and $\Sigma_2 = (\mathcal{Z}, \mathcal{U}, \mathcal{Y}, h, g)$. Let $(\chi, \leq)$, $(\mathcal{Z}_E, \leq)$, and $(\mathcal{L}, \leq)$ be partial orders with $\mathcal{U} \subseteq \chi$, $\mathcal{Z} \subseteq \mathcal{Z}_E$, and $\chi \times \mathcal{Z}_E \subseteq \mathcal{L}$. The system extension is defined as $\tilde{\Sigma} = (\mathcal{L}, \mathcal{Y}, \tilde{F}, \tilde{G})$, in which*

(i) $\tilde{F} : \mathcal{L} \to \mathcal{L}$ *with* $\tilde{F}|_{\mathcal{U}\times\mathcal{Z}} = (f', h')$ and $\mathcal{L} - (\mathcal{U} \times \mathcal{Z})$ *is invariant under* $\tilde{F}$;

(ii) $\tilde{G} : \mathcal{L} \to \mathcal{Y}$ *with* $\tilde{G}|_{\mathcal{U}\times\mathcal{Z}} = g'$;

(iii) $\tilde{\Sigma}|_{\chi\times\mathcal{Z}_E} = \tilde{\Sigma}_1 \circ_f \tilde{\Sigma}_2$, in which $\tilde{\Sigma}_1 = (\chi, \mathcal{Y}, \chi, \tilde{f}, \mathrm{id}_1)$ *and* $\tilde{\Sigma}_2 = (\mathcal{Z}_E, \chi, \mathcal{Y}, \tilde{h}, \tilde{g})$, *with* $\tilde{f}|_{\mathcal{U}\times\mathcal{Y}} = f$, $\tilde{h}|_{\mathcal{Z}\times\mathcal{U}} = h$, *and* $\tilde{g}|_{\mathcal{Z}\times\mathcal{U}} = g$;

(iv) *the partial order $(\mathcal{L}, \leq)$ is closed with respect to $\chi \times \mathcal{Z}_E$.*

The $\tilde{\Sigma}$-transition sets thus take the following form. For any $y_1, y_2 \in \mathcal{Y}$, $T_{(y_1,y_2)}(\tilde{\Sigma}) = \{w \in \chi \mid \exists\, z \in \mathcal{Z}$ such that $y_2 = \tilde{g}(\tilde{F}(w, z))$ and $y_1 = \tilde{g}(w, z)\}$. The $\tilde{\Sigma}$-transition sets correspond to the set of all possible values of $w \in \chi$ compatible with two consecutive outputs of the extended system $\tilde{\Sigma}$. The *output set* is again given by $O_y(k) := T_{(y(k),y(k+1))}(\tilde{\Sigma})$. The next definition introduces the notion of interval compatibility of the tuple $(\tilde{\Sigma}_1, \tilde{\Sigma}, (\chi, \leq))$.

Definition 16.8.5 *The tuple $(\tilde{\Sigma}_1, (\chi, \leq))$ is said to be interval compatible if the following are verified*

(i) for any $y_1, y_2 \in \mathcal{Y}$, we have that $T_{y_1,y_2}(\tilde{\Sigma}) = [\wedge T_{y_1,y_2}(\tilde{\Sigma}), \vee T_{y_1,y_2}(\tilde{\Sigma})]$;

(ii) the extension $\tilde{\Sigma}_1$ is such that $\tilde{f} : (T_{y_1,y_2}(\tilde{\Sigma}), y_1) \to [\tilde{f}(\bigwedge T_{y_1,y_2}(\tilde{\Sigma}), y_1), \tilde{f}(\bigvee T_{y_1,y_2}(\tilde{\Sigma}), y_1)]$ is an order isomorphism.

In order to determine the set of variable values in $\mathcal{L}$ of the extended system that are compatible with an output pair y_1, y_2 and with a set of possible discrete variable values $[w_1, w_2] \subseteq T_{y_1,y_2}(\tilde{\Sigma})$, we introduce the notion of induced output set.

Definition 16.8.6 *Consider the system $\tilde{\Sigma} = (\mathcal{L}, \mathcal{Y}, \tilde{F}, \tilde{g})$ and a transition set $T_{(y_1,y_2)}(\tilde{\Sigma})$ for some $y_1, y_2 \in \mathcal{Y}$. For any $w_1, w_2 \in T_{(y_1,y_2)}(\tilde{\Sigma})$ with $w_1 \leq w_2$, the sets*

$$I_{(y_1,y_2)}^{[w_1,w_2]} = \{q \in \mathcal{L} \mid \pi_1 \circ a_L(q) \geq w_1,\ \pi_1 \circ a_U(q) \leq w_2,\ y_2 = \tilde{g}(\tilde{F}(q)),\ \textit{and } y_1 = \tilde{g}(q)\}$$

are named the induced output sets of $\tilde{\Sigma}$ induced by an interval $[w_1, w_2] \subseteq T_{y_1,y_2}(\tilde{\Sigma})$.

The meaning of an induced output set is the following. The set $I^{[w_1,w_2]}_{(y_1,y_2)}$ is the set of all possible values of $q \in \mathcal{L}$ that are compatible with two output measurements y_1, y_2 *and* whose upper and lower approximations in $\chi \times \mathcal{Z}_E$ have the discrete component contained in the set $[w_1, w_2]$. One can easily verify that if $[w_1, w_2] \subseteq T_{y_1,y_2}(\tilde{\Sigma})$, then $\{(\alpha, z) \mid g(\alpha, z) = y_1 \text{ and } g(f(\alpha, y_1), h(\alpha, z)) = y_2\}$ with $\alpha \in [w_1, w_2]$ is contained in $I^{[w_1,w_2]}_{(y_1,y_2)}$. Next, a definition similar to interval compatibility is introduced for the induced output sets.

Definition 16.8.7 *The pair $(\tilde{\Sigma}, (\mathcal{L}, \leq))$ is said to be induced interval compatible if $(\tilde{\Sigma}_1, (\chi, \leq))$ is interval compatible and for any $[w_1, w_2] \subseteq T_{y_1,y_2}(\tilde{\Sigma})$ for $y_1, y_2 \in \mathcal{Y}$, we have that*

(i) $\tilde{F} : ([\bigwedge I^{[w_1,w_2]}_{(y_1,y_2)}, \bigvee I^{[w_1,w_2]}_{(y_1,y_2)}]) \to [\tilde{F}(\bigwedge I^{[w_1,w_2]}_{(y_1,y_2)}), \tilde{F}(\bigvee I^{[w_1,w_2]}_{(y_1,y_2)})]$ *is order preserving;*

(ii) $\tilde{F} : ([\bigwedge I^{[\alpha,\alpha]}_{(y_1,y_2)}, \bigvee I^{[\alpha,\alpha]}_{(y_1,y_2)}]) \to [\tilde{F}(\bigwedge I^{[\alpha,\alpha]}_{(y_1,y_2)}), \tilde{F}(\bigvee I^{[\alpha,\alpha]}_{(y_1,y_2)})]$ *is an order isomorphism;*

(iii) for any $[w_1, w_2] \subseteq T_{y_1,y_2}(\tilde{\Sigma})$, *we have that* $d\left(\pi_2 \circ a_L \circ \tilde{F}(\bigwedge I^{[w_1,w_2]}_{(y_1,y_2)}), \pi_2 \circ a_U \circ \tilde{F}(\bigvee I^{[w_1,w_2]}_{(y_1,y_2)})\right) \leq \gamma(|[w_1, w_2]|)$, *for some distance function 'd,' and* $\gamma : \mathbb{N} \to \mathbb{R}$ *a monotonic function of its argument.*

Item (i) and (ii) of this definition require that the extended function $\tilde{F}$ has order preserving properties on the induced output sets. Item (iii) establishes that the distance between the lower and upper bounds of the interval sublattice in $(\mathcal{Z}_E, \leq)$ induced by an interval $[w_1, w_2] \in \chi$ is bounded by a monotonic function of the size of $[w_1, w_2]$. When $(y_1, y_2) = (y(k), y(k+1))$ in the above definitions, in which $\{y(k)\}_{k\in\mathbb{N}}$ is an output sequence of Σ, we will use the notation $(y(k), y(k+1)) := Y(k)$ so that $I^{[w_1,w_2]}_{y(k),y(k+1)} = I^{[w_1,w_2]}_{Y(k)}$. A solution to Problem 16.8.1 is proposed in the following theorem.

Theorem 16.8.8 *Let $\{y(k)\}_{k\in\mathbb{N}}$ be the output sequence corresponding to an execution of Σ. Let $\hat{\Sigma}_1$ be as in Theorem 16.5.6. Consider the additional system with input $\hat{\Sigma}_2 = (\mathcal{L} \times \mathcal{L}, \chi \times \chi \times \mathcal{Y} \times \mathcal{Y}, \mathcal{Z}_E \times \mathcal{Z}_E, (f_3, f_4), (g_1, g_2, g_3, g_4))$ with $f_3 : \mathcal{L} \times \chi \times \chi \times \mathcal{Y} \times \mathcal{Y} \to \mathcal{L}$, $f_4 : \mathcal{L} \times \chi \times \chi \times \mathcal{Y} \times \mathcal{Y} \to \mathcal{L}$, $g_3 : \mathcal{L} \to \mathcal{Z}_E$, and $g_4 : \mathcal{L} \to \mathcal{Z}_E$ given by*

$$f_3\,(q_L(k), L(k), U(k), y(k), y(k+1)) = \tilde{F}\left(q_L(k) \curlyvee \bigwedge I^{[L^*(k),U^*(k)]}_{Y(k)}\right) \quad (16.28)$$

$$f_4\,(q_U(k), L(k), U(k), y(k), y(k+1)) = \tilde{F}\left(q_U(k) \curlywedge \bigvee I^{[L^*(k),U^*(k)]}_{Y(k)}\right) \quad (16.29)$$

$$g_3(q_L(k)) = \pi_2 \circ a_L(q_L(k)) \quad (16.30)$$

$$g_4(q_U(k)) = \pi_2 \circ a_U(q_U(k)) \quad (16.31)$$

in which $L(k), U(k), y(k), y(k+1)$ is the output of $\hat{\Sigma}_1$, $L^*(k) = \bigwedge O_y(k) \curlyvee L(k)$, $U^*(k) = \bigvee O_y(k) \curlywedge U(k)$. If Σ is independent discrete state observable and $(\tilde{\Sigma}, (\mathcal{L}, \leq))$ is induced interval compatible, $\hat{\Sigma}_1 \circ_c \hat{\Sigma}_2$ solves Problem 16.8.1.

Proof. Properties (i)–(iii) follow directly from Theorem 16.5.6. The proof of (i')–(iii') proceeds as follows. The proof of (i') exploits the order preserving properties of $\tilde{F}$, a_L,

a_U, and π_2. The proof of (ii') exploits the property of induced order compatibility and the definition of distance on a partial order. The proof of (iii') uses directly the observability of system Σ.

Proof of (i'). We use induction argument on k. Initially, $q_L(0) = \bigwedge \mathcal{L}$ and $q_U(0) = \bigvee \mathcal{L}$. Therefore, we have that $q_L(0) \leq (\alpha(0), z(0)) \leq q_U(0)$. Next, we show that if $q_L(k) \leq (\alpha(k),\ z(k)) \leq q_U(k)$ then also $q_L(k+1) \leq (\alpha(k+1), z(k+1)) \leq q_U(k+1)$. Since $\alpha(k) \in [L^*(k), U^*(k)] \subseteq T_{y(k),y(k+1)}(\tilde{\Sigma})$ and also $(\alpha(k), z(k)) \in \{(\alpha, z) \mid g(\alpha, z) = y(k)$ and $g(f(\alpha, y(k)),\ h(\alpha, z)) = y(k+1)\}$, it follows that $(\alpha(k), z(k)) \in I_{Y(k)}^{[L^*(k),U^*(k)]}$. As a consequence, we have that

$$q_L(k) \curlyvee \bigwedge I_{Y(k)}^{[L^*(k),U^*(k)]} \leq (\alpha(k), z(k)) \leq q_U(k) \curlywedge \bigvee I_{Y(k)}^{[L^*(k),U^*(k)]}.$$

By property (i) of Definition 16.8.7 (order preserving property) it follows that $\tilde{F}(q_L(k) \curlyvee \bigwedge I_Y^{[L^*,U^*]}) \leq (\alpha(k+1), z(k+1)) \leq \tilde{F}(q_U(k) \curlywedge \bigvee I_Y^{[L^*,U^*]})$, which in turn implies by Equations (16.28) and (16.29) that $q_L(k+1) \leq (\alpha(k+1), z(k+1)) \leq q_U(k+1)$. We are thus left to show that $q_L(k) \leq (\alpha(k), z(k))$ implies that $\pi_2 \circ a_L(q_L(k)) \leq z(k)$ and that $(\alpha(k), z(k)) \leq q_U(k)$ implies that $z(k) \leq \pi_2 \circ a_U(q_U(k))$. These are true as $\pi_2 \circ a_L$ and $\pi_2 \circ a_U$ are order preserving maps, $\pi_2 \circ a_L(\alpha(k), z(k)) = z(k)$, and $\pi_2 \circ a_U(\alpha(k), z(k)) = z(k)$.

Proof of (ii'). Since $\tilde{F}$ is order preserving on the induced transition sets, we have (neglecting the dependence on k) that

$$\tilde{F}(\bigwedge I_Y^{[L^*,U^*]}) \leq \tilde{F}(q_L \curlyvee \bigwedge I_Y^{[L*,U^*]}) \text{ and } \tilde{F}(\bigvee I_Y^{[L^*,U^*]}) \geq \tilde{F}(q_L \curlywedge \bigvee I_Y^{[L*,U^*]}).$$

Since $\pi_2 \circ a_L$ and $\pi_2 \circ a_U$ are also order preserving, by using property (iii) of the distance function (Definition 16.3.4), we have that

$$d\left(\pi_2 \circ a_L \circ \tilde{F}(q_L \curlyvee \bigwedge I_Y^{[L^*,U^*]}), \pi_2 \circ a_U \circ \tilde{F}(q_U \curlywedge \bigvee I_Y^{[L^*,U^*]})\right) \leq d\left(\pi_2 \circ a_L \circ \tilde{F}(\bigwedge I_Y^{[L^*,U^*]}), \pi_2 \circ a_U \circ \tilde{F}(\bigvee I_Y^{[L^*,U^*]})\right).$$

By property (iii) of Definition 16.8.7, we have that

$$d\left(\pi_2 \circ a_L \circ \tilde{F}(\bigwedge I_Y^{[L^*,U^*]}), \pi_2 \circ a_U \circ \tilde{F}(\bigvee I_Y^{[L^*,U^*]})\right) \leq \gamma([L^*, U^*]). \qquad (16.32)$$

Since $\tilde{f}([L^*, U^*], y) = [\tilde{f}(L^*, y), \tilde{f}(U^*, y)]$, $\tilde{f}(L^*(k), y(k)) = L(k+1)$, and $\tilde{f}(U^*(k), y(k)) = U(k+1)$, we have by the order isomorphism property of $\tilde{f}$ that $|\tilde{f}([L^*, U^*], y(k))| = |[L^*, U^*]| = |[L(k+1), U(k+1)]|$. Thus $\gamma(|[L^*, U^*]|) = \gamma([[L(k+1), U(k+1)]])$. This along with equation (16.32) completes the proof.

Proof of (iii'). For $k > k_0$, $L'(k) = \alpha(k) = U'(k)$ because $[L(k), U(k)] \cap \mathcal{U} = \alpha(k)$. As a consequence, $q_{L'}(k+1) = \tilde{F}(q_{L'}(k) \curlyvee \bigwedge I_{Y(k)}^{[\alpha(k),\alpha(k)]})$ and $q_{U'}(k+1) = \tilde{F}(q_{U'}(k) \curlywedge \bigvee I_{Y(k)}^{[\alpha(k),\alpha(k)]})$. By property (ii) of Definition 16.8.7, it follows that for any $k > k_0$ we have that for any $q' \in [q_{L'}(k+1), q_{U'}(k+1)]$ there is $q \in [q_{L'}(k), q_{U'}(k)]$ such that $q' = \tilde{F}(q)$. Also, $\mathcal{L} - (\mathcal{U} \times \mathcal{Z})$ is invariant under $\tilde{F}$ and $\tilde{F}|_{\mathcal{U}\times\mathcal{Z}} = (f', h')$, Therefore, it is also true that for any $(\alpha', z') \in [q'_L(k+1), q'_U(k+1)] \cap (\mathcal{U} \times \mathcal{Z})$ there is $(\alpha, z) \in$

$[q_{L'}(k), q_{U'}(k)] \cap (\mathcal{U} \times \mathcal{Z})$ such that $(\alpha', z') = (f(\alpha, y(k)), h(\alpha, z))$. In addition, we have that such (α, z) is in the induced transition set, that is, $(\alpha, z) \in I_{Y(k)}^{[\alpha(k),\alpha(k)]}$. Thus in turn implies that $g(\alpha, z) = y(k)$. Since this is true for any $k \geq k_0$, if for any k we have that $[q'_L(k), q'_U(k)] \cap (\mathcal{U} \times \mathcal{Z})$ contains more than one element, it means that there are at least two executions of Σ, $\sigma_1 \neq \sigma_2$, such that $g(\sigma_1) = g(\sigma_2)$. This contradicts observability of Σ. Thus, it must be that there is $k'_0 > k_0$ such that for $k \geq k'_0$ we have that $[q'_L(k), q'_U(k)] \cap (\mathcal{U} \times \mathcal{Z}) = (\alpha(k), z(k))$. As a consequence, by virtue of equations (16.26,16.27,16.30,16.31) we also have that $z_{L'}(k) = z_{U'}(k)$ for all $k \geq k'_0$.

16.8.1 RoboFlag Drill with continuous dynamics

A version of the RoboFlag Drill system of Section 16.2 is considered in which now the defender robots have partially measured second order dynamics instead of having fully measured first order dynamics. To describe this dynamics, each defender robot motion is denoted by variable $z_i \in \mathbb{R}^2$, in which $z_{i,1}$ is the position. We consider the problem of estimating the current assignment α and continuous variables z_i given measured positions $z_{i,1}$ of the defender robots. The function $h : \mathcal{U} \times \mathcal{Z} \to \mathcal{Z}$ that updates the z variables is given by

$$\begin{aligned} z'_{i,1} &= (1-\beta)z_{i,1} - \beta z_{i,2} + 2\beta x_{\alpha_i} \\ z'_{i,2} &= (1-\lambda)z_{i,2} + \lambda x_{\alpha_i} \end{aligned} \tag{16.33}$$

for any i. The set $\mathcal{Z}$ is such that $z_{i,1} \in [x_i, x_{i+1}]$ and $z_{i,2} \in [x_i, x_{i+1}]$ for any i, which is guaranteed if β and λ are assumed sufficiently small. This means that each defender moves toward the x position of the assigned attacker with second order damped dynamics. The continuous variables are $z = (z_{1,1}, z_{1,2}, \ldots, z_{N,1}, z_{N,2}) \in \mathcal{Z}$, with output $g(z) = (z_{1,1}, \ldots, z_{N,1}) \in \mathcal{Y}$. Thus, $\Sigma_1 = (\mathcal{U}, \mathcal{Y}, \mathcal{U}, f, \mathrm{id}_1)$ and $\Sigma_2 = (\mathcal{Z}, \mathcal{U}, h, g)$, in which f and $\mathcal{U}$ are the same as in Section 16.2 and h is now given by Equation (16.33). The overall system is given by the feedback interconnection of Σ_1 with Σ_2, that is, $\Sigma = \Sigma_1 \circ_f \Sigma_2$. The partial order $(\chi, \leq)$ and extension $\tilde{f}$ are the same as the ones defined in Equations (16.9) and (16.10). We define $\mathcal{L} = \chi \times \mathcal{Z}$ and $\tilde{F} = (\tilde{f}, \tilde{h})$, in which $\tilde{h} : \chi \times \mathcal{Z} \to \mathcal{Z}$ is the same as h in Equation (16.33) but with $x \in \chi$ in place of $\alpha \in \mathcal{U}$. The partial order in $\mathcal{Z}$ is established by $z^a \leq z^b$ for $z^a, z^b \in \mathcal{Z}$ if $z^a_{i,2} \leq z^b_{i,2}$. One can verify that the order on each $z_{i,2}$ is preserved by the dynamics in Equation (16.33). In fact, if $z^{(1)}_{i,2} < z^{(2)}_{i,2}$ and $w^{(1)}_i \leq w^{(2)}_i$ then $(1-\lambda)z^{(1)}_{i,2} + \lambda x_{w^{(1)}_i} \leq (1-\lambda)z^{(2)}_{i,2} + \lambda x_{w^{(2)}_i}$ because $x_{w^{(1)}_i} \leq x_{w^{(2)}_i}$ whenever $w^{(1)}_i \leq w^{(2)}_i$, and because $(1-\lambda) > 0$. With such partial order on $\mathcal{Z}$, one can show that the assumptions of Theorem 16.8.8 are verified. The estimator in Theorem 16.8.8 has been implemented. For the continuous state estimator, let $z_L, z_U \in \mathbb{R}^N$ represent the lower and upper bound of $(z_{1,2}, \ldots, z_{N,2})$. Figure 16.6 illustrates the estimator performance, in which $W(k) = \sum_{i=1}^N |m_i(k)|$, where $|m_i(k)|$ is the cardinality of the sets $m_i(k)$ that are the sets of possible α_i for each component obtained from the sets $[L_i, U_i]$ by removing iteratively a singleton occurring at component i by all other components. When $[L(k), U(k)] \cap \mathrm{perm}(N)$ has converged to α, then $m_i(k) = \alpha_i(k)$. The distance function for $z, x \in \mathbb{R}^N$ is defined by $d(x, z) = \sum_{i=1}^N \mathrm{abs}(z_i - x_i)$. The function $V(k)$ in the plot

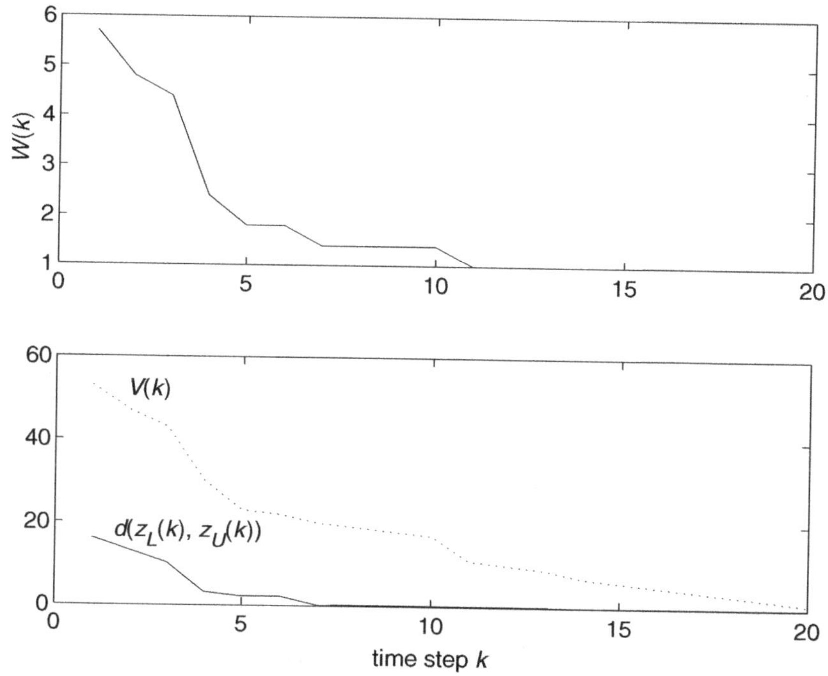

Figure 16.6 Estimator performance with $N = 10$ agents.

is defined by $V(k) = \frac{1}{2}\sum_{i=1}^{N}(x_{U_i(k)} - x_{L_i(k)})$, and it is always non increasing. A possible candidate for γ is thus $\gamma(|[L,U]|) := |[L,U]|\sup_i|x_{U_i} - x_{L_i}|$. Note that even if the discrete state has not converged yet, the continuous state estimation error after $k = 8$ is close to zero.

16.9 CONCLUSION

In this work, we have presented an approach to state estimation in decision and control systems, which reduces complexity by using partial order theory. The developed algorithms enjoy scalability properties that are substantial in multi-agent systems. This has been done for estimating the discrete state once the continuous state is measured and for estimating both discrete and continuous state when an estimator in cascade form is possible. Partial order theory has proved to be a useful tool borrowed from theoretical computer science to address this issue, and it was nicely merged with classical control theory to reach our goal. The question of how to deal with the estimation problem for both the continuous and the discrete state when an estimator in cascade form is not possible will be addressed in our future work. In particular, we will take advantage of the 'cooperative' and 'competitive' nature of the multi-agent systems under study that naturally imposes order preserving dynamics in suitable partial orders. The dynamic feedback control problem will also be addressed in our future work by using partial order theory. For the case of the RoboFlag Drill, we will design a dynamic controller for the attackers, which on the basis of the state estimates, decides the next action to take in order to win the game. These are mainly synthesis issues, however, we will also explore how partial order theory can be employed to solve analysis problems such as the computation of escape sets and controlled invariant sets with low computational load.

ACKNOWLEDGEMENT

Research supported by AFOSR/MURI grant #F49620-01-1-0361.

REFERENCES

Abramsky S and Jung A 1994 Domain Theory. In *Handbook of Logic in Computer Science* (S. Abramsky, DM Gabbay and TSE Maibaum EDs), Clarendon Press, Oxford, pp. 1–168.

Alessandri A and Coletta P 2001 Design of Luenberger observer for a class of hybrid linear systems. In *Lecture Notes in Computer Science* 2034, (M D Di Benedetto, A Sangiovanni-Vincentelli Eds.) Springer Verlag, Berlind. pp. 7–18.

Alessandri A and Coletta P 2003 Design of observer for switched discrete-time linear systems. *American Control Conference*, pp. 2785–2790.

Balluchi A, Benvenuti L, Benedetto MDD and Sangiovanni-Vincentelli A 2002 Design of observers for hybrid systems. In *Lecture Notes in Computer Science* 2289, (C J Tomlin and M R Greenstreet Eds.) Springer Verlag, Berlin. pp. 76–89.

Balluchi A, Benvenuti L, Benedetto MDD and Sangiovanni-Vincentelli A 2003 Observability of hybrid systems. In *Conf. on Decision and Control*, pp. 1159–1164.

Bemporad A, Ferrari-Trecate G and Morari M 1999 Observability and controllability of piecewise affine and hybrid systems. *IEEE Transactions on Automatic Control* **45**, 1864–1876.

Bui HH, Venkatesh S and West G 2002 Policy recognition in the abstract hidden Markov model. *Journal of Artificial Intelligence Research* **17**, 451–499.

Caines PE and Wang S 1995 Cocolog: A conditional observer and controller logic for finite machines. *SIAM J. Control and Optimization* **34**, 1687–1715.

Caines PE and Wei Y 1996 The hierarchical lattices of a finite machine. *Systems and Control Letters* **25**, 257–263.

Caines PE, Greiner R and Wang S 1991 Classical and logic-based dynamic observers for finite automata. *IMA J. of Mathematical Control and Information,* pp. 45–80.

Cassandra AR, Kaelbling LP and Littman ML 1994 Acting optimally in partially observable stochastic domains. In *Proc. 12th Conference on Artificial Intelligence*, pp. 1023–1028, Seattle, WA.

Collins RT, Lipton AJ, Fujiyoshi H and Kanade T 2001 Algorithms for cooperative multisensor surveillance. In *Proceedings of the IEEE* **89**(10), 1456–1477.

Costa EF and do Val JBR 2002 On the observability and detectability of continuous-time jump linear systems. *SIAM Journal on Control and Optimization* **41**, 1295–1314.

Cousot P and Cousot R 1977 Abstract interpretation: A unified lattice model for static analysis of programs by construction or approximation of fixpoints. In *Conference Record of the Sixth Annual ACM Symposium on Principles of Programming Languages*, pp. 238–252, Los Angeles.

D'Andrea R, Murray RM, Adams JA, Hayes AT, Campbell M and Chaudry A 2003 The RoboFlag Game. In *American Control Conference*, pp. 661–666.

Davey BA and Priesteley HA 2002 *Introduction to Lattices and Order*. Cambridge University Press, Cambridge.

DelVecchio D and Murray RM 2004 Existence of discrete state estimators for hybrid systems on a lattice. In *43rd Conf. on Decision and Control*, pp. 1–6.

Diaz M, Juanole G and Courtiat J 1994 Observer: a concept for formal on-line validation of distributed systems. *IEEE Trans. on Software Engineering* **20**, 900–913.

Godefroid P 1996 Partial-order methods for the verification of concurrent systems. In *Lecture notes in Computer Science*. Springer, New York.

Kalman RE 1960 A new approach to linear filtering and prediction problems. *Transactions of the ASME – Journal of Basic Engineering* **82**(Series D), 35–45.

Klavins E 2003 A formal model of a multi-robot control and communication task. In *Conf. on Decision and Control*, pp. 4133–4139, Hawaii.

Klavins E and Murray RM 2004 Distributed algorithms for cooperative control. *Pervasive Computing* **3**, 56–65.

Luenberger DG 1971 An introduction to observers. *IEEE Transactions on Automatic Control* **AC-16:6**, 596–602.

Oishi M, Hwang I and Tomlin C 2003 Immediate observability of discrete event systems with application to user-interface design. In *Conf. on Decision and Control*, pp. 2665–2672, Hawaii.

Ozveren CM and Willsky AS 1990 Observability of discrete event dynamic systems. *IEEE Transactions on Automatic Control* **35**(7), 797–806.

Ramadge PJ 1986 Observability of discrete event systems. In *Conf. on Decision and Control*, pp. 1108–1112, Athens, Greece.

Rudie K, Lafortune S and Ling F 2003 Minimal communication in a distributed discrete event system. *IEEE Trans. on Automatic Control* **48**, 957–975.

Santis ED, Benedetto MDD and Pola G 2003 On observability and detectability of continuous-time linear switching systems. In *Conf. on Decision and Control*, pp. 5777–5782.

Vidal R, Chiuso A and Soatto S 2002 Observability and identifiability of jump linear systems. In *Conf. on Decision and Control*, pp. 3614–3619, Las Vegas.

Vidal R, Chiuso A, Soatto S and Sastry S 2003 Observability of linear hybrid systems. In *Lecture Notes in Computer Science* 2623, (O Maler and A Pnueli Eds.) Springer Verlag, New York, pp. 526–539.

17

Abstraction-based command and control with patch models

V. G. Rao, S. Goldfarb and R. D'Andrea

17.1 INTRODUCTION

In this chapter, we consider the problem of complexity management in modern warfare, particularly as it concerns cooperative control (with humans in the loop) of assets such as autonomous vehicles, employed against an adversary. Our approach to this broad problem is based on systematic abstraction of battlespace information into models that can be tailored to be relevant to the local situation. Decision-makers using such models, through appropriate software and connected through appropriate communication mechanisms, can provide, we believe, an information infrastructure that can manage large-scale complexity. In military terms, the objective is to achieve a system-wide condition of 'shared situation awareness' (SSA) (Endsley 1995; Nofi 2000) that allows greater responsiveness to events on the battlefield. The specific instantiation of this approach described here – the method of patch models – is adapted to represent fast-changing spatio-temporal information, concerning mobile friendly and enemy assets. The approach, however, can be generally applied in broader contexts, and we indicate elsewhere in this chapter the guiding principles that should be employed in the development of such abstraction-based systems.

Our objective is to provide an expository overview of work that has previously been described in technical detail elsewhere (Rao and D'Andrea 2006; Rao *et al.* 2006), and to discuss several important emerging problems for which we offer promising attacks.

For motivation, consider the following prototypical modern battlefield narrative. An enemy threat is detected by a large-area surveillance aircraft, and tracked coarsely. Shortly, a lower-flying autonomous surveillance aircraft, with greater resolution, arrives to loiter on the scene and track the target more closely. Meanwhile an autonomous attack aircraft is diverted from a different mission. Once the attack aircraft arrives on the scene, there is a short period of joint surveillance, followed by an attack. The mission is completed when the surveillance aircraft takes pictures of the effect of the strike, an operation usually

Cooperative Control of Distributed Multi-Agent Systems edited by Jeff S. Shamma

referred to as 'battle damage assessment,' or BDA. The entire sequence of events takes twenty minutes.

Consider the problem of realizing the capabilities required to make the above scenario a routine one. Viewed in isolation, it may appear that the capabilities already exist. Certainly, situations very similar to this have already occurred. Imagine, however, this level of responsiveness and success for almost all threats detected anywhere in a large battlespace, and imagine further, that the detection rate is high, due to highly capable surveillance systems. The nature of the problem then, becomes clear, since we must now consider the infrastructure of capabilities that would cause such a scenario to play out in nearly *every* instance where it was necessary, against the backdrop of a large number of similar scenarios already in progress. It is this infrastructure that concerns us in this chapter.

Such a hypothetical future command and control system would be a highly complex entity. It has been noted (Simon 1996) that such complex systems typically comprise nearly-decomposable components. Our hypothetical system must be factored, or partitioned, into component capabilities that, one hopes, constitute a nearly exhaustive and almost mutually exclusive set. The work in this book was driven by one relatively reasonable partitioning of the overall problem into four such capabilities: distributed control and computation, management of adversarial interactions, management of uncertain evolution, and complexity management. This chapter concerns the fourth aspect, 'complexity management.'

Complexity management, in our view, is a problem dimension that is orthogonal to all other dimensions. In every dimension, complexity arises from at least two sources. The first is the effect of scaling up functions and subsystems. What, for instance, happens to a resource allocation system when it deals with hundreds or thousands of resources instead of dozens? This effect is largely a computational-complexity effect, and one seeks efficient, preferably distributed methods for solving problems known to be intrinsically hard (in the sense of NP completeness). The work of Carla Gomes[1] and her colleagues on recognizing and exploiting the phenomena of phase transitions and backdoors in NP-complete problems is an appropriate approach for this aspect of complexity.

The second source is the increased heterogeneity, richer ontology and increased communication complexity (in a semantic sense) that accompanies scaling to broader contexts of activity. Can a scheduler developed for aircraft in the scenario above be adapted to the presence of other battlespace elements such as spacecraft and missiles? Can a new kind of video camera used on a new variant of a surveillance system seamlessly feed its output into the battlespace management software used on a JSTARS aircraft? Such issues are the focus of 'complexity management' as understood in this chapter.

Our approach to such problems is based on four guiding principles. The first is that the *representation* of information necessary to make decisions is more fundamental than the decision-making processes themselves. Second, the representations must be amenable to sharing across a variety of decision-processes (such as resource allocation and mission profile selection). Third, the representations must lend themselves to consistent, shared use by both human and artificial decision-makers. Fourth, the representations must be applicable to a variety of decision-making contexts, ranging from ground-level high-speed tactical contexts to slower-changing strategic contexts much further upstream in the decision-making hierarchy.

[1] http://www.cs.cornell.edu/gomes/

These four principles have been applied in the development of the methodology of patch models described here. In Section 17.2, we provide an overview of the modeling framework. In Section 17.3, we consider procedures for realization and verification of the models, with particular attention to the latter. In Section 17.4, we consider human and artificial decision-making with patch models. In Section 17.5, we consider the issues involved in the collaborative use of such models by a hierarchy of decision-makers. In Section 17.6, we present our conclusions. Throughout, we indicate open problems where they appear in the course of the development.

17.2 OVERVIEW OF PATCH MODELS

We begin by considering how we might represent a battlefield situation involving several friendly, neutral and enemy moving entities from the point of view of an individual decision-maker. This decision-maker might be either a human performing a well-defined function (such as allocation of aircraft to targets, or air-traffic control), or an ambiguous function such as reviewing status information and evaluating various what-if scenarios to develop or select a course of action. In the battlefield of the future, the decision-maker might well be a software program (with appropriate override mechanisms for supervising humans). Both human and artificial decision-makers may either be onboard the mobile resources they control, or at remote locations. All these decision-makers, performing different functions in different contexts require representations of the battlespace suited to their needs, and patch models serve this purpose. We provide here a simplified description of these models, which accommodate information that can be linked to primitive *spatio-temporal* information. A more detailed technical exposition with a detailed simulation case study can be found in Rao and D'Andrea (2006).

The information content of a battlespace is derived from a (possibly very large) set of independent noisy variables that must be measured, such as vehicle positions. Call this information *primitive* information. We restrict ourselves in this work to spatio-temporal primitive information. We suppose that there is a set $\mathcal{A}$ of primitive agents (such as vehicles) $\{a_1, \ldots, a_{\mathbf{q}}\}$, each of which is described by a state vector. Each state vector must contain spatio-temporal elements such as vehicle positions and velocities that allows the information to be organized spatio temporally. Each state vector may also include different, agent-specific non-spatio-temporal elements such as camera-state or weapon-state. We may also include design-dependent abstract variables such as 'damage state.' We stack these individual agent state vectors and write the detailed domain dynamics as follows:

$$
\begin{aligned}
\dot{\overline{x}} &= f(\overline{x}, \overline{u}, \overline{w}), \\
\overline{y} &= h(\overline{x}, \overline{v}), \\
\overline{u} &= g(\overline{y}), \\
p_i(\overline{x}, \overline{u}) &\geq 0,\ i = 1, 2, \ldots,
\end{aligned}
\tag{17.1}
$$

where $\overline{x}$, $\overline{u}$ and $\overline{y}$ are the global state, control and observation vectors respectively, and the p_i represent constraints in a standard form. The noise processes $\overline{w}$ and $\overline{v}$ are noise processes capturing the effects of imperfect surveillance capabilities and vehicle control

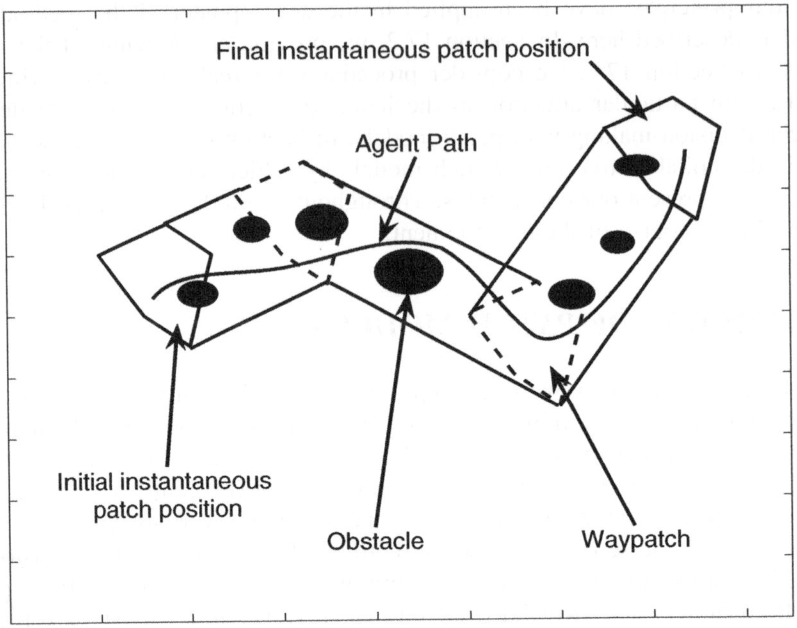

Figure 17.1 Primitive spatio-temporal information (an agent path) and its representation with a patch.

capabilities. Clearly, individual decision-makers cannot observe $\overline{y}$ completely or influence all elements of $\overline{u}$ equally. These decentralization conditions must be modeled through constraints on the allowable forms of the control and observation functions, f and g which we will elaborate upon in later sections.

The primitive information is clearly vast in scope as well as quantity, and its evolution is described by complex dynamics. Much of this information is irrelevant to most decision-making roles. We cannot, however, merely omit states, as is done in model-order reduction in classical control, since in some cases high *accuracy* and *precision* may be irrelevant for a decision, but coarse information may still be required. The first priority in modeling is therefore to select relevant information and to aggregate it into more abstract building blocks with relevance-dependent precision. We achieve this by introducing a construct called a patch. A patch is a moving convex figure, (here we restrict ourselves to polygons), as illustrated in Figure 17.1, that moves along a piece-wise linear path in two-dimensional space, defined as follows:

$$P = \{R, T_1, \ldots T_{\mathbf{j}}\}, \tag{17.2}$$

$$R = \begin{pmatrix} x_1 & y_1 \\ \cdots & \cdots \\ x_{\mathbf{n}} & y_{\mathbf{n}} \end{pmatrix}, \tag{17.3}$$

$$T_j = \begin{bmatrix} x_j & y_j & \lambda_j & t_j \end{bmatrix}. \tag{17.4}$$

In the above, R is a root polygon and the T_i, $i \leq \mathbf{i}$, (called *waypatches*) represent affine transformations, each comprising a new translated position, a scaling factor and a time. This information structure is a patch. As shown in Figure 17.1, visually a patch is a series of piece-wise linear bands in space (or a tube in space-time). The spatial 'band' is referred to as the *trace of the patch*, $P([t_0, t_f])$, while the instantaneous position of the transformed root polygon is denoted $P(t)$. Note that the term 'patch' refers to the entire trajectory of the root polygon, and is thus an integral rather than a differential representation primitive.

A patch is meant to represent one or more agents over an interval of time of interest in a manner that aggregates individual agent information, and simplifies their dynamic behavior. For example, a sufficiently large triangle that is assumed to move at constant velocity in a straight line between start and end points requires just 14 independent pieces of primitive information (a root polygon and two waypatches, velocity being implicit), but can subsume an arbitrary number of agents with complex dynamics at an appropriate scale. The cost of using a patch as a representational primitive in place of primitive data is that there is loss of precision in both observation and control. For a given decision-maker, some primitive information will require representation with very precise, tight patches that move in ways that very closely conform to the actual motion of a small number of agents, while other parts of the primitive information set may be lumped together to form a very coarse and slowly varying representation that is sufficient for the objective at hand. We therefore construct *patch models* out of individual patches as follows. We define a set of patches, $M = \{P_1, \ldots, P_\mathbf{i}\}$ to be a patch model, and associate the individual patches with agents (assumed to be the loci of all primitive information) with a *representation function*:

$$h : \mathcal{A} \rightarrow M \equiv \{P_1, \ldots, P_\mathbf{i}\}. \tag{17.5}$$

Next, consider information relevant to decision-making that is *not* based on primitive measured data. This can be of two sorts. First, a decision-maker may anticipate more than one predicted evolution of the situation. Second, a decision-maker may also maintain a collection of possible courses of action, even if we anticipate only one evolving scenario. We must therefore maintain multiple *possible world* models. These models are all patch models defined exactly as above, with a specific model designated as the current working hypothesis model at any given time, and used for control. Such a collection of models is called a *view*:

$$V = \{M_1, \ldots, M_\mathbf{k}\} \tag{17.6}$$

A view is an information structure that represents both actual and possible scenarios and is designed to support the sort of 'what-if' reasoning that is required in more complex decision-making. Finally, we may conceptualize an entire command and control system, comprising several decision-makers, as a system of interconnected views, $\{V_1, V_2, \ldots, V_n\}$ for all n decision makers in the hierarchy. Both humans and artificial decision-makers observe and influence the battlefield through computer-maintained views comprising patch models. The topology of interconnection of this set of views describes the architecture of the command and control system.

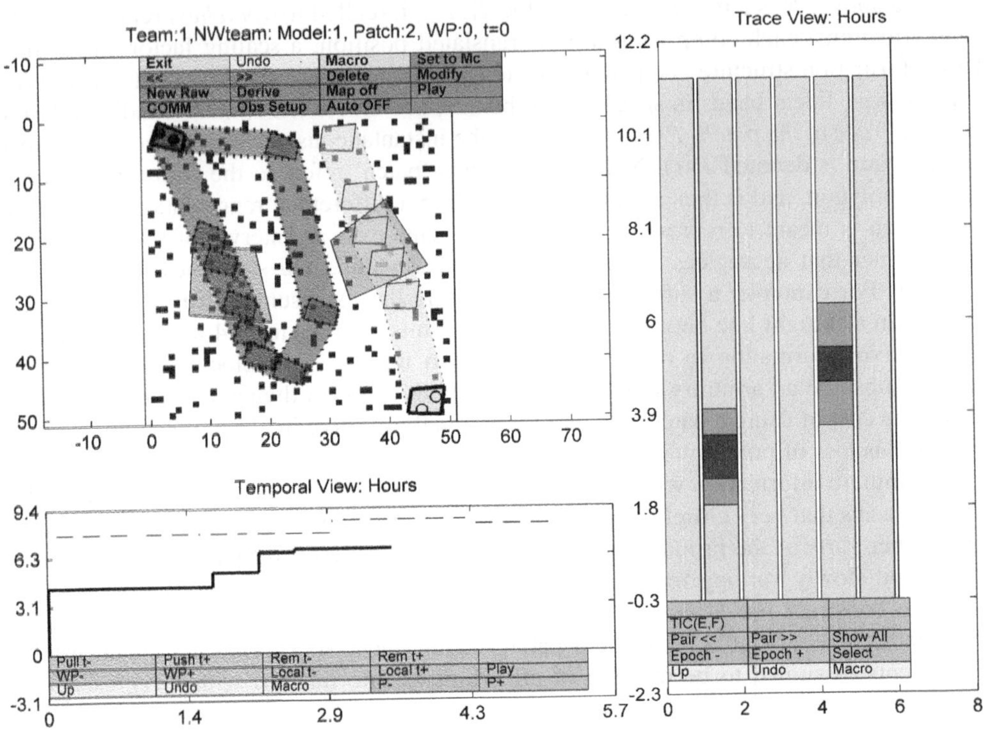

Figure 17.2 *Patchworks:* a simulation environment based on patch models. The top left pane shows a spatial view of four patches; the bottom left pane shows the velocities of the two moving patches. The right pane shows the qualitative state vector and is described in Section 17.5.

Patch models, as described, are an enabling mechanism that can be used in a great variety of ways, once appropriate software has been designed to computationally represent and manipulate them. We have developed a simulation environment for distributed command and control of simulated point vehicles called *Patchworks*. A screen shot of this software is shown in Figure 17.2. The software constructs visual renderings of a patch model, allowing humans to reason about and manipulate the models visually. Artificial decision-makers must directly work with the data structures comprising the model. A thorough overview of the features and capabilities of this software, illustrated through a detailed combat mission simulation, can be found elsewhere (Rao and D'Andrea 2006).

Since the purpose of this chapter is to understand fundamental themes and important open problems, we restrict ourselves to three particularly illuminating lines of investigation in the next three sections: realization and verification, decision-making, and hierarchical control. These three areas contain a rich set of open problems that require further investigation, and the purpose of these sections is to motivate these problems through results that have already been obtained, indicate how they might be formally posed, and suggest promising approaches.

17.3 REALIZATION AND VERIFICATION

Given a patch model formally set up as described in the previous section, we require mechanisms for controlling agent actions through the models and verifying the models through observations of primitive variables. These are the *realization* and *verification* problems respectively.

Realization is the problem of finding a path for an agent that maintains the truth of the patch representing it in the model maintained by the controlling decision-maker. Call this patch the *ego patch*. Our current architecture supports only single ego patches for simplicity, but multiple ego patches can be supported at the cost of greater software complexity. The elements of the realization problem are illustrated in Figure 17.1. The path indicated for the vehicle must avoid small obstacles while staying within the instantaneous position of the patch. A weaker form of control requires the vehicle to stay with the trace of the patch, and within the instantaneous patches only at end times, thus allowing in-mission delays. These problems are referred to as differential and integral patch realization problems respectively, and practical methods for solving them are discussed in detail in Rao *et al.* (2006). Realization problems are standard robotics path-planning problems with time-varying state constraints, but the special structure of the problem leads to many useful heuristics for computational efficiency. We do not discuss the basic problem of realization further in this overview, since the techniques are relatively well known. We discuss briefly the problem of *hierarchical* realization, which is significantly different and an open problem.

Consider the situation where an agent is controlled by a hierarchy of supervisors through a reporting chain, rather than through a single ego patch. This corresponds to a hierarchy of patch models, where each model has a *unique* parent model. Supervisors achieve control by editing patches in their models, with changes being cascaded down the hierarchy (as constraints on the corresponding patches in lower-level models). More intuitively, we can consider patches in lower-level models to be virtual agents themselves, being controlled by higher-level models. The hierarchy of patches representing the agents controlled by the ego patch of the lowest-level model must satisfy a consistency condition defined as follows. Let $M_0, M_1, M_2, \ldots, M_k$ be a hierarchy of k models, with M_0 being the ground-level model and $P_e \in M_0$ being the ego patch. Let $\mathcal{A}_e = h_0^{-1}(P_e)$ be the set of agents controlled by P_e. Then, for all t:

$$h_0(\mathcal{A}_e) \subset h_1(\mathcal{A}_e) \subset \ldots \subset h_k(\mathcal{A}_e), \tag{17.7}$$

where the sequence of inclusions correspond to geometric inclusion in space-time. Note that the condition involves proper inclusion, since if a downstream patch is equal to an upstream patch and one assumes that supervising models may be changed continuously by a decision-maker, one would require a physically impossible zero-lag connection between the two models to maintain model truth at all levels.

It might appear that the hierarchical version of the problem is no more than a series of progressively constrained path planning problems. There is a crucial difference, however: the 'virtual agents' or intermediate patches being non-physical entities allows them to move in more unconstrained ways than physical agents. In particular, they can move

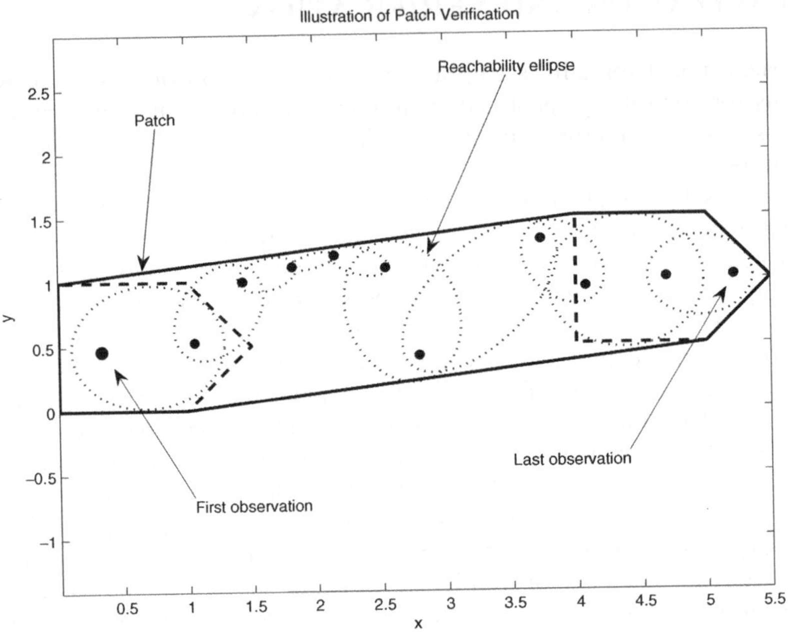

Figure 17.3 Patch Verification with noisy, intermittent, non-sequential observations.

'through' obstacles below a particular size, and through other patches, under appropriate conditions. This makes their motion much less constrained than motion in the robotics paradigm, where rigid-body obstacle avoidance is central. At this stage, solving this type of under-constrained path planning problem this remains an interesting open problem. Techniques drawn from fluid dynamics or optics, rather than robotics, and based on metaphors such as friction, viscosity and refraction, applied to patches, may be the appropriate ones.

Now consider the inverse problem of patch verification. Recall that decentralization assumptions must be modeled as constraints on the observations available to a given decision-maker. We require, however, a more general notion than observability as it is typically understood in control theory. Observability is a useful conceptual model for a static set of information-availability patterns – a given state variable is either available or not available to a decision-maker based on a single 'observability' characteristic of that state within the system. When a state may be observed through a set of non-sequential, variably delayed, sparsely scattered (in time) and noisy measurements, a more appropriate notion is that of the 'information pattern' introduced by Witsenhausen (Witsenhausen 1971) and used extensively in the nonlinear stochastic control literature. Within this framework, we do not attempt to work through a separate representation of observation capabilities (such as the observation matrix, C, used in continuous LTI theory). We work, instead, directly with measurement sets. Introducing this approach is beyond the scope of this article, so we restrict ourselves to an example that illustrates the underlying principles and resultant problem formulations.

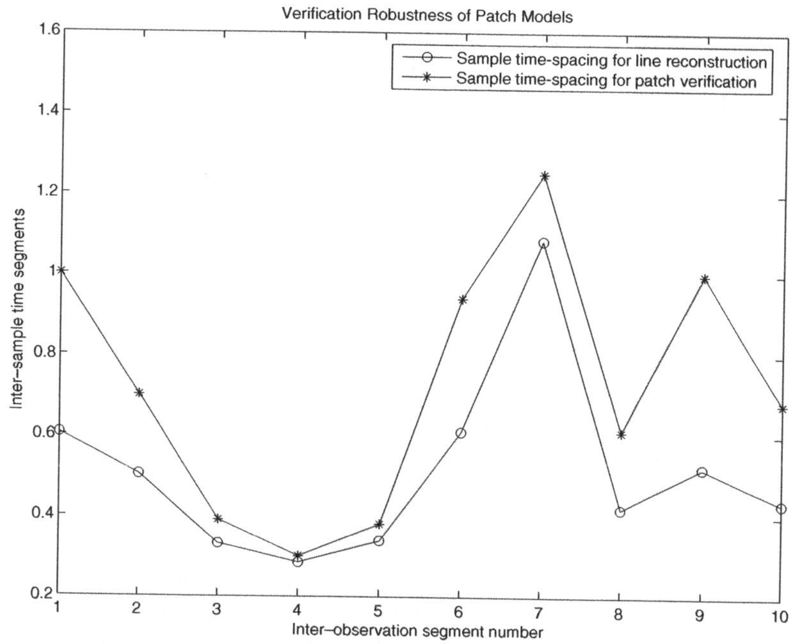

Figure 17.4 Interaction of patch geometry and sampling intervals.

Let, at time t, a set $Y_{ij}(t)$ of observations of agent i be available to decision-maker j. Assume that j knows that the maximum velocity of i is v_i. Let $P(t)$ be the patch representing i in the patch model maintained by j. We suppose Y to be of the general form:

$$Y(t) = \begin{bmatrix} t_1 & x_1 & y_1 \\ t_2 & x_2 & y_2 \\ \vdots & \vdots & \vdots \\ t_k & x_k & y_k \\ \vdots & \vdots & \vdots \\ t_n & x_n & y_n \end{bmatrix}, \tag{17.8}$$

where $t_k < t$ and the x and y variables are two-dimensional location coordinates. Let $t_{\min} = \min\{t_1, \ldots, t_n\}$ and $t_{\max} = \max\{t_1, \ldots, t_n\}$. Given this data, we wish to verify that

$$(x(t), y(t)) \in P(t) \text{ for all } t \in (t_{\min}, t_{\max}). \tag{17.9}$$

We may also be interested in the weaker form of verification given by the conditions:

$$\begin{aligned} &(x(t), y(t)) \in P([t_{\min}, t]) \text{ for all } t \in (t_{\min}, t_{\max}). \\ &(x(t_{\min}), y(t_{\min})) \in P(t_{\min}) \\ &(x(t_{\max}), y(t_{\max})) \in P(t_{\max}) \end{aligned} \tag{17.10}$$

The former condition asserts that the agent stayed within the instantaneous patch location at all times, while the latter asserts that the agent stayed within the trace of the instantaneous moving patch in the interior of its trajectory, and within the instantaneous patches at the first and last observations. The latter set of conditions accommodates in-transit delays and speed-ups so long as they are compensated for. We will focus on the latter since it is geometrically more intuitive. The former case merely leads to faster sampling requirements.

Figure 17.3 shows a set of 11 observations corresponding to a moving agent represented by the pentagonal patch. To verify conditions 17.10 we require that the reachable regions between observations be contained with the trace of the patch (the long band between the start and end pentagon positions). Some straightforward geometry reveals that the reachable set for agent i between consecutive observations is an elliptical region with foci at the observation locations, (x_k, y_k) and (x_{k+1}, y_{k+1}), and semi-major axis given by $v_i(t_{k+1} - t_k)/2$ (for noisy observations, the reachable set is the union of a set of ellipses whose foci are at the boundaries of the uncertainty regions). If, as shown, all 10 of the ellipses obtained between the observations lies within $P([t_{\min}, t_{\max}])$, one can conclude that the first of the conditions in 17.10 is satisfied, with the other two conditions being trivial to check.

Now note a feature of Figure 17.3: all 10 ellipses are tangent to the patch trace boundary. The trajectory and sampling points were constructed specially to achieve this condition, which can be interpreted usefully by asking the following question. Given a patch and a set of spatial sampling locations on the trajectory (the last two columns of Equation 17.8), what sets of corresponding observation times lead to verification and non-verification respectively? More generally, given that we know the maximum agent velocity (v_i), how closely can we reconstruct its trajectory using a finite set of samples? One extreme is easy to identify. If the sample locations and time are related by:

$$||(x_{k+1}, y_{k+1}) - (x_k, y_k)||_2 = v_i(t_{k+1} - t_k) \tag{17.11}$$

it is clear that only the straight-line paths between observation points could have been taken by the agent. The reachability ellipses in this case collapse to a poly-line. The set of observation times corresponding to this condition is then, in a sense, a maximal set of samples for the time period, since it allows full recovery of the path, with further sampling being unnecessary if no noise is present. The condition illustrated in Figure 17.3 is the other extreme. Given the spatial observation locations, the sequence of ellipses that are *just* contained within the patch trace correspond to the minimal set of samples for the time period (or the slowest average sampling). In other words, were the samples any further apart in time, the patch would not be verifiable. Another useful way to understand this extreme is to note that if a patch does not tangentially touch all reachability sets, it is either less precise than it could be or more precise than the data warrants.

Figure 17.4 illustrates this relation between the minimal and maximal time-separations between observations for the set of spatial sampling locations and patch geometry in Figure 17.3. Where the two curves are highly separated, patch verification is much easier than full trajectory reconstruction. Where the two curves are close or touch, patch verification requires sampling of the same order as is required for trajectory reconstruction.

We have assumed in this discussion that sampling locations are provided, and considered the implications of various associated time vectors. The interpretation here is that given the same spatial geometry of the patch and path, the validity of the patch with respect to slower moving vehicles (longer inter-observation times) is easier to verify than that of faster vehicles moving along the same path. For faster vehicles, one must either observe at more points along the path, in order to shrink the reachability ellipses, or expand the patches themselves. Many interesting lines of investigation emerge from the phenomenology of patch verification. For instance, one could pose optimization problems to select the fewest possible sampling locations for a given patch and path, or alternately, attempt to find the smallest patch (in the sense of a metric such as area, aspect ratio, or number of waypatches) that can be verified with a set of observations considered *a posteriori*. In a real-time setting, we may also pose the problem of where and when to request an observation from an available surveillance resource in order to verify an *a priori* model that extends further in time than available data (i.e. is predictive with respect to observed entities, or imperative with respect to a controlled resource). At a more theoretical level, an interesting question is: what is the relationship between the sampling phenomenology discussed here, and sampling theorems of various sorts?

Next, we continue the development of these ideas by asking, given a sound representational framework of patch models and adequate realization and verification mechanisms, how can we actually use these models for control?

17.4 HUMAN AND ARTIFICIAL DECISION-MAKING

Human decision-makers can apply powerful visual cognition mechanisms to analyze situations, form intentions and issue imperatives through visual manipulation of patch models via GUIs such as in Figure 17.2. It is reasonable to assume that patch models represented through appropriate software usefully, if reductively, mirror the corresponding spatio-temporal mental models (Endsley 2000), or 'situation awareness' that provides the basis for human decision-making. For humans therefore, patch models can be a low-friction mechanism for quickly capturing spatio-temporal beliefs, objectives, ideas and intentions with respect to controlled or observed entities. Given sufficiently powerful realization and verification mechanisms, these intentions can also be directly translated into action. The important open problems concerning the human aspect of patch-model based decision making are beyond the scope of this overview, since they involve issues in cognitive psychology.

The construction of artificial decision-makers capable of working with patch models, whether they are labeled AI systems or software agents, presents an interesting challenge. Since such systems do not have the powerful visual faculties that humans do, they must work directly with patch models in the form of data structures. Automated geometric or diagrammatic reasoning of the complexity that patch models support, unfortunately, is much too difficult for state-of-the-art AI techniques to handle. (Sacks and Doyle 1992) provide a useful discussion of the issues. AI-based techniques are largely limited to automated reasoning and planning (Traverso *et al.* 2004) within the confines of first-order predicate logic, over finite ranges of discretized variables. Automated reasoning about models (such as patch models) constructed with geometric primitives requires reduction of these models to logic-based declarative models.

The tools we use for achieving this reduction are a pair of axiomatic systems for representing time and space respectively, and developed in the AI community. The temporal interval calculus (TIC) (Allen 1983, 1991) represents time through a time-segment primitive, $\tau = [t_1, t_2]$, rather than as a point-set, and analyzes temporal relations through a set of 13 primitive *qualitative* dyadic relations between two time intervals, $c(\tau_1, \tau_2)$, capturing conditions ranging from 'before' through 'during' to 'after.' A related system for representing primitive *regions* in space (simply-connected convex sets in the simplest case), the region connection calculus (RCC) (Randell *et al.* 1992) has also been developed, and comprises eight primitive dyadic relations, $C(A, B)$, ranging from 'disconnected from' through 'partially overlapping' to 'proper part' (the fewer primitive relations result from the absence of a past/future asymmetry in space).

Using TIC and RCC, one can construct from a patch model M a qualitative state vector, $\overline{X}(t)$, whose elements are the pairwise instantaneous region-connection states, $C(P_i(t), P_j(t))$, of all the instantaneous patches in a model, M. Each element of this vector evolves in a piece-wise constant manner. Each such piece-wise constant segment is called an epoch, and the interaction of epochs can be modeled with TIC. The right pane in Figure 17.2 shows the qualitative state for a finite future window in time. Each vertical bar is an element of the state, and each uniformly-shaded portion of each bar is an epoch.

The qualitative state $\overline{X}$ over a window of time captures the condition of a patch model (with loss of quantitative geometric information) and is sufficient to support very flexible kinds of planning and scheduling. Here we consider a general reasoning problem: that of generating behavior rules for instantaneous patch motion that lead to a desired qualitative state. Since humans manipulate patch models to achieve control over agents primarily by editing the spatio-temporal geometry of patches, automation of this capability is fundamental to the construction of artificial decision makers.

The behavior design problem for a set of patches is defined as follows. Let M be a patch model, containing instantaneous patches, $P(t_0), \ldots, P_n(t_0)$ at t_0. For simplicity, replace the continuous time variable t with the discrete time index k and assume that the geometric state of the model is governed by:

$$\overline{x}_i\,(k+1) \equiv \overline{x}_i\,(k) + \overline{v}_i\,(k) \tag{17.12}$$

where $\overline{x}_i\,(k) = (x_i\,(k)\,, y_i\,(k))$ is the mean of the vertex coordinates of $P_i\,(k)$ and $\overline{v}_i$, the velocity, is the control variable. For this system of evolving patches, we specify goals via an equivalence class of qualitative goal states, $\mathcal{X}^* \equiv \{\overline{X}_1^*, \overline{X}_2^*, \ldots, \overline{X}_m^*\}$. This allows us to formally state the behavior rule design problem:

Behavior Rule Design Problem: Determine a set of behavior rules $g_1, g_2, \ldots, g_n$ for the n agents, such that $\overline{v}_i\,(k) = g_i\left(\overline{x}_1, \overline{x}_2, \ldots, \overline{x}_n, \overline{X}, k\right)$ and

$$\lim_{k \to k^*} \overline{X}_i\,(k) = \overline{X}_i^* \tag{17.13}$$

for some $\overline{X}_i^* \in \mathcal{X}^*$ and some $k^* < \infty$.

We note four features of this formulation. First, we note that as defined above, the goal set $\mathcal{X}$ may allow geometric configurations that are not intended. Second, we note that the reason an equivalence class of states is required, rather than a single goal state

$\overline{X}^*$, is that in general we may not want to distinguish among different ways of achieving the same geometry (for example, if a configuration of a ring of n polygons is required, we may be agnostic with respect to the $(n-1)!$ different ways of realizing it). Third, we note that even though the *goals* are formulated in terms of $\overline{X}$, the behavior *rules* are allowed to depend on quantitative (position/velocity) variables. Forcing the rules to be functions of only the qualitative state ($v_i = g_i(\overline{X})$) in general leads to a very impoverished design space. Finally, note that the formulation only concerns the achievement of an *instantaneous* state represented through RCC. Automated manipulation along the temporal dimension alone is also possible (and somewhat simpler), and an example of this, involving rescheduling two patches (adjusting waypatch times) to restore a violated TIC constraint, is in Rao and D'Andrea (2006). Simultaneously achieving non-trivial spatial and temporal configurations, however, is a much harder problem. The behavior rule design problem is, in the general case, equivalent to theorem proving with joint RCC and TIC descriptions. Reasoning with RCC and TIC in general is known to be undecidable and reasoning with decidable subsets can be NP-complete (Cohn and Hazarika 2001). Developing tractable reasoning mechanisms with a subset useful for decision-support with patch models is therefore an important open problem.

Here we illustrate the general features of the problem with an example. Details of this form of behavior design through both specially-constructed rules and general-purpose computational mechanisms such as simulated annealing is discussed in Goldfarb *et al.* (2006).

17.4.1 Example: the surround behavior

The intent of the behavior 'surround' is to achieve a condition where all terminal instantaneous patches completely surround a static target patch P_T. The desired geometric configuration is an annular configuration of shapes along the perimeter of P_T, as shown in Figure 17.5. We attempt to model this intent as follows. For the set of instantaneous patches $P_1(k), P_2(k), \ldots P_n(k)$, we require that every $P_i, i \in [1, n], P_i \neq P_T$, overlap the target patch and two other patches (all non-disconnected relations being PO, or partially overlapping, in RCC terms). The equivalence class of goal states thus defined includes what is intended: every configuration that satisfies the geometric intent. A little thought will show, however, that the goal requirement lets in other unintended configurations as well. This is the case in general, and cannot be avoided, since it is an effect of the reductive nature of the representation with respect to the patch model.

In Basic Surround, each polygon has *at least* two neighbors and is in relation PO with target polygon P_T. In Refined Surround, each polygon has *exactly* two neighbors and is in relation PO with target polygon P_T. We omit the formal definition of the goal state equivalence class and the associated conditions on the qualitative state vector X for brevity.

We solved the problem using flocking rules designed specifically for this problem (details are described in Goldfarb *et al.* 2006). The successful configuration shown in Figure 17.5 is a an element $\overline{X}^*$ of the solution set $\mathcal{X}^*$, and this configuration was frequently achieved with appropriate tuning. An alternative undesirable configuration, admitted by the definition of the goal state for Basic Surround, is also shown in Figure 17.5, and can also appear in simulation.

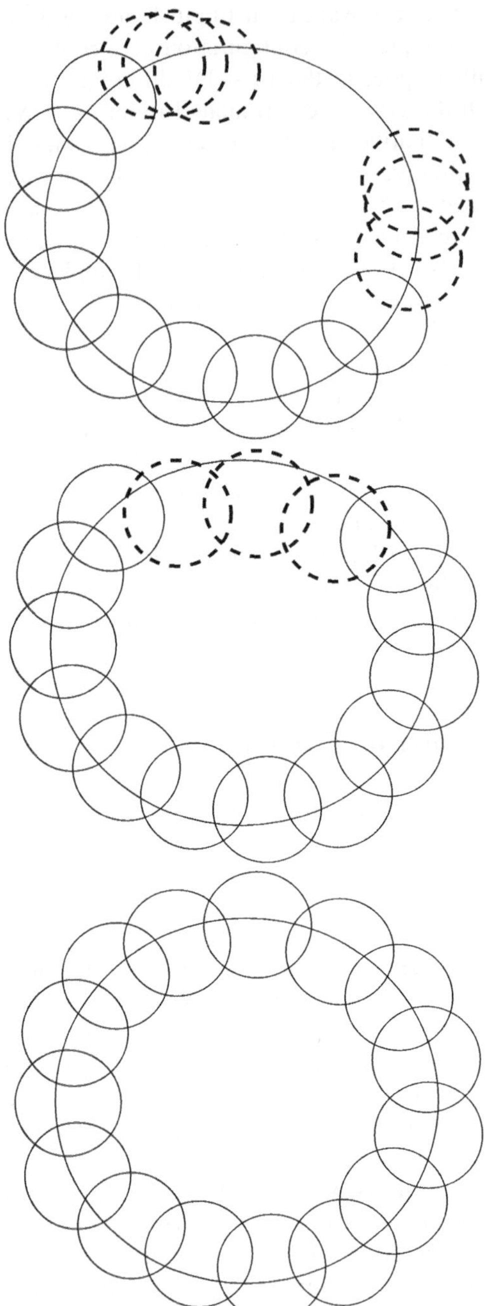

Figure 17.5 Comparison of solution states for Basic Surround (top), Refined Surround (middle), and the intended configuration (bottom). Both Basic and Refined Surround constraints require that each agent be in relation PO with the target. Basic Surround constraints also require that each agent be in relation PO with *at least* two other agents, and Refined Surround constraints require that each agent be in relation PO with *exactly* two other agents.

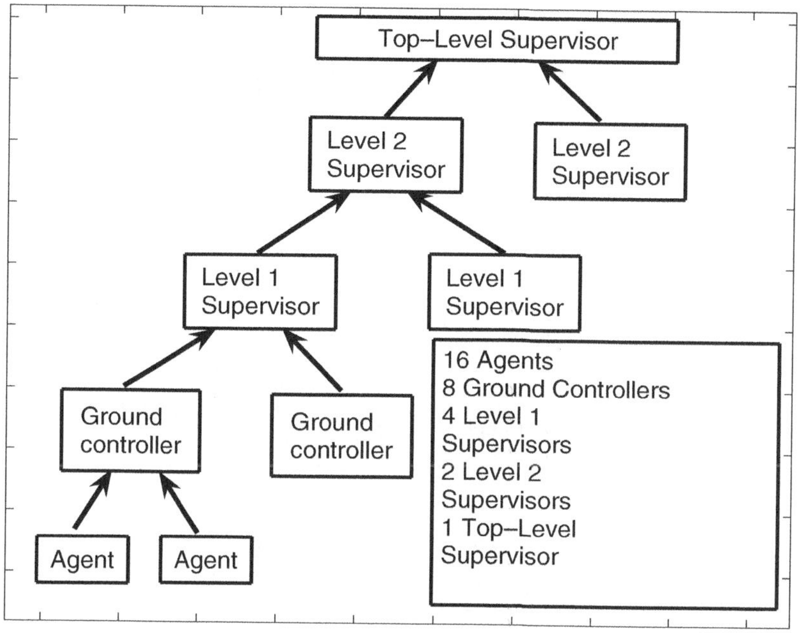

Figure 17.6 Hierarchy of decision-makers in example of 16-vehicle binary reporting tree.

17.5 HIERARCHICAL CONTROL

Consider now the problem of multiple hierarchically-organized decision-makers attempting to cooperatively observe and/or control a set of vehicles through patch models. The set of vehicles is partitioned into a set of teams, each of which is controlled (or in the case of enemy surveillance, tracked) by the ego patch $P_e(t)$ of exactly one patch model M corresponding to one decision maker, the *ground level* decision-maker. Assume, for the sake of simplicity that all decision-makers have similar patch models (i.e. they partition the agent set in the same way, such that there is one to one correspondence between the patches representing a given pair of agents for any pair of models. In the notation of 17.2, $h_m(q_i) = h_m(q_j) \Leftrightarrow h_n(q_i) = h_n(q_j)$ for any pair of models, M_m and M_n and any pair of agents q_i and q_j).

For a particular team, comprising, say agents $\{q_1, q_2\}$, let this decision-maker be G. Now let G maintain a reporting relation with respect to a parent decision-maker, designated $Par(G)$, defined as follows. For agents $\{q_1, q_2\}$, G periodically informs $Par(G)$ of their location by sharing its current ego patch, $P(t)$. For all agents *except* $\{q_1, q_2\}$, $Par(G)$ provides G with updated positions by sharing the relevant patches.

If patches are being updated in real time, to reflect unpredictable movements, the patch representing $\{q_1, q_2\}$ in $Par(G)$ must contain P_e. In fact, if $Par(G)$ is to maintain a robust model of the locations of $\{q_1, q_2\}$, the inclusion must be *strict*, with a tolerance dependent on the frequency of updates and the maximum velocities of the vehicles. If this

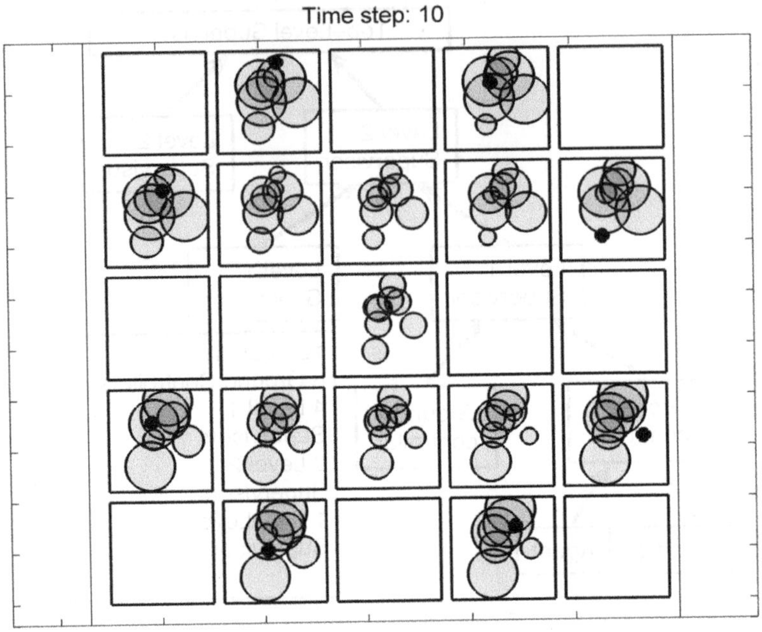

Figure 17.7 Views of all 15 decision-makers. The central view is that of the top-level supervisor; the ones above and below it level 2 views; the four placed diagonally around the center are level 1 views, and the 8 on the rim are ground views. Dark patches in the outer rim are ego patches in ground-level models.

strict inclusion is maintained with sufficiently large tolerances throughout the reporting hierarchy, we can go even further and note that the rate at which $Par(G)$ reports to *its* parent, $Par(Par(G))$, with updates on the positions of $\{q_1, q_2\}$, is lower than the frequency with which it *receives* such reports. This is not surprising, since given strict inclusions, one may expect that some updates in lower-level patches will not violate the strict inclusion condition, and therefore require no perturbation of the parent model at all, leading to that particular update propagating no further.

Before we proceed further, it will be useful to introduce an example that we will use as a running example. The situation is depicted in Figure 17.6 as follows. Sixteen vehicles, divided into eight teams of two each, are being controlled by a hierarchy of 15 decision-makers. There are eight ground level decision-makers, four one level above, two at the third level and one top-level decision maker. In Figure 17.7, the eight views arranged in the outer rim of the square correspond to ground views, while the central view is that of the top-level decision-maker. In this case, the strict inclusion conditions were satisfied by constructing the initial instantaneous circular patches as follows. The radius of the patch representing team j in the model at node i is given by:

$$r(i, j) = r_0 \lambda^{K(i,j)},$$

$$K = \begin{bmatrix} 0 & 0 & 0 & 0 & 0 & 0 & 0 & 0 \\ 1 & 1 & 1 & 1 & -1 & -1 & -1 & -1 \\ -1 & -1 & -1 & -1 & 1 & 1 & 1 & 1 \\ 2 & 2 & 0 & 0 & -2 & -2 & -2 & -2 \\ 0 & 0 & 2 & 2 & -2 & -2 & -2 & -2 \\ -2 & -2 & -2 & -2 & 2 & 2 & 0 & 0 \\ -2 & -2 & -2 & -2 & 0 & 0 & 2 & 2 \\ 3 & 1 & -1 & -1 & -3 & -3 & -3 & -3 \\ 1 & 3 & -1 & -1 & -3 & -3 & -3 & -3 \\ -1 & -1 & 3 & 1 & -3 & -3 & -3 & -3 \\ -1 & -1 & 1 & 3 & -3 & -3 & -3 & -3 \\ -3 & -3 & -3 & -3 & 3 & 1 & -1 & -1 \\ -3 & -3 & -3 & -3 & 1 & 3 & -1 & -1 \\ -3 & -3 & -3 & -3 & -1 & -1 & 3 & 1 \\ -3 & -3 & -3 & -3 & -1 & -1 & 1 & 3 \end{bmatrix} \tag{17.14}$$

For a strict hierarchy of k levels, m ground-level agents and exactly p reports per decision-maker (p physical agents for ground-level decision makers), and a total of N decision makers, the following relations must hold:

$$p^{k+1} = m,$$

$$k = \log_p m - 1, \tag{17.15}$$

$$N = \sum_{i=0}^{k} p^i.$$

(17.16)

In general, we may conjecture that the number of levels in the hierarchy will be in a logarithmic relationship with the number of agents being commanded, with the base of the logarithm being of the order of the average number of reports per decision-maker. If the degree distribution in the reporting hierarchy is other than a narrow normal distribution, however (such as a scale-free or small-world distribution), it is not clear what forms these relationships can assume.

For our example, the nominal command for each team is to stay in its initial position, but we imagine that they are all moving to evade enemy pursuers, necessitating updates in models propagating bottom-up and then top-down all through the hierarchy. We imagine, further, that all patches are circular at any given instant, and imagine that patches are updated whenever a new observation comes in, by adding a waypatch designed to maintain verifiability as discussed in Section 17.3. The circles representing a given pair of agents therefore must increase in radius upstream through the chain of parent decision-makers, and then *continue* to increase down from the top of the hierarchy towards *other* ground level decision-makers. The precision of representation therefore, decreases monotonically in the network as a function of degrees of separation from the ground-level reporting node.

Now imagine that all sensor signals are lost at time $t = 0$, and consider what must happen next in order to maintain the truth of all models for $t > 0$. This is a simple way to probe the effect of increasingly fast domain dynamics and external events. Clearly, since patch models are a conservative, worst-case representation, the ground level instantaneous patches at the last observation instant must be grown at a rate of at least $\dot{r}(t) = v_{\max}$, where r is the radius of a particular ego patch, and $v_{\max}$ is the maximum vehicle/agent velocity. Patches that are further away in the reporting web must also grow at a similar rate to maintain the inclusion relation. We now analyze the evolution of the 15 models in this example over time from an information-content perspective.

17.5.1 Information content and situation awareness

How do we quantify the information content of a patch model? This question must be addressed before we approach such interesting matters as the relative situation awareness at various levels in the hierarchy. Consider the case of point agents, with spatio-temporal location being the only state of interest. In this case, a decision-maker may be said to possess complete information about an agent if the patch representing a given agent collapses to a one-dimensional curve coincident with the agent's trajectory in space-time. At the other extreme, if an agent is represented by a patch that is larger than the overall global area (or universal set) of interest (in this case the square 'battlefield' represented in each of the non-empty views in Figure 17.7, to which all activity is restricted), the patch contains no useful information at all, since the information that the agent is in the battlefield at all times may be considered *a priori* and tautological. In between, so long as a patch does not cover the universal set, the patch contains *some* information. This notion of information content is clearly restricted to spatio-temporal location information and assumes the existence of a suitable bounded universal set to which all agents are known to be restricted. In all realistic cases, such a set will exist (though it might be as large as the entire surface of the earth).

Relating this notion to a formal information measure such as Shannon entropy is nontrivial, and remains an open problem. A measure called *residual area* is, however, very useful within the context of patch models, and we analyze the example from the previous subsection in terms of this measure. We define the residual area as:

$$r_A(P(t)) = 1 - ||\Sigma \cap (\Sigma - P(t))||/||\Sigma||. \tag{17.17}$$

Note that the measure is an instantaneous one. The measure is the normalized area of the portion of the set difference that is in the universal set, Σ (the norm notation indicating area). Note that strictly speaking, Σ is not a universal set in the usual sense, since parts of a patch can be outside it. While this issue can be fixed with additional notation, we will restrict ourselves to the present notation for clarity, since it does not cause confusion for the limited discussion presented here.

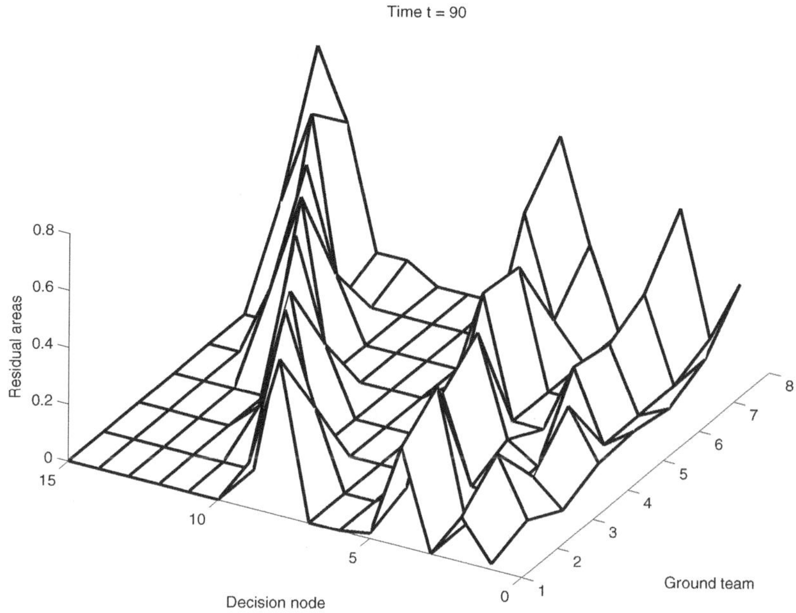

Figure 17.8 Information content in hierarchy, measured through residual areas.

Now, let us analyze the information content in the various models in the example over time. First, consider the total information at a given instant in time. Figure 17.8 shows the residual areas for each of the 8 patches for each of the 15 decision-makers at time $t = 90$, or approximately two-fifths of the way through the 240 time-step simulation. Decision-makers 8-15 are ground-level, 4-7 are level 1, 2-3 are level 2 and 1 is the top-level decision maker. Note the three lines of peaks. From left to right, these correspond to the content of the ground-level, level 1 and level 2 supervisors. The information content of the top-level supervisor corresponds to the line of partial peaks along the line for decision-maker 1. The left-most line of peaks, forming a diagonal from decision-makers 8-15 and patches 1-8 shows that ground-level decision-makers have high information on their ego patches and almost no information elsewhere. The information content spreads out as one proceeds up the hierarchy. Figure 17.9 shows the same information in contour form, this time for 4 time slices at times $t = 10, 90, 160$ and 210. Note that information degrades first in the top decision-maker's view, and then progressively down the hierarchy, with the ground-level decision-makers being the last to lose all information. Note also that if one views the contour maps as a set of vertical slices of width 1, 2, 4 and 8, the amount of concentration of content along a diagonal in each slice indicates the degree of localization of information.

These two views of the evolution might suggest that as sensing and communication start to lag behind domain evolution, supervisors possess decreasing amounts of information. Consider an alternate view of the example. We sum the residual areas in each model and normalize, to get at the *total* information content per model. We then average the information content across all models at a given level in the hierarchy. The result is a measure of total information content at a given level. Figure 17.10 plots these quantities

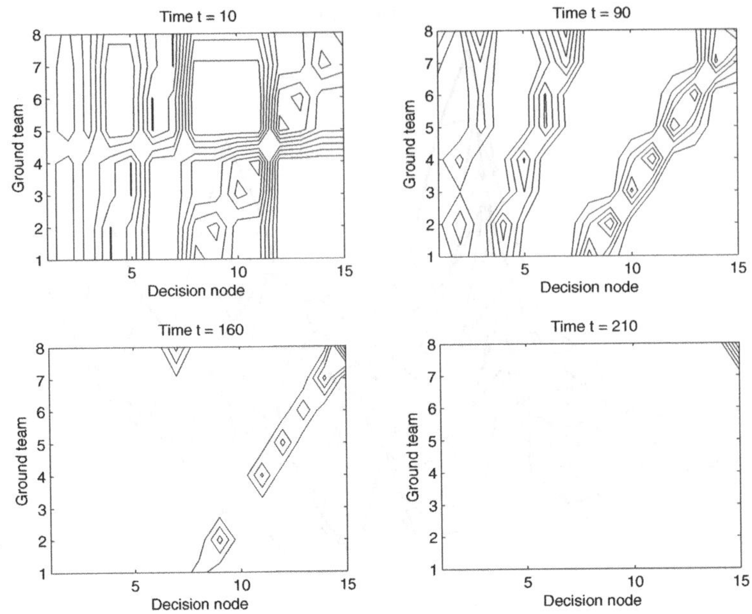

Figure 17.9 Contours of information content, measured by residual area, at four time points.

over time. Note the interesting behavior of the four curves: their *order*, defined as the order of ordinates at a given abscissa (time) point is completely reversed during the mission. This suggests that during epochs when sensing and reporting can keep up with the domain dynamics, supervisors do in fact possess more *total* information and are thus enabled to make better broad decisions. Once information degrades below a particular point, however, reports know more than supervisors. If we interpret informed higher-level decision making as strategic, and lower-level decision-making as tactical, we can conclude that as things speed up, informed tactical responsiveness is retained longer than informed strategic responsiveness. Interpreted another way, this suggests that increasing sensing and reporting frequency by leveraging improved computing and communication increases the horizon of informed decision-making and responsiveness at *all* levels. If we assume that being better informed leads to decision making that is both *faster* and better, then this simulation is a demonstration of the validity of one of the key hypotheses of network-centric warfare (DoD 2001): that better networks enable better shared situation awareness (SSA), leading to faster responsiveness. Of course, this demonstration is an empirical one, and restricted, moreover, to the context of a very specific construction involving patch models. We can hope, however, that more generalized information models of C2 systems analyzed in a similar fashion might yield more general and powerful insights.

We conclude our dissection of this example with one final view of the data. Consider again the aggregate information evolution, but this time viewing each patch individually at a representative node at each level, yielding $4 \times 8 = 32$ curves. Figure 17.11 shows this evolution for the residual areas. The key point to note is the increasing asymmetry of the evolution of the information across patches as one goes down the hierarchy. This neatly captures an intuitive idea: that local information must degrade much more slowly than

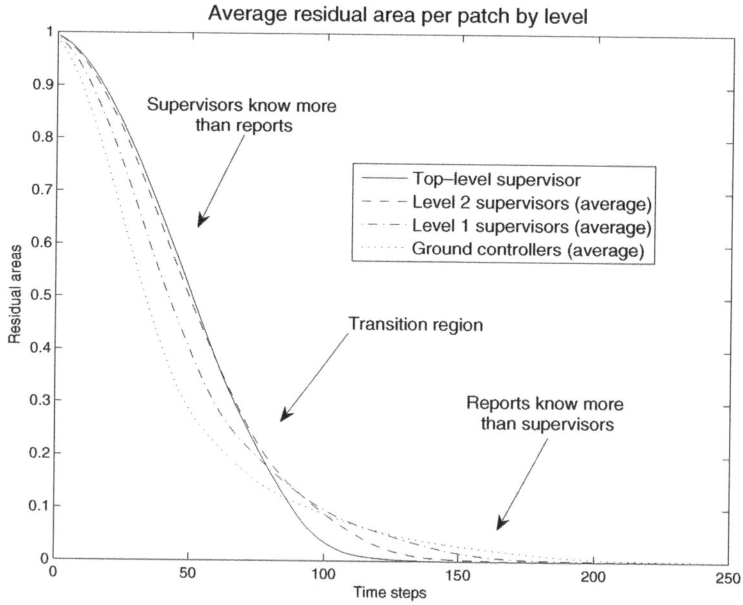

Figure 17.10 Relative information content at supervisory and ground-level decision nodes over time. Note that initially, supervisors have more information, but it degrades faster than at lower levels.

information that is distant in the reporting network. The value of having a supervisory role is that even though the total information may eventually be less than that of lower-level models, the *symmetry* is maintained through time. This balanced degradation may have some value in decision-making even beyond the point where the total information content starts to lag behind. Investigating this kind of 'obsolete big picture' information and its uses is an open problem.

17.6 CONCLUSION

The scenario we used to motivate this work in the introduction to the chapter, a version of the concept of operation referred to as *cooperative strike*, is in several ways, as old as warfare itself. An enemy threat is detected, a course of action selected, resources allocated and a mission executed. Though modern military terminology would describe this sequence of events as a 'kill chain' (comprising *find, fix, track, target, engage* and *assess* stages), a very similar description might be applied to a Bronze Age battle involving chariots and archers.

The similarity is superficial primarily for one reason: the sheer scale and speed at which the decision-making is enabled and completed have no counterpart in pre-modern warfare. The technical enablers are reliable high-bandwidth communications in the battlefield, significant computing power at every decision node in the system and sophisticated software capabilities that are generally bucketed under the term 'autonomy.' Employed against an adversary without similar technology, the asymmetry in capabilities can be so

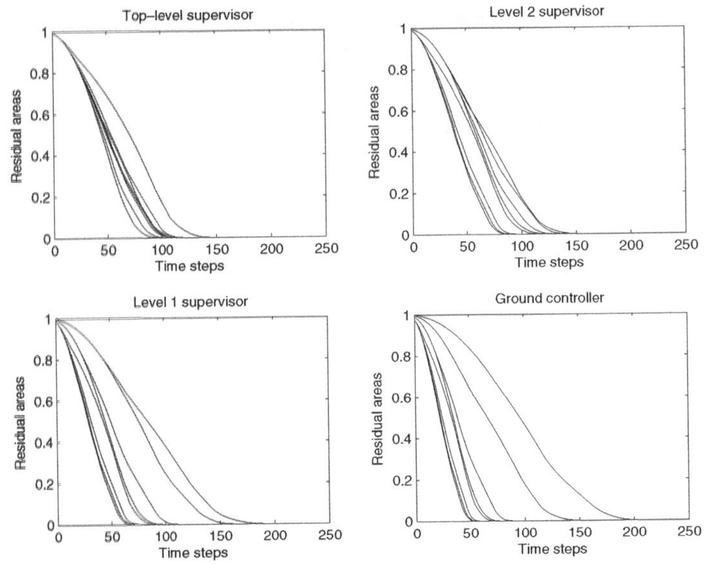

Figure 17.11 Asymmetric degradation of information at various levels. Note that lower-level decision-makers have greater asymmetry, leading to localization of knowledge, even though they retain useful knowledge longer.

vast that the resulting battlefield situations are unprecedented, even though one might say that the only change has been a speed-up of the building block processes of battle. Information technology asymmetrically employed in the battlefield can lead to a discontinuous change in the nature of combat, even though the underlying changes in decision-making speed may be smooth (Albert *et al.* 2001). This observation has led to a new military doctrine called 'Information Superiority' and a corresponding prescriptive concept of warfare called 'network-centric warfare' (NCW) (Albert *et al.* 2000; Cebrowski 1999; DoD 2001). NCW as it is understood (and partially applied) today largely concerns the impact of communication and computing power on the battlefield. The impact of autonomy, however, is not yet fully understood, since autonomy technology is a software capability that still lags significantly behind the hardware capabilities. Autonomy, arguably the central element of this work, is a more delayed aspect of the impact of information technology, since it refers to capabilities manifested in advanced software, rather than improved hardware.

The typical problems of modern warfare tend to defy compact description. The work in this volume, for instance, was motivated by the unwieldy phrase, 'cooperative control of distributed autonomous vehicles in adversarial environments' (often with the qualifying phrase, 'with humans in the loop,' appended). The description is not redundant, however, and each element is integral to the intent of the description. The complexity of the description is simply a function of the all-encompassing nature of the source – information technology – of the most pressing modern research challenges (and opportunities). Since the evolution of information technology is impacting every aspect of military operations, the underlying research imperative is to reinvent the nature of warfare.

Patch models represent one approach – a representation-driven one – for managing the impact of information technology on combat technology. The sharable (among humans and artificial decision-makers) and interoperable nature (among decision processes) of the models enable clean, modular architectures, while their capacity for supporting contextual relevance-based abstractions make them suitable for managing information overload. While patch models themselves are a simplified framework, largely for dealing with simulated agents displaying a restricted subset of the behaviors that occur in battle, we expect that any larger-scale implementation effort will benefit from the principles that drove (and were validated by) this work.

REFERENCES

Albert DS, Garstka JJ and Stein FP 2000 *Network Centric Warfare: Developing and Leveraging Information Superiority (2nd Ed.).* CCRP, Washington, DC.

Albert DS, Garstka JJ, Hayes RE and Signori D 2001 *Understanding Information Age Warfare.* CCRP.

Allen JF 1983 Maintaining knowledge about temporal intervals. *Communications of the ACM* **26**(11), 832–843.

Allen JF 1991 Planning as temporal reasoning. In *Proc. Second International Conference on the Principles of Knowledge Representation and Reasoning*, Cambridge, MA, pp. 3–14.

Cebrowski AK 1999 Network-centric warfare: An emerging military response to the information age. In *Command and Control Research and Technology Symposium.*

Cohn AG and Hazarika SM 2001 Qualitative spatial representation and reasoning: An overview. *Fundamenta Informaticae* **43**, 2–32.

DoD 2001 *Network-Centric Warfare: DoD Report to Congress.* Department of Defense, Washington, DC.

Endsley MR 1995 Towards a theory of situation awareness in dynamic systems. *Human Factors* **37**(1), 32–64.

Endsley MR 2000 Situation models: an avenue to the modeling of mental models. In *Proc. 14th Triennial Congress of the Intl. Ergonomics Assoc.*

Goldfarb S, Rao VG and D'Andrea R 2006 Agent-based modeling with polygon primitives for aerospace applications *AIAA Modeling and Simulation Technologies Conference and Exhibit*, Keystone, CO. AIAA Paper 2002–6469.

Nofi AA 2000 Defining and measuring shared situation awareness. Technical Report CRM D0002895.A1, Center for Naval Analysis, Alexandria, VA.

Randell D, Cui Z and Cohn AG 1992 A spatial logic based on regions and connection. In *Proc. 3rd Intl. Conf. on Knowledge Representation and Reasoning*, pp. 165–176.

Rao VG and D'Andrea R 2006 Patch models and their applications to multi-vehicle command and control. In *Proc. American Control Conference*, pp. 4958–4963, Minneapolis, MN.

Rao VG, Wongpiromsarn T, Ho T, Chung K and D'Andrea R 2006 Encapsulated motion planning for abstraction-based control of multivehicle systems. In *American Control Conference*, pp. 2995–3000, Minneapolis, MN.

Sacks EP and Doyle J 1992 Prolegomena to any future qualitative physics. *Computational Intelligence* **8**(2), 187–209.

Simon H 1996 *The Sciences of the Artificial* (Third Edition). MIT Press, Cambridge, MA.

Traverso P, Ghallab M and Nau D 2004 *Automated Planning: Theory and Practice*. Morgan Kauffman, New York.

Witsenhausen HS 1971 Separation of estimation and control for discrete time systems. In *Proceedings of the IEEE* **59**(11), 1557–1565.

Index

Cooperative Control of Distributed Multi-Agent Systems edited by Jeff S. Shamma

Printed and bound by CPI Group (UK) Ltd, Croydon, CR0 4YY
16/09/2026
14963389-0005